Advances in Experimental Medicine and Biology

Volume 1514

Series Editors

Nima Rezaei , Research Center for Immunodeficiencies,
Children's Medical Center
Tehran University of Medical Sciences
Tehran, Iran

Avia Rosenhouse-Dantsker, Department of Chemistry
University of Illinois at Chicago
Chicago, IL, USA

Robert Gerlai, Department of Psychology
University of Toronto
Mississauga, ON, Canada

Anthony M. Pedley

Editor

Supramolecular Protein Assemblies In Cells

 Springer

Editor
Anthony M. Pedley
Department of Biochemistry and Molecular Biology
Roy J. and Lucille A. Carver College of Medicine, Holden Comprehensive Cancer Center,
Fraternal Order of Eagles Diabetes Research Center, University of Iowa
Iowa City, Iowa, USA

ISSN 0065-2598   ISSN 2214-8019   (electronic)
Advances in Experimental Medicine and Biology
ISBN 978-3-032-26628-6   ISBN 978-3-032-26629-3   (eBook)
https://doi.org/10.1007/978-3-032-26629-3

This Springer imprint is published by the registered company Springer Nature Switzerland AG
The registered company address is: Gewerbestrasse 11, 6330 Cham, Switzerland

If disposing of this product, please recycle the paper.

Preface

From an early age, we learn that cells are densely packed with proteins, lipids, genetic material, and metabolites. This crowded and heterogeneous landscape poses a fundamental, yet largely unanswered question: How do cells efficiently coordinate the complex array of biological processes necessary for life? Recently, answers to this question have emphasized the importance of supramolecular protein assemblies in regulating these activities by generating unique biochemical environments to enhance the efficiencies of signal transmission, enzymatic activities, and metabolic channeling.

This volume is a result of the growing desire to highlight emerging examples of such assemblies and showcase the innovative approaches being used to study them. By no means is this book intended to be comprehensive. The first several chapters illustrate the prevalence and importance of protein assemblies in membrane organization, protein import, and intracellular signaling. Later chapters address the organization of higher ordered protein complexes associated with the transfer of genetic information through DNA replication and repair as well as transcription. Finally, the volume concludes with a series of chapters focused on the compartmentalization of metabolic enzymes ranging from stable megadalton complexes like the pyruvate dehydrogenase complex to dynamic structures like metabolons.

As scientists, we are at a thrilling time to study spatial and temporal cellular biochemistry. Rapid advances in technologies have started to provide molecular level insights into these higher ordered assemblies, and as a result, have started to transform the way we view the regulation of biological processes. I hope that this book generates new questions, cultivates long-term collaborations, and inspires a new generation of scientists to explore this dynamic and constantly evolving field.

This book would not have been possible without the contributions of talented researchers who provided insightful and engaging chapters. I am grateful for their efforts in shaping this volume. I would also like to thank the members of my research group, past and present, not only for their patience as I work to complete this book but also for all their work in advancing our understanding of metabolic condensates. I wish each of you continued success in your future endeavors, both in and out of the laboratory. To my family and friends: I know I often "geek out" over concepts that might not make much sense to you, but your encouragement and well-timed distractions have kept me grounded and motivated. Finally, I want to acknowledge my previous research mentors—Stephen Benkovic (Penn State University), V. Jo Davisson (Purdue University), and Laurie Witucki (Grand Valley State University). Thank you for introducing me to the fascinating world of multi-protein assemblies and for your guidance, inspiration, and generosity in providing me opportunities to explore this exciting area of research.

Iowa City, Iowa, USA   Anthony M. Pedley

Contents

1 Introduction... 1
Stephen J. Benkovic

**2 Structural Insights into Ordered Multicomponent Assemblies
in Cell Junctions** 7
Oliver James Harrison, Priyanka Mathews, and Julia Brasch

3 Supramolecular Assemblies Drive Activation of Inflammasomes ... 41
Zhangfei Shen, Chen Wang, and Tian-Min Fu

**4 Conformational Dynamics and Allostery in the cAMP-Protein
Kinase A Signalosome**.. 55
Varun Venkatakrishnan and Ganesh S. Anand

**5 The TOM Complex in the Outer Membrane of Mitochondria:
A Supramolecular Assembly for Protein Import**.................. 85
Stephan Nussberger, Robin Ghosh, and Shuo Wang

6 Dynamic Assemblies in Genome Maintenance.................... 113
Paras Gaur and Maria Spies

**7 PCNA Macromolecular Complexes: PCNA Serves as a Molecular
Hub Regulating Multiple Cellular Processes Inside and Outside
of the Nucleus** 157
Hamsini Kala, Dana Abou Abbas, Linda H. Malkas,
and Robert J. Hickey

**8 Regulation of the Anti-termination RNA Transcription Complex
by Lon-Mediated Lambda N Degradation**...................... 189
Marianita Castro, Sanghyuk Lee, and Irene Lee

**9 The Pyruvate Dehydrogenase Complex: A 90-Year-Old Enigma
Shaping the Future of Structural Enzymology** 207
Toni K. Träger, Fotis L. Kyrilis, Greg Kafetzopoulos,
Christian Tüting, and Panagiotis L. Kastritis

10 Enzyme Assemblies in Nucleotide Metabolism: Structure, Regulation, and Disease Implications 255
Jack P. Boylan, Timothy D. Iles, Alexis Nguyen,
and Anthony M. Pedley

11 Mammalian Respiratory Chain Complex Assemblies and Their Links to Mitochondria Stress-Induced Human Diseases 299
Runyu Guo and Maojun Yang

Index 331

Chapter 1
Introduction

Stephen J. Benkovic

Abstract Owing to a lengthy career, I occupied a ringside seat to watch the evolution of our understanding of enzymatic catalysis. The field has progressed from primitive bioorganic models of enzyme active sites, which guided physical organic-based studies on how enzymes hydrolyze esters and amides, as well as laborious purifications of individual enzymes from animal organs, to investigations of multienzyme complexes, as featured in this text. My intent in this short essay is to provide background for the chapters in this volume. What papers and texts might have inspired the research reported within them? Where was the field at that time, and where might it be going?

Keywords Bioorganic chemistry · Enzymology · Multi-enzyme complexes · Enzyme compartmentalization · Replisome · Metabolons · Purinosome

Owing to a lengthy career, I occupied a ringside seat to watch the evolution of our understanding of enzymatic catalysis. The field has progressed from primitive bioorganic models of enzyme active sites, which guided physical organic-based studies on how enzymes hydrolyze esters and amides, as well as laborious purifications of individual enzymes from animal organs, to investigations of multienzyme complexes, as featured in this text. My intent in this short essay is to provide background for the chapters in this volume. What papers and texts might have inspired the research reported within them? Where was the field at that time, and where might it be going?

The term initially used for this nascent field was bioorganic chemistry, where chemists created molecular facsimiles of catalytically active sites of enzymes using small organic molecules. Their models were based on early X-ray crystallographic structures of serine proteases, which revealed their active sites and their now

S. J. Benkovic (✉)
Department of Chemistry, The Pennsylvania State University, University Park, PA, USA
e-mail: sjb1@psu.edu

A. M. Pedley (ed.), *Supramolecular Protein Assemblies In Cells*, Advances in Experimental Medicine and Biology 1514,
https://doi.org/10.1007/978-3-032-26629-3_1

familiar catalytic triad. These constructs provided an estimate of how the rates of enzymatic reactions could be accelerated by the proximity of a bound substrate to catalytic residues within an active site, which in turn could function as nucleophiles or general acid-base catalysts. In retrospect, early attempts to imitate active sites employed cyclodextrins, micelles, and simple polymers with a repetitive side chain were constructs of synthetic biology and are now seen in contemporary research.

Increased availability of X-ray crystallographic structures, as a result of the widespread use of recombinant protein expression, drove in-depth studies of the mechanism of action of a variety of monomeric enzymes. The development of clever mechanistic probes for these enzymes was particularly noteworthy. Those probes included measurements of substrate specificity, pH dependency, and isotope effects on key enzymatic steps, leading to the generation of complete reaction cycles for triosephosphate isomerase and dihydrofolate reductase. Other kinetic methods soon followed, such as stop flow, T-jump, and pulse chase methods that teased out evidence for transient intermediates. Ingenious methods to synthesize chiral phosphoryl or thiophosphoryl substrates were developed and applied to a variety of phosphoryl transfer reactions to deduce from the stereochemical outcomes the nature of the transfer process. This resulted in questions; Is the reaction direct or does it proceed through an intermediate?

With this increasing knowledge of the mechanism of action of specific enzymes, one could now imagine how such information might be used in the creation of more potent, specific inhibitors and as potential chemotherapies. This information was embodied in the design of suicide inhibitors, transition state analogs, and irreversible and covalent inhibitors to design new pharmaceuticals. In my view, these studies were the precursors for what is now termed chemical biology.

Advancements in recombinant DNA technology and genomic sequencing provided the opportunity for studying molecular features of enzymes and their dynamics. The appearance of site-specific mutagenesis validated the earlier assumptions of which residues were important for catalysis and spread from the active site to the protein itself, as it was recognized that all residues contributed to catalytic activity through their conformational mobility. This flexibility to mold active sites was demonstrated by NMR relaxation studies and by sophisticated molecular dynamics simulations. Recent studies confirmed the importance of conformational flexibility through probes such as hydrogen-deuterium (H/D) exchange, Stokes shift and knowledge-based (X-ray structures) conformational ensembles. Allostery became widely accepted in the field and rationalized in terms of conformational changes that were transmitted through the protein structure. This quickly extended beyond the conformational dynamics of a single enzyme and soon protein allostery was being applied to multifunctional assemblies such as fatty acid synthase and the pyruvate dehydrogenase complex.

The annotation of protein databases opened the floodgates for the study of a plethora of complex enzyme-based macromolecular assemblies that included DNA replication and repair foci and metabolons. I dedicated countless hours to studying how individual enzymes behave in the context of these structures (Benkovic 2021). For the last four decades, my laboratory studied the function of the eight proteins that constitute the T4 DNA replisome Fig. 1.1a. This work soon progressed into looking into the coordinated actions of components within the various subcomplexes and eventually the entire replisome. During this time, our understanding of this system greatly benefited from new techniques such as single-molecule total internal reflection fluorescence and molecular tweezers (Benkovic and Spiering 2017). I also searched for additional evidence to support a decades-old observation of a complex between transformylase enzymes in de novo purine biosynthesis. Coinciding with the routine use of fluorescence microscopy to investigate enzyme dynamics in live cells, we discovered that not only the two transformylase enzymes associate into a dynamic assembly but so do the other four pathway enzymes Fig. 1.1b (An et al. 2008). This supramolecular protein assembly, the purinosome, has been a poster child for the study of many analogous assemblies comprised of metabolic enzymes and is comprehensively reviewed in Chap. 10.

In my own pursuits with studying these higher-ordered assemblies, several questions naturally arose—What is the origin of the signals that foster the formation of such assemblies within a cell? How broad are the reaction types for which such complexes are responsible? If they carry out a multistep process, is it through a coordinated or channeled mechanism?

This volume presents three primary areas where supramolecular protein assemblies contribute to cellular function: cell-cell communication and signaling, genetic information transfer and repair, and cellular metabolism. Speculative answers to these questions are discussed within various chapters of the book. Further perusal of the chapters reveals how much of our current understanding of these processes depends on technology. Fluorescence microscopy has enabled us to visualize the dynamics of proteins and their assemblies inside cells in real time. Structural insights have been guided by the discovery and advances in electron microscopy (cryogenic, scanning, and transmission) and cryo-tomography. And, our understanding of the complex architecture and dynamics has benefited from mass spectrometry (chemical cross-linking, native mass spectrometry, H/D exchange) and computational simulations.

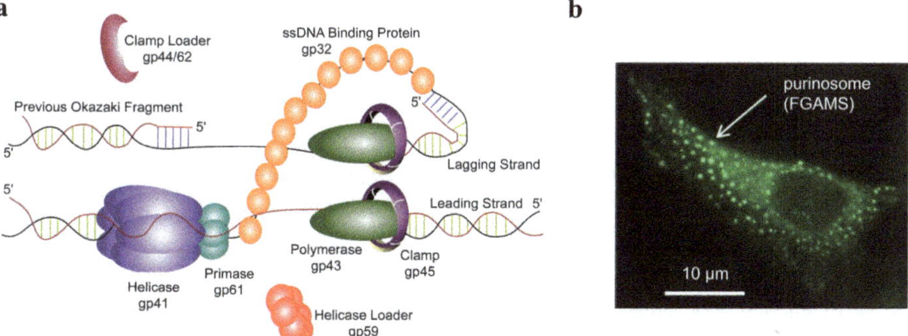

Fig. 1.1 The DNA replisome and purinosome metabolon assemblies. (**a**) The replisome figure was modified from Spiering et al. (2008) and Noble et al. (licensed under CC BY 4.0) (2015). (**b**) The purinosome figure was modified from An et al. and used with permission from *Science* (2008)

With these technologies in hand, what are the foremost questions that remain? Where might the study of supramolecular protein assemblies go? I conclude by offering a list of questions to help direct those researchers working in this area and challenge them to develop and use innovative strategies and technologies to seek out answers to these questions.

- How do enzymes within supramolecular protein assemblies find each other? Is it through diffusion, or are they restricted to spaces by cytoskeletal elements? Given the crowded confines of the cells, is it possible that proteins necessary for these complexes are synthesized proximal to where they will function? This is especially true for the metabolons, where their dependency on cofactors is furnished by nearby organelles such as the mitochondria.
- If signaling triggers the complex assembly, then what is the nature of the signaling? Is it driven by post-translational modifications such as phosphorylation, ubiquitination, acetylation, etc.? How is the signaling reversed to dissociate the complex?
- What is the actual architecture of the complexes, such as their stoichiometry and subcellular localization? In the case of those assemblies that exist as liquid-like condensates, which involve many molecules, understanding the exact composition and the stoichiometry currently appears to be tenuous.
- Critical to the regulation of these assemblies is the presence of a scaffolding molecule(s). How can a scaffold be modified to trigger complex changes? What are the structural, biochemical, and biological consequences of this modification? While computational modeling has provided some tools to look at the dynamics of large complexes. I am intrigued by the possibility that advanced modeling by machine learning and artificial intelligence algorithms might assist.
- Is there communication between distinct supramolecular protein assemblies? For example, if metabolic pathways are carried out by specific metabolons in liquid-like droplets, then do individual condensates contact one another under certain conditions? What is the spatial topology of a cell that has many metabolons, where each catalyzes a specific metabolic pathway? Does this communication result in a more regulated and efficient process?

References

An S, et al. Reversible compartmentalization of de novo purine biosynthetic complexes in living cells. Science. 2008;320(5872):103–6.

Benkovic SJ. From bioorganic models to cells. Annu Rev Biochem. 2021;90:57–76.

Benkovic SJ, Spiering MM. Understanding DNA replication by the bacteriophage T4 replisome. J Biol Chem. 2017;292(45):18434–42.

Noble E, Spiering MM, Benkovic SJ. Coordinated DNA replication by the bacteriophage T4 replisome. Viruses. 2015;7(6):3186–200.

Spiering MM, Nelson SW, Benkovic SJ. Repetitive lagging strand DNA synthesis by the bacteriophage T4 replisome. Mol Biosyst. 2008;4(11):1070–4.

Readers who desire more information about the evolution of enzymology from bioorganic molecules to multi-enzyme assemblies can find them in the following:

Bell RP. Acid-base catalysis. Oxford & Clarendon Press; 1941.

Bernhard S. The structure and function of enzymes. W. A. Benjamin, Inc.; 1968.

Blakely RL, Benkovic SJ. Folates and pterins: chemistry and biochemistry of folates, vol. 1. Wiley; 1984.

Blakely RL, Benkovic SJ. Folates and pterins: chemistry and biochemistry of pterins, vol. 2. Wiley; 1985.

Boyer PD, editor. The enzymes, vol. I–XV. 3rd ed. Academic Press; 1971.

Breslow R. The mechanism of thiamine action: predictions from model experiments. Ann N Y Acad Sci. 1962;98(2):445–52. Wiley-Blackwell.

Cantor CR, Schimmel PR. Biophysical chemistry part III: the behavior of biological macromolecules. W. H. Freeman & Company; 1980.

Cleland WW, O'Leary MH, Northrop DB. Isotope effects on enzyme catalyzed reactions. University Park Press; 1977.

Copeland RA. Enzymes: a practical introduction to structure, mechanism, and data analysis. 2nd ed. Wiley-VCH; 2000.

Craik CS, Fletterick R, Matthews CR, Wells J. Protein & pharmaceutical engineering. Vol. UCLA symposium of molecular & cellular biology, vol. 110. Wiley-Liss; 1990.

Du S, et al. Conformational ensembles reveal the origins of serine protease catalysis. Science. 2025;387(6735):eado5068.

Dugas H, Penney C. Bioorganic chemistry: a chemical approach to enzyme action. Springer; 1981.

Eckstein F, Lilley DMJ. Nucleic acids & molecular biology, vol. 5. Springer; 1991.

Fersht A. Enzyme structure and mechanism. W H Freeman & Company; 1977.

Gao S, Klinman JP. Functional roles of enzyme dynamics in accelerating active site chemistry: emerging techniques and changing concepts. Curr Opin Struct Biol. 2022;75:102434.

Hammes GG. Enzyme catalysis and regulation. Academic Press; 1982.

Hammes GG. Spectroscopy for the biological sciences. Wiley-Intersciences; 2005.

Hammes GG. Physical chemistry for the biological sciences. Wiley; 2007.

Hecht SM, editor. Bioorganic chemistry: nucleic acids. Oxford University Press; 1986.

Jencks WP. Catalysis in chemistry and enzymology. McGraw Hill; 1969.

Kornberg A. DNA replication. W. H. Freeman & Co.; 1974/1980.

Mahler HR, Cordes EH. Biological chemistry. 2nd ed. Harper & Row; 1966/1971.

Page MI, editor. The chemistry of enzyme action. Vols. New comprehensive biochemistry, vol. 6. Elsevier; 1984.

Page MI, Williams A. Enzyme mechanisms. Royal Society of Chemistry; 1987.

Sheraga HA. Protein structure. Volume 1: molecular biology. Academic Press; 1961.

Steinhart J, Reynolds JA. Multiple equilibria in proteins. Academic Press; 1969.

Westley J. Enzymic catalysis. Harper & Row; 1969.

Wilcox CS, Hamilton AD. Molecular design and bioorganic catalysis. Kluwer Academic Publishers; 1996.

Wu R, editor. Recombinant DNA methodology II. Academic Press; 1992.

Chapter 2
Structural Insights into Ordered Multicomponent Assemblies in Cell Junctions

Oliver James Harrison (ID), Priyanka Mathews (ID), and Julia Brasch (ID)

Abstract Cell junctions are essential structures of metazoan tissues that maintain cohesion and integrity and permit coordinated tissue rearrangements in development. Vertebrate junctions include stable structures, like adherens junctions, desmosomes, and focal adhesions, and transient, ultrastructurally less-defined complexes that mediate recognition and signaling. All are mediated by assemblies of cell adhesion and recognition proteins at the cell surface and cytoplasmic proteins that mediate signaling or linkage to the cytoskeleton. These multiple components must organize together to form large functional intercellular structures. Structural methods have provided deep insights into how these large structures assemble and revealed roles for intrinsic propensities of adhesion and recognition protein ectodomains to form ordered assemblies on membranes, often mirrored by their intracellular components. In this chapter, we discuss the structural and molecular principles underlying the assembly and organization of these large complexes and focus in detail on three important cadherin-mediated junctions: adherens junctions, desmosomes, and neuronal self-recognition complexes formed by clustered protocadherins. These examples highlight the highly specific and organized arrangements that contribute to junction assembly and how these mechanisms are closely tuned to the biological roles of specific junctions.

Keywords Cell adhesion · Cell junction · Cadherin · Clustered protocadherin · Delta protocadherin · Desmosome · Adherens junction · X-ray crystallography · Cryo-EM · Cryo-ET · Protein assembly · Membrane assembly · Protein interaction · Structure · Structural biology

O. J. Harrison · P. Mathews · J. Brasch (✉)
Department of Biochemistry, University of Utah School of Medicine,
Salt Lake City, UT, USA
e-mail: julia.brasch@biochem.utah.edu

© The Author(s), under exclusive license to Springer Nature Switzerland AG 2026
A. M. Pedley (ed.), *Supramolecular Protein Assemblies In Cells*, Advances in
Experimental Medicine and Biology 1514,
https://doi.org/10.1007/978-3-032-26629-3_2

2.1 Introduction

Cellular junctions are fundamental structures for metazoan life. They provide a link between adjacent cells or between cells and the extracellular matrix to permit cohesion within multicellular structures and allow coordinated movements in tissues throughout development. Cell junctions are diverse in form and function and are formed by a correspondingly diverse set of adhesion proteins at the cell surface and associated proteins below the membrane. Fundamentally, cell junctions represent massive transcellular assemblies of multiple proteins whose properties influence a wide range of events at the tissue and cell level, including migration, patterning, and sorting during development and signaling and homeostasis in a wide range of tissues. Their importance is reflected in their roles in human disease as major regulators of tumor progression and metastasis (Hazan et al. 2004; Hamidi and Ivaska 2018; Yu et al. 2019), neurodevelopmental and behavioral disorders (Laszlo and Lele 2022; Flaherty and Maniatis 2020; Taylor et al. 2020; Jaudon et al. 2021), skin and heart disorders (Nitoiu et al. 2014; Al-Jassar et al. 2013; Waschke 2008), and musculoskeletal diseases (Brancaccio 2019; Goody et al. 2015). This chapter focuses on the principles underlying the assembly of these large structures and, in particular, how specific molecular assembly properties of adhesion proteins can contribute to their organization and order.

2.2 Cell Junctions Are Diverse Centers of Adhesion and Signaling

Cells in vertebrate solid tissues engage in multiple types of cell junctions (Fig. 2.1a), including tight junctions (Citi et al. 2024) that seal the subapical spaces between cells, gap junctions (Mese et al. 2007) that allow exchange of small molecules, adherens junctions (Meng and Takeichi 2009) and desmosomes (Perl et al. 2024) that promote general tissue cohesion and mechanical coupling, and focal adhesions (Legerstee and Houtsmuller 2021) and hemidesmosomes (Walko et al. 2015) that link cells to the extracellular matrix. Together, these junctions are responsible for the mechanical integrity of solid tissues and are vital to the processes of morphogenesis and tissue organization in development (Fig. 2.1b) (Takeichi 2014; Harris and Tepass 2010; Honig and Shapiro 2020; Punovuori et al. 2021). In the nervous system, similar stable cell-cell junctions form in the adhesive region of synapses and can be involved in target recognition and neural circuit formation (Jontes 2018). In addition, more transient junctional structures can mediate dynamic adhesion, migration, and recognition processes between cells to shape the nervous system and other tissues (Hirano and Takeichi 2012; Takeichi 2014). This chapter will deal in detail primarily with the molecular interactions that underlie the assembly of adherens junctions, the fundamental junctions mediating cell-cell adhesion; desmosomes, which preserve tissue integrity, especially in tissues under mechanical stress;

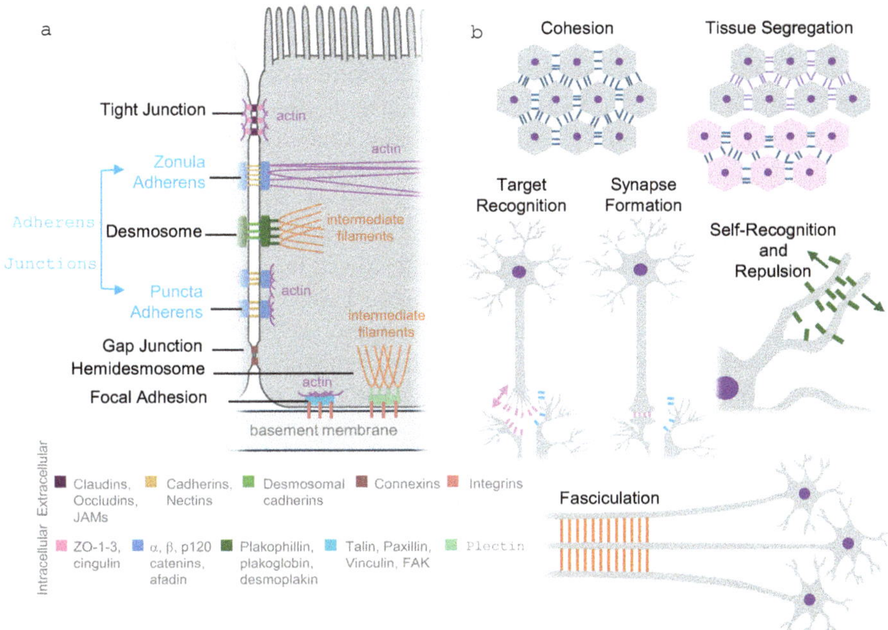

Fig. 2.1 Overview of vertebrate cellular junctions and major tissue patterning roles. (**a**) Schematic showing the general arrangement of cellular junctions in a vertebrate simple polarized epithelium. Tight junctions, zonula adherens junctions, and desmosomes form a tripartite arrangement in the subapical region of the lateral membranes. Desmosomes and punctate adherens junctions are also distributed throughout the lateral membranes, as are gap junctions. Basal attachment is mediated by focal adhesions and hemidesmosomes. For each junction, the core transmembrane, cytoplasmic, and cytoskeletal constituents are detailed in the legend. JAMs: junctional adhesion molecules; ZO-1-3: zonula occludens protein 1–3; FAK: focal adhesion kinase. (**b**) Schematic views showing selected major roles of adhesion in tissue patterning. Cells are shown in gray or pink with purple nuclei; lines between cells in the *upper panels* indicate interactions of adhesion molecules; adhesion molecules themselves are shown as rectangles with compatible binding pairs colored identically

and dynamic neuronal contacts mediated by protocadherins, which can not only link neurons together but also recognize and repel sister neurites in self-recognition.

2.2.1   Cell Junctions Are Multicomponent Structures

Cellular junctions are composed of multiple proteins, including cell surface adhesion and recognition proteins that mediate intercellular interactions and cytoplasmic proteins that mediate signaling, turnover, and cytoskeletal association. At the surface, a majority of cell-cell adhesion molecules belong to two superfamilies: the immunoglobulin superfamily (IgSF) and the cadherin superfamily. Both are large and highly diverse and comprise transmembrane or GPI-linked proteins that present adhesive extracellular regions at the cell surface for interaction. IgSF adhesion proteins (reviewed in Aricescu and Jones 2007) include nectins, DIP/Dprs, and nephrins and contain extracellular immunoglobulin domains that mediate adhesion through protein-protein interactions with themselves or other ligand proteins. Cadherins (reviewed in Brasch et al. (2012)), which are the group that will be described in most detail in this chapter, contain tandem extracellular cadherin (EC) domains with characteristic sequence motifs involved in the binding of three structurally critical calcium ions between each pair of successive domains. Cadherins mediate calcium-dependent adhesion and recognition through their EC domains, usually between identical cadherins or similar members of the cadherin superfamily, though interactions with other proteins have been described (Kilshaw 1999; Whittard et al. 2002). The large cadherin superfamily contains numerous subfamilies that can be grouped by differences in the number and arrangement of EC domains and by the presence of different functional motifs in the cytoplasmic regions (Hulpiau and Van Roy 2009; Nollet et al. 2000). Major subfamilies of cadherins include classical cadherins, which are the core transmembrane components of adherens junctions; desmosomal cadherins, which have a comparable role in desmosomes; and protocadherins, which function in the organization of the nervous system, most notably in self-recognition by neurites.

Below the membrane, cadherins and other cell adhesion molecules are often linked to the cytoskeleton via interactions of their intracellular domains with adapter proteins that in turn bind or modify cytoskeletal components (Pokutta and Weis 2007; Acharya and Yap 2016). Linkage to actin allows for coordinated cell movements through actin filament polymerization and depolymerization and provides resistance to mechanical stress, which allows tissues to reshape without losing their cohesion. Adherens junctions, which are linked to the actin cytoskeleton, provide sufficient cohesion to resist forces during morphogenesis, while desmosomes, which are linked to intermediate filaments, provide additional robustness that is critical in vertebrate skin and cardiac muscle, where physical stresses are pronounced (Hegazy et al. 2022). The classical cadherin and desmosomal cadherin subfamilies associate with different members of the catenin family of adaptor proteins to mediate their linkages to actin and intermediate filaments, respectively. Outside of the cadherin family, diverse adapter proteins occupy analogous roles (see Fig. 2.1a), notably afadin for nectin-mediated cell-cell adhesion and talin for integrin-mediated cell-matrix adhesion (Goult et al. 2018; Mandai et al. 2013). Interestingly, adhesion molecule-cytoskeleton linkages can be mechanosensitive: alpha-catenin and talin can undergo well-characterized force-dependent conformational changes that expose regions that can recruit other adapter proteins to further solidify the coupling of extracellular adhesion and intracellular scaffolds (Yao et al. 2014).

While adapter proteins like catenins provide physical linkage of the scaffold of one cell to that of a neighboring cell, numerous other cytoplasmic proteins are associated with cell-cell junctions and can function in junction regulation, signaling, dynamic modulation of the cytoskeleton, and cell polarity (Zaidel-Bar 2013; Zaidel-Bar et al. 2007). Some surface proteins that function primarily in signaling and recognition rather than adhesion, such as clustered protocadherins, associate primarily with cytoplasmic proteins that are involved in signaling upon membrane contact and appear to dynamically regulate, but not stably associate with, the cytoskeleton (Pancho et al. 2020).

2.2.2 Molecular Properties of Adhesion Proteins Drive Junction Assembly and Function

Cellular junctions are thus built from complex arrays of interacting proteins that must organize to form functional intercellular structures. Among these proteins, cell surface adhesion molecules occupy a unique role in providing the bridge between adjacent membranes and potentially the contact-dependent initial steps of overall junction assembly or signaling. Structural and biophysical studies of adhesion and recognition proteins have uncovered critical molecular properties that allow them to carry out their function.

2.2.2.1 Adhesive Interactions Provide the Specific Glue

At the most basic level, cell adhesion molecules must engage in adhesive *trans* binding across the intermembrane space through protein:protein interactions involving their ectodomains. A hallmark of adhesive *trans* interactions is their high degree of specificity. *Trans* binding is referred to as homophilic when it occurs between identical proteins, and heterophilic when it occurs between two different proteins, either belonging to different families or representing different subtypes within a family (Fig. 2.2). Both modes of binding are common in adhesion molecules, with molecular specificity encoded in surface residues that form the *trans* binding interfaces. Adhesive specificity is also frequently more complicated, with members of a subfamily engaging in both homo- and heterophilic interactions mediated through the same interface with varying affinities. Adhesive specificity of adhesion proteins can underlie tissue patterning in multiple ways (see Fig. 2.1b) (Honig and Shapiro 2020). Whole tissues can express different homophilic adhesion proteins to permit or drive their segregation. This is observed for classical cadherins during vertebrate neurulation when the neural plate, a region that gives rise to the adult nervous system, downregulates epithelial (E)-cadherin and begins to express neural (N)-cadherin as it segregates from the surrounding ectoderm that continues to express the epithelial form (Stepniak et al. 2009; Radice et al. 1997). More generally, switching of classical cadherin expression from E- to N- is a hallmark of epithelial-mesenchymal transitions that occur elsewhere in development and disease (Punovuori et al. 2021). Homophilic specificity can also direct more specific targeting of cells expressing

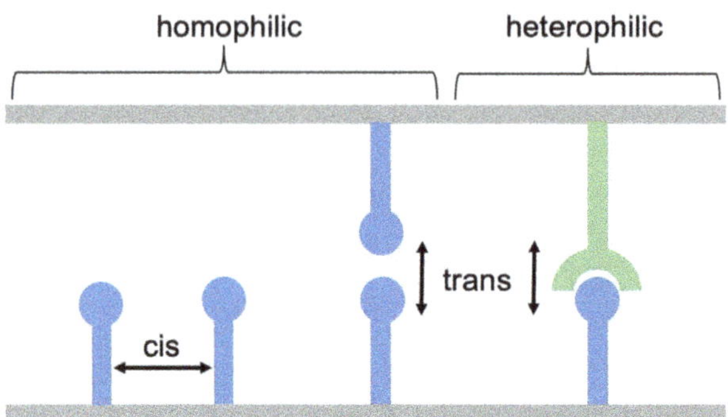

Fig. 2.2 Binding interactions of adhesion proteins. Schematic showing binding interactions (arrows) between cell surface adhesion and recognition proteins. *Trans* interactions form between molecules on opposing membranes from different cells or from different processes of one cell. *Cis* interactions form between adjacent proteins on the same membrane. The specificity of binding interactions can be homophilic, where identical molecules bind, or heterophilic, where the binding partners are nonidentical and may represent different subtypes or isoforms within a protein family or members of different families

matching adhesion proteins, as is observed in neural circuit and motor pool formation mediated by the type II cadherin family (Basu et al. 2018; Brasch et al. 2018; Patel et al. 2006; Price et al. 2002) and for organization of retinal cone cells in the fly eye by classical cadherins (Hayashi and Carthew 2004). Heterophilic adhesive specificity can mediate associations between nonidentical cells or tissues expressing the respective binding partners. This is observed for the nectin family of adhesion proteins whose preferential heterophilic interactions are necessary for heterotypic cell-cell interactions during ear and eye development, nervous system organization, and sperm maturation (Harrison et al. 2012; Togashi et al. 2011; Okabe et al. 2004; Ozaki-Kuroda et al. 2002). More complex arrangements directed by the differential homophilic and heterophilic affinities of nectins are strikingly observed in the inner ear organ of Corti, where a precise checkerboard arrangement of sensory hair cells and supporting cells develops through maximization of high-affinity heterophilic nectin interactions at the expense of weaker homophilic interactions (Togashi et al. 2011). More recently, the DIP/Dpr family of insect IgSF cell adhesion molecules was shown to display a close relationship between affinities of homophilic and heterophilic adhesive *trans* interactions and their effects on wiring of the Drosophila visual system (Cosmanescu et al. 2018; Xu et al. 2018). In addition to binding preference, different affinities of homophilic binding of individual subtypes within a family (Honig and Shapiro 2020) or differences in surface expression levels of a single adhesion molecule (Godt and Tepass 1998) have also been suggested or shown to affect tissue patterning. The combined effects of binding specificity, molecular affinities, and expression levels were incorporated in Steinberg's differential adhesion hypothesis (Steinberg 1970; Foty and Steinberg 2005; Honig and Shapiro 2020), which proposes that tissues reorganize to maximize mutual adhesive forces that arise from the combination of these factors. This hypothesis explains some cases of tissue patterning well (e.g., Godt and Tepass 1998; Steinberg 2007), but in other cases, the linkage of adhesion to cytoskeletal cortical tension needs to be considered to explain tissue behaviors, as accounted for in the later differential interfacial tension hypothesis (Brodland 2002; Foty and Steinberg 2013).

2.2.2.2 Lateral Interactions Promote Junction Assembly

In addition to *trans* interactions, lateral *cis* interactions can form between cell adhesion molecules on the same cell surface (Fig. 2.2). They potentially allow individual *trans* interactions to integrate into ordered assemblies at cell contact sites. Such lateral assemblies can have effects on adhesive strength and on the ability of cell adhesion and recognition molecules to signal, recruiting not only the adhesion molecules themselves but also associated intracellular components. The effective adhesive strength at a contact site is determined by avidity, which is the product of the binding affinities of individual interactions and their multiplicity (valency) over the contact region (Carman and Springer 2003). Binding affinities of adhesion molecule *trans* interactions are frequently quite weak, with K_D values spanning the micromolar range for vertebrate classical and related cadherins (Katsamba et al. 2009; Brasch et al. 2011, 2018; Vendome et al. 2014; Chappuis-Flament et al. 2001; Harrison et al. 2016). Weak individual affinities may have an advantage in enhancing the regulatability and reversibility of adhesive interactions (Gumbiner 2005; Shapiro and Weis 2009; Brasch et al. 2012). The combination of organized *cis* and *trans* interactions, however, can amplify the weak individual *trans* interactions to increase avidity and provide overall adhesion strong enough to influence cell shape and resist morphogenetic forces. Classical and desmosomal cadherins represent good examples of adhesion proteins with weak *trans* interactions (Harrison et al. 2016; Katsamba et al. 2009; Syed et al. 2002) that are nonetheless able to mediate strong adhesion by assembling into ordered intermembrane structures in adherens junctions and desmosomes, respectively (see Sects. 2.3 and 2.4).

Signaling in response to cell recognition or adhesion, termed outside-in signaling, requires a mechanism to communicate changes induced by binding in the extracellular region to the intracellular domains where signal activation occurs. Some adhesion proteins, such as adhesion GPCRs, contain large seven-pass transmembrane domains that allow extracellular changes induced by ligand binding to alter the relative apposition of the transmembrane helices to activate coupled signaling proteins below the membrane, though precise mechanisms of adhesion GPCR signaling are still under investigation (Vizurraga et al. 2020). Integrins, a large family of cell-matrix and cell-cell adhesion proteins, have a single-pass transmembrane region but function as α/β-heterodimers on the cell surface. This heterodimer architecture allows them to adopt multiple conformational states that bidirectionally couple the structure of their extracellular ligand binding domains with the apposition of their transmembrane and cytoplasmic signaling domains (Luo et al. 2007). The majority of adhesion and recognition proteins, however, have simple single-pass transmembrane regions and lack a constitutive dimer structure analogous to that of integrins. In these proteins, lateral interactions can play a critical role in coupling adhesive binding with signaling. A combination of *trans* and *cis* binding can elaborate continuous structures, including lattices that extend in two dimensions at membrane contact regions, as seen for classical and desmosomal cadherins (Harrison et al. 2011; Sikora et al. 2020) and vertebrate Dscams (Guo et al. 2021), or linear zipper-like arrangements as seen for clustered protocadherins (Brasch et al. 2019), ephrins (Seiradake et al. 2010), and multiple members of the immunoglobulin superfamily (Aricescu and Jones 2007). In each case, the formation of ordered assemblies of extracellular regions necessarily induces ordered or semi-ordered apposition of transmembrane and cytoplasmic domains that can affect their signaling. The relationship between extracellular and intracellular ordered assembly has been strikingly visualized for desmosome junctions in human skin using cryo-ET (Al-Amoudi et al. 2007, 2011; see Sect. 2.4).

In the following sections, we will discuss the mechanism of assembly for three classes of critical cadherin adhesion and recognition molecules. The mechanisms described reflect a combination of insights from methods such as cellular imaging, biophysical characterization, bioinformatics, x-ray crystallography, cryo-EM, and cryo-electron tomography (cryo-ET) of native or reconstituted junctions.

2.3 Spontaneous Ectodomain Assembly in Vertebrate Adherens Junctions

Vertebrate adherens junctions are present in essentially all vertebrate solid tissues from the earliest stages of development into adulthood. They take on ultrastructurally distinct forms in different tissues (Niessen and Gottardi 2008; Yap et al. 1997a). In simple polarized epithelia like the intestinal epithelium, adherens junctions form a belt-like zonula adherens (see Fig. 2.1a) that rings the sub-apical region of the lateral membranes of adjacent cells, whereas in heart muscle, adherens junctions are arranged as broad ribbon-like fascia adherens that connect cardiac myocytes end-to-end at intercalated disc regions (Nielsen et al. 2023). In addition, smaller spot-like punctate adherens junctions are observed along the whole lateral membranes of epithelial and non-epithelial cells (Fig. 2.3a) (Indra et al. 2018; Kametani and Takeichi 2007). All adherens junctions have a common core structure comprising classical cadherins that form intercellular connections between the membranes, and adapter proteins α-,β-, and p120 catenins that together associate with the cadherin cytoplasmic tail and actin filaments to allow adherens junctions to link the internal scaffolds of adjacent cells (Fig. 2.3b).

There are around 20 classical cadherins encoded in vertebrate genomes that are divided into type I classical cadherins (e.g., epithelial E-cadherin, neural N-cadherin, retinal R-cadherin, placental P-cadherin, and Xenopus-specific compaction C-cadherin) and type II classical cadherins (e.g., cadherins-5–12, 18–20, 22, and

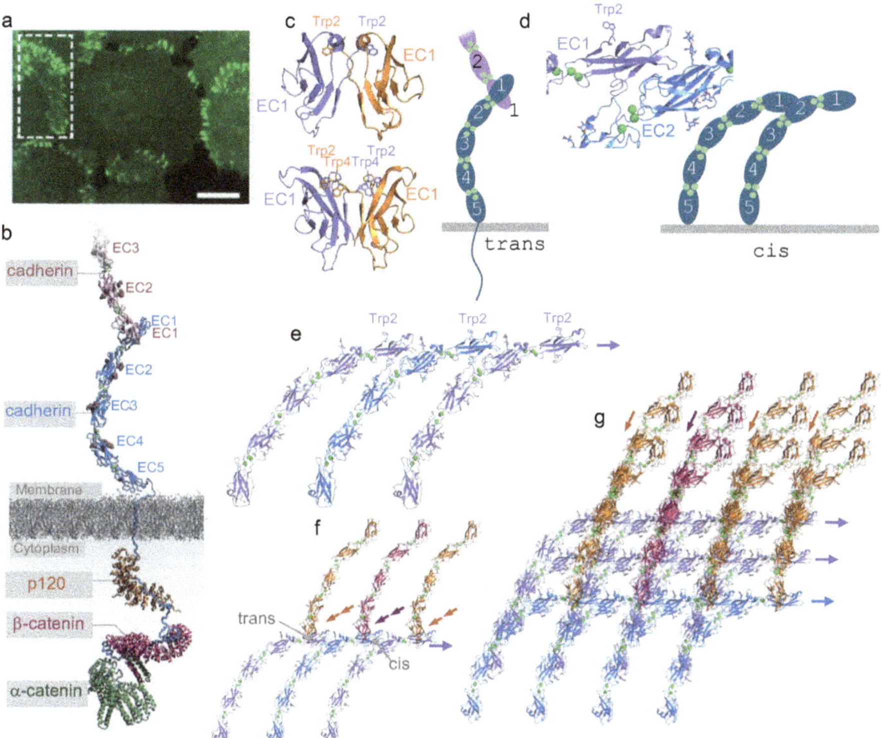

Fig. 2.3 Extracellular molecular assembly in vertebrate adherens junctions. (**a**) Fluorescence microscopy of A431 epithelial cells expressing GFP-tagged E-cadherin. E-cadherin is concentrated at cell-cell contacts and adopts a punctate distribution between lateral membranes (boxed). Scale bar 10 μm. (Adapted from Troyanovsky et al. (2021)). (**b**) Schematic showing the core classical cadherin adhesive complex assembled from experimental structures of E-cadherin ectodomain (PDB:3Q2V) in blue, with adhesive binding partner in pink, and adapter proteins p120-catenin (PDB:3L6X), β-catenin (PDB:1I7W), and α-catenin (PDB:1DOW, 4IGG, 4K1N). p120 and β-catenin bind directly to the cadherin cytoplasmic domain (blue), while α-catenin associates indirectly by binding β-catenin. E-cadherin transmembrane and cytoplasmic regions of unknown structure that link the known components are modelled. Calcium ions shown as green spheres; glycans as dark rose spheres. (Adapted from Takeichi (2014)). (**c**) Detailed view of type I (*top*) and type II (*bottom*) classical cadherin *trans* interfaces from crystal structures of N-cadherin EC1–5 (PDB: 3Q2W) and cadherin-22 EC1–2 (PDB: 6CG7). EC1 domains of interacting protomers are shown in blue and orange ribbon representation with Trp side chains anchoring the strand-swapped interaction shown as sticks. *Right panel* shows a schematic representation of the *trans* dimer between complete molecules. (**d**) Detailed view of the type I cadherin *cis* interface from the crystal structure of N-cadherin EC1–5 (PDB:3Q2W). Interacting EC1 and EC2 domains, providing the concave and convex faces, respectively, are shown in light and dark blue. *Right panel* shows schematic of complete ectodomains interacting in *cis*. (**e**) Linear array of three cadherin ectodomains interacting via the *cis* interface, from the crystal lattice of N-cadherin EC1–5 (PDB: 3Q2W). (**f**) Linear *cis* array of three cadherins (blue) engaged in *trans* interactions with cadherins from an opposing membrane (orange/magenta) that belong to *cis* arrays oriented approximately perpendicular to those in blue (arrows). (**g**) Molecular assembly formed by combined *cis* and *trans* interfaces, where linear arrays intersect to form a grid-like pattern that extends in two dimensions with membrane-proximal EC5 domains on parallel membrane planes. From the crystal lattice of N-cadherin EC1–5 (PDB: 3Q2W)

24) based on sequence (Nollet et al. 2000). All share a common domain structure comprising an ectodomain of five EC domains, numbered EC1–5 starting from the membrane-distal N-terminus; a single transmembrane region; and a cytoplasmic domain containing binding motifs for catenin adapter proteins (Fig. 2.3b). Type I cadherins are very broadly expressed throughout whole tissues, while type II cadherins are expressed in more fine-grained patterns restricted to individual cell types or regions, particularly in the central nervous system, where they can be responsible for neural circuit assembly (Fig. 2.1b) (Sanes and Zipursky 2020). Adherens junction formation is best characterized for type I classical cadherins, likely due to their wide expression patterns and presence in ultrastructurally discernible adherens junctions, such as the zonula adherens and fascia adherens. Type II cadherins associate with the cytoskeleton through the same adapter proteins, so are likely to also form adherens junctions (Kiener et al. 2006), but these are less well-studied. The best characterized type II cadherin junctions are those of vascular endothelial (VE) cadherin, a somewhat divergent type II cadherin (Brasch et al. 2011), which is found exclusively in the junctions of endothelial cells lining blood vessels (Uehara 2006; Kametani and Takeichi 2007).

2.3.1 Adhesive Trans Interactions in Adherens Junctions

Extensive structural and mutagenesis studies have identified the *trans* interface that mediates classical cadherin adhesive interactions (Tamura et al. 1998; Shapiro et al. 1995; Parisini et al. 2007; Boggon et al. 2002; Harrison et al. 2011; Brasch et al. 2011, 2018; Patel et al. 2006; Vendome et al. 2014; Dalle Vedove et al. 2015; Kudo et al. 2016). Adhesion is mediated by a strand swapping mechanism in which the N-terminal A-strand of the cadherin EC1 domain disengages from its own domain in which it can self-dock and instead docks into a conserved hydrophobic pocket of a partner cadherin, which reciprocates the interaction (Fig. 2.3c). The docking is anchored by a single conserved tryptophan residue (Trp2) in the A-strand of type I classical cadherins or by two such residues (Trp2 and Trp4) in type II cadherins (Fig. 2.3c). While the *trans* interaction positions EC1 domains approximately in parallel, overall curvature of the cadherin ectodomains allows it to link cadherins that emanate from opposing cell surfaces (Fig. 2.3c). Type I and type II classical cadherins do not interact with each other in trans, owing to the differences in their swapped elements and pocket residues, which can allow them to function orthogonally (Patel et al. 2006; Brasch et al. 2011, 2018; Shimoyama et al. 2000). Within each group, the different subtypes of classical cadherins show specificity in their *trans* interactions that may be critical for their roles in cell sorting and tissue organization (Honig and Shapiro 2020; Sanes and Zipursky 2020). All classical cadherins can bind homophilically, which permits their general role in tissue cohesion, and a preference for homophilic binding observed for many classical cadherins is thought to be important for permitting or driving the separation of tissues expressing different cadherin subtypes (Hayashi and Takeichi 2015; Takeichi 2014; Honig and Shapiro 2020) and for targeting of neurons with matching cadherins (Basu et al. 2018). In addition, heterophilic *trans* binding between different individual cadherins occurs in specific combinations to give a wide range of potential heterophilic interactions that may also play roles in tissue organization (Katsamba et al. 2009; Basu et al. 2018; Brasch et al. 2018; Shimoyama et al. 2000).

2.3.2 *Lateral Assembly of Classical Cadherins in Adherens Junctions*

It is notable that the binding affinities for individual classical cadherin *trans* interactions are comparatively weak in the micromolar to hundred micromolar range (Katsamba et al. 2009; Brasch et al. 2011, 2018; Vendome et al. 2014; Chappuis-Flament et al. 2001). Nonetheless, these interactions mediate strong intercellular cohesion owing to the multiplicity of such interactions between contacting membranes (for more detail, see, e.g., Honig and Shapiro 2020). Early studies showed that cadherins accumulate and cluster rapidly at sites of adhesive cell-cell contact (Vasioukhin et al. 2000; Adams et al. 1996) and that such clustering could contribute to strengthening their adhesive force (Yap et al. 1997b; Brieher et al. 1996). A role for the cadherin cytoplasmic domain and adaptor proteins in clustering and junction assembly was established (Yap et al. 1998; Nagafuchi and Takeichi 1988). Later, however, it was found that the ectodomain of type I classical cadherins itself could also mediate assembly of cadherin molecules into specific ordered extracellular assemblies through the combination of *trans* and specific *cis* interactions. Extracellular *cis* interactions could not be detected for ectodomains in solution but were revealed by X-ray crystallography (Boggon et al. 2002; Harrison et al. 2011). The method likely allowed very weak *cis* interactions to be observed due to high local concentrations of cadherin during crystallization. Topologically identical *cis* interfaces were observed in crystal lattices first of full-ectodomain C-cadherin (Boggon et al. 2002) then of N- and E-cadherin (Harrison et al. 2011). The *cis* interactions are formed through an interface between a concave surface on the EC1 domain opposite the *trans* binding site and a convex face at the base of the EC2 domain close to the EC2–3 junction (Fig. 2.3d). The *cis* interface positions the curved cadherin ectodomains in parallel and allows for propagation of a linear array (Fig. 2.3d, e). Critically, orthogonal orientation of the *trans* and *cis* interfaces allows a two-dimensional lattice of cadherin ectodomains to assemble in which linear arrays from the respective putative membrane planes intersect (Fig. 2.3f, g). The assemblies observed in the crystal lattices provided a potential mechanism by which cadherins could assemble in an ectodomain-driven manner between cell membranes.

Identification of a specific potential *cis* interface allowed its biological role to be probed by targeted mutagenesis in a cellular context. Ablation of the interface in transfected A431D cells showed a modest effect on full-length cadherin, whereby fluorescent cadherins were observed to become more distributed over the cell surface and concentrated less efficiently in cell junctions (Harrison et al. 2011). The effect was strikingly more pronounced when cadherins lacking their cytoplasmic domains were used, such that they depended entirely on their ectodomains for assembly (Harrison et al. 2011). In these experiments, the *cis* interface mutant cadherins failed to form junctions and were evenly distributed over the cell surfaces, even though the *trans* interface was shown to be fully functional in solution. Subsequent mutagenesis studies in a variety of systems have also shown that aspects of adherens junction formation and function depend on the weak extracellular *cis* interactions identified by X-ray crystallography (Thompson et al. 2020; Strale et al. 2015). Molecular simulations suggest that assembly of cadherins in adherens junctions could be driven by the combination of specific ectodomain *cis* and *trans* interactions, in addition to clustering via the cytoplasmic domains (Wu et al. 2011; Chen et al. 2021). Specifically, they reveal that assembly of cadherin ectodomains into ordered *cis-trans* lattices may occur as a cooperative process by which nascent adherens junctions could nucleate upon first *trans* contact (Wu et al. 2011; Chen et al. 2021). Initial *trans* binding favors formation of *cis* interactions by constraining the cadherin ectodomains and reducing the entropic cost of forming *cis* bonds, which are potentially too weak to form efficiently between flexible free monomers. Such extracellular mechanisms of spontaneous nanoscale assembly upon contact are suggested to be of potentially broad importance in adhesion and signaling processes at the cell surface (Boni et al. 2022).

On the scale of whole mature adherens junctions, ordered ectodomain-mediated assemblies of type I classical cadherins appear to represent nanostructures within larger cellular adherens junctions, in some cases interspersed with clusters of nectin adhesive complexes (Strale et al. 2015; Indra et al. 2013, 2018). Ordered extracellular cadherin density is not observed across overall junctions in electron micrographs (Le Bihan et al. 2015; Rayns et al. 1969), suggesting that the ordered nanoclusters undergo wider, less uniformly ordered, assembly into full adherens junctions through distinct, likely cytoplasmic, mechanisms. Recent studies have shown that the intermolecular spacing of the extracellular lattice is compatible with submembrane assembly of core catenin components of the adhesive complex, whereas additional proteins, such as the cell polarity proteins erbin and scribble, are likely to associate outside of the nanocluster regions (Troyanovsky et al. 2021).

Notably, type II classical cadherins, which are closely related to type I cadherins and form a similar strand-swapped *trans* dimer, did not contain the type I cadherin *cis* interface in any of their numerous EC1–2 and EC1–3 crystal structures (Brasch et al. 2018; Patel et al. 2006). This absence suggests that the junctional assembly of this subfamily differs from that of type I cadherins and involves distinct interactions. Their mechanisms of assembly remain to be defined, as does their overall comparability to junctions formed by type I cadherins.

2.4 Extensive Ordered Assembly Throughout Whole Desmosome Junctions

Desmosomes are a major class of vertebrate adhesive junctions that link the intermediate filament networks of adjacent cells in a wide variety of solid tissues, where they are present alongside actin-based adherens junctions. Functionally, desmosomes appear to be particularly required to provide strong intercellular cohesion in tissues that experience high mechanical stress, such as skin and heart muscle. Phenotypes arising from experimental ablation of desmosomes in mice or from inherited mutations in human disease predominantly affect these tissues, leading to severe skin fragility or heart arrhythmias (Kowalczyk and Green 2013; Koch et al. 1997; Krusche et al. 2011; Chidgey et al. 2001; Chen et al. 2008; Al-Jassar et al. 2013). Desmosomes are also the target of autoantibodies in pemphigus, a group of well-characterized autoimmune diseases in which the disassembly of desmosome junctions by autoantibodies results in widespread blistering and loss of barrier function in skin and mucosae (Spindler and Waschke 2018; Kasperkiewicz et al. 2017).

The ultrastructure of desmosomes is highly characteristic in electron micrographs (Fig. 2.4a). Desmosomes appear as large (up to 1 µm) (Garrod and Chidgey 2008) "spot-weld" like junctions with a dense submembrane plaque composed of adapter proteins and associated intermediate filaments, separable into inner and outer dense regions. Between the apposed membranes, a constant intermembrane spacing is maintained that appears slightly wider than that of adherens junctions (~25–35 nm vs 15–30 nm) (Desai et al. 2009; Al-Amoudi et al. 2007; Mcnutt and Weinstein 1973; Farquhar and Palade 1963; Fawcett and Mcnutt 1969; Miyaguchi 2000; Sikora et al. 2020). Within this, a dense accumulation of adhesion molecule extracellular domains is observed, which, in some preparations, has a characteristic dense midline and evidence of periodic order that persists throughout the entire width of the junction (Fig. 2.4a). As expected from their dense and highly ordered structure, mature desmosomes can be extremely stable and are thought to adopt a "hyper-adhesive" state in which they persist for long periods before being disassembled by internalization of the whole structure (Garrod and Tabernero 2014).

2.4.1 Specialized Cadherins Bridge the Desmosome Intercellular Space

Adhesion in desmosomes is mediated by members of two cadherin subfamilies, desmocollins (1–3) and desmogleins (1–4), that are related to classical cadherins in their domain organization (Fig. 2.4b) but which associate with distinct cytoplasmic catenin adapter proteins that link to the intermediate filament system (Nollet et al. 2000). Individual desmocollins and desmogleins differ in their expression patterns, with desmoglein-2 and desmocollin-2 being widely expressed throughout solid tissues, including cardiac muscle, and the remaining subtypes being more restricted to

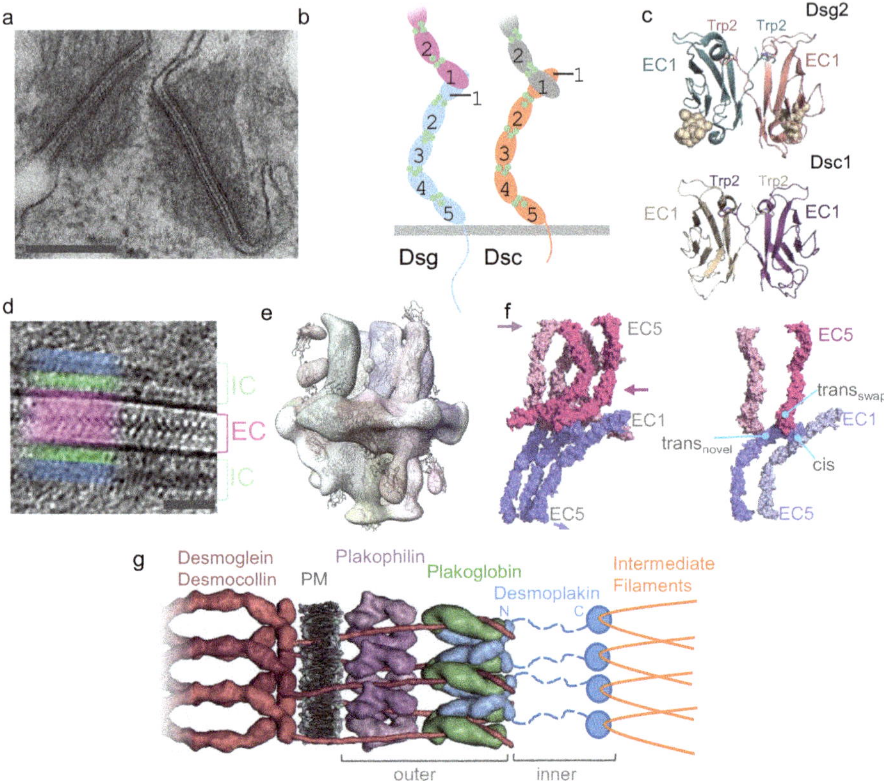

Fig. 2.4 Ordered assembly of desmosome extracellular and intracellular regions. (**a**) Transmission electron micrograph view of desmosomes in human skin. Note the extensive parallel membranes, electron-dense cytoplasmic plaques, and intermembrane density, including a midline and evidence of regular arrangement of extracellular adhesive regions. Scale bar 250 nm. (Adapted from McGrath (1999)). (**b**) Schematic showing domain organization of desmogleins (blue) and desmocollins (orange) engaged in *trans* binding with partners shown in purple and gray. Dsg: Desmoglein; Dsc: Desmocollin. (**c**) Detailed view of strand-swapped *trans* interfaces of desmoglein-2 and desmocollin-1 from EC1–5 fragment crystal structures (PDB:5ERD,5IRY). Ribbon representation is shown colored by protomer; anchoring Trp2 residues are shown as sticks. Glycans shown as wheat spheres. (**d**) Cryo-EM projection image of a vitrified desmosome from human skin showing highly ordered intermembrane cadherin density (pink) with correspondingly ordered density in the cytoplasmic outer dense plaque, in which plakophilin (green) and plakoglobin (blue) layers can be discerned. Scale bar 40 nm. (Adapted from Al-Amoudi et al. (2011)). (**e**) Cryo-ET density map from subtomogram averaging of purified mouse liver desmosomes into which molecular models of desmosomal cadherin ectodomains are fitted. (Adapted from Sikora et al. (2020)). (**f**) Molecular models of the assembly of desmosomal cadherin ectodomains in purified mouse liver desmosomes from flexible fitting to cryo-ET density (PDB: 7A7D). Molecules from opposing membranes are colored blue and magenta and engage in a network of strand-swapped *trans* dimers, classical cadherin-like *cis* interactions, and novel *trans* dimers (see text). (**g**) Integrated model of desmosome molecular organization above and below the membrane derived from fitting stuctures of desmosomal cadherin ectodomains (red), plakophilin (purple, PDB:1XM9), plakoglobin (green, modelled on β-catenin, PDB: 2Z6G0), and desmoplakin (blue, modelled on spectrin, PDB:1U4Q, with PDB: 1DOW used to model the interaction with plakoglobin) to densities from cryo-ET of human skin desmosomes. Regions of unknown structure are modeled (cadherin transmembrane and cytoplasmic regions) or shown schematically (desmoplakin C-terminus and intermediate filaments). (Adapted from Al-Amoudi et al. (2011))

stratified epithelia, such as the skin, where their precise expression varies by layer (Perl et al. 2024). X-ray crystallography and mutagenesis approaches have revealed that both types of desmosomal cadherins can form strand-swapped *trans* dimers anchored by docking of a single tryptophan (Trp2) as observed for type I classical cadherins (Fig. 2.4c) (Harrison et al. 2016; Schinner et al. 2022; Lowndes et al. 2014; Nie et al. 2011). Like classical cadherins, desmosomal cadherins have a curved overall ectodomain structure that allows the *trans* dimers to link cadherins from opposing cells. Notably, this curvature appears to be particularly flexible for desmosomal cadherins owing to incomplete calcium ion coordination in the EC3–4 junction, a feature that could impart additional robustness to the linkages by allowing them to flex under stress (Harrison et al. 2016; Tariq et al. 2015). Notably, SPR and bead aggregation experiments show that heterophilic strand-swapped *trans* binding between desmocollins and desmogleins is of markedly higher affinity than either of the respective homophilic *trans* bonds, suggesting that desmosomes could be heterophilic structures (Harrison et al. 2016). Nevertheless, homophilic as well as heterophilic *trans* binding interactions were readily detected in single-molecule force experiments (Shafraz et al. 2018; Priest et al. 2019; Lowndes et al. 2014) and homophilic adducts were also found in a cross-linking study of living cells (Nie et al. 2011). Both desmocollins and desmogleins have essential roles in proper desmosomal adhesion (Marcozzi et al. 1998; Koch et al. 1997; Chidgey et al. 2001), yet it remains an open question as to how their roles differ and how they are arranged with respect to each other in the junction.

2.4.2 *Unweaving the Interactions That Maintain Molecular Order in Desmosomes*

The dense and highly ordered intercellular density of desmosomes in electron micrographs is highly suggestive of assembly of desmosomal cadherins into a specific lattice structure that would likely involve *cis* interactions along with the characterized *trans* dimers. Unlike those of classical cadherins, crystal structures of full ectodomains of several desmogleins and a desmocollin did not reveal conserved candidate *cis* interfaces that could orient the cadherins in parallel, possibly reflecting a requirement for a mixture of subtypes not present in the crystallization experiments (Harrison et al. 2016). Major strides in understanding the assembly of desmosomes came from the methods of electron tomography (ET) and cryo-electron tomography (cryo-ET) (He et al. 2003; Sikora et al. 2020; Al-Amoudi et al. 2007, 2011). These methods involve the collection of low-dose electron microscopy images from a sample as it is physically tilted through a range of angles to provide 3D density information. Resolution is typically lower than single-particle cryo-EM but can approach subnanometer through subtomogram averaging (Ni et al. 2022; Zivanov et al. 2022; Castano-Diez et al. 2012). A specific strength of cryo-ET is that it can be readily applied to native and near-native frozen samples to provide in situ molecular-resolution information.

The first ET study of desmosomes examined neonatal mouse skin sections prepared from freeze-substituted and resin-embedded tissue (He et al. 2003). Tomographic reconstructions of intercellular regions of the mouse skin desmosomes readily identified elongated curved densities corresponding closely to the overall classical/desmosomal cadherin ectodomain structures from x-ray crystallography. In addition, EC1-mediated *trans* interactions were also discernible from the densities and were compatible in orientation with strand-swapped dimers. Outside of the *trans* interactions, however, little overall order was observed in the arrangement of desmosomal cadherins, likely due to the stresses of the plastic embedding and sectioning process. These limitations were overcome in two subsequent cryo-ET studies. The first examined vitrified native desmosomes in human skin to allow preservation of the tissue as close as possible to its native state (Al-Amoudi et al. 2007, 2011). This revealed a striking level of order in the desmosomal intercellular space in which the dense midline and periodic lattice-like arrangement of desmosomal cadherin extracellular domains, including their characteristic curvature, were observed (Fig. 2.4d). Sub-tomogram averaging between similar regions of the desmosomes maximized resolution to around 30 Å.

Fitting of available cadherin ectodomain structures into the density suggested a lattice model distinct from that observed for classical cadherins. In the model, strand-swapped dimers are formed, and cadherins on the same membrane are arranged in linear arrays as observed for classical cadherins; however, adjacent arrays are antiparallel with respect to each other and are not maintained by the classical cadherin *cis* interface, instead interacting in a zipper-like fashion via their EC1 domains. The model was modified in the second study by the same group that examined partially purified, vitrified rat liver desmosomes using cryo-ET augmented by constrained molecular dynamics for flexible model fitting (Fig. 2.4e) (Sikora et al. 2020). While still supporting an antiparallel arrangement of cadherin arrays, the modified model suggested that three distinct interactions formed the lattice (Fig. 2.4f): the strand-swap *trans* interaction; a *cis* interface between EC1 and EC2, as seen for classical cadherins; and a novel *trans* interface. The latter is proposed to form between the same EC1 and EC2 regions that mediate the *cis* interface, but in a radically different interface orientation that links cadherins from opposing membranes. Molecular simulations suggested that this *trans-cis-trans* lattice could act in a truss-like manner to maintain desmosome stiffness in the face of forces perpendicular to the membrane (Sikora et al. 2020). While other models of cadherin assembly could be compatible with the observed densities from cryo-ET(Fig. 2.4e), given the modest resolution (~26 Å), the current model provided the best explanation for known biophysical properties of desmosomes compared to a range of alternate models tested in the molecular dynamics simulations (Sikora et al. 2020). The resolution also precluded differentiation of the two types of desmosomal cadherins thought to be present in the lattice (desmoglein-2 and desmocollin-2), given their close similarity in overall ectodomain structure (Harrison et al. 2016). However, the physiological relevance of the current model and the roles of different cadherin types are now amenable to future testing by targeted mutagenesis since the model makes clear predictions of interface regions.

2.4.3 Molecular Order Extends to Desmosome Cytoplasmic Regions

Below the membrane, cryo-ET has also provided insights into how the apparent order of the desmosome extracellular structure is reflected in the arrangement of intracellular components (Fig. 2.4d, g) (Al-Amoudi et al. 2011). Desmocollins and desmogleins bind to adapter proteins plakophilin and plakoglobin directly via their cytoplasmic domains and indirectly to the large armadillo protein desmoplakin that links them to intermediate filaments. The dense arrangement of these elements produces the desmosomal plaque observed in electron micrographs (see Fig. 2.4a, d). Comprehensive mapping by immunogold labelling showed that plakophilin, plakoglobin, the cadherin cytoplasmic regions, and the desmoplakin N-terminus form the "outer" dense plaque close to the membrane, while the "inner" dense plaque, further from the membrane, contains the desmoplakin C-terminus and intermediate filaments (North et al. 1999). Al-Amoudi and colleagues built on this map using cryo-ET density of the outer dense plaque from subtomogram averaging of desmosomes in vitreous cryo-sections of human skin (Al-Amoudi et al. 2011). This allowed the outer dense plaque to be divided into plakophilin and plakoglobin layers separated by a narrow electron-lucent region (Fig. 2.4d). Critically, both layers displayed highly ordered density with a periodicity corresponding closely to that seen in the extracellular region (~70 Å). Fitting of available molecular structures into the density suggested a model for how plakophilin could be arranged in the first layer (Fig. 2.4g), with each cadherin cytoplasmic domain binding one plakophilin monomer. The best-fitting model for the plakoglobin layer contained a repeating V-shaped unit containing a plakoglobin monomer and the N-terminal plakin domain of desmoplakin (Fig. 2.4g). Further from the membrane, the inner dense plaque did not display appreciable order, suggesting that the flexibility of the long desmoplakin molecule "washes out" order at these long distances.

Along with subsequent analyses (Dean and Mattheyses 2022; Pasani et al. 2024), the immunogold and cryo-ET studies provide a remarkably comprehensive model for an ordered arrangement of desmosomal cadherins and their adapter proteins that extends from the intercellular space to approximately 250 Å below the membrane (Fig. 2.4d, g). The propagation of order across the membrane suggests that extracellular and intracellular structures in desmosomes could be interdependent. An extensive literature suggests that desmosome formation and maintenance are influenced by the phosphorylation state of the cadherin cytoplasmic tails and adapter proteins (Kimura et al. 2012; Spindler and Waschke 2018). Whether order and disorder in the extracellular region can also induce changes in cytoplasmic assembly remains to be investigated.

2.5 Specialized Protocadherin Signaling and Adhesion Complexes in the Nervous System

Vertebrate nervous systems represent highly complex arrangements of millions to billions of individual neurons that depend on precise intercellular connections for their coordinated function. At a fundamental level, neural development depends on a range of adhesion and recognition mechanisms that include axon pathfinding, target recognition, and self-avoidance (see Fig. 2.1b). The latter mechanism is a basic, but critical, requirement for allowing neurons to extend neurites throughout their territory without aberrant formation of autapses when sister neurites from the same cell meet (Fig. 2.5a). A critical component of self-avoidance is the ability of individual neurons to distinguish "self" from "non-self" using cell surface cues; only recognition of self should activate an avoidance pathway. Cell surface adhesion and recognition molecules can provide neurons with unique surface identities; however, this requires a very large number of distinguishable recognition codes to minimize the probability that different neurons share the same complement and are erroneously recognized as "self." In insect nervous systems, the Dscam family of cell adhesion molecules carries out this role (Hattori et al. 2008). Transcripts of the Dscam1 gene are alternatively spliced in three regions encoding parts of the extracellular recognition domains to generate a repertoire of >19,000 individual specificities sufficient to mediate self-recognition and avoidance (Wojtowicz et al. 2004). In vertebrates, the Dscam family does not undergo extensive alternative splicing and consequently does not appear to be generally responsible for self-recognition and avoidance (Guo et al. 2021). Instead, vertebrate genomes contain a large, specialized cluster of cadherin genes that fulfil this role.

2.5.1 A Protocadherin Gene Cluster that Can Encode Neuronal Identity

The clustered protocadherin family is one of the largest cadherin subfamilies and contains 50–60 isoforms in mammals. They are encoded by a large, regulated gene-cluster composed of three subclusters, α, β, and γ (Wu and Maniatis 1999). Each subcluster contains a close tandem arrangement of multiple variable exons, each with its own promoter, that encode extracellular and transmembrane anchor regions of different protocadherin isoforms. Clustered protocadherin isoforms are stochastically expressed in individual neurons by alternative promoter choice and, for α and γ clusters, splicing to downstream constant cytoplasmic region exons (Canzio and Maniatis 2019; Zipursky and Sanes 2010). Each encoded clustered protocadherin contains a variable extracellular binding domain that diverges in sequence from other isoforms while retaining a common EC1–6 structure, coupled with a constant cytoplasmic signaling domain (Fig. 2.5b). Critically, up to 15 clustered protocadherin isoforms are expressed per neuron, depending on neuron type (Yagi 2012; Lv

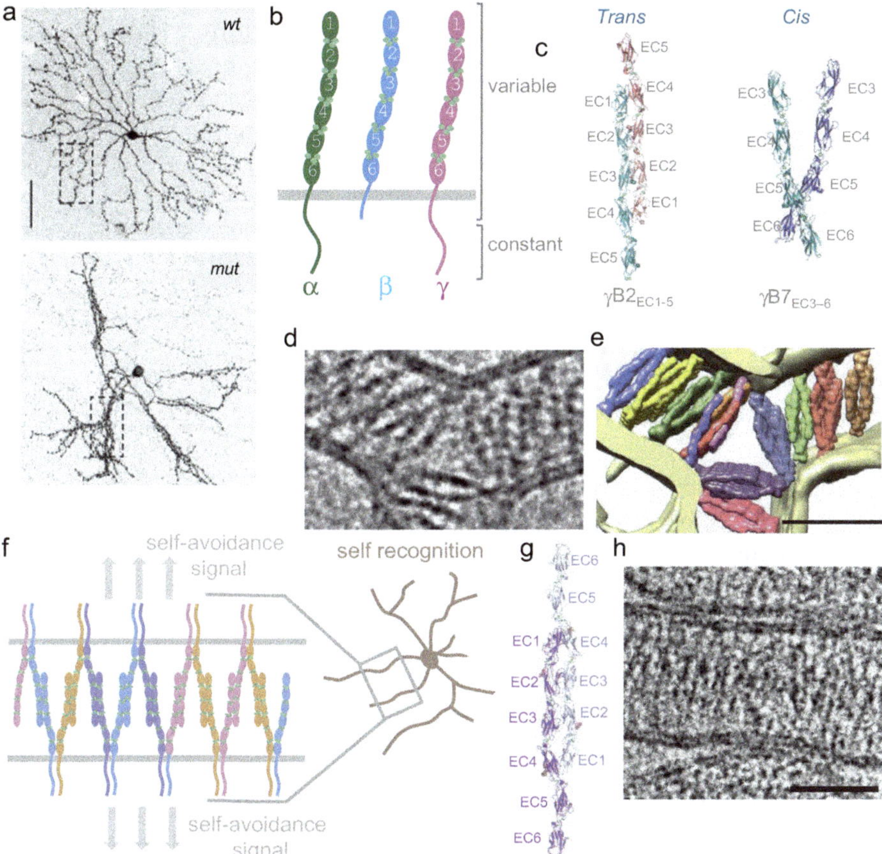

Fig. 2.5 Assembly of clustered protocadherin signaling complexes for neuronal self-recognition. (**a**) Light microscopy image of a starburst amacrine cell from wild-type (*top*) and protocadherin gamma cluster knockout mice (*bottom*). Self-avoidance is lost in the mutant. Scale bar 50 μm. (Adapted from Lefebvre et al. (2012)). (**b**) Schematic showing domain organization of α, β, and γ clustered protocadherins. Calcium ions shown in green. (**c**) Crystal structures showing *trans* and *cis* interfaces of clustered protocadherins, colored by protomer. From PDB:5T9T, 5V5X. (**d**) Slice from a tomographic reconstruction showing assembly of purified protocadherin γB6 ectodomains between coated liposome membranes. Scale bar for (**d**) and (**e**): 350 Å. (**e**) Segmented tomogram view of protocadherin yB6 assembly with membranes colored in wheat and assembled zippers differentiated by color. (Panels (**e**) and (**f**) adapted from Brasch et al. (2019)). (**f**) Model for self-recognition and avoidance by clustered protocadherins. Contacting neurites from a single neuron express identical surface protocadherin complements, allowing extensive zippers to assemble and initiate repulsion (see text). (**g**) Crystal structure showing the δ protocadherin 8.1 trans dimer from Xenopus. From PDB: 6VG1. Glycans shown as wheat spheres. (**h**) Slice from a tomographic reconstruction showing assembly of purified non-clustered δ1 protocadherin ectodomains between coated liposome membranes. Note the lack of zipper-like structures compared to the clustered protocadherin in panel (**d**). Scale bar: 400 Å. (Adapted from Harrison et al. (2020))

et al. 2022), including up to five divergent "c-type" forms that are more uniformly expressed across neurons (Yagi 2012; Canzio and Maniatis 2019). This combinatorial stochastic expression has the potential to generate an enormous number of individual neuron "identities" for self-recognition. Indeed, a central role for clustered protocadherins in self-avoidance was established in functional studies in which deletion of the gamma subcluster in mice disrupted dendritic self-avoidance in cerebellar Purkinje cells and retinal starburst amacrine cells (Lefebvre et al. 2012; Ing-Esteves and Lefebvre 2024) (Fig. 2.5a) and deletion of all three subclusters leads to clumping of olfactory sensory neurons (Mountoufaris et al. 2017).

2.5.2   Zipper-Like Assembly of Clustered Protocadherins Drives Self-Recognition

Encoding of neuronal identity by a cell-specific repertoire of up to 15 clustered protocadherin isoforms requires high binding specificity and a mechanism to coordinate recognition as a combinatorial unit so only complements that fully match can generate a repulsion signal. Clustered protocadherins have been found to engage in a novel assembly mechanism that exhibits these features and could facilitate their role in neuronal self-recognition. Their ectodomains form two molecular interactions that were observed in crystal structures of protocadherin fragments: a *trans* interaction mediated by the membrane distal EC1–4 domains (Goodman et al. 2016; Nicoludis et al. 2016), and a *cis* interaction between the membrane proximal EC5–6 domains (Goodman et al. 2017) (Fig. 2.5c). The *trans* dimer is an extensive antiparallel interaction formed by complete overlap of the EC1–4 regions stabilized by symmetric EC1:EC4 and EC2:EC3 contacts (Fig. 2.5c). Critically, the *trans* interaction is found to be strictly homophilic in biophysical and cell aggregation experiments (Goodman et al. 2022; Thu et al. 2014) such that each protocadherin isoform can bind in *trans* only to itself, consistent with the need for high specificity. The *cis* interaction is an asymmetric "V"-shaped dimer formed close to the membrane between EC5 of one molecule and EC5–6 of a neighbor (Fig. 2.5c). In contrast to the *trans* dimer, this interaction is highly promiscuous: each isoform can bind in *cis* to most other isoforms tested, with a preference for heterophilic over homophilic *cis* bonds in some cases (Goodman et al. 2022).

The *cis* and *trans* interfaces of clustered protocadherins are well-separated and could in principle form together in two possible arrangements: a closed tetramer where two *cis* dimers bind to each other via two *trans* bonds to satisfy all interfaces; or an open zipper where a *cis* dimer engages two opposing *cis* dimers, one through each arm, leaving *trans* interfaces open for extension of a linear zipper. Notably, bioinformatic analysis suggested that the linear zipper model would provide the properties necessary to encode single-cell identity (Rubinstein et al. 2015), and such an arrangement was directly observed in a low-resolution crystal structure of protocadherin γB4 (Brasch et al. 2019). However, some support for the closed tetramer model came from solution biophysics and single-particle cryo-EM experiments that suggested discrete tetramers could form, albeit with some strain (Brasch et al. 2019). To resolve the conflict, a liposome reconstitution system was used to tether purified protocadherin ectodomains to liposomes to mimic the plasma membrane environment for cryo-ET imaging of contact sites (Brasch et al. 2019). In this system, ectodomains of protocadherin yB6 were discernible in reconstructed tomograms to EC domain-level resolution (Fig. 2.5d), allowing the architecture of the contact sites to be traced by segmentation of the map (Fig. 2.5e). The ectodomains assembled into extensive linear zippers containing upwards of 20 molecules at liposome contact sites, providing clear evidence for the zipper model in a membrane context. Multiple zippers were observed at each contact site and could pack closely together, suggesting potential for higher-order assemblies of zippers (Brasch et al. 2019; Boni et al. 2022), though no conserved contacts were observed (Brasch et al. 2019).

The zipper model observed in cryo-ET provides a basis for how neuronal identity can be encoded by complements of protocadherins (Fig. 2.5f). In the model, when neurites from the same cell make contact, their identical protocadherin complements allow elongated zippers to form through alternating homophilic *trans* interactions and promiscuous *cis* interactions. These zippers then trigger signaling pathways that initiate repulsion to guide the neurites away from aberrant contact. When neurites from different cells make contact, zippers can begin to form in the same way between any matched protocadherin isoforms but are truncated upon *cis* incorporation of an isoform with no matching *trans* binding partner on the other neurite. These truncated zippers do not reach the threshold to trigger a repulsion signal, and other synaptic adhesion receptors can then enable synapse formation. It remains to be shown in targeted mutagenesis experiments that the zipper model underlies self-avoidance in native neurons but experiments with transfected myeloid leukemia K562 cells show that combinatorial recognition specificity depends on the integrity of the *cis* interface (Wiseglass et al. 2024). The intracellular signaling pathways linking zipper elongation with repulsion mechanisms also remain to be elucidated. Clustered protocadherin intracellular domains are implicated in the activation of Wnt and WAVE cascades, in addition to association with other signaling pathways, often in a subtype-specific manner (Pancho et al. 2020; Schalm et al. 2010; Mah and Weiner 2017). Currently, the importance of these pathways for repulsion remains unclear, especially since engagement of clustered protocadherins can also activate other cellular effects such as homophilic cell adhesion in some cell types (Lefebvre 2017; Molumby et al. 2016).

2.5.3 *Nonclustered Protocadherins Diverge in Function and Mechanism*

While clustered protocadherins play a major role in neuronal self-avoidance, members of a related "non-clustered" family of δ protocadherins have a range of largely adhesive functions and lack comparably complex genomic organization and stochastic expression (Light and Jontes 2017).δ protocadherins are highly expressed in nervous tissue and are linked to a female-limited form of epilepsy (Dibbens et al. 2008) and other neurodevelopmental disorders (Light and Jontes 2017). There are 11 isoforms in humans that can be subdivided into δ1 and δ2 subfamilies, with seven and six consecutive EC domains, respectively (Hulpiau and Van Roy 2009; Kahr et al. 2013). They form topologically the same *trans* interface as clustered protocadherins (Fig. 2.5g)(Cooper et al. 2016; Modak and Sotomayor 2019; Harrison et al. 2020; Nicoludis et al. 2016) but diverge in other mechanistic features. While their adhesive binding is homophilic (Bisogni et al. 2018; Harrison et al. 2020) it is not as strict as that of clustered protocadherins; some heterophilic interactions are detectable in vitro within the δ1 and δ2 subfamilies (Harrison et al. 2020). This "looser" specificity likely reflects the difference in biological roles and expression patterns of non-clustered protocadherins. Notably, cell aggregation experiments using combinatorial expression of different δ protocadherins, mimicking native expression found in neurons (Bisogni et al. 2018; Mincheva-Tasheva et al. 2024), show that single mismatches do not uniformly cause complete cell segregation as is seen for clustered protocadherins (Thu et al. 2014). Instead, partially mismatched δ protocadherin complements produce partial or complete intermixing of transfected cells in certain combinations that appear to reflect quantitative contributions to adhesive strength by individual subtypes (Bisogni et al. 2018; Mincheva-Tasheva et al. 2024; Pederick et al. 2018). δ protocadherins also do not appear to adopt the same *cis* interactions or zipper-like assemblies formed by clustered protocadherins, either in crystal structures, biophysical experiments, or at liposome contact sites visualized by cryo-ET (Fig. 2.5h) (Harrison et al. 2020; Hudson et al. 2021). Given their role in adhesion, the δ protocadherins are likely to laterally cluster at cell contacts (Light and Jontes 2017; Pederick et al. 2018; Bisogni et al. 2018), but the mechanism by which they assemble has yet to be defined and may be mediated by the transmembrane and/or the intracellular domains and binding partners.

2.6 Summary

Assembly mechanisms have emerged as a critical component of adhesion and recognition. The assemblies characterized in detail so far include those described above and numerous others, such as Eph:Ephrin signaling complexes (Seiradake et al. 2010), immunoglobulin superfamily receptors in the immunological synapse (Dustin 2014), netrin-neogenin-RGM complexes in neuronal guidance (Robinson et al. 2021), nephrins and nephs in the kidney slit diaphragm (Birtasu et al. 2023) and large molecular synaptic assemblies of adhesion GPCRs and their ligands (Li et al. 2023; Lu et al. 2015; Jackson et al. 2015; Ranaivoson et al. 2015). Mechanisms of *cis* and *trans* interactions are highly diverse and, as is seen for the examples highlighted in this review, are exquisitely tuned in their characteristics to support their biological roles.

2.6.1 Modes of Ordered Assembly Are Adapted to Specific Biological Functions

Type I cadherins form linear *cis*-arrays that intersect arrays from the other cell membrane through *trans* interactions (Harrison et al. 2011), allowing initial clustering of the cadherins into puncta upon cell:cell contact. The specificity and affinity of the individual *trans* interactions allow these cadherins to participate and drive cell sorting and tissue patterning, while the geometry of the *cis* interaction translates individual molecular *trans* binding events into organized assembly at the cell membrane. The resulting two-dimensional assemblies may increase the avidity of adhesion and provide a basis for signaling and eventual assembly of mature junctions linked to the cytoskeleton. While type I classical cadherins appear to form multiple microscale lattices within larger and less ordered physiological adherens junctions, desmosomal cadherins display remarkable extracellular order even at the ultrastructural level of whole desmosomes. The extensive and extraordinarily ordered assembly of desmosomes extends to their intracellular binding partners and likely reflects their special physiological role in maintaining intercellular cohesion in the face of strong mechanical forces. For clustered protocadherins, their primary role in neuronal self-recognition requires that they form extremely specific *trans* interactions in combination with promiscuous *cis* interactions to allow entire cellular complements of clustered protocadherins to be integrated in unidimensional polymeric zippers in which only perfectly matched complements of sister dendrites permit polymerization and signaling. Even highly specialized functions can have a direct molecular basis that is discernible in junctional assemblies. The recently characterized structural organization of nephrin and neph cell adhesion proteins in the kidney slit diaphragm suggests a sieve-like extracellular assembly permeated by gaps of regular size (Birtasu et al. 2023; Ozkan et al. 2014). Slit diaphragm junctions form between the foot processes of podocytes that completely wrap fenestrated glomerular capillaries and are responsible for filtration to prevent loss of serum proteins into the glomerular space (Martin and Jones 2018). The ordered, sieve-like intermembrane assembly in the slit diaphragm, therefore, likely constitutes a molecular-scale filter that mediates this critical function (Birtasu et al. 2023).

2.6.2 Structural Methods at the Frontier of Junctional Assembly Characterization

In the systems described in this chapter, identification of specific *trans* and *cis* interfaces and visualization of their combination in membrane junctions has been critical in defining assembly mechanisms. X-ray crystallography can play a valuable role in identifying the molecular interfaces, since it routinely provides atomic resolution and allows even weak interactions to be observed in the high-concentration environment of the crystal. Adherens junctions provide a good example of a cell surface assembly mechanism that was revealed primarily by x-ray crystallography, which provided sufficient resolution to identify specific interfaces for targeting by mutation and sufficient sensitivity to observe very weak *cis* interactions. Notably, the classical cadherin *cis* interface was present in the full ectodomain crystal lattices of three type I cadherin subtypes and was also observed when prior structures of truncated EC1–2 fragments were reexamined (Harrison et al. 2011). This illustrates that, in general, interfaces that have a physiological role are expected to be observed in crystal lattices containing the relevant regions since even weak interfaces are likely to favorably compete with small nonphysiological crystal contacts. In some cases, the crystallization conditions (which are typically arrived at through random screening) can specifically prevent physiologically relevant interfaces from forming due to extremes of pH or solvent components. Nonetheless, repeated observation of topologically identical interfaces across different crystal forms and/or different subtypes or orthologs provides a good indicator of their potential physiological relevance for testing by targeted mutagenesis. By the same token, absence of a putative interface from multiple crystal lattices can provide evidence that the interface is not relevant to a protein family, as was the case for type II cadherins that do not appear to share the type I cadherin assembly mechanism.

The relatively simple homomeric assembly of classical cadherins allowed not only individual interfaces but also their combination into complete assemblies to be directly observed using X-ray crystallography. For more complex multicomponent systems, Cryo-EM and cryo-ET have provided powerful insights by visualizing the relevant adhesion proteins in a context where they can assemble, e.g., on artificial membranes or by visualizing whole physiological junctions in situ. Cryo-EM can be used to provide 2D information on ordered assemblies—as was done for type I classical cadherins (Harrison et al. 2011), and potentially to derive high-resolution 3D structures from ordered assemblies by a common single particle processing pipeline (Schirra et al. 2023), which takes advantage of the random orientation of individual, separate assemblies in the sample. Cryo-ET presents an alternative approach to obtaining 3D information for assemblies formed between membranes. Individual tomograms containing one or more junctions contain sufficient 3D information to allow for segmentation, a process by which individual proteins and membranes can be demarcated and differentiated from each other to allow their arrangement in the assembly to be determined at protein-level and even domain-level as for clustered protocadherins (Brasch et al. 2019). In addition, higher resolution can be derived for very ordered assemblies through subtomogram averaging of multiple subregions of the assembly, which was successfully applied to the desmosome (Al-Amoudi et al. 2007, 2011; Birtasu et al. 2023; Sikora et al. 2020) and, more recently, the kidney slit diaphragm (Birtasu et al. 2023). As for many structural approaches, a confounding factor can be the inherent flexibility of adhesion proteins, which tend to have elongated structures and require a degree of mechanical flexibility for their function. These properties can limit the achievable resolution obtainable from subtomogram averaging, highlighting the importance of integrative approaches that use other sources of information, such as fragment structures and biophysics, and that assess multiple potential models computationally (Birtasu et al. 2023; Sikora et al.

2020). Perhaps the most powerful feature of cryo-ET is that it can reveal both the extracellular and intracellular assembly of native junctions in situ. This approach is rapidly developing and should be capable of revealing the full extent to which macromolecular assemblies within and outside the cell adopt ordered structures.

Acknowledgments This work was supported by National Institutes of Health grant R35GM150961.

Competing Interests The authors have no conflicts of interest to declare that are relevant to the content of this chapter.

References

Acharya BR, Yap AS. Cell–cell adhesion and the cytoskeleton. In: Encyclopedia of cell biology. Oxford: Elsevier; 2016. p. 704–12.

Adams CL, Nelson WJ, Smith SJ. Quantitative analysis of cadherin-catenin-actin reorganization during development of cell-cell adhesion. J Cell Biol. 1996;135:1899–911.

Al-Amoudi A, Diez DC, Betts MJ, Frangakis AS. The molecular architecture of cadherins in native epidermal desmosomes. Nature. 2007;450:832–7.

Al-Amoudi A, Castano-Diez D, Devos DP, Russell RB, Johnson GT, Frangakis AS. The three-dimensional molecular structure of the desmosomal plaque. Proc Natl Acad Sci USA. 2011;108:6480–5.

Al-Jassar C, Bikker H, Overduin M, Chidgey M. Mechanistic basis of desmosome-targeted diseases. J Mol Biol. 2013;425:4006–22.

Aricescu AR, Jones EY. Immunoglobulin superfamily cell adhesion molecules: zippers and signals. Curr Opin Cell Biol. 2007;19:543–50.

Basu R, Duan X, Taylor MR, Martin EA, Muralidhar S, Wang Y, Gangi-Wellman L, Das SC, Yamagata M, West PJ, Sanes JR, Williams ME. Heterophilic type II cadherins are required for high-magnitude synaptic potentiation in the hippocampus. Neuron. 2018;98:658–68.

Birtasu AN, Wieland K, Ermel UH, Rahm JV, Scheffer MP, Flottmann B, Heilemann M, Grahammer F, Frangakis AS. The molecular architecture of the kidney slit diaphragm. bioRxiv. 2023. https://doi.org/10.1101/2023.10.27.564405.

Bisogni AJ, Ghazanfar S, Williams EO, Marsh HM, Yang JY, Lin DM. Tuning of delta-protocadherin adhesion through combinatorial diversity. elife. 2018;7:e41050.

Boggon TJ, Murray J, Chappuis-Flament S, Wong E, Gumbiner BM, Shapiro L. C-cadherin ectodomain structure and implications for cell adhesion mechanisms. Science. 2002;296:1308–13.

Boni N, Shapiro L, Honig B, Wu Y, Rubinstein R. On the formation of ordered protein assemblies in cell-cell interfaces. Proc Natl Acad Sci USA. 2022;119:e2206175119.

Brancaccio A. A molecular overview of the primary dystroglycanopathies. J Cell Mol Med. 2019;23:3058–62.

Brasch J, Harrison OJ, Ahlsen G, Carnally SM, Henderson RM, Honig B, Shapiro L. Structure and binding mechanism of vascular endothelial cadherin: a divergent classical cadherin. J Mol Biol. 2011;408:57–73.

Brasch J, Harrison OJ, Honig B, Shapiro L. Thinking outside the cell: how cadherins drive adhesion. Trends Cell Biol. 2012;22:299–310.

Brasch J, Katsamba PS, Harrison OJ, Ahlsen G, Troyanovsky RB, Indra I, Kaczynska A, Kaeser B, Troyanovsky S, Honig B, Shapiro L. Homophilic and heterophilic interactions of type II cadherins identify specificity groups underlying cell-adhesive behavior. Cell Rep. 2018;23:1840–52.

Brasch J, Goodman KM, Noble AJ, Rapp M, Mannepalli S, Bahna F, Dandey VP, Bepler T, Berger B, Maniatis T, Potter CS, Carragher B, Honig B, Shapiro L. Visualization of clustered protocadherin neuronal self-recognition complexes. Nature. 2019;569:280–3.

Brieher WM, Yap AS, Gumbiner BM. Lateral dimerization is required for the homophilic binding activity of C-cadherin. J Cell Biol. 1996;135:487–96.

Brodland GW. The differential interfacial tension hypothesis (DITH): a comprehensive theory for the self-rearrangement of embryonic cells and tissues. J Biomech Eng. 2002;124:188–97.

Canzio D, Maniatis T. The generation of a protocadherin cell-surface recognition code for neural circuit assembly. Curr Opin Neurobiol. 2019;59:213–20.

Carman CV, Springer TA. Integrin avidity regulation: are changes in affinity and conformation underemphasized? Curr Opin Cell Biol. 2003;15:547–56.

Castano-Diez D, Kudryashev M, Arheit M, Stahlberg H. Dynamo: a flexible, user-friendly development tool for subtomogram averaging of cryo-EM data in high-performance computing environments. J Struct Biol. 2012;178:139–51.

Chappuis-Flament S, Wong E, Hicks LD, Kay CM, Gumbiner BM. Multiple cadherin extracellular repeats mediate homophilic binding and adhesion. J Cell Biol. 2001;154:231–43.

Chen J, Den Z, Koch PJ. Loss of desmocollin 3 in mice leads to epidermal blistering. J Cell Sci. 2008;121:2844–9.

Chen Y, Brasch J, Harrison OJ, Bidone TC. Computational model of E-cadherin clustering under force. Biophys J. 2021;120:4944–54.

Chidgey M, Brakebusch C, Gustafsson E, Cruchley A, Hail C, Kirk S, Merritt A, North A, Tselepis C, Hewitt J, Byrne C, Fassler R, Garrod D. Mice lacking desmocollin 1 show epidermal fragility accompanied by barrier defects and abnormal differentiation. J Cell Biol. 2001;155:821–32.

Citi S, Fromm M, Furuse M, Gonzalez-Mariscal L, Nusrat A, Tsukita S, Turner JR. A short guide to the tight junction. J Cell Sci. 2024;137:jcs261776.

Cooper SR, Jontes JD, Sotomayor M. Structural determinants of adhesion by Protocadherin-19 and implications for its role in epilepsy. elife. 2016;5:e18529.

Cosmanescu F, Katsamba PS, Sergeeva AP, Ahlsen G, Patel SD, Brewer JJ, Tan L, Xu S, Xiao Q, Nagarkar-Jaiswal S, Nern A, Bellen HJ, Zipursky SL, Honig B, Shapiro L. Neuron-subtype-specific expression, interaction affinities, and specificity determinants of DIP/Dpr cell recognition proteins. Neuron. 2018;100:1385–1400.e6.

Dalle Vedove A, Lucarelli AP, Nardone V, Matino A, Parisini E. The X-ray structure of human P-cadherin EC1-EC2 in a closed conformation provides insight into the type I cadherin dimerization pathway. Acta Crystallogr F Struct Biol Commun. 2015;71:371–80.

Dean WF, Mattheyses AL. Defining domain-specific orientational order in the desmosomal cadherins. Biophys J. 2022;121:4325–41.

Desai BV, Harmon RM, Green KJ. Desmosomes at a glance. J Cell Sci. 2009;122:4401–7.

Dibbens LM, Tarpey PS, Hynes K, Bayly MA, Scheffer IE, Smith R, Bomar J, Sutton E, Vandeleur L, Shoubridge C, Edkins S, Turner SJ, Stevens C, O'Meara S, Tofts C, Barthorpe S, Buck G, Cole J, Halliday K, Jones D, Lee R, Madison M, Mironenko T, Varian J, West S, Widaa S, Wray P, Teague J, Dicks E, Butler A, Menzies A, Jenkinson A, Shepherd R, Gusella JF, Afawi Z, Mazarib A, Neufeld MY, Kivity S, Lev D, Lerman-Sagie T, Korczyn AD, Derry CP, Sutherland GR, Friend K, Shaw M, Corbett M, Kim HG, Geschwind DH, Thomas P, Haan E, Ryan S, Mckee S, Berkovic SF, Futreal PA, Stratton MR, Mulley JC, Gecz J. X-linked protocadherin 19 mutations cause female-limited epilepsy and cognitive impairment. Nat Genet. 2008;40:776–81.

Dustin ML. The immunological synapse. Cancer Immunol Res. 2014;2:1023–33.

Farquhar MG, Palade GE. Junctional complexes in various epithelia. J Cell Biol. 1963;17:375–412.

Fawcett DW, Mcnutt NS. The ultrastructure of the cat myocardium. I. Ventricular papillary muscle. J Cell Biol. 1969;42:1–45.

Flaherty E, Maniatis T. The role of clustered protocadherins in neurodevelopment and neuropsychiatric diseases. Curr Opin Genet Dev. 2020;65:144–50.

Foty RA, Steinberg MS. The differential adhesion hypothesis: a direct evaluation. Dev Biol. 2005;278:255–63.

Foty RA, Steinberg MS. Differential adhesion in model systems. Wiley Interdiscip Rev Dev Biol. 2013;2:631–45.

Garrod D, Chidgey M. Desmosome structure, composition and function. Biochim Biophys Acta. 2008;1778:572–87.

Garrod D, Tabernero L. Hyper-adhesion: a unique property of desmosomes. Cell Commun Adhes. 2014;21:249–56.

Godt D, Tepass U. Drosophila oocyte localization is mediated by differential cadherin-based adhesion. Nature. 1998;395:387–91.

Goodman KM, Rubinstein R, Thu CA, Mannepalli S, Bahna F, Ahlsen G, Rittenhouse C, Maniatis T, Honig B, Shapiro L. gamma-Protocadherin structural diversity and functional implications. elife. 2016;5:e20930.

Goodman KM, Rubinstein R, Dan H, Bahna F, Mannepalli S, Ahlsen G, Aye Thu C, Sampogna RV, Maniatis T, Honig B, Shapiro L. Protocadherin cis-dimer architecture and recognition unit diversity. Proc Natl Acad Sci USA. 2017;114:E9829–37.

Goodman KM, Katsamba PS, Rubinstein R, Ahlsen G, Bahna F, Mannepalli S, Dan H, Sampogna RV, Shapiro L, Honig B. How clustered protocadherin binding specificity is tuned for neuronal self-/nonself-recognition. elife. 2022;11:e72416.

Goody MF, Sher RB, Henry CA. Hanging on for the ride: adhesion to the extracellular matrix mediates cellular responses in skeletal muscle morphogenesis and disease. Dev Biol. 2015;401:75–91.

Goult BT, Yan J, Schwartz MA. Talin as a mechanosensitive signaling hub. J Cell Biol. 2018;217:3776–84.

Gumbiner BM. Regulation of cadherin-mediated adhesion in morphogenesis. Nat Rev Mol Cell Biol. 2005;6:622–34.

Guo L, Wu Y, Chang H, Zhang Z, Tang H, Yu Y, Xin L, Liu Y, He Y. Structure of cell-cell adhesion mediated by the Down syndrome cell adhesion molecule. Proc Natl Acad Sci USA. 2021;118:e2022442118.

Hamidi H, Ivaska J. Every step of the way: integrins in cancer progression and metastasis. Nat Rev Cancer. 2018;18:533–48.

Harris TJ, Tepass U. Adherens junctions: from molecules to morphogenesis. Nat Rev Mol Cell Biol. 2010;11:502–14.

Harrison OJ, Jin X, Hong S, Bahna F, Ahlsen G, Brasch J, Wu Y, Vendome J, Felsovalyi K, Hampton CM, Troyanovsky RB, Ben-Shaul A, Frank J, Troyanovsky SM, Shapiro L, Honig B. The extracellular architecture of adherens junctions revealed by crystal structures of type I cadherins. Structure. 2011;19:244–56.

Harrison OJ, Vendome J, Brasch J, Jin X, Hong S, Katsamba PS, Ahlsen G, Troyanovsky RB, Troyanovsky SM, Honig B, Shapiro L. Nectin ectodomain structures reveal a canonical adhesive interface. Nat Struct Mol Biol. 2012;19:906–15.

Harrison OJ, Brasch J, Lasso G, Katsamba PS, Ahlsen G, Honig B, Shapiro L. Structural basis of adhesive binding by desmocollins and desmogleins. Proc Natl Acad Sci USA. 2016;113:7160–5.

Harrison OJ, Brasch J, Katsamba PS, Ahlsen G, Noble AJ, Dan H, Sampogna RV, Potter CS, Carragher B, Honig B, Shapiro L. Family-wide structural and biophysical analysis of binding interactions among non-clustered delta-protocadherins. Cell Rep. 2020;30:2655–2671.e7.

Hattori D, Millard SS, Wojtowicz WM, Zipursky SL. Dscam-mediated cell recognition regulates neural circuit formation. Annu Rev Cell Dev Biol. 2008;24:597–620.

Hayashi T, Carthew RW. Surface mechanics mediate pattern formation in the developing retina. Nature. 2004;431:647–52.

Hayashi S, Takeichi M. Emerging roles of protocadherins: from self-avoidance to enhancement of motility. J Cell Sci. 2015;128:1455–64.

Hazan RB, Qiao R, Keren R, Badano I, Suyama K. Cadherin switch in tumor progression. Ann N Y Acad Sci. 2004;1014:155–63.

He W, Cowin P, Stokes DL. Untangling desmosomal knots with electron tomography. Science. 2003;302:109–13.

Hegazy M, Perl AL, Svoboda SA, Green KJ. Desmosomal cadherins in health and disease. Annu Rev Pathol. 2022;17:47–72.

Hirano S, Takeichi M. Cadherins in brain morphogenesis and wiring. Physiol Rev. 2012;92:597–634.

Honig B, Shapiro L. Adhesion protein structure, molecular affinities, and principles of cell-cell recognition. Cell. 2020;181:520–35.

Hudson JD, Tamilselvan E, Sotomayor M, Cooper SR. A complete Protocadherin-19 ectodomain model for evaluating epilepsy-causing mutations and potential protein interaction sites. Structure. 2021;29:1128–1143.e4.

Hulpiau P, Van Roy F. Molecular evolution of the cadherin superfamily. Int J Biochem Cell Biol. 2009;41:349–69.

Indra I, Hong S, Troyanovsky R, Kormos B, Troyanovsky S. The adherens junction: a mosaic of cadherin and nectin clusters bundled by actin filaments. J Invest Dermatol. 2013;133:2546–54.

Indra I, Choi J, Chen CS, Troyanovsky RB, Shapiro L, Honig B, Troyanovsky SM. Spatial and temporal organization of cadherin in punctate adherens junctions. Proc Natl Acad Sci USA. 2018;115:E4406–15.

Ing-Esteves S, Lefebvre JL. Gamma-protocadherins regulate dendrite self-recognition and dynamics to drive self-avoidance. Curr Biol. 2024;34:4224–4239.e4.

Jackson VA, Del Toro D, Carrasquero M, Roversi P, Harlos K, Klein R, Seiradake E. Structural basis of latrophilin-FLRT interaction. Structure. 2015;23:774–81.

Jaudon F, Thalhammer A, Cingolani LA. Integrin adhesion in brain assembly: from molecular structure to neuropsychiatric disorders. Eur J Neurosci. 2021;53:3831–50.

Jontes JD. The cadherin superfamily in neural circuit assembly. Cold Spring Harb Perspect Biol. 2018;10:a029306.

Kahr I, Vandepoele K, Van Roy F. Delta-protocadherins in health and disease. Prog Mol Biol Transl Sci. 2013;116:169–92.

Kametani Y, Takeichi M. Basal-to-apical cadherin flow at cell junctions. Nat Cell Biol. 2007;9:92–8.

Kasperkiewicz M, Ellebrecht CT, Takahashi H, Yamagami J, Zillikens D, Payne AS, Amagai M. Pemphigus. Nat Rev Dis Primers. 2017;3:17026.

Katsamba P, Carroll K, Ahlsen G, Bahna F, Vendome J, Posy S, Rajebhosale M, Price S, Jessell TM, Ben-Shaul A, Shapiro L, Honig BH. Linking molecular affinity and cellular specificity in cadherin-mediated adhesion. Proc Natl Acad Sci USA. 2009;106:11594–9.

Kiener HP, Stipp CS, Allen PG, Higgins JM, Brenner MB. The cadherin-11 cytoplasmic juxtamembrane domain promotes alpha-catenin turnover at adherens junctions and intercellular motility. Mol Biol Cell. 2006;17:2366–76.

Kilshaw PJ. Alpha E beta 7. Mol Pathol. 1999;52:203–7.

Kimura TE, Merritt AJ, Lock FR, Eckert JJ, Fleming TP, Garrod DR. Desmosomal adhesiveness is developmentally regulated in the mouse embryo and modulated during trophectoderm migration. Dev Biol. 2012;369:286–97.

Koch PJ, Mahoney MG, Ishikawa H, Pulkkinen L, Uitto J, Shultz L, Murphy GF, Whitaker-Menezes D, Stanley JR. Targeted disruption of the pemphigus vulgaris antigen (desmoglein 3) gene in mice causes loss of keratinocyte cell adhesion with a phenotype similar to pemphigus vulgaris. J Cell Biol. 1997;137:1091–102.

Kowalczyk AP, Green KJ. Structure, function, and regulation of desmosomes. Prog Mol Biol Transl Sci. 2013;116:95–118.

Krusche CA, Holthofer B, Hofe V, Van De Sandt AM, Eshkind L, Bockamp E, Merx MW, Kant S, Windoffer R, Leube RE. Desmoglein 2 mutant mice develop cardiac fibrosis and dilation. Basic Res Cardiol. 2011;106:617–33.

Kudo S, Caaveiro JM, Tsumoto K. Adhesive dimerization of human P-cadherin catalyzed by a chaperone-like mechanism. Structure. 2016;24:1523–36.

Laszlo ZI, Lele Z. Flying under the radar: CDH2 (N-cadherin), an important hub molecule in neurodevelopmental and neurodegenerative diseases. Front Neurosci. 2022;16:972059.

Le Bihan O, Decossas M, Gontier E, Gerbod-Giannone MC, Lambert O. Visualization of adherent cell monolayers by cryo-electron microscopy: a snapshot of endothelial adherens junctions. J Struct Biol. 2015;192:470–7.

Lefebvre JL. Neuronal territory formation by the atypical cadherins and clustered protocadherins. Semin Cell Dev Biol. 2017;69:111–21.

Lefebvre JL, Kostadinov D, Chen WV, Maniatis T, Sanes JR. Protocadherins mediate dendritic self-avoidance in the mammalian nervous system. Nature. 2012;488:517–21.

Legerstee K, Houtsmuller AB. A layered view on focal adhesions. Biology (Basel). 2021;10:1189.

Li J, Bandekar SJ, Arac D. The structure of fly Teneurin-m reveals an asymmetric self-assembly that allows expansion into zippers. EMBO Rep. 2023;24:e56728.

Light SEW, Jontes JD. delta-Protocadherins: organizers of neural circuit assembly. Semin Cell Dev Biol. 2017;69:83–90.

Lowndes M, Rakshit S, Shafraz O, Borghi N, Harmon RM, Green KJ, Sivasankar S, Nelson WJ. Different roles of cadherins in the assembly and structural integrity of the desmosome complex. J Cell Sci. 2014;127:2339–50.

Lu YC, Nazarko OV, Sando R 3rd, Salzman GS, Li NS, Sudhof TC, Arac D. Structural basis of latrophilin-FLRT-UNC5 interaction in cell adhesion. Structure. 2015;23:1678–91.

Luo BH, Carman CV, Springer TA. Structural basis of integrin regulation and signaling. Annu Rev Immunol. 2007;25:619–47.

Lv X, Li S, Li J, Yu XY, Ge X, Li B, Hu S, Lin Y, Zhang S, Yang J, Zhang X, Yan J, Joyner AL, Shi H, Wu Q, Shi SH. Patterned cPCDH expression regulates the fine organization of the neocortex. Nature. 2022;612:503–11.

Mah KM, Weiner JA. Regulation of Wnt signaling by protocadherins. Semin Cell Dev Biol. 2017;69:158–71.

Mandai K, Rikitake Y, Shimono Y, Takai Y. Afadin/AF-6 and canoe: roles in cell adhesion and beyond. Prog Mol Biol Transl Sci. 2013;116:433–54.

Marcozzi C, Burdett ID, Buxton RS, Magee AI. Coexpression of both types of desmosomal cadherin and plakoglobin confers strong intercellular adhesion. J Cell Sci. 1998;111(Pt 4):495–509.

Martin CE, Jones N. Nephrin signaling in the podocyte: an updated view of signal regulation at the slit diaphragm and beyond. Front Endocrinol (Lausanne). 2018;9:302.

Mcgrath JA. Hereditary diseases of desmosomes. J Dermatol Sci. 1999;20:85–91.

Mcnutt NS, Weinstein RS. Membrane ultrastructure at mammalian intercellular junctions. Prog Biophys Mol Biol. 1973;26:45–101.

Meng W, Takeichi M. Adherens junction: molecular architecture and regulation. Cold Spring Harb Perspect Biol. 2009;1:a002899.

Mese G, Richard G, White TW. Gap junctions: basic structure and function. J Invest Dermatol. 2007;127:2516–24.

Mincheva-Tasheva S, Pfitzner C, Kumar R, Kurtsdotter I, Scherer M, Ritchie T, Muhr J, Gecz J, Thomas PQ. Mapping combinatorial expression of non-clustered protocadherins in the developing brain identifies novel PCDH19-mediated cell adhesion properties. Open Biol. 2024;14:230383.

Miyaguchi K. Ultrastructure of the zonula adherens revealed by rapid-freeze deep-etching. J Struct Biol. 2000;132:169–78.

Modak D, Sotomayor M. Identification of an adhesive interface for the non-clustered delta1 protocadherin-1 involved in respiratory diseases. Commun Biol. 2019;2:354.

Molumby MJ, Keeler AB, Weiner JA. Homophilic protocadherin cell-cell interactions promote dendrite complexity. Cell Rep. 2016;15:1037–50.

Mountoufaris G, Chen WV, Hirabayashi Y, O'Keeffe S, Chevee M, Nwakeze CL, Polleux F, Maniatis T. Multicluster Pcdh diversity is required for mouse olfactory neural circuit assembly. Science. 2017;356:411–4.

Nagafuchi A, Takeichi M. Cell binding function of E-cadherin is regulated by the cytoplasmic domain. EMBO J. 1988;7:3679–84.

Ni T, Frosio T, Mendonca L, Sheng Y, Clare D, Himes BA, Zhang P. High-resolution in situ structure determination by cryo-electron tomography and subtomogram averaging using emClarity. Nat Protoc. 2022;17:421–44.

Nicoludis JM, Vogt BE, Green AG, Scharfe CP, Marks DS, Gaudet R. Antiparallel protocadherin homodimers use distinct affinity- and specificity-mediating regions in cadherin repeats 1-4. elife. 2016;5:e18449.

Nie Z, Merritt A, Rouhi-Parkouhi M, Tabernero L, Garrod D. Membrane-impermeable cross-linking provides evidence for homophilic, isoform-specific binding of desmosomal cadherins in epithelial cells. J Biol Chem. 2011;286:2143–54.

Nielsen MS, Van Opbergen CJM, Van Veen TAB, Delmar M. The intercalated disc: a unique organelle for electromechanical synchrony in cardiomyocytes. Physiol Rev. 2023;103:2271–319.

Niessen CM, Gottardi CJ. Molecular components of the adherens junction. Biochim Biophys Acta. 2008;1778:562–71.

Nitoiu D, Etheridge SL, Kelsell DP. Insights into desmosome biology from inherited human skin disease and cardiocutaneous syndromes. Cell Commun Adhes. 2014;21:129–40.

Nollet F, Kools P, Van Roy F. Phylogenetic analysis of the cadherin superfamily allows identification of six major subfamilies besides several solitary members. J Mol Biol. 2000;299:551–72.

North AJ, Bardsley WG, Hyam J, Bornslaeger EA, Cordingley HC, Trinnaman B, Hatzfeld M, Green KJ, Magee AI, Garrod DR. Molecular map of the desmosomal plaque. J Cell Sci. 1999;112(Pt 23):4325–36.

Okabe N, Shimizu K, Ozaki-Kuroda K, Nakanishi H, Morimoto K, Takeuchi M, Katsumaru H, Murakami F, Takai Y. Contacts between the commissural axons and the floor plate cells are mediated by nectins. Dev Biol. 2004;273:244–56.

Ozaki-Kuroda K, Nakanishi H, Ohta H, Tanaka H, Kurihara H, Mueller S, Irie K, Ikeda W, Sakai T, Wimmer E, Nishimune Y, Takai Y. Nectin couples cell-cell adhesion and the actin scaffold at heterotypic testicular junctions. Curr Biol. 2002;12:1145–50.

Ozkan E, Chia PH, Wang RR, Goriatcheva N, Borek D, Otwinowski Z, Walz T, Shen K, Garcia KC. Extracellular architecture of the SYG-1/SYG-2 adhesion complex instructs synaptogenesis. Cell. 2014;156:482–94.

Pancho A, Aerts T, Mitsogiannis MD, Seuntjens E. Protocadherins at the crossroad of signaling pathways. Front Mol Neurosci. 2020;13:117.

Parisini E, Higgins JM, Liu JH, Brenner MB, Wang JH. The crystal structure of human E-cadherin domains 1 and 2, and comparison with other cadherins in the context of adhesion mechanism. J Mol Biol. 2007;373:401–11.

Pasani S, Menon KS, Viswanath S. The molecular architecture of the desmosomal outer dense plaque by integrative structural modeling. Protein Sci. 2024;33:e5217.

Patel SD, Ciatto C, Chen CP, Bahna F, Rajebhosale M, Arkus N, Schieren I, Jessell TM, Honig B, Price SR, Shapiro L. Type II cadherin ectodomain structures: implications for classical cadherin specificity. Cell. 2006;124:1255–68.

Pederick DT, Richards KL, Piltz SG, Kumar R, Mincheva-Tasheva S, Mandelstam SA, Dale RC, Scheffer IE, Gecz J, Petrou S, Hughes JN, Thomas PQ. Abnormal cell sorting underlies the unique X-linked inheritance of PCDH19 epilepsy. Neuron. 2018;97:59–66.e5.

Perl AL, Pokorny JL, Green KJ. Desmosomes at a glance. J Cell Sci. 2024;137:jcs261899.

Pokutta S, Weis WI. Structure and mechanism of cadherins and catenins in cell-cell contacts. Annu Rev Cell Dev Biol. 2007;23:237–61.

Price SR, De Marco Garcia NV, Ranscht B, Jessell TM. Regulation of motor neuron pool sorting by differential expression of type II cadherins. Cell. 2002;109:205–16.

Priest AV, Koirala R, Sivasankar S. Single-molecule studies of classical and desmosomal cadherin adhesion. Curr Opin Biomed Eng. 2019;12:43–50.

Punovuori K, Malaguti M, Lowell S. Cadherins in early neural development. Cell Mol Life Sci. 2021;78:4435–50.

Radice GL, Rayburn H, Matsunami H, Knudsen KA, Takeichi M, Hynes RO. Developmental defects in mouse embryos lacking N-cadherin. Dev Biol. 1997;181:64–78.

Ranaivoson FM, Liu Q, Martini F, Bergami F, Von Daake S, Li S, Lee D, Demeler B, Hendrickson WA, Comoletti D. Structural and mechanistic insights into the latrophilin3-FLRT3 complex that mediates glutamatergic synapse development. Structure. 2015;23:1665–77.

Rayns DG, Simpson FO, Ledingham JM. Ultrastructure of desmosomes in mammalian intercalated disc; appearances after lanthanum treatment. J Cell Biol. 1969;42:322–6.

Robinson RA, Griffiths SC, Van De Haar LL, Malinauskas T, Van Battum EY, Zelina P, Schwab RA, Karia D, Malinauskaite L, Brignani S, Van Den Munkhof MH, Dudukcu O, De Ruiter AA, Van Den Heuvel DMA, Bishop B, Elegheert J, Aricescu AR, Pasterkamp RJ, Siebold C. Simultaneous binding of guidance cues NET1 and RGM blocks extracellular NEO1 signaling. Cell. 2021;184:2103–2120.e31.

Rubinstein R, Thu CA, Goodman KM, Wolcott HN, Bahna F, Mannepalli S, Ahlsen G, Chevee M, Halim A, Clausen H, Maniatis T, Shapiro L, Honig B. Molecular logic of neuronal self-recognition through protocadherin domain interactions. Cell. 2015;163:629–42.

Sanes JR, Zipursky SL. Synaptic specificity, recognition molecules, and assembly of neural circuits. Cell. 2020;181:536–56.

Schalm SS, Ballif BA, Buchanan SM, Phillips GR, Maniatis T. Phosphorylation of protocadherin proteins by the receptor tyrosine kinase Ret. Proc Natl Acad Sci USA. 2010;107:13894–9.

Schinner C, Xu L, Franz H, Zimmermann A, Wanuske MT, Rathod M, Hanns P, Geier F, Pelczar P, Liang Y, Lorenz V, Studle C, Maly PI, Kauferstein S, Beckmann BM, Sheikh F, Kuster GM, Spindler V. Defective desmosomal adhesion causes arrhythmogenic cardiomyopathy by involving an integrin-alphaVbeta6/TGF-beta signaling cascade. Circulation. 2022;146:1610–26.

Schirra RT, Dos Santos NFB, Zadrozny KK, Kucharska I, Ganser-Pornillos BK, Pornillos O. A molecular switch modulates assembly and host factor binding of the HIV-1 capsid. Nat Struct Mol Biol. 2023;30:383–90.

Seiradake E, Harlos K, Sutton G, Aricescu AR, Jones EY. An extracellular steric seeding mechanism for Eph-ephrin signaling platform assembly. Nat Struct Mol Biol. 2010;17:398–402.

Shafraz O, Rubsam M, Stahley SN, Caldara AL, Kowalczyk AP, Niessen CM, Sivasankar S. E-cadherin binds to desmoglein to facilitate desmosome assembly. elife. 2018;7:e37629.

Shapiro L, Weis WI. Structure and biochemistry of cadherins and catenins. Cold Spring Harb Perspect Biol. 2009;1:a003053.

Shapiro L, Fannon AM, Kwong PD, Thompson A, Lehmann MS, Grubel G, Legrand JF, Als-Nielsen J, Colman DR, Hendrickson WA. Structural basis of cell-cell adhesion by cadherins. Nature. 1995;374:327–37.

Shimoyama Y, Tsujimoto G, Kitajima M, Natori M. Identification of three human type-II classic cadherins and frequent heterophilic interactions between different subclasses of type-II classic cadherins. Biochem J. 2000;349:159–67.

Sikora M, Ermel UH, Seybold A, Kunz M, Calloni G, Reitz J, Vabulas RM, Hummer G, Frangakis AS. Desmosome architecture derived from molecular dynamics simulations and cryo-electron tomography. Proc Natl Acad Sci USA. 2020;117:27132–40.

Spindler V, Waschke J. Pemphigus – a disease of desmosome dysfunction caused by multiple mechanisms. Front Immunol. 2018;9:136.

Steinberg MS. Does differential adhesion govern self-assembly processes in histogenesis? Equilibrium configurations and the emergence of a hierarchy among populations of embryonic cells. J Exp Zool. 1970;173:395–433.

Steinberg MS. Differential adhesion in morphogenesis: a modern view. Curr Opin Genet Dev. 2007;17:281–6.

Stepniak E, Radice GL, Vasioukhin V. Adhesive and signaling functions of cadherins and catenins in vertebrate development. Cold Spring Harb Perspect Biol. 2009;1:a002949.

Strale PO, Duchesne L, Peyret G, Montel L, Nguyen T, Png E, Tampe R, Troyanovsky S, Henon S, Ladoux B, Mege RM. The formation of ordered nanoclusters controls cadherin anchoring to actin and cell-cell contact fluidity. J Cell Biol. 2015;210:1033.

Syed SE, Trinnaman B, Martin S, Major S, Hutchinson J, Magee AI. Molecular interactions between desmosomal cadherins. Biochem J. 2002;362:317–27.

Takeichi M. Dynamic contacts: rearranging adherens junctions to drive epithelial remodelling. Nat Rev Mol Cell Biol. 2014;15:397–410.

Tamura K, Shan WS, Hendrickson WA, Colman DR, Shapiro L. Structure-function analysis of cell adhesion by neural (N-) cadherin. Neuron. 1998;20:1153–63.

Tariq H, Bella J, Jowitt TA, Holmes DF, Rouhi M, Nie Z, Baldock C, Garrod D, Tabernero L. Cadherin flexibility provides a key difference between desmosomes and adherens junctions. Proc Natl Acad Sci USA. 2015;112:5395–400.

Taylor MR, Martin EA, Sinnen B, Trilokekar R, Ranza E, Antonarakis SE, Williams ME. Kirrel3-mediated synapse formation is attenuated by disease-associated missense variants. J Neurosci. 2020;40:5376–88.

Thompson CJ, Su Z, Vu VH, Wu Y, Leckband DE, Schwartz DK. Cadherin clusters stabilized by a combination of specific and nonspecific cis-interactions. elife. 2020;9:e59035.

Thu CA, Chen WV, Rubinstein R, Chevee M, Wolcott HN, Felsovalyi KO, Tapia JC, Shapiro L, Honig B, Maniatis T. Single-cell identity generated by combinatorial homophilic interactions between alpha, beta, and gamma protocadherins. Cell. 2014;158:1045–59.

Togashi H, Kominami K, Waseda M, Komura H, Miyoshi J, Takeichi M, Takai Y. Nectins establish a checkerboard-like cellular pattern in the auditory epithelium. Science. 2011;333:1144–7.

Troyanovsky RB, Sergeeva AP, Indra I, Chen CS, Kato R, Shapiro L, Honig B, Troyanovsky SM. Sorting of cadherin-catenin-associated proteins into individual clusters. Proc Natl Acad Sci USA. 2021;118:e2105550118.

Uehara K. Distribution of adherens junction mediated by VE-cadherin complex in rat spleen sinus endothelial cells. Cell Tissue Res. 2006;323:417–24.

Vasioukhin V, Bauer C, Yin M, Fuchs E. Directed actin polymerization is the driving force for epithelial cell-cell adhesion. Cell. 2000;100:209–19.

Vendome J, Felsovalyi K, Song H, Yang Z, Jin X, Brasch J, Harrison OJ, Ahlsen G, Bahna F, Kaczynska A, Katsamba PS, Edmond D, Hubbell WL, Shapiro L, Honig B. Structural and energetic determinants of adhesive binding specificity in type I cadherins. Proc Natl Acad Sci USA. 2014;111:E4175–84.

Vizurraga A, Adhikari R, Yeung J, Yu M, Tall GG. Mechanisms of adhesion G protein-coupled receptor activation. J Biol Chem. 2020;295:14065–83.

Walko G, Castanon MJ, Wiche G. Molecular architecture and function of the hemidesmosome. Cell Tissue Res. 2015;360:529–44.

Waschke J. The desmosome and pemphigus. Histochem Cell Biol. 2008;130:21–54.

Whittard JD, Craig SE, Mould AP, Koch A, Pertz O, Engel J, Humphries MJ. E-cadherin is a ligand for integrin alpha2beta1. Matrix Biol. 2002;21:525–32.

Wiseglass G, Boni N, Smorodinsky-Atias K, Rubinstein R. Clustered protocadherin cis-interactions are required for combinatorial cell-cell recognition underlying neuronal self-avoidance. Proc Natl Acad Sci USA. 2024;121:e2319829121.

Wojtowicz WM, Flanagan JJ, Millard SS, Zipursky SL, Clemens JC. Alternative splicing of Drosophila Dscam generates axon guidance receptors that exhibit isoform-specific homophilic binding. Cell. 2004;118:619–33.

Wu Q, Maniatis T. A striking organization of a large family of human neural cadherin-like cell adhesion genes. Cell. 1999;97:779–90.

Wu Y, Vendome J, Shapiro L, Ben-Shaul A, Honig B. Transforming binding affinities from three dimensions to two with application to cadherin clustering. Nature. 2011;475:510-U107.

Xu S, Xiao Q, Cosmanescu F, Sergeeva AP, Yoo J, Lin Y, Katsamba PS, Ahlsen G, Kaufman J, Linaval NT, Lee PT, Bellen HJ, Shapiro L, Honig B, Tan L, Zipursky SL. Interactions between the Ig-superfamily proteins DIP-alpha and Dpr6/10 regulate assembly of neural circuits. Neuron. 2018;100:1369–1384.e6.

Yagi T. Molecular codes for neuronal individuality and cell assembly in the brain. Front Mol Neurosci. 2012;5:45.

Yao M, Qiu W, Liu R, Efremov AK, Cong P, Seddiki R, Payre M, Lim CT, Ladoux B, Mege RM, Yan J. Force-dependent conformational switch of alpha-catenin controls vinculin binding. Nat Commun. 2014;5:4525.

Yap AS, Brieher WM, Gumbiner BM. Molecular and functional analysis of cadherin-based adherens junctions. Annu Rev Cell Dev Biol. 1997a;13:119–46.

Yap AS, Brieher WM, Pruschy M, Gumbiner BM. Lateral clustering of the adhesive ectodomain: a fundamental determinant of cadherin function. Curr Biol. 1997b;7:308–15.

Yap AS, Niessen CM, Gumbiner BM. The juxtamembrane region of the cadherin cytoplasmic tail supports lateral clustering, adhesive strengthening, and interaction with p120ctn. J Cell Biol. 1998;141:779–89.

Yu W, Yang L, Li T, Zhang Y. Cadherin signaling in cancer: its functions and role as a therapeutic target. Front Oncol. 2019;9:989.

Zaidel-Bar R. Cadherin adhesome at a glance. J Cell Sci. 2013;126:373–8.

Zaidel-Bar R, Itzkovitz S, Ma'ayan A, Iyengar R, Geiger B. Functional atlas of the integrin adhesome. Nat Cell Biol. 2007;9:858–67.

Zipursky SL, Sanes JR. Chemoaffinity revisited: Dscams, protocadherins, and neural circuit assembly. Cell. 2010;143:343–53.

Zivanov J, Oton J, Ke Z, Von Kugelgen A, Pyle E, Qu K, Morado D, Castano-Diez D, Zanetti G, Bharat TAM, Briggs JAG, Scheres SHW. A Bayesian approach to single-particle electron cryo-tomography in RELION-4.0. elife. 2022;11:e83724.

Chapter 3
Supramolecular Assemblies Drive Activation of Inflammasomes

Zhangfei Shen, Chen Wang, and Tian-Min Fu

Abstract Inflammasomes are large cytosolic protein complexes that detect and respond to pathogenic and endogenous danger signals, triggering critical innate immune defenses. This chapter outlines the structural and mechanistic features of canonical inflammasomes, focusing on NLRP1, NLRP3, NAIP–NLRC4, NLRP6, and AIM2. Canonical inflammasomes often comprise a sensor protein (e.g., an NLR or AIM2), the adaptor ASC, and pro-caspase-1, which are assembled upon ligand binding. NLR sensors detect diverse signals, ranging from bacterial toxins and lipopolysaccharides to crystals and ATP, while AIM2 recognizes cytosolic double-stranded DNA. Once activated, these inflammasomes facilitate caspase-1-dependent maturation of IL-1β and IL-18, as well as cleavage of gasdermin D, culminating in pyroptotic cell death and pro-inflammatory cytokine release. Structural insights reveal that inflammasome activation transitions from an autoinhibited state to filamentous or ring-shaped supramolecular assemblies. Despite ligand diversity and varying activation pathways, inflammasomes share unifying principles in their oligomerization-driven activation. Elucidating these mechanisms advances our understanding of how cells mount rapid, robust immune responses and paves the way for developing therapeutic interventions in inflammatory and infectious diseases.

Keywords Inflammasome · Innate immunity · Pyroptosis · Noncanonical inflammasome · Caspase · NLR family · NLRP1 · NLRP3 · NAIP · NLRC4 · NLRP6 · AIM2 · GSDMD · PAMP · DAMP

Zhangfei Shen and Chen Wang contributed equally with all other contributors.

Z. Shen · C. Wang · T.-M. Fu (✉)
Department of Pathology, UMass Chan Medical School, Worcester, MA, USA
e-mail: Tianmin.Fu@umassmed.edu

© The Author(s), under exclusive license to Springer Nature Switzerland AG 2026   41
A. M. Pedley (ed.), *Supramolecular Protein Assemblies In Cells*, Advances in Experimental Medicine and Biology 1514,
https://doi.org/10.1007/978-3-032-26629-3_3

3.1 Overview of Inflammasomes

The innate immune system employs multiple families of pattern recognition receptors (PRRs) to sense intracellular or extracellular danger signals and trigger immune responses (Yin et al. 2015). PRRs recognize pathogen-associated molecular patterns (PAMPs), damage-associated molecular patterns (DAMPs), or other stimuli that disrupt cellular homeostasis (Janeway Jr. 1989; Li and Wu 2021). Among these immune sensors, inflammasomes are cytoplasmic supramolecular protein complexes essential for innate immune responses (Martinon et al. 2002; Fu et al. 2024). Upon detecting cytosolic immune signals, inflammasomes assemble into large, multiprotein complexes that mediate the maturation and secretion of pro-inflammatory cytokines (Martinon et al. 2002; Chou et al. 2023; Zheng et al. 2020). Indeed, the identification of inflammasomes is based on the study of the maturation of the inflammatory cytokine IL-1beta (Martinon et al. 2002). In the 1980s, studies revealed that the secreted mature IL-1b is generated by cleaving the pro-IL1beta (Auron et al. 1984; Van Damme et al. 1985). In 1992, caspase-1 was identified as the enzyme responsible for converting pro-IL1beta into mature IL-1beta (Cerretti et al. 1992; Thornberry et al. 1992). To identify the molecular machinery for activating caspase-1, the group of the late J. Tschopp found the first inflammasome, NLRP1, that can form a large complex to promote the activation of caspase-1 (Martinon et al. 2002).

Inflammasomes are generally classified into canonical and noncanonical types (Fu et al. 2024). Canonical inflammasomes consist of sensors, the adaptor protein ASC (apoptosis-associated speck-like protein containing a CARD), and the effector protein pro-caspase-1. Many sensor proteins belong to the nucleotide-binding and oligomerization domain (NOD) leucine-rich repeat containing (NLR) family (Chou et al. 2023). Upon activation by PAMPs or DAMPs, the sensor proteins recruit ASC to nucleate a higher-order oligomerization platform that, in turn, recruits and activates caspase-1 (Fu et al. 2024). This leads to cleavage of pro-IL-1β and pro-IL-18 into their mature forms and processing of gasdermin D (GSDMD), resulting in GSDMD pore formation for cytokine secretion and pyroptosis (Ding et al. 2016; Liu et al. 2016; Xia et al. 2021a). In contrast, noncanonical inflammasomes involve caspases-4/5 (human) or caspase-11 (mouse) with an N-terminal CARD domain and a C-terminal caspase domain, which function as both sensors and effectors and are directly activated by cytosolic lipopolysaccharides (LPS) from Gram-negative bacteria (Kayagaki et al. 2011, 2015; Shi et al. 2014; Vigano et al. 2015). This activation triggers cleavage of GSDMD to drive pyroptosis, bypassing IL-1β or IL-18 maturation (Kayagaki et al. 2015).

The largest family of canonical inflammasomes in mammals is formed by NLRs, which share a conserved domain arrangement (Fig. 3.1a–d). In general, NLRs contain an N-terminal variable domain for downstream signaling, a central nucleotide-binding domain (NBD), also referred to as the NACHT domain, and a C-terminal leucine-rich repeat (LRR) domain that senses stimuli to activate inflammatory responses (Meunier and Broz 2017). The NACHT domain includes four

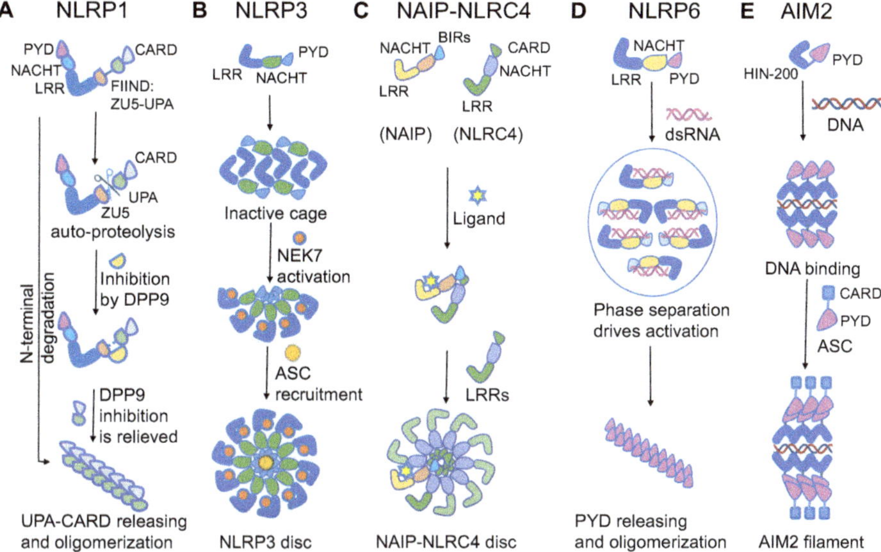

Fig. 3.1 Overview of canonical inflammasomes. Assembly and activation of canonical inflammasomes, including NLRP1 (**a**), NLRP3 (**b**), NAIP-NLRC4 (**c**), NLRP6 (**d**) and AIM2 (**e**). PYD, pyrin domain; NACHT, nucleotide-binding and oligomerization domain; LRR, leucine-rich repeat; FIIND, function-to-find domain; CARD, caspase recruitment domain; DPP9, dipeptidyl peptidase 9; VbP, Val-boro-pro (DPP9 inhibitor); NEK7, NIMA-related kinase 7; ASC, apoptosis-associated speck-like protein containing a CARD,; BIR, baculoviral inhibitor of apoptosis repeat; NAIP, neuronal apoptosis inhibitory protein; NBD, nucleotide-binding domain. Detailed descriptions of the assembly pathways are presented in the main text

subdomains—NBD, helical domain 1 (HD1), winged-helix domain (WHD), and helical domain 2 (HD2)—and undergoes nucleotide-driven conformational changes and self-oligomerization. Based on the variable N-terminal domains, human NLRs can be classified into five subfamilies: NLRA (acidic transactivating domain-containing), NLRB (baculovirus IAP repeat-containing; also known as NAIP), NLRC (CARD-containing), NLRP (PYD-containing), and NLRX (with an uncharacterized N-terminal domain) (Chou et al. 2023; Meunier and Broz 2017; Sundaram et al. 2024).

Aside from NLR inflammasomes, a unique receptor called absent in melanoma 2 (AIM2) detects cytosolic double-stranded DNA (dsDNA) (Hornung et al. 2009). AIM2 features an N-terminal PYD domain and a C-terminal HIN domain (Fig. 3.1e). The HIN domain binds dsDNA, leading to AIM2 oligomerization and inflammasome activation (Yin et al. 2015; Jin et al. 2012).

In this chapter, we summarize current knowledge regarding the structure, assembly, regulation, and function of canonical inflammasomes, including NLRP1, NLRP3, NAIP–NLRC4, NLRP6, and AIM2 (Fig. 3.1).

3.2   NLRP1 Inflammasome

NLRP1 was the first described inflammasome sensor in the NLR family (Martinon et al. 2002). Human NLRP1 features an N-terminal PYD domain, a central NACHT domain, and a C-terminal LRR domain followed by a unique function-to-find domain (FIIND) and a CARD domain (Fig. 3.2a). In contrast, mice encode three NLRP1 homologs—NLRP1a, NLRP1b, and NLRP1c—all lacking the N-terminal PYD domain (Boyden and Dietrich 2006; Sastalla et al. 2013). The FIIND domain comprises the subdomains ZU5 and UPA (D'Osualdo et al. 2011; Finger et al. 2012).

NLRP1 activation requires proteolytic or ubiquitin-mediated degradation of its N-terminal region (Chui et al. 2019; Sandstrom et al. 2019). For example, mouse NLRP1b can be cleaved by anthrax lethal factor and tobacco etch virus (TEV)

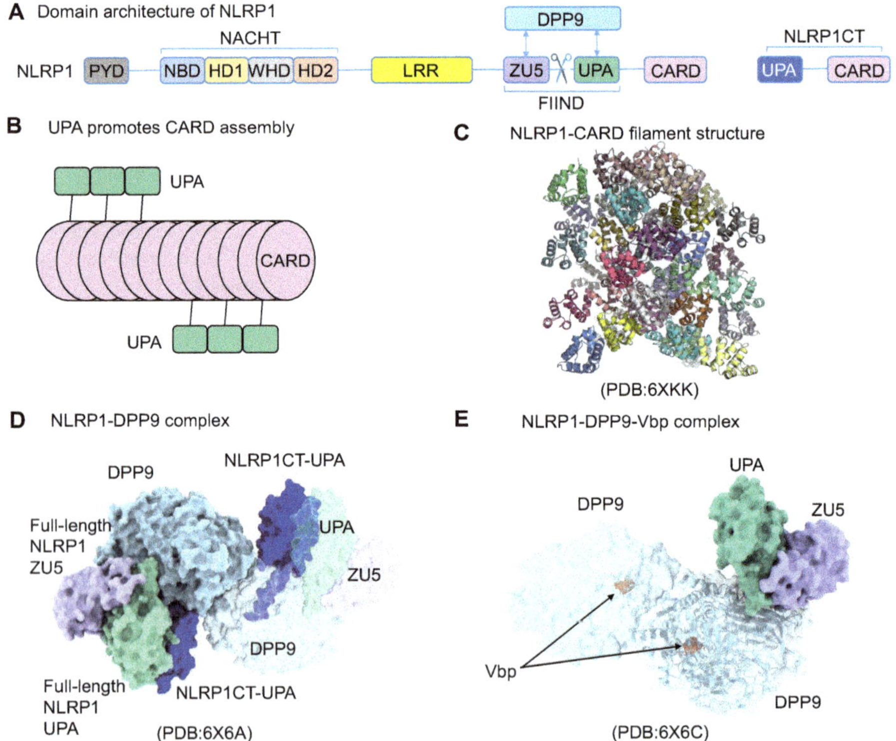

Fig. 3.2 Assembly of the NLRP1 inflammasome. (**a**) Schematic domain architecture of NLRP1. (**b**) Schematic arrangement of NLRP1 UPA domains promotes the NLRP1 CARD domains to form a filament. (**c**) Cryo-EM structure of NLRP1 CARD filament with each subunit colored individually (PDB: 6XKK). (**d**) Structure of NLRP1 and DPP9 ternary complex (PDB:6X6A). A dimeric complex contains two copies of DPP9, the FIIND domain (ZU5 and UPA) of full-length NLRP1 and the UPA of a processed NLRP1 C-terminal fragment. (**e**) Structure of dimeric DPP9 bound with VbP (Val-boro-pro, DPP9 inhibitor) in complex with the FIIND domain of NLRP1 (PDB: 6X6C)

protease, or ubiquitinated by the Shigella ligase IpaH7.8, while human NLRP1 is targeted by various picornaviral 3C proteases (Chui et al. 2019; Sandstrom et al. 2019; Robinson et al. 2020; Tsu et al. 2021). The N-terminal fragment is cleaved in the FIIND domain, which results in liberating the C-terminal UPA–CARD region for oligomerization and activation (Sandstrom et al. 2019) (Fig. 3.2a). The UPA–UPA interaction drives the formation of the oligomeric NLRP1 CARD filament (Fig. 3.2b, c), which then recruits and activates caspase-1 (Gong et al. 2021; Robert Hollingsworth et al. 2021).

Depletion or inhibition of dipeptidyl peptidases DPP8 and DPP9 also modulates NLRP1 activity (Okondo et al. 2017, 2018; Gai et al. 2019). Cryo-EM structures of human or rat NLRP1 bound to DPP9 revealed that each DPP9 dimer associates with one full-length NLRP1 and one cleaved UPA–CARD fragment (Hollingsworth et al. 2021; Huang et al. 2021) (Fig. 3.2d). In these complexes, the cleaved UPA–CARD tail inserts into the DPP9 active site, preventing release of the active fragment. Inhibitors like Val-boro-Pro (VbP) compete with this tail, displacing the UPA–CARD fragment from DPP9 (Hollingsworth et al. 2021) (Fig. 3.2e). When the amount of proteolytically cleaved UPA–CARD exceeds available DPP9 complexes, the free C-terminal fragments accumulate and oligomerize to form functional inflammasomes (Hollingsworth et al. 2021).

3.3 NLRP3 Inflammasome

NLRP3 is a prototypical NLR-family sensor that recognizes a wide range of PAMPs and DAMPs, including pore-forming toxins, LPS, crystalline particles, ATP, and ion flux (Fu and Wu 2023). NLRP3 consists of an N-terminal PYD domain that recruits ASC, a central NACHT domain that mediates ATP-dependent oligomerization, and a C-terminal LRR domain that contributes to ligand sensing (Fig. 3.3a). Under

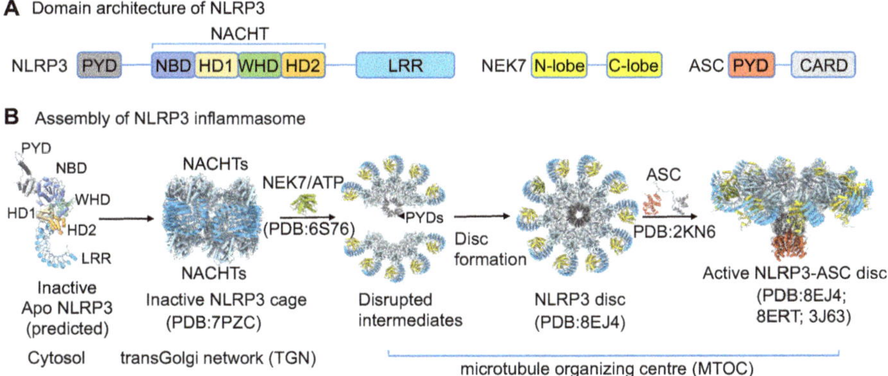

Fig. 3.3 Assembly of the NLRP3 inflammasome. (**a**) Schematic domain architecture of NLRP3, NEK7 and ASC. (**b**) Assembly and activation of the NLRP3 inflammasome

resting conditions, NLRP3 is expressed at low levels (Bauernfeind et al. 2009). Upon pathogen infection or inflammatory cues, NLRP3 transcription is upregulated ("priming"), similar to NLRP6 but in contrast to NLRC4 or AIM2 (Fu et al. 2024).

Following priming, NLRP3 activation involves its trafficking from the trans-Golgi network (TGN) to the microtubule organizing center (MTOC), recruitment of the kinase NEK7, and subsequent assembly of an NLRP3–ASC–caspase-1 complex (He et al. 2016; Shi et al. 2016) (Fig. 3.3a). NLRP3 in the apo state forms cage-like oligomers associated with the TGN membrane, representing an autoinhibited conformation (Andreeva et al. 2021; Hochheiser et al. 2022). Structural analysis revealed that human NLRP3 forms decameric cages, whereas mouse NLRP3 assembles into dodecamers to hexadecamers (Andreeva et al. 2021; Hochheiser et al. 2022). Two layers of NLRP3 rings are held together by LRR–LRR interactions, although the PYD domains are poorly resolved in these structures (Fig. 3.3b).

To activate NLRP3, these cages disperse as NLRP3 moves from the TGN to the MTOC, aided by dynein and the adaptor HDAC6 (Magupalli et al. 2020). At the MTOC, NEK7 is enriched and binds to NLRP3, causing a major conformational transition from the cage-like state to a ring-like active oligomer (Xiao et al. 2023). The NLRP3(ΔPYD)-NEK7 structure provides the first glimpse of how NEK7 interacts with NLRP3 (Sharif et al. 2019). The NACHT domain and LRR domain of NLRP3 form a tight grip around the C-terminal lobe of NEK7 (Fig. 3.3b). A recent cryo-EM structure of NLRP3–NEK7–ASC shows a ring of approximately 10–11 NLRP3 protomers with NEK7 interacting at the periphery and PYD filaments protruding orthogonally for ASC recruitment (Xiao et al. 2023) (Fig. 3.3b). Large subdomain rearrangements within the NACHT domain drive this transition, while NEK7 and the LRR domains stabilize the ring but do not directly participate in NLRP3 oligomerization (Xiao et al. 2023) (Fig. 3.3b). This mechanism highlights how NLRP3 transitions from an autoinhibited cage state at the TGN to an active ring-like oligomer at the MTOC (Fig. 3.3b).

3.4 NAIP–NLRC4 Inflammasome

The NAIP–NLRC4 inflammasome involves two NLR-family members, NAIP (the sensor) and NLRC4 (the adaptor) (Kofoed and Vance 2011; Zhao et al. 2011). Humans possess a single NAIP, whereas mice have seven, with NAIP3 identified as a pseudogene (Sundaram et al. 2024). NAIPs recognize bacterial components such as flagellin and the type III secretion system (T3SS) needle or rod proteins from pathogens, including *Salmonella typhimurium*, *Escherichia coli*, *Shigella flexneri*, and *Burkholderia* species (Zhao et al. 2011; Yang et al. 2013; Grandjean et al. 2017; Reyes Ruiz et al. 2017). Upon detecting PAMPs, NAIPs recruit NLRC4 for inflammasome assembly.

Structurally, both NAIPs and NLRC4 contain an N-terminal signaling domain (BIR in NAIPs, CARD in NLRC4), a central NACHT domain (NBD, HD1, WHD, HD2), and C-terminal LRR domains (Hu et al. 2015; Zhang et al. 2015; Tenthorey

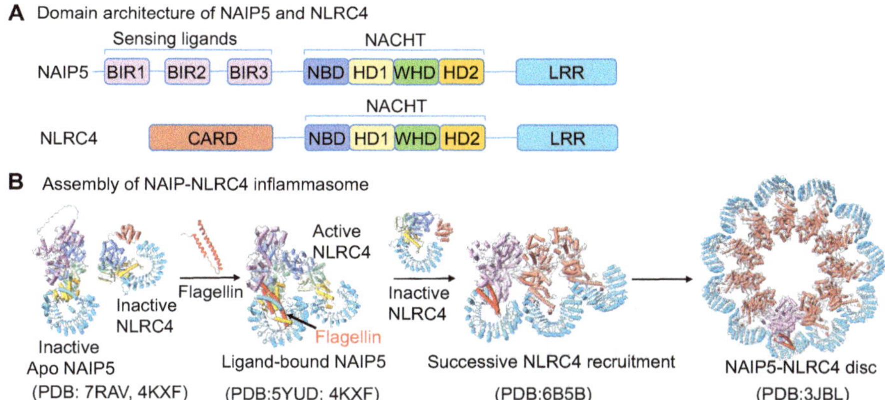

Fig. 3.4 Assembly of the NAIP5-NLRC4 inflammasome. (**a**) Schematic domain architecture of NAIP5 and NLRC4. (**b**) Assembly and activation of the NAIP5-NLRC4 inflammasome. Flagellin in red is the ligand of NAIP5 and can bind NAIP5 to trigger the assembly of the NAIP5-NLRC4 inflammasome

et al. 2017) (Fig. 3.4a). In the NAIP–NLRC4 inflammasome, the ligand is bound by NAIP, which then nucleates NLRC4 oligomerization (Hu et al. 2015; Zhang et al. 2015). For instance, NAIP2 specifically recognizes the T3SS rod protein PrgJ, while NAIP5 detects flagellin (Zhang et al. 2015; Tenthorey et al. 2017) (Fig. 3.4b). Upon ligand binding, NAIPs undergo conformational rearrangements that stabilize key subdomains and facilitate NLRC4 recruitment (Tenthorey et al. 2017; Paidimuddala et al. 2023a, b).

In the apo state, NLRC4 is autoinhibited (Hu et al. 2013). Once recruited by ligand-bound NAIPs, it oligomerizes into a ring-shaped structure (Hu et al. 2015; Zhang et al. 2015). Cryo-EM analysis of PrgJ–NAIP2–NLRC4 complexes shows a disc-like assembly with one NAIP2 and 10–12 NLRC4 protomers, indicating that NAIP2 nucleates NLRC4 polymerization (Zhang et al. 2015) (Fig. 3.4b). This transition involves major conformational changes within the NACHT domain and between the NACHT and LRR domains (Fig. 3.4b). Similarly, flagellin-activated NAIP5 initiates formation of a comparable NLRC4 oligomer (Tenthorey et al. 2017) (Fig. 3.4b). The CARD domains of oligomerized NLRC4 then form a helical structure that recruits caspase-1 via CARD–CARD interactions, enabling cytokine maturation (Li et al. 2018).

3.5 NLRP6 Inflammasome

NLRP6 is particularly important within the gastrointestinal tract (Chen et al. 2011; Elinav et al. 2011; Normand et al. 2011). Its activation can be triggered by microbial ligands such as lipopolysaccharide (LPS), lipoteichoic acid (LTA), and microbial

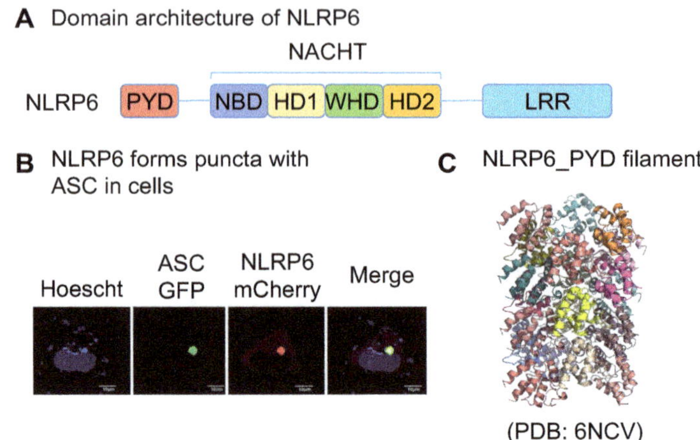

Fig. 3.5 Assembly of the NLRP6 inflammasome. (**a**) Schematic domain architecture of NLRP6. (**b**) NLRP6 and ASC form puncta at cytosol in cells. NLRP6 is fused with a mCherry while ASC is fused with a GFP. DNA is stained by Hoechst. (**c**) Cryo-EM structure of NLRP6 PYD filament with each subunit colored individually (PDB: 6NCV)

RNA, as well as by metabolic signals and type I interferons (IFNs) (Ghimire et al. 2018; Barnett et al. 2023; Wang et al. 2015). Structurally, NLRP6 has an N-terminal PYD, a central NACHT domain, and a C-terminal LRR, resembling NLRP3 (Fig. 3.5a). Besides forming inflammasomes, NLRP6 has additional roles, such as regulating mucus production in colonic goblet cells (Wlodarska et al. 2014).

Biochemical analyses indicate that NLRP6 can recognize RNA and undergo liquid–liquid phase separation (LLPS) that promotes its oligomerization and activation (Shen et al. 2021) (Fig. 3.5b). The NACHT and LRR domains of NLRP6 contain positively charged patches critical for RNA binding. DHX15, a DExH-box helicase, also enhances NLRP6 LLPS, facilitating the integration of signals from dsRNA and LTA (Shen et al. 2021). Structurally, NLRP6 filaments are primarily driven by PYD–PYD interactions, with NACHT and LRR domains surrounding a central PYD filament (Shen et al. 2019) (Fig. 3.5c). This filament scaffold recruits ASC and activates caspase-1, analogous to other canonical inflammasomes (Fu et al. 2024; Xia et al. 2021b). Thus, LLPS is emerging as a key mechanism for driving NLRP6 oligomerization and function.

3.6   AIM2 Inflammasome

AIM2 was first identified as a tumor suppressor in melanomas (DeYoung et al. 1997; Chen et al. 2006). In 2009, AIM2 was shown to recognize cytosolic dsDNA and form an inflammasome (Hornung et al. 2009; Burckstummer et al. 2009; Fernandes-Alnemri et al. 2009). Structurally, it contains an N-terminal PYD domain and a C-terminal HIN (hematopoietic expression, IFN-inducible, and nuclear localization with 200–amino acid repeat) domain (Jin et al. 2013) (Fig. 3.6a). The HIN domain binds dsDNA in a sequence-independent manner, whereas the PYD recruits ASC (Jin et al. 2012). AIM2 is the founding member of the AIM2-like receptor (ALR) family, which includes four members in humans and 14 in mice (Wang and Yin 2017). All ALRs share a similar domain arrangement to AIM2, except for mouse p202, which lacks the PYD domain and serves as a dominant negative regulator of immune defense (Wang and Yin 2017).

Apo AIM2 is believed to be autoinhibited by intramolecular PYD–HIN interactions (Wang and Yin 2017). Upon recognizing cytosolic dsDNA, the HIN domain binds the DNA backbone through electrostatic interactions (Jin et al. 2012). This exposes the PYD domain for higher-order oligomerization, driving inflammasome formation (Lu et al. 2015) (Fig. 3.6b, c). DNA activates AIM2 in a length-dependent manner with a minimal length of 70–80 bp in cells (Jin et al. 2013; Morrone et al. 2015). In vitro footprint assays showed that the minimal length of DNA for AIM2 binding is 8–9 bp, while the isolated AIM2 HIN domain can effectively bind to 20 bp dsDNA with an affinity of ~30 nM (Wang and Yin 2017). Structural analysis showed that the HIN domain of AIM2 is composed of two OB-fold subdomains with highly positively charged surfaces, which engage B-form dsDNA through electrostatic interactions (Jin et al. 2012). Therefore, AIM2 recognizes dsDNA in a sequence-independent manner. Although the PYD domain of AIM2 is not involved in DNA binding, it contributes to the cooperative clustering of AIM2 on dsDNA through its intrinsic capability of self-association (Wang and Yin 2017). Together, cooperative clustering of AIM2 along extended dsDNA promotes PYD self-association and inflammasome activation (Fig. 3.6c).

The AIM2 inflammasome is regulated by cellular or pathogen proteins through diverse mechanisms. For example, cells express some inhibitor proteins to AIM2, including PYD-only proteins (POPs) and CARD-only proteins (COPs) (Khare et al.

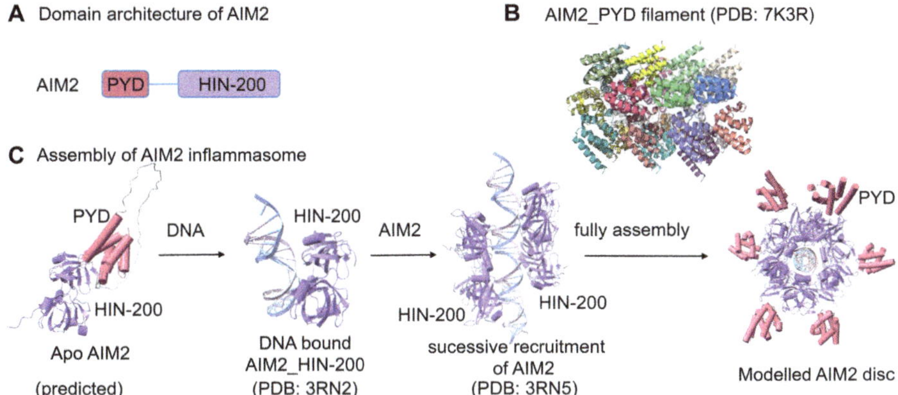

Fig. 3.6 Assembly of the AIM2 inflammasome. (**a**) Schematic domain architecture of AIM2. (**b**) Cryo-EM structure of AIM2 PYD filament with subunits color-coded (PDB: 7K3R). (**c**) DNA sensing and activation of the AIM2 inflammasome

2014; Johnston et al. 2005). These proteins can inhibit AIM2 by interfering with the interactions between AIM2 and ASC, thereby preventing AIM2 inflammasome assembly (Khare et al. 2014; Johnston et al. 2005). Pathogens have also evolved mechanisms to evade AIM2 detection or suppress its activation. For instance, *Francisella tularensis* uses the CRISPR-Cas system to prevent AIM2 recognition, and various viral proteins can inhibit AIM2 oligomerization and activation (Sampson et al. 2014).

3.7 Conclusion

Inflammasomes are cytosolic multiprotein complexes that serve as critical sensors in innate immunity. Over the past decade, numerous advances have revealed how these complexes recognize ligands, assemble, and become activated. This chapter focused on several well-characterized inflammasomes (NLRP1, NLRP3, NAIP–NLRC4, NLRP6, and AIM2), highlighting both their unique and shared activation features. Despite diverse ligands and regulatory mechanisms, these inflammasomes converge on common principles: (1) they exist in an autoinhibited apo state, (2) ligand detection or signal reception initiates oligomerization-driven activation, and (3) the formation of supramolecular assemblies is integral to the activation process. Looking ahead, many critical questions remain unanswered in the study of inflammasomes. One challenge is understanding how NLRP3 integrates diverse upstream signals to trigger inflammasome assembly. What molecular mechanisms govern its activation in response to such a wide array of stimuli? Additionally, the regulation of inflammasome activity occurs within a complex and dynamic network—how do cellular pathways fine-tune this process to balance immunity and prevent pathological inflammation? Lastly, with inflammasomes implicated in numerous inflammatory diseases, the development of targeted therapeutics remains a major goal. How can we design molecules to modulate inflammasome activation without compromising host defense? Ongoing research will continue to illuminate these fundamental aspects, deepening our understanding of inflammasomes as crucial immune sensors and paving the way for innovative therapeutic strategies to manage inflammasome-associated diseases.

Conflict of Interest Statement The authors have no conflicts of interest to declare that are relevant to the content of this chapter.

References

Andreeva L, et al. NLRP3 cages revealed by full-length mouse NLRP3 structure control pathway activation. Cell. 2021;184:6299–312 e6222. https://doi.org/10.1016/j.cell.2021.11.011.

Auron PE, et al. Nucleotide sequence of human monocyte interleukin 1 precursor cDNA. Proc Natl Acad Sci USA. 1984;81:7907–11. https://doi.org/10.1073/pnas.81.24.7907.

Barnett KC, Li S, Liang K, Ting JP. A 360 degrees view of the inflammasome: mechanisms of activation, cell death, and diseases. Cell. 2023;186:2288–312. https://doi.org/10.1016/j.cell.2023.04.025.

Bauernfeind FG, et al. Cutting edge: NF-kappaB activating pattern recognition and cytokine receptors license NLRP3 inflammasome activation by regulating NLRP3 expression. J Immunol. 2009;183:787–91. https://doi.org/10.4049/jimmunol.0901363.

Boyden ED, Dietrich WF. Nalp1b controls mouse macrophage susceptibility to anthrax lethal toxin. Nat Genet. 2006;38:240–4. https://doi.org/10.1038/ng1724.

Burckstummer T, et al. An orthogonal proteomic-genomic screen identifies AIM2 as a cytoplasmic DNA sensor for the inflammasome. Nat Immunol. 2009;10:266–72. https://doi.org/10.1038/ni.1702.

Cerretti DP, et al. Molecular cloning of the interleukin-1 beta converting enzyme. Science. 1992;256:97–100. https://doi.org/10.1126/science.1373520.

Chen IF, et al. AIM2 suppresses human breast cancer cell proliferation in vitro and mammary tumor growth in a mouse model. Mol Cancer Ther. 2006;5:1–7. https://doi.org/10.1158/1535-7163.MCT-05-0310.

Chen GY, Liu M, Wang F, Bertin J, Nunez G. A functional role for Nlrp6 in intestinal inflammation and tumorigenesis. J Immunol. 2011;186:7187–94. https://doi.org/10.4049/jimmunol.1100412.

Chou WC, Jha S, Linhoff MW, Ting JP. The NLR gene family: from discovery to present day. Nat Rev Immunol. 2023;23:635–54. https://doi.org/10.1038/s41577-023-00849-x.

Chui AJ, et al. N-terminal degradation activates the NLRP1B inflammasome. Science. 2019;364:82–5. https://doi.org/10.1126/science.aau1208.

D'Osualdo A, et al. CARD8 and NLRP1 undergo autoproteolytic processing through a ZU5-like domain. PLoS One. 2011;6:e27396. https://doi.org/10.1371/journal.pone.0027396.

DeYoung KL, et al. Cloning a novel member of the human interferon-inducible gene family associated with control of tumorigenicity in a model of human melanoma. Oncogene. 1997;15:453–7. https://doi.org/10.1038/sj.onc.1201206.

Ding J, et al. Pore-forming activity and structural autoinhibition of the gasdermin family. Nature. 2016;535:111–6. https://doi.org/10.1038/nature18590.

Elinav E, et al. NLRP6 inflammasome regulates colonic microbial ecology and risk for colitis. Cell. 2011;145:745–57. https://doi.org/10.1016/j.cell.2011.04.022.

Fernandes-Alnemri T, Yu JW, Datta P, Wu J, Alnemri ES. AIM2 activates the inflammasome and cell death in response to cytoplasmic DNA. Nature. 2009;458:509–13. https://doi.org/10.1038/nature07710.

Finger JN, et al. Autolytic proteolysis within the function to find domain (FIIND) is required for NLRP1 inflammasome activity. J Biol Chem. 2012;287:25030–7. https://doi.org/10.1074/jbc.M112.378323.

Fu J, Wu H. Structural mechanisms of NLRP3 inflammasome assembly and activation. Annu Rev Immunol. 2023;41:301–16. https://doi.org/10.1146/annurev-immunol-081022-021207.

Fu J, Schroder K, Wu H. Mechanistic insights from inflammasome structures. Nat Rev Immunol. 2024;24:518–35. https://doi.org/10.1038/s41577-024-00995-w.

Gai K, et al. DPP8/9 inhibitors are universal activators of functional NLRP1 alleles. Cell Death Dis. 2019;10:587. https://doi.org/10.1038/s41419-019-1817-5.

Ghimire L, et al. NLRP6 negatively regulates pulmonary host defense in Gram-positive bacterial infection through modulating neutrophil recruitment and function. PLoS Pathog. 2018;14:e1007308. https://doi.org/10.1371/journal.ppat.1007308.

Gong Q, et al. Structural basis for distinct inflammasome complex assembly by human NLRP1 and CARD8. Nat Commun. 2021;12:188. https://doi.org/10.1038/s41467-020-20319-5.

Grandjean T, et al. The human NAIP-NLRC4-inflammasome senses the Pseudomonas aeruginosa T3SS inner-rod protein. Int Immunol. 2017;29:377–84. https://doi.org/10.1093/intimm/dxx047.

He Y, Zeng MY, Yang D, Motro B, Nunez G. NEK7 is an essential mediator of NLRP3 activation downstream of potassium efflux. Nature. 2016;530:354–7. https://doi.org/10.1038/nature16959.

Hochheiser IV, et al. Structure of the NLRP3 decamer bound to the cytokine release inhibitor CRID3. Nature. 2022;604:184–9. https://doi.org/10.1038/s41586-022-04467-w.

Hollingsworth LR, et al. DPP9 sequesters the C terminus of NLRP1 to repress inflammasome activation. Nature. 2021;592:778–83. https://doi.org/10.1038/s41586-021-03350-4.

Hornung V, et al. AIM2 recognizes cytosolic dsDNA and forms a caspase-1-activating inflammasome with ASC. Nature. 2009;458:514–8. https://doi.org/10.1038/nature07725.

Hu Z, et al. Crystal structure of NLRC4 reveals its autoinhibition mechanism. Science. 2013;341:172–5. https://doi.org/10.1126/science.1236381.

Hu Z, et al. Structural and biochemical basis for induced self-propagation of NLRC4. Science. 2015;350:399–404. https://doi.org/10.1126/science.aac5489.

Huang M, et al. Structural and biochemical mechanisms of NLRP1 inhibition by DPP9. Nature. 2021;592:773–7. https://doi.org/10.1038/s41586-021-03320-w.

Janeway CA Jr. Approaching the asymptote? Evolution and revolution in immunology. Cold Spring Harb Symp Quant Biol. 1989;54(Pt 1):1–13. https://doi.org/10.1101/sqb.1989.054.01.003.

Jin T, et al. Structures of the HIN domain:DNA complexes reveal ligand binding and activation mechanisms of the AIM2 inflammasome and IFI16 receptor. Immunity. 2012;36:561–71. https://doi.org/10.1016/j.immuni.2012.02.014.

Jin T, Perry A, Smith P, Jiang J, Xiao TS. Structure of the absent in melanoma 2 (AIM2) pyrin domain provides insights into the mechanisms of AIM2 autoinhibition and inflammasome assembly. J Biol Chem. 2013;288:13225–35. https://doi.org/10.1074/jbc.M113.468033.

Johnston JB, et al. A poxvirus-encoded pyrin domain protein interacts with ASC-1 to inhibit host inflammatory and apoptotic responses to infection. Immunity. 2005;23:587–98. https://doi.org/10.1016/j.immuni.2005.10.003.

Kayagaki N, et al. Non-canonical inflammasome activation targets caspase-11. Nature. 2011;479:117–21. https://doi.org/10.1038/nature10558.

Kayagaki N, et al. Caspase-11 cleaves gasdermin D for non-canonical inflammasome signalling. Nature. 2015;526:666–71. https://doi.org/10.1038/nature15541.

Khare S, et al. The PYRIN domain-only protein POP3 inhibits ALR inflammasomes and regulates responses to infection with DNA viruses. Nat Immunol. 2014;15:343–53. https://doi.org/10.1038/ni.2829.

Kofoed EM, Vance RE. Innate immune recognition of bacterial ligands by NAIPs determines inflammasome specificity. Nature. 2011;477:592–5. https://doi.org/10.1038/nature10394.

Li D, Wu M. Pattern recognition receptors in health and diseases. Signal Transduct Target Ther. 2021;6:291. https://doi.org/10.1038/s41392-021-00687-0.

Li Y, et al. Cryo-EM structures of ASC and NLRC4 CARD filaments reveal a unified mechanism of nucleation and activation of caspase-1. Proc Natl Acad Sci USA. 2018;115:10845–52. https://doi.org/10.1073/pnas.1810524115.

Liu X, et al. Inflammasome-activated gasdermin D causes pyroptosis by forming membrane pores. Nature. 2016;535:153–8. https://doi.org/10.1038/nature18629.

Lu A, et al. Plasticity in PYD assembly revealed by cryo-EM structure of the PYD filament of AIM2. Cell Discov. 2015;1:15013-. https://doi.org/10.1038/celldisc.2015.13.

Magupalli VG, et al. HDAC6 mediates an aggresome-like mechanism for NLRP3 and pyrin inflammasome activation. Science. 2020;369:eaas8995. https://doi.org/10.1126/science.aas8995.

Martinon F, Burns K, Tschopp J. The inflammasome: a molecular platform triggering activation of inflammatory caspases and processing of proIL-beta. Mol Cell. 2002;10:417–26. https://doi.org/10.1016/s1097-2765(02)00599-3.

Meunier E, Broz P. Evolutionary convergence and divergence in NLR function and structure. Trends Immunol. 2017;38:744–57. https://doi.org/10.1016/j.it.2017.04.005.

Morrone SR, et al. Assembly-driven activation of the AIM2 foreign-dsDNA sensor provides a polymerization template for downstream ASC. Nat Commun. 2015;6:7827. https://doi.org/10.1038/ncomms8827.

Normand S, et al. Nod-like receptor pyrin domain-containing protein 6 (NLRP6) controls epithelial self-renewal and colorectal carcinogenesis upon injury. Proc Natl Acad Sci USA. 2011;108:9601–6. https://doi.org/10.1073/pnas.1100981108.

Okondo MC, et al. DPP8 and DPP9 inhibition induces pro-caspase-1-dependent monocyte and macrophage pyroptosis. Nat Chem Biol. 2017;13:46–53. https://doi.org/10.1038/nchembio.2229.

Okondo MC, et al. Inhibition of Dpp8/9 activates the Nlrp1b inflammasome. Cell Chem Biol. 2018;25:262–7 e265. https://doi.org/10.1016/j.chembiol.2017.12.013.

Paidimuddala B, et al. Mechanism of NAIP-NLRC4 inflammasome activation revealed by cryo-EM structure of unliganded NAIP5. Nat Struct Mol Biol. 2023a;30:159–66. https://doi.org/10.1038/s41594-022-00889-2.

Paidimuddala B, Cao J, Zhang L. Structural basis for flagellin-induced NAIP5 activation. Sci Adv. 2023b;9:eadi8539. https://doi.org/10.1126/sciadv.adi8539.

Reyes Ruiz VM, et al. Broad detection of bacterial type III secretion system and flagellin proteins by the human NAIP/NLRC4 inflammasome. Proc Natl Acad Sci USA. 2017;114:13242–7. https://doi.org/10.1073/pnas.1710433114.

Robert Hollingsworth L, et al. Mechanism of filament formation in UPA-promoted CARD8 and NLRP1 inflammasomes. Nat Commun. 2021;12:189. https://doi.org/10.1038/s41467-020-20320-y.

Robinson KS, et al. Enteroviral 3C protease activates the human NLRP1 inflammasome in airway epithelia. Science. 2020;370:eaay2002. https://doi.org/10.1126/science.aay2002.

Sampson TR, et al. A CRISPR-Cas system enhances envelope integrity mediating antibiotic resistance and inflammasome evasion. Proc Natl Acad Sci USA. 2014;111:11163–8. https://doi.org/10.1073/pnas.1323025111.

Sandstrom A, et al. Functional degradation: a mechanism of NLRP1 inflammasome activation by diverse pathogen enzymes. Science. 2019;364:eaau1330. https://doi.org/10.1126/science.aau1330.

Sastalla I, et al. Transcriptional analysis of the three Nlrp1 paralogs in mice. BMC Genomics. 2013;14:188. https://doi.org/10.1186/1471-2164-14-188.

Sharif H, et al. Structural mechanism for NEK7-licensed activation of NLRP3 inflammasome. Nature. 2019;570:338–43. https://doi.org/10.1038/s41586-019-1295-z.

Shen C, et al. Molecular mechanism for NLRP6 inflammasome assembly and activation. Proc Natl Acad Sci USA. 2019;116:2052–7. https://doi.org/10.1073/pnas.1817221116.

Shen C, et al. Phase separation drives RNA virus-induced activation of the NLRP6 inflammasome. Cell. 2021;184:5759–74 e5720. https://doi.org/10.1016/j.cell.2021.09.032.

Shi J, et al. Inflammatory caspases are innate immune receptors for intracellular LPS. Nature. 2014;514:187–92. https://doi.org/10.1038/nature13683.

Shi H, et al. NLRP3 activation and mitosis are mutually exclusive events coordinated by NEK7, a new inflammasome component. Nat Immunol. 2016;17:250–8. https://doi.org/10.1038/ni.3333.

Sundaram B, Tweedell RE, Prasanth Kumar S, Kanneganti TD. The NLR family of innate immune and cell death sensors. Immunity. 2024;57:674–99. https://doi.org/10.1016/j.immuni.2024.03.012.

Tenthorey JL, et al. The structural basis of flagellin detection by NAIP5: a strategy to limit pathogen immune evasion. Science. 2017;358:888–93. https://doi.org/10.1126/science.aao1140.

Thornberry NA, et al. A novel heterodimeric cysteine protease is required for interleukin-1 beta processing in monocytes. Nature. 1992;356:768–74. https://doi.org/10.1038/356768a0.

Tsu BV, et al. Diverse viral proteases activate the NLRP1 inflammasome. Elife. 2021;10:e60609. https://doi.org/10.7554/eLife.60609.

Van Damme J, et al. Homogeneous interferon-inducing 22K factor is related to endogenous pyrogen and interleukin-1. Nature. 1985;314:266–8. https://doi.org/10.1038/314266a0.

Vigano E, et al. Human caspase-4 and caspase-5 regulate the one-step non-canonical inflammasome activation in monocytes. Nat Commun. 2015;6:8761. https://doi.org/10.1038/ncomms9761.

Wang B, Yin Q. AIM2 inflammasome activation and regulation: a structural perspective. J Struct Biol. 2017;200:279–82. https://doi.org/10.1016/j.jsb.2017.08.001.

Wang P, et al. Nlrp6 regulates intestinal antiviral innate immunity. Science. 2015;350:826–30. https://doi.org/10.1126/science.aab3145.

Wlodarska M, et al. NLRP6 inflammasome orchestrates the colonic host-microbial interface by regulating goblet cell mucus secretion. Cell. 2014;156:1045–59. https://doi.org/10.1016/j.cell.2014.01.026.

Xia S, et al. Gasdermin D pore structure reveals preferential release of mature interleukin-1. Nature. 2021a;593:607–11. https://doi.org/10.1038/s41586-021-03478-3.

Xia S, Chen Z, Shen C, Fu TM. Higher-order assemblies in immune signaling: supramolecular complexes and phase separation. Protein Cell. 2021b;12:680–94. https://doi.org/10.1007/s13238-021-00839-6.

Xiao L, Magupalli VG, Wu H. Cryo-EM structures of the active NLRP3 inflammasome disc. Nature. 2023;613:595–600. https://doi.org/10.1038/s41586-022-05570-8.

Yang J, Zhao Y, Shi J, Shao F. Human NAIP and mouse NAIP1 recognize bacterial type III secretion needle protein for inflammasome activation. Proc Natl Acad Sci USA. 2013;110:14408–13. https://doi.org/10.1073/pnas.1306376110.

Yin Q, Fu TM, Li J, Wu H. Structural biology of innate immunity. Annu Rev Immunol. 2015;33:393–416. https://doi.org/10.1146/annurev-immunol-032414-112258.

Zhang L, et al. Cryo-EM structure of the activated NAIP2-NLRC4 inflammasome reveals nucleated polymerization. Science. 2015;350:404–9. https://doi.org/10.1126/science.aac5789.

Zhao Y, et al. The NLRC4 inflammasome receptors for bacterial flagellin and type III secretion apparatus. Nature. 2011;477:596–600. https://doi.org/10.1038/nature10510.

Zheng D, Liwinski T, Elinav E. Inflammasome activation and regulation: toward a better understanding of complex mechanisms. Cell Discov. 2020;6:36. https://doi.org/10.1038/s41421-020-0167-x.

Chapter 4
Conformational Dynamics and Allostery in the cAMP-Protein Kinase A Signalosome

Varun Venkatakrishnan and Ganesh S. Anand

Abstract Protein kinase A (PKA) is a master kinase that controls numerous cellular processes via phosphorylation. PKA, or cAMP-dependent protein kinase, is regulated by the second messenger 3′5′ cyclic adenosine monophosphate (cAMP) generated in response to upstream G-protein-coupled receptor stimulation. Intracellularly, PKA is organized into multiprotein complexes that bring together the regulatory and catalytic elements of PKA, termed signalosomes. cAMP-PKA signalosomes exert spatiotemporal control of PKA activity by precisely regulating activation and termination phases of PKA signaling. While cAMP binds to PKA, triggering signal activation, phosphodiesterases catalyze cAMP degradation to terminate PKA signaling. Amide-hydrogen/deuterium exchange mass spectrometry (HDXMS), a powerful biophysical technique, has captured the conformational dynamics of regulation in PKA signaling. This has revealed allosteric mechanisms of PKA regulation and identified key pathway intermediates in the activation and termination phases of PKA. This chapter describes the components that constitute the PKA signalosome and the allosteric mechanisms of PKA regulation as revealed by HDXMS.

Keywords Protein kinase A · Phosphodiesterase · cAMP · Second messenger signaling · Spatiotemporal regulation · HDXMS · Protein dynamics · Allostery

V. Venkatakrishnan
Department of Chemistry, The Pennsylvania State University, University Park, PA, USA
e-mail: vxv5106@psu.edu

G. S. Anand (✉)
Department of Chemistry, The Pennsylvania State University, University Park, PA, USA

Department of Biochemistry and Molecular Biology, The Pennsylvania State University, University Park, PA, USA

The Huck Institutes of the Life Sciences, The Pennsylvania State University, University Park, PA, USA
e-mail: gsa5089@psu.edu

A. M. Pedley (ed.), *Supramolecular Protein Assemblies In Cells*, Advances in Experimental Medicine and Biology 1514,
https://doi.org/10.1007/978-3-032-26629-3_4

4.1 Introduction

Phosphorylation of proteins is among the most essential and ubiquitous mechanisms of signal transduction and intracellular communication (Hunter 2012; Li et al. 2013). Phosphorylation has major implications for numerous cellular processes, ranging from cell growth and development to metabolism and cell death. The human kinome, comprising 518 protein kinases and 20 lipid kinases, encodes ~2% of all genes and catalyzes the phosphotransfer reaction of the γ-phosphate group from ATP to target proteins, most preferably at S/T/Y residues (Kole et al. 1988; Zhang et al. 2021; Manning et al. 2002). Among these is protein kinase A (PKA), a "master kinase" that catalyzes the phosphorylation of >100 substrates at S/T residues to control the expression of multiple genes and regulate intracellular metabolism (Shabb 2001; Sassone-Corsi 2012). PKA belongs to the AGC family of protein kinases that includes 60 kinases that share highly conserved catalytic domains. PKA is unique among AGC kinases in that its regulatory (R) and catalytic (C) domains exist on separate protein subunits (Fig. 4.1) (Taylor et al. 2012). As the

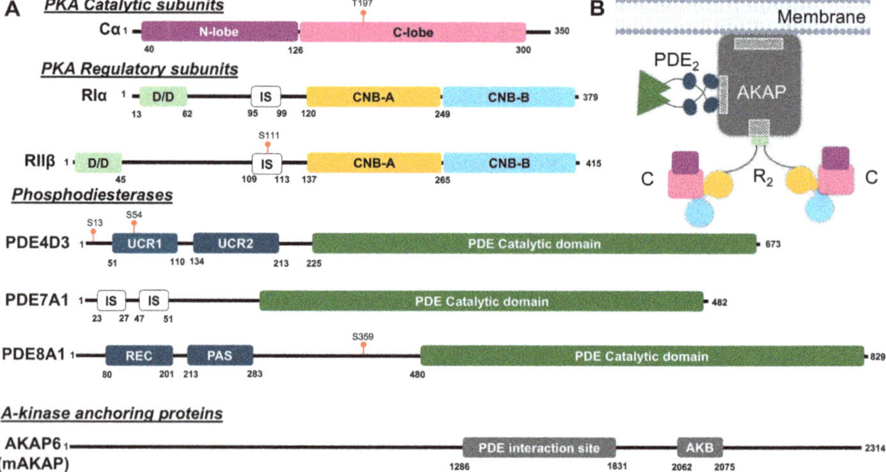

Fig. 4.1 Domain organization of the PKA signalosomal complex. (**a**) Domain organization of the components of the PKA signalosome. A representative protein from each class of PKA signaling proteins is shown. Cα is representative of PKA catalytic subunits (Cβ, Cγ). RIα is representative of type I R-isoforms which includes RIβ. RIIβ is representative of type II R-isoforms which includes RIIα. A representative PDE from each cAMP-specific PDE is also shown. PDE4D3 is a long PDE isoform representative of the PDE4 family, PDE7A1 is representative of the PDE7 family, and PDE8A1 is representative of the PDE8 family. AKAP6 (mAKAP) is also shown to be representative of the AKAPs that bind tether PDEs and PKA to the sarcoplasmic reticulum or nuclear membrane. N-lobe-N-terminal lobe, C-lobe-C-terminal lobe, D/D-docking/dimerization domain, IS-PKA inhibitory site, CNB-A-cyclic nucleotide binding domain A, CNB-B-cyclic nucleotide binding domain B, UCR1-upstream conserved region 1, UCR2-upstream conserved region 2, REC-receiver domain, PAS-PER-ARNT-SIMs domain. PKA Phosphorylation sites are highlighted on R-subunits and PDEs. Phosphorylation at site T197 is necessary for kinase activation by an upstream kinase, PDK1. (**b**) Architecture of the heteroheptameric PKA signalosome localized to a membrane. All proteins are color-coded based on the domain organization. The dimerization states of the proteins are given in the subscript

first kinase whose structure was solved at high resolution, the PKA C-subunit is considered a prototype for describing phosphotransfer (Reikhardt and Shabanov 2020).

PKA is a major intracellular target for the second messenger 3′5′cyclic adenosine monophosphate (cAMP) that is generated by Adenylyl cyclases (AC) upon upstream hormonal stimulation of G-protein coupled receptors (GPCRs) (Sassone-Corsi 2012). PKA, therefore, functions to relay hormonal messages (e.g., adrenaline, glucagon) to downstream intracellular targets to elicit highly specific responses (Surdo et al. 2017). Mammalian PKA is encoded by two major C-subunit isoforms (Cα, Cβ) with four nonredundant isoforms of R-subunits (RIα, RIβ, RIIα, and RIIβ) (Fig. 4.1a) (Skalhegg and Tasken 2000; Amieux and McKnight 2002). Additional C-subunit isoforms, such as Cγ and PRKX, are relatively low in abundance, and their expression is highly selective across cell types (Li et al. 2002).

PKA activity is spatiotemporally regulated by activation and termination phases to complete one round of PKA signaling (Fig. 4.2) (Walsh and Van Patten 1994). In the basal inactive state, PKA is a tetrameric holoenzyme complex composed of a dimer of R-subunits complexed to two C-subunits (R$_2$:C$_2$) (Taylor et al. 2012; Kim

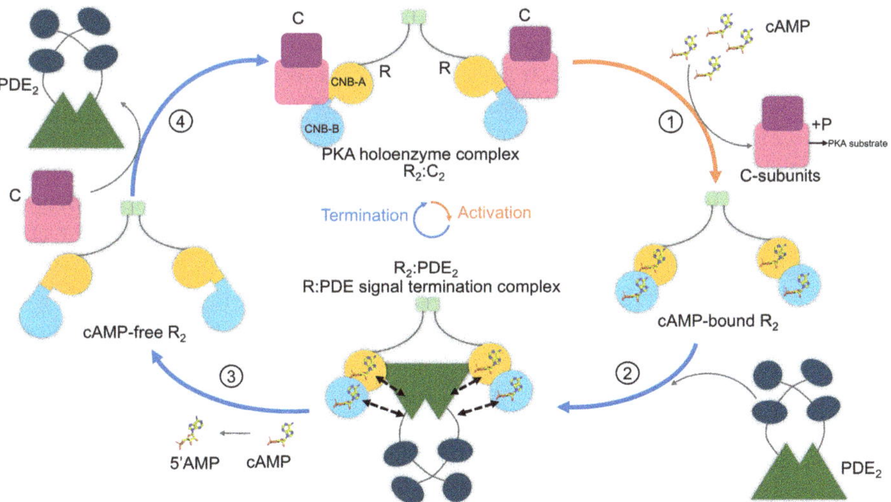

Fig. 4.2 Activation and termination phases of PKA signaling. In the inactive state, PKA is a heterotetrameric holoenzyme complex R$_2$:C$_2$. Step 1: cAMP generated by upstream hormonal stimulation binds to the four cAMP binding sites (CNBs) on dimeric R-subunits, promoting a substrate-assisted release of C-subunits. Step 2: PDEs are recruited to the R-subunits to form a transient signal termination complex. Step 3: Bound cAMP is hydrolyzed by PDEs, generating product 5′AMP and cAMP-free R subunits. Step 4: C-subunits reassociate with cAMP-free R-subunits to reset PKA signaling and reform the PKA holoenzyme complex. In dimeric R-subunits (R$_2$), CNB-A-orange circle, CNB-B-sky blue circle, D/D- pale green square, linker-grey lines. In C-subunits, N-lobe-violet square, C-lobe-pink square. cAMP, AMP-yellow stick representation. In dimeric PDEs (PDE$_2$), catalytic domain-green triangle, N-terminal regulatory domains-navy blue circles, linkers-grey lines. Orange arrow, activation phases; blue arrow, termination phase. The dimerization states of the proteins are given in the subscript

et al. 2007; Lu et al. 2019). cAMP binding to the R-subunits promotes the release of C-subunits, which phosphorylate downstream targets (Skalhegg and Tasken 2000; Das et al. 2007). The precise and timed termination of PKA is mediated by cAMP-specific phosphodiesterases (PDEs) that catalyze the degradation of cAMP that is bound to R-subunits (Moorthy et al. 2011a; Krishnamurthy et al. 2013; Conti and Beavo 2007). This primes R-subunits for reassociation with C-subunits to regenerate the PKA holoenzyme complex to reset PKA for a subsequent round of activation. Intracellular timing of PKA signaling ranges from 2 to 5 min for each activation and termination phase (Anton et al. 2022; Borner et al. 2011). The scaffold proteins—A-kinase anchoring proteins (AKAPs)—colocalize PDE and PKA in close proximity to each other and tether PKA to GPCRs (Wong and Scott 2004; Smith et al. 2013). This spatially localizes PKA signaling to a 30–60 nm range around GPCRs and is a major determinant in the hormonal specificity of PKA signaling (Anton et al. 2022). Together, the R-subunits, C-subunits, PDEs, and AKAPs make up the PKA signalosome complex that spatiotemporally regulates cAMP-PKA signaling (Fig. 4.1b) (Baillie et al. 2005; Hoshi et al. 2005; Steinberg and Brunton 2001). Dysregulation of PKA is implicated in a myriad of diseases such as diabetes mellitus, cardiovascular disorders, acrodysostosis, Carney complex, Cushing's syndrome, and multiple cancers (Kirschner et al. 2000; Linglart et al. 2012; Rhayem et al. 2015; Kammer et al. 1996; Walker et al. 2019).

Compartmentalization of PKA signaling was first proposed in the 1980s when specific GPCRs were observed to selectively activate either type I or II R-subunits in cardiomyocytes (Hayes and Brunton 1982; Buxton and Brunton 1983). Since then, numerous PKA compartments have been identified across multiple cell types (Surdo et al. 2017; Anton et al. 2022; Baillie et al. 2005; Steinberg and Brunton 2001; Asirvatham et al. 2004; Pare et al. 2005; Lynch et al. 2005; Lehnart et al. 2005; Lohmann et al. 1984). This led to the hypothesis that cAMP acted locally as opposed to the prevailing dogma at the time that cAMP was a highly diffusible messenger. The remarkable discovery of the localization of a PDE4 isoform to the plasma membrane in mice identified a plausible mechanism for restricting the diffusion of cAMP (Shakur et al. 1993). The parallel discovery of AKAPs identified a scaffolding protein that could integrate cAMP, PKA, PDEs, and GPCRs within a three-dimensional nanodomain spanning 30–60 nm (Anton et al. 2022; Lohmann et al. 1984; Theurkauf and Vallee 1982).

R-subunits are central regulators of cAMP-PKA signaling and undergo large conformational changes as they transition between activation and termination phases by mediating differential interactions with cAMP, C-subunits, PDEs, and AKAPs (Moorthy et al. 2011a; Herberg et al. 1994). This makes the PKA regulatory pathway a highly dynamic signaling process. Amide-hydrogen deuterium exchange mass spectrometry (HDXMS) is a powerful analytical technique that has captured the conformational dynamics of the PKA signalosome across activation and termination phases. This has also revealed that allostery is a key feature of PKA regulation. In this chapter, we discuss the application of HDXMS in uncovering the molecular mechanisms of spatiotemporal regulation of PKA signaling. First, we describe the structural biology of the components that make up the fundamental PKA signalosome. This will be followed by a brief description of HDXMS and its merits. Finally, we summarize the key findings made using HDXMS, discussing the dynamics and allostery of the activation and termination phases of PKA.

4.2 Structural Biology of the PKA Signalosome

Multiple regulatory elements (R-subunits, PDEs, and AKAPs) associate with the C-subunits to constitute PKA signaling islands or signalosomes, which exert spatio-temporal control on PKA (Smith et al. 2013; Baillie et al. 2005; Steinberg and Brunton 2001). The fundamental PKA signalosome is a dynamic hetero-oligomeric complex with variable molecular weights ranging from ~200 to ~800 kDa (Smith et al. 2013; Hoshi et al. 2005). This variability in molecular weights is attributable to the diversity of PDEs and AKAPs, which differentially colocalize with PKA (Neves-Zaph 2017). In addition to compartmentalizing PKA near GPCRs and ACs, AKAPs can also dock accessory proteins, including phosphatases and other kinases (Anton et al. 2022; Redden and Dodge-Kafka 2011). The differential expression of PKA, PDE, and AKAP isoforms thus leads to diverse combinations of PKA signalosomes. There is a lesser degree of variability among the isoforms of R- and C-subunits, leading to two types of PKA holoenzymes, types I and II, each with two isoforms, respectively. Type I PKA holoenzyme isoforms consist of $RI\alpha_2{:}C\alpha_2$ and $RI\beta_2{:}C\beta_2$, while type II PKA holoenzyme isoforms consist of $RII\alpha_2{:}C\alpha_2$ and $RII\beta_2{:}C\beta_2$ (Skalhegg and Tasken 2000). The overall oligomeric state of the PKA signalosome is a heteroheptamer complex made up of a dimer of R-subunits, two monomeric C subunits, a dimer of PDEs, and a monomeric AKAP (Fig. 4.1b). In this section, we describe the structural basis for the functioning of the individual components of the PKA signalosome.

4.2.1 Catalytic Subunits

PKA C-subunit is a ~40 kDa (350 amino acids) monomeric protein that contains the prototypical AGC kinase domain and catalyzes phosphotransfer from ATP to characteristic -RRX[S/T]φ- motifs (X-any, φ-hydrophobic) present in >100 substrates, including type II R-isoforms (Shabb 2001; Reikhardt and Shabanov 2020; Shoji et al. 1981; Soberg et al. 2013). The canonical kinase fold of the C-subunit is organized into N-terminal (N-lobe) and C-terminal (C-lobe) lobes and is highly conserved across C-subunit isoforms (>90% sequence homology) (Fig. 4.1a) (McClendon et al. 2014). The N-terminal of the C-subunit can be myristoylated at a glycine residue to enable anchoring to cell membranes, exerting additional spatial regulation on PKA (Gangal et al. 1999). The N-lobe spans residues 40–126 and consists of three α-helices and five β-turns and is responsible for ATP binding (along with 2 Mg^{2+} ions) (Zheng et al. 1993; Herberg et al. 1999). Within this region is the glycine-rich loop (G-loop) that contains a conserved cluster of glycines (G50, G52, G55) to interact with ATP. The G-loop acts as a lid, facilitating ATP entry and ADP exit within the catalytic pocket, and stabilizes the triple phosphate moieties of ATP. A critical salt-bridge between K72 in the β-subdomain and E91 in the αC-helix is a hallmark of kinase activation upon ATP binding (Tsigelny et al. 1999). This couples ATP binding with conformational changes at the β-III strand, leading to the closure of the kinase pocket since the αC-helix is positioned between both lobes and forms a hinge.

The larger C-lobe (residues 127–300) is an all-α helical subdomain containing seven contiguous α-helices that are primarily involved in substrate binding and phosphotransfer. The C-lobe contains the activation loop, which is phosphorylated by the upstream kinase PDK1 at T197, leading to PKA activation. The activation loop is a characteristic feature of all AGC kinases. The phosphate moiety of pT197 forms hydrogen bonds with H87 in the α-C helix of the N-lobe, further stabilizing the kinase fold in addition to the K72-E91 salt bridge (Steichen et al. 2012). Kinase activation results in the formation of two spines, the catalytic and regulatory spines, which are anchored by the core hydrophobic αF-helix (McClendon et al. 2014). Upon activation, the N- and C-lobes close to form a clamp consisting of negatively charged glutamates, which enables the docking of the substrate phosphorylation site. The HRD motif in the catalytic loop mediates phosphotransfer of the γ-phosphate of ATP to the substrate. Regulation of C-subunit activity by R-subunits is achieved by mimicking the substrate phosphorylation sites (Kim et al. 2007). In type I PKA holoenzymes, closure of the Cα catalytic pocket upon ATP binding significantly enhances the binding affinity for RIα by >100-fold (K_d < 0.04 nM), whereas in type II PKA holoenzymes, the R:C interactions are not significantly impacted by ATP occupancy (Herberg et al. 1994, 1996; Anand et al. 2007).

4.2.2 Regulatory Subunits

The dimeric R-subunits are central to the regulation of PKA signaling. The R-subunits bind to C-subunits and are the receptors for cAMP during signal activation. During signal termination, the R-subunits interact with PDEs. They also interact with AKAPs, which spatially compartmentalize PKA signaling. Each R monomer is organized into an N-terminal docking/dimerization domain (D/D) followed by a disordered linker segment and two tandem cAMP-binding domains (CNB-A or CNB-B) (Fig. 4.1a). The D/D domain on each protomer spans 45–60 residues in the N-terminal and forms the primary interface for interprotomer interactions via formation of a four-helix bundle (Banky et al. 2003). The D/D also serves as the interaction site for the AKAPs, where an antiparallel helix that forms the interprotomer interface interacts with a conserved ~14–18 residue helix found in AKAPs (Carr et al. 1991; Newlon et al. 1999). RII-isoforms interact more preferentially with AKAPs than RI-isoforms, attributable to differences in the sequences of D/D domains. Additionally, RI-isoforms are a covalent dimer under oxidizing conditions as the D/D interface contains an inter-chain disulfide bridge (C17–C37 in RIα) (Leon et al. 1997). This is followed by the disordered interdomain linker, which is ~60–90 residues in length. The length of the disordered linker is a key distinguishing feature between RI and RII-isoforms, wherein RII-isoforms have longer linker regions spanning ~90 residues. These linkers contribute to the large hydrodynamic radius of R and are likely to increase the spatial range of PKA signaling (Herberg et al. 1994; Smith et al. 2017). The linkers have been demonstrated to contribute to phase separation in RIα, revealing a mechanism for sequestration of cAMP with RIα liquid droplets (Zhang et al. 2020). The linkers also contain the C-subunit interaction site (autoinhibitory site), which in RI-subunits is a pseudosubstrate site (-RRG[A/G]φ-) whereas in RII-subunits, this site is a true substrate site (RRX[S/T]φ-), reflecting differential regulation of C-subunit across various R isoforms (Kim et al. 2007; Anand et al. 2007; Diskar et al. 2007; Martin et al. 2007; Wu et al. 2007; Zhang et al. 2012).

C-terminal to the disordered linkers are the tandem cyclic nucleotide binding domains A (CNB-A) and B (CNB-B). The R-subunits are unique among cAMP binding proteins as they are the only proteins that possess a tandem cAMP-binding domain on a single protein subunit (Kannan et al. 2007). This is in contrast to other cAMP binding proteins, such as the prototypical DNA binding protein CAP and the ion channel HCN, which only possess one CNB site on each protein subunit (Weber et al. 1982). Each CNB spans ~120 residues and is organized into an eight β-stranded barrel, making up two β-subdomains and three noncontiguous α-helices. The β-barrel forms a protective cage, where cAMP is buried, limiting its passive release. Across both R chains, there are four CNBs, which mediate high-affinity interactions with cAMP (K_d ~ 1–300 nM) (Herberg et al. 1994; Adams et al. 1991; Mongillo et al. 2004; Doskeland and Ogreid 1984). Each CNB possesses a phosphate binding cassette (PBC), which is a conserved 13-residue motif that mediates highly specific, high-affinity interactions with cAMP (Berman et al. 2005). Each PBC contains a conserved arginine that interacts with the negatively charged exocyclic phosphate moiety of cAMP and a glutamate whose side chain interacts with the ribose moiety. An additional π-π interaction is facilitated by a distal hydrophobic residue with the adenine ring of cAMP. This hydrophobic capping residue is key for allosteric communication between both CNBs upon cAMP occupancy. The tandem CNBs are connected by an interdomain helix (αB:C helix of CNB-A) that undergoes conformational changes depending on cAMP occupancy at the CNBs. In the absence of cAMP, crystal structures of the PKA holoenzyme have shown that the αB/C helix forms an extended and rigid helix separating both CNBs and adopting a dumbbell-like conformation or H-conformation (Kim et al. 2007). When cAMP is occupied at both CNBs, the αB/C helix breaks to form a separate αB helix and a kinked αC helix, leading to the B-conformation of R-subunits, where both CNBs move toward each other (Su et al. 1995). CNB-A is the primary interaction site for C-subunits, whereas CNB-B is considered the "gatekeeper" for PKA activation since cAMP binds to CNB-B on each chain before accessing CNB-A (Das et al. 2007; Berman et al. 2005). The C-terminal of R-subunits possesses a PDZ domain interaction sequence.

RIα is a ~84 kDa protein (~42 kDa monomer) that is considered the prototypical R isoform as it is conserved across all eukaryotes. In mammals, RIα is the most abundant isoform expressed across all cell types and can compensate for the loss of expression of any other R isoforms (Amieux et al. 1997). Complete knockout of RIα in mice (prkar1a$^{-/-}$) conferred embryonic lethality attributable to a failure of cardiac morphogenesis, highlighting the essential role of RIα in development (Amieux and McKnight 2002). Furthermore, RIα possesses the highest affinity for cAMP among the R-isoforms (K_d ~ 1–50 nM) (Herberg et al. 1994; Doskeland and Ogreid 1984). Together, this makes RIα the most essential PKA isoform. RIβ is primarily expressed in the central nervous system and is associated with learning and memory (Marbach et al. 2021). RII isoforms are larger than RI (~90 kDa or ~45 kDa monomer) and are primarily localized near membranes. RIIα is expressed in myocytes and is essential for cardiac contractility, while RIIβ is the most abundant isoform in the liver and is associated with metabolic signaling (Pare et al. 2005; Mantovani et al. 2009; McKnight et al. 1998). All these contribute to the functional diversity across R-isoforms and drive differential regulation of PKA.

4.2.3 PDEs, the Terminators of cAMP Signaling

Cyclic nucleotide phosphodiesterases (PDEs) are a major class of dimeric proteins that catalyze the hydrolysis of exocyclic phosphate moieties of cyclic nucleotides to generate nucleotide monophosphates. There are 11 mammalian PDE families, out of which three are specific to cAMP (PDE4, PDE7, and PDE8), while five exhibit dual specificity for cAMP and cGMP (PDE1, PDE2, PDE3, PDE10, and PDE11) and the remaining 3 are cGMP specific (PDE5, PDE6, and PDE9) (Conti and Beavo 2007; Neves-Zaph 2017; Francis et al. 2001; Omori and Kotera 2007; Marchmont and Houslay 1980). Multiple isoforms exist with splice variants for each PDE family, leading to broad structural and functional diversity of PDEs. Since this chapter is on cAMP-PKA signaling, we will focus only on cAMP-specific PDEs and their role in the termination and reset of PKA signaling, although dual-specific PDEs are also important for the regulation of cAMP degradation. cAMP-specific PDEs catalyze the hydrolysis of cAMP to 5′adenosine monophosphate (5′AMP) to terminate PKA signaling. PDEs are organized into one or more N-terminal regulatory domains followed by a conserved C-terminal catalytic domain, although some isoforms and families do not possess N-terminal regulatory domains (Fig. 4.1a). The N-terminal regions of PDEs contain accessory and regulatory domains that vary significantly across PDE families and participate in the dimerization and AKAP binding (Bolger et al. 2015). However, the primary dimerization motif (H-loop) is present in the C-terminal catalytic domains, which are highly conserved across PDEs (~50% sequence homology). This forms a noncontiguous dual dimerization interface.

The catalytic domains (~350 amino acids) of PDE are essential to the catalytic activity and contain all α-helical domains of 16–18 α-helices. In vitro biophysical studies on full-length PDEs have been challenging due to their low levels of expression and their propensity to aggregate when overexpressed. However, the truncated catalytic domains of multiple PDEs have been successfully expressed and purified, yielding high-resolution structures, although PDE8 catalytic domains still form inclusion bodies, necessitating a tedious unfolding and refolding process (Yan et al. 2009; Wang et al. 2008; Huai et al. 2003a, 2004a, b; Iffland et al. 2005; Scapin et al. 2004; Sung et al. 2003). The catalytic pocket within the PDE catalytic domain contains two metal ions, Mg^{2+} and Zn^{2+}, which are coordinated by multiple histidine residues (Huai et al. 2003b). These are important for precisely positioning the exocyclic phosphate of cAMP and catalyzing cAMP hydrolysis. The adenine ring is stabilized by residues in α-helix 15 through π-π interactions with a conserved hydrophobic residue, along with a conserved glutamine residue that mediates H-bonding with the primary amine group of the adenine moiety (Zhang et al. 2004). This glutamine is proposed to be the primary determinant for the specificity of adenine over guanine. Additional substrate binding is also contributed by the M-loop between α-helix 14 and 15, which spans a highly disordered region. The H-loop (spanning α-helix 8 and 9) forms a part of the dimerization motif, is also a part of the catalytic pocket, and is involved in inhibitor selectivity (Wang et al. 2005, 2008).

The PDE4 subfamily is the most abundantly expressed cAMP-specific PDE intracellularly. The expression of PDE7 and PDE8 is more selective and higher in specific cell types, and their functions are still being characterized. Although PDE7 is the least abundant of the PDEs, it possesses the highest catalytic activity among the cAMP-specific PDEs, with a $k_{cat} = 3.9$ s^{-1} (Wang et al. 2005). This is followed by PDE8, which possesses a $k_{cat} = 2$ s^{-1} (Wang et al. 2008). PDE7 and PDE8 are considered high-affinity cAMP-specific PDEs. PDE4 has a $k_{cat} = 1.6$ s^{-1} and displays moderate affinity for cAMP. This is representative of the functional diversity across PDEs at a biochemical level and will likely extend broadly to the various PKA signalosomes across the cell.

The N-terminal domains vary across PDE families and have regulatory functions and mediate interactions with other proteins. Alternate splicing leads to short PDE isoforms that lack one or more N-terminal domains. PDE4 has two N-terminal domains: UCR1 and UCR2 (upstream conserved regions 1 and 2) (Beard et al. 2000). UCR2 binds to the catalytic domain of PDE4 and disrupts PDE catalytic activity. Negative feedback phosphorylation by PKA C-subunits at UCR1 (S54 in PDE4D3) releases the catalytic domain, enabling activating cAMP hydrolysis cAMP to terminate PKA signaling (MacKenzie et al. 2002). C-subunit-mediated phosphorylation at an additional site (S13) on PDE4D3 initiates AKAP (mAKAP) binding (Carlisle Michel et al. 2004). A similar feedback mechanism occurs in PDE8A, which has an N-terminal Receiver domain followed by a PAS domain. PKA phosphorylation at S359 between the PAS and catalytic domains significantly increases PDE8 activity (Brown et al. 2012). While PDE7A does not possess additional domains, it regulates PKA via tandem pseudosubstrate sites (Han et al. 2006).

Interactions between PDEs and C-subunits represent negative feedback in PKA signaling. The N-terminal regions of PDEs possess a targeting signal enabling them to interact with AKAPs, thereby spatially integrating PKA and PDE signaling (Asirvatham et al. 2004; Lynch et al. 2005; Yarwood et al. 1999). Notably, the dual-specific PDE3 also colocalizes with PKA via AKAPs (Mongillo et al. 2004; Movsesian et al. 1991; Beca et al. 2013). Tethering to AKAPs enables the PDE catalytic domains to participate in cAMP hydrolysis from the R-subunits to terminate cAMP-PKA signaling (Moorthy et al. 2011a; Krishnamurthy et al. 2013, 2014; Tulsian et al. 2017).

4.2.4   AKAPs in Spatial Regulation of PKA Signaling

AKAPs (A-kinase anchoring proteins) are molecular scaffolding proteins that localize PKA to specific subcellular compartments and are critical for the spatial regulation of PKA signaling. The microtubule binding protein MAP 2 was the first AKAP discovered, as it co-eluted with RII subunits in brain extract (Lohmann et al. 1984). Since then, more than 50 structurally diverse AKAPs have been identified, all sharing a common PKA docking site (A-kinase binding motif or AKB) (Wong and Scott 2004). Each AKAP also possesses its own localization signal directing it toward specific subcellular compartments. The consensus PKA interface on AKAPs is a 14–18 residue amphipathic helix that interacts with the D/D domains in dimeric R-subunits (Carr et al. 1991; Newlon et al. 1997). The AKB helix binds across the D/D domain of R-subunits (Newlon et al. 1999). Structural differences between the D/D domains in type I and II R-subunits are a major determinant for AKAP specificity (Banky et al. 2003; Burns et al. 2003). Most AKAPs show specificity for type II R over type I R while some also show dual specificity for both types of R-subunits (D-AKAP) (Banky et al. 2003; Burns et al. 2003). This reflects the intracellular diversity of PKA localization mediated by AKAPs. Additionally, AKAPs also serve as docking sites for PDEs, thereby integrating activation and termination components of PKA. Of all the PDEs, PDE4 isoforms preferentially interact with multiple AKAPs. For example, the PDE4D3-mAKAP-RIIα signalosome is essential for maintaining cardiac contractile function in cardiomyocytes (Bedioune et al. 2018). PDE8 has been demonstrated to interact with AKAP12 in cardiomyocytes, although PDE4 also interacts with the same AKAP. PDE7 interacts with AKAP MTG16b in the Golgi apparatus of T-lymphocytes (Asirvatham et al. 2004; Qasim et al. 2024). AKAPs thus spatially integrate the activation and termination components of the PKA signalosome. Additionally, AKAPs interact with other kinases as well as phosphatases that further localize PKA signaling and promote cross talk (Redden and Dodge-Kafka 2011).

4.2.5   PKA Regulation Beyond the Signalosome

Although PKA primarily signals via cAMP and the R-subunits, numerous proteins have been shown to mediate interactions with the different components of PKA, driving non-canonical regulation of PKA. PKA signaling can extend far beyond the range of AKAPs, wherein C-subunits activated in the cytoplasm can translocate to the nucleus and phosphorylate proteins, such as the transcription factor CREB (Gonzalez and Montminy 1989). Nuclear translocation of active C-subunits requires an accessory protein that possesses a nuclear import sequence known as A-kinase interacting proteins or AKIPs (Sastri et al. 2005). AKIPs interact with the N-terminal αA-helix of C-subunits and enable efficient trafficking of the C-subunits to the nucleus. There are no R-subunits in the nucleus to regulate PKA nuclear activity. Instead, a small protein (~8 kDa) known as PKI contains a pseudosubstrate site (RRGAI) that binds to and potently inhibits C-subunit activity (Zheng et al. 1993; Wen et al. 1995). PKI also contains a nuclear export signal that facilitates the exit of C-subunits to the cytoplasm, where it can return to its corresponding signalosome.

Another example is the multilevel cross talk between hedgehog signaling and type I PKA (RIα$_2$:Cα$_2$). In the primary cilia, the orphan GPCR GPR161 has an AKAP site on its cytoplasmic tail that preferentially binds Riα (Bachmann et al. 2016). Localizing PKA near GPR161 inhibits hedgehog signaling by Cα-mediated phosphorylation of the transcription factor GLI, thereby preventing its nuclear translocation. Parallelly, Smoothened (SMO), the major downstream effector of hedgehog (Hh) possesses a PKA pseudosubstrate (-RRGAI-) site in its own cytoplasmic tail domain, which, upon activation by Hh, sequesters Cα, inhibiting GLI phosphorylation and thereby promoting its nuclear translocation (Arveseth et al. 2021; Happ et al. 2022). Like SMO, the N-terminal disordered segment of PDE7A1 possesses two consensus pseudosubstrate sites (RRGAI), which can bind and inhibit Cα. PDE7A1 is an example of a non-catalytic mode of PKA deactivation by a PDE (Han et al. 2006). A cAMP-independent mechanism of PKA activation has been shown to occur via interactions with the RING ubiquitin ligase PJA2, which also functions as an AKAP (Lignitto et al. 2011). Upon phosphorylation by Cβ, PJA2 ubiquitinates RIIβ, priming its degradation, leading to irreversible and sustained activation of Cβ. Beyond AKAPs, RIα possesses a polyproline helix (PPPPNP) that has been demonstrated to interact with SH3 domain containing proteins such as GRB2, forming a basis for cross talk with EGFR signaling (Tortora et al. 1997). RAF1 kinase, a downstream component in the EGFR pathway, interacts with PDE8A, indicating the presence of an AKAP-independent PDE8A-PKA signalosome integrated via EGFR signaling (Brown et al. 2013).

4.3 Amide Hydrogen/Deuterium Exchange Mass Spectrometry

Amide hydrogen/deuterium exchange mass spectrometry (HDXMS) is a powerful analytical technique for mapping conformational dynamics and ensemble behavior of proteins using isotopic deuterium (Englander and Kallenbach 1983). HDMXS reports on intrinsic protein dynamics, ligand binding, protein-protein interactions, and allosteric changes (Katta and Chait 1991). In the deuterium exchange reaction, backbone amide hydrogens present in the peptide bond in proteins are stably exchanged with their heavy isotope, deuterium (Bai et al. 1993). The regions of the protein that have more ordered H-bonding networks, such as α-helices and β-sheets or biomolecular interfaces, will incorporate deuterium at a slower rate than disordered regions that have weaker H-bonding networks. When coupled with bottom-up mass spectrometry analysis, protein dynamics can be quantified at peptide-level resolution (Masson et al. 2019; Hoofnagle et al. 2003). A reaction time course in the order of seconds to minutes ($t = 0.5$–100 min) is typically performed to resolve localized dynamics across different regions of a protein. HDXMS thus provides temporal resolution of protein dynamics. Conventionally, comparative HDXMS analysis across various protein perturbation states is performed. Previously, X-ray crystallography (XRC) was the primary technique used for structural and biophysical characterization of PKA (Kim et al. 2005, 2007; Lu et al. 2019; Zheng et al. 1993; Wu et al. 2004, 2007; Zhang et al. 2012; Su et al. 1995; Bruystens et al. 2014). XRC provided high-resolution structures of PKA complexes, revealing biomolecular interfaces, ligand binding, and amino acid orientations. However, this only captured static snapshots of end-state conformations of protein complexes and was unable to report on protein dynamics, disordered regions, and reveal transient biomolecular interactions. These limitations are overcome by HDXMS although the complementation of HDXMS with XRC presents a more powerful approach.

The components of the PKA signalosome undergo major conformational changes as they cycle between active and inactive states. HDXMS has been successfully applied to map the activation and termination phases of PKA, revealing allosteric regulation of PKA and capturing conformational intermediates in this dynamic pathway (Figs. 4.2 and 4.3a). Since the R-subunit is critical for PKA regulation, these studies have majorly focused on the conformational transitions of the R-subunit across activation and termination phases.

4.4 HDXMS Mapping Dynamics of PKA Activation Phase

The activation phase of PKA entails a transition from the basal PKA holoenzyme complex (R_2:C_2) to dissociated C-subunits and cAMP-saturated R-subunits (Figs. 4.2 and 4.3a). In this section, we summarize the key findings made by HDXMS studies focused on the activation phase of PKA.

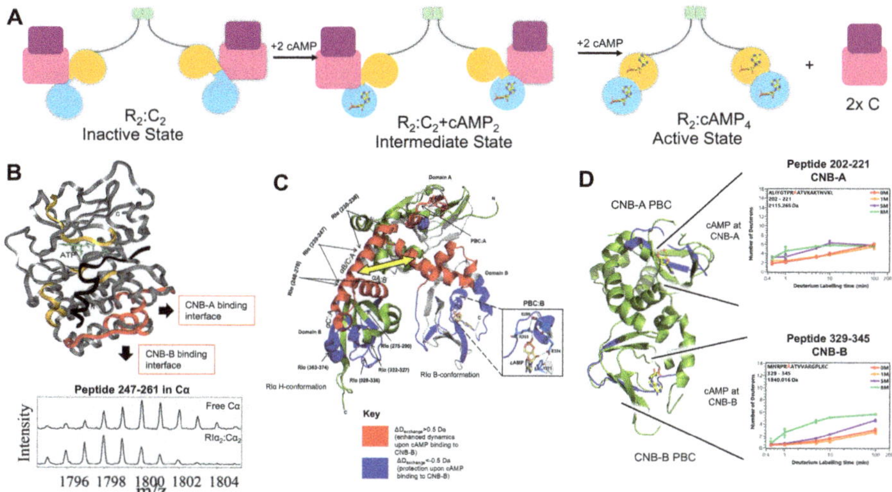

Fig. 4.3 Dynamics of PKA signal activation by cAMP. (**a**) Sequence of cAMP activation via conformational intermediates. Refer to Figs. 4.1 and 4.2 for the key. (**b**) Deuterium exchange differences (HDXMS) in Cα upon RIα binding are mapped onto the crystal structure of Cα in ribbon representation (PDB:1ATP). The αG-H loop (highlighted in red) is stabilized by interactions with RIα CNBs, while the glycine-rich loop and β5-αD (gold) are stabilized by the PKA pseudosubstrate site of RIα (black). ATP is also shown (green sticks). Bottom-HDXMS mass spectral envelope of peptide 247–261 in the αG-H loop of Cα that is critical for CNB interactions. A mass shift to a lower m/z species is observed when RIα is bound. The x-axis gives the m/z range while the y-axis gives the relative intensity. (Figure obtained with permission from Anand et al. (2003). Copyright (2003) National Academy of Sciences, U.S.A.). (**c**) HDXMS analysis of mutant PKA holoenzymes RIα R209K:Cα upon binding of cAMP to CNB-B alone in RIα reveals enhanced dynamics (regions colored red) in the interdomain αB/C helix, accompanied by deuterium exchange protection at the PBC (cAMP binding pocket) of CNB-B (colored in blue). These changes overlaid onto the structure of cAMP-free RIα (green, PDB-2QCS) and cAMP-bound RIα (grey, PDB-1RGS), which are aligned to each other relative to CNB-A. The yellow arrow indicates the conformational transitions of the αB/C helix between cAMP-free and cAMP-bound RIα states. $\Delta D_{\text{exchange}}$ represents the difference in deuterium exchange between cAMP-free and a single cAMP-bound holoenzyme. Negative difference confers (<−0.5 Da) deuterium exchange protection while a positive difference (>0.5 Da) confers enhanced dynamics upon cAMP binding. The PBC of CNB-B is shown in the inset. (Figure obtained with permission from Moorthy et al. (2011b)). (**d**) Crystal structure of monomeric cAMP-saturated RIα (green) with the PBC (blue) and cAMP (yellow sticks) of each CNB highlighted (PDB:1RGS). Right-HDXMS uptake plot comparing the effect of different urea concentrations (0, 1, 5, and 8 M) on cAMP release from each of the CNBs from cAMP-saturated RIα across different deuterium labeling times. Plots for two representative peptides shown, peptide 202–221 from CNB-A PBC and peptide 329–345 from CNB-B PBC. X-axis: number of deuterons (Da), Y-axis: Deuterium labeling time (min). Higher deuterium exchange at lower urea concentrations correlates with weaker interactions with cAMP at the respective CNBs. (Figure obtained with permission from Chandramohan et al. (2017))

4.4.1   Isoform-Specific Allostery in the Basal PKA Holoenzyme

In the basal inactive state, PKA is maintained as a high-affinity holoenzyme complex $R_2:C_2$ with a $K_d \sim 0.04$ nM (Herberg et al. 1994, 1996). The primary site for stable C-subunit interactions on RIα/RIIβ is isolated to the autoinhibitory site, along with a part of the α-A helix of CNB-A, which stably binds the C-subunit even when the tandem CNB-B domain is deleted by mutagenesis. Electrostatic interactions stabilize this interface as a cluster of negatively charged residues in the catalytic cleft of the C-subunit interacts with the tandem arginine residues in the autoinhibitory site of the R-subunits. Upon binding of CNB-A alone (truncated at 244 in the αB/C helix), stabilization occurs at the bottom of the C-lobe spanning αG, H, and I helices in the C-subunit (Anand et al. 2002, 2003; Hamuro et al. 2004). These residues span the CNB-B interacting sites on the C-subunit, suggesting allosteric coupling of both CNBs via the C-lobe of C-subunits. This was confirmed by analysis of full-length $RIα_2:Cα_2$ complexes, where a similar magnitude stabilization was observed in the bottom of the C-lobe in Cα (Fig. 4.3b) (Anand et al. 2003). A larger magnitude of deuterium exchange protection was observed at the αG, H helices in RIIβ:Cβ complexes, indicating stronger interactions between CNB-B of RII relative to RI isoforms (Anand et al. 2007). This was the first application of HDXMS-guided computational docked modeling of R:C complexes. Such conformational changes have been difficult to visualize in crystal structures of R:C complexes, which do not reveal significant structural differences between the two isoforms.

ATP binding to C-subunits is also a major determinant in the strength of R:C interactions. RIα:Cα formed weaker complexes in the absence of ATP, whereas the binding strengths of RIIβ:Cα complexes remained unaltered in the absence of ATP binding (Anand et al. 2007). This is attributable to the difference between the C-subunit autoinhibitory sites across both R isoforms, where RIIβ has a true substrate site instead of a pseudosubstrate. Compensatory mechanisms have been proposed as the basis for this lack of ATP dependency in RII holoenzymes. ATP binding is thought to stabilize the RIIβ:C interaction while autophosphorylation of the substrate site is assumed to weaken RIIβ:C interactions.

4.4.2   Cooperativity and Allostery in PKA Activation by cAMP

The complete activation of PKA requires binding of four cAMP molecules to the CNB sites in the dimeric R-subunit (Fig. 4.3a). This process has been best studied for RIα and can be extended to other R isoforms. cAMP binds RIα at very high affinities ($K_d \sim 1$–60 nM), promoting dissociation of C-subunits (Herberg et al. 1994; Doskeland and Ogreid 1984). While both CNBs share high sequence homology, they differ in the binding affinities for cAMP. CNB-B is considered the "gatekeeper" for PKA activation and has a higher cAMP binding affinity than CNB-A (Das et al. 2007; Berman et al. 2005). A variation of HDXMS combined with titrated urea denaturation has been utilized to determine the relative strengths of the cAMP binding determinants across both CNBs (Chandramohan et al. 2017). This revealed that the ribose (E324) and phosphate -interacting (R333) residues in CNB-B mediated the strongest contacts with cAMP, which did not release cAMP even under high denaturing conditions (5 M urea), whereas the equivalent contacts on CNB-A (E200, R209) were lost at lower denaturant concentrations (1–5 M urea) (Fig. 4.3d).

During activation, cAMP binds first to CNB-B, which is accompanied by a drastic conformational change priming cAMP to bind to the CNB-A site (Fig. 4.3c). This leads to cooperative cAMP binding at CNB-A with Hill coefficients ~1.4–1.8. The basis for this positive cooperativity is W260 in CNB-B, which functions as the adenine capping residue for cAMP at CNB-A and thus allosterically couples both domains. The interdomain B/C helix connecting both CNB-A and CNB-B on each chain undergoes a large conformational change upon cAMP binding, forming a kinked and twisted helix. HDMXS analysis of mutant RIα R209K:Cα holoenzymes, which have ablated cAMP binding at CNB-A, revealed that cAMP binding to CNB-B alone was insufficient to dissociate C (Moorthy et al. 2011b). However, this enhanced the dynamics of CNB-B, which occupied the bottom of the Cα in crystal structures, identifying a previously unknown conformational intermediate in the activation phase of PKA. This is attributable to the dynamic αB/C helix, which undergoes large conformational changes due to cAMP binding at the CNBs (Fig. 4.3c). Binding of cAMP to CNB-A significantly weakens interactions between RIα and Cα, reducing the binding affinity 5000-fold. This unleashes the C-subunit for substrate phosphorylation, representing complete activation. However, in the absence of a physiological substrate, Cα reassociates back with cAMP-bound Riα (Anand et al. 2002). Under these conditions, Cα accelerates the dissociation of cAMP from CNB-A to restore its high binding affinity for the pseudosubstrate and CNB-A. This is representative of an oscillating bidirectional allostery, where cAMP binding to CNB-A dissociates Cα, while subsequent Cα binding to cAMP-bound CNB-A promotes dissociation of cAMP. Here, HDXMS identified an allosteric network propagating through the αA-helix in CNB-A of Riα, which is the interaction interface with Cα. This is likely to be a reversible equilibrium process in the absence of external drivers such as PKA substrates or cAMP-degrading PDEs.

4.5 HDXMS for Mapping PKA Termination

Once activated, PKA needs to be terminated in a timely manner to reset PKA for subsequent cycles of activation and termination. While it has been well known for decades that PDEs colocalize with PKA via AKAPs, exactly how PDEs terminate PKA signaling was not understood. The mechanisms by which PDEs drive termination by dissociating cAMP from the R-subunits are now only being uncovered. HDXMS has been the primary tool for mapping the termination phase of PKA. The major focus on PKA termination has been on the type I PKA isoform (RIα). In this section, we summarize the key findings by HDXMS studies that have revealed novel mechanisms of PKA termination.

4.5.1 Dynamics of the PKA Signal Termination Complex

To terminate PKA, PDEs mediate interactions with RIα to weaken its interactions with cAMP. This is necessary due to the high binding affinities between RIα and cAMP that hinder the passive release of cAMP. The first evidence of a direct interaction between RIα and PDEs was demonstrated for the PDE RegA from the slime mold *D. discoideum* (Moorthy et al. 2011a). HDXMS together with pulldown assays and peptide microarrays revealed that the catalytic domain of RegA preferentially interacted with RIα, forming a signal termination complex with a binding affinity of 0.1–10 μM (Krishnamurthy et al. 2013). Since then, the catalytic domains of mammalian PDEs, such as PDE2, PDE5, and PDE8, have all been shown to interact with Riα (Krishnamurthy et al. 2014). Among these, PDE8 is the only cAMP-specific PDE and shares the highest sequence homology with RegA. Signal termination has been best characterized by HDMXS for RIα:PDE8 signal termination complex, which is a critical intermediate in the signal termination pathway (Krishnamurthy et al. 2014; Tulsian et al. 2017, 2020; Venkatakrishnan et al. 2023).

The primary PDE8 interaction site on RIα is localized to the β-subdomains of each of the CNBs. The β-subdomains on each CNB form a cage with an extensive H-bonding network that shields cAMP from degradation (Kannan et al. 2007). PDE8 interacts with this site, weakening the binding determinants for cAMP across both CNBs. The RIα interaction site around PDE8 was isolated to the PDE8 catalytic site and a distal site corresponding to a 3–10 helix of PDE8 (residues 685–710) (Krishnamurthy et al. 2014). A structural model of the RIα:PDE8 complex generated by HDXMS and molecular docking revealed that each CNB directly interfaced with the catalytic sites in the PDE8 dimer coupling both the active sites across the proteins (Fig. 4.4a). This reveals a cAMP channel between both active sites wherein cAMP is trafficked directly from the CNB PBC to the catalytic pocket of PDE8 without having the need to dissociate into the surrounding microenvironment. This model also points to a heterohexameric RIα:PDE8 complex organization where two PDE8 dimers engage all four CNBs in dimeric RIα, forming an $RI\alpha_2{:}PDE8_4$ complex (Tulsian et al. 2017, 2020, 2021) (Fig. 4.4c). Since only a dimer of PDE localizes with PKA via AKAPs, this indicates that hydrolysis occurs one chain at a time.

4.5.2 PDEs Catalyze Stepwise Release of cAMP from R-Subunits

Termination of PKA signaling is dependent on the release of cAMP from the RIα. HDXMS analysis of PDE8-mediated cAMP release revealed that release occurs first from CNB-A (half-life ~3 min) followed by CNB-B (half-life ~8 min) (Tulsian et al. 2017; Venkatakrishnan et al. 2023). This rate is likely to be enhanced by feedback mechanisms on full-length PDEs such as activation by C-subunit phosphorylation on PDE accessory domains. Upon cAMP release, the PBCs within each CNB

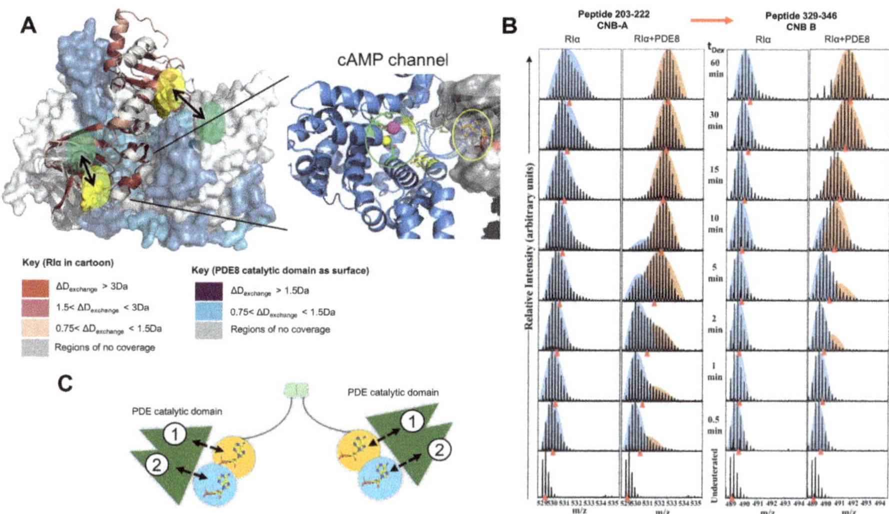

Fig. 4.4 Mechanism of PKA signal termination by PDEs. (**a**) Structural model of RIα:PDE8 signal termination complex generated by HDXMS guided molecular docking showing an RIα monomer complexed with a PDE8 dimer. Differences in deuterium exchange ($\Delta D_{exchange}$) between free proteins (RIα, PDE8) and proteins in complex are overlaid onto the docking model. Regions showing deuterium exchange protection in RIα (cartoon representation, PDB:1RGS) are highlighted in shades of red, while deuterium exchange protection in PDE8 catalytic domain (surface representation, PDB:3ECN) is given in shades of blue. Inset-cAMP channel connecting CNB PBC of RIα (yellow circle, surface representation) to the PDE8 catalytic site is shown (green circle, cartoon representation). (Figure obtained with permission from Krishnamurthy et al. (2014)). (**b**) Reaction monitoring of PDE8-mediated cAMP release from RIα revealed by bimodal mass spectral kinetics of CNB PBC peptides (203–222 in CNB-A and 329–346 in CNB-B). Comparative analysis of mass spectra of cAMP-bound RIα with and without PDE8 across an HDXMS time course ($t = 0.5$–60 min). Blue spectral envelope, cAMP-bound low deuterium exchanging population; orange envelope, cAMP-free high exchanging population. With time, the high-exchanging species becomes more abundant in the presence of PDE8, while the low-exchanging species is lost. This occurs stepwise, first at CNB-A ($t_{1/2} \sim 8$ min) followed by CNB-B ($t_{1/2} \sim 10$ min). X-axis represents m/z, Y-axis represents relative intensity. The red arrows indicate the average centroids for each mass spectral envelope. (Figure obtained with permission from Tulsian et al. (2017)). (**c**) Sequence of cAMP release overlaid onto a model of RIα$_2$:PDE8$_4$, revealing cAMP is released from CNB-A first, followed by CNB-B. Key is as shown in Figs. 4.1 and 4.2

become more dynamic. This is also reflective of the relative binding strengths of cAMP across each CNB site, where CNB-A mediates weaker interactions with cAMP than CNB-B. Mutational analysis reveals that cAMP needs to be released from CNB-A first before CNB-B can be accessed by PDE8 (Tulsian et al. 2020) (Fig. 4.4b, c). This is tied to the cooperativity between both domains, which are coupled by the CNB-A capping residue W260 and the interdomain B/C helix, whose conformation is dependent on cAMP occupancy at the CNBs. In R-subunits from lower eukaryotes such as *D. discoideum*, the sequence of cAMP release is proposed to occur at CNB-B first followed by CNB-A (Krishnamurthy et al. 2015). The *D. discoideum* homolog for RIα is DictR, while RegA is the corresponding

PDE. A major difference between RIα and DictR is the oligomerization state, where RIα is a dimer and DictR is a monomer, only possessing both CNB sites. Interchain dynamics in the dimeric RIα are a likely determinant in the sequence of cAMP release. The release of cAMP from the R-subunit promotes reassociation with C to reform the PKA holoenzyme complex and reset PKA and completes the PKA regulatory cycle.

4.5.3 cAMP Channeling and Processivity in Signal Termination

Hormonal GPCR stimulation is accompanied by an amplification of second messenger cAMP (Sassone-Corsi 2012; Pizzoni et al. 2024). The compartmentalization of PDEs near GPCRs controls the passive diffusion of cAMP and localizes the effect of cAMP within the range of signalosomes (Anton et al. 2022). Although only four cAMP molecules are required for complete activation of PKA, a large molar excess of cAMP remains in solution following activation. To squelch excess cAMP, RIα activates cAMP hydrolysis for PDE8 by > 10x termed as processivity (Tulsian et al. 2017) (Fig. 4.5a). Processivity forms the basis for molecular adaptation by ensuring that temporal control of PKA is maintained across varying levels of cAMP. This is tied to the mechanism of cAMP channeling by the RIα:PDE8 complex, where CNBs on the RIα and hydrolysis sites on PDE8 are integrated (Krishnamurthy et al. 2014). Substrate channeling is a classical feature of metabolic signaling enzyme cascades that accelerates substrate turnover, similar to the activation of PDE8 catalysis (Spivey and Ovadi 1999). During processivity, excess cAMP associates with RIα, following which it is trafficked to the PDE8 catalytic site for accelerated hydrolysis (Fig. 4.4b). The M-loop of PDE8 (748–764) is likely to facilitate entry of cAMP and exit of AMP, serving as a portal in the termination complex. Due to the transient nature of the RIα:PDE8 complex, it is still unclear whether processivity contributes to the allosteric activation of PDE8 solely or occurs completely via cAMP channeling or is a combination of both processes. A high-resolution structure of the dynamic signal termination complex will enable better interpretation of the HDXMS analyses.

4.5.4 Dysregulation of PKA Termination in Diseases

Dysregulation of PKA signaling is implicated in multiple systemic diseases (Bruystens et al. 2014; Forlino et al. 2014; Cho-Chung et al. 1995; Bruystens et al. 2016). Mutations in the congenital bone disorder acrodysostosis (ACRDYS) are localized to RIα and PDE4 (Linglart et al. 2012; Elli et al. 2016). This indicates that

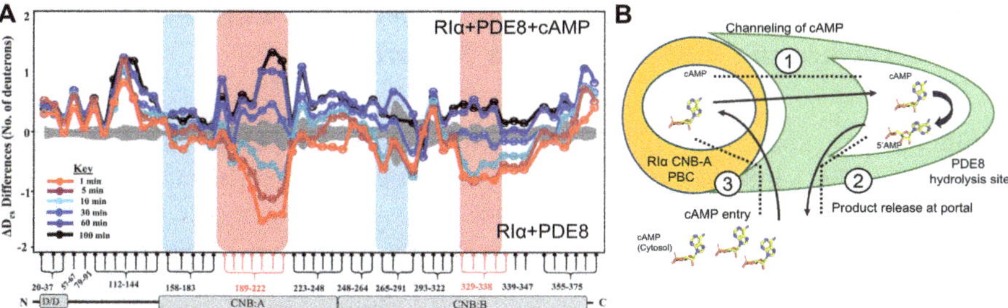

Fig. 4.5 cAMP channeling and processivity by signal termination complexes. (**a**) Deuterium exchange difference plot comparing dynamics of RIα in the presence of PDE8 with RIα in the presence of PDE8 and excess cAMP (330 μM), revealing processive hydrolysis of cAMP. Positive difference correlates to enhanced dynamics in RIα, whereas a negative difference correlates to decreased dynamics in RIα (with PDE8) in the presence of excess cAMP, respectively. Each peptide is shown as a point in the plot spanning the domains of Riα, which are highlighted in the X-axis from N- to C-terminal while the Y-axis corresponds to the difference in deuterium exchange ($\Delta D_{exchange}$). Each time point ($t = 1$–100 min) is color-coded as given. The PDE8 interaction site across both CNBs is highlighted by the blue box, while the PBC of each CNB is enclosed in the red box. At lower time points ($t = 1$–10 min) deuterium exchange protection is observed at both PBC sites, revealing the association of excess cAMP to RIα, while increased dynamics is observed at higher time points ($t = 30$–100 min) represented by the loss of cAMP by PDE8 action. (Figure obtained with permission from Tulsian et al. (2017)). (**b**) Simplified model of cAMP channel between a CNB-A site (orange) and a PDE8 catalytic domain (green), revealing the mechanism of cAMP channeling. Step 1: cAMP from the CNB-A PBC is trafficked to the PDE8 catalytic site, following which it is hydrolyzed to cAMP. This first step represents the release of hydrolysis of cAMP originally bound to RIα. Step 2: product 5′AMP is released from the catalytic pocket. Step 3: excess cAMP from the cytosol enters the channel into the PBC, wherein it is trafficked to the PDE8 catalytic pocket for processive hydrolysis. This three-step process continues until all the cAMP in the cytosol is hydrolyzed

the disease pathogenesis of ACRDYS is tied to PKA termination. Classically, ACRDYS mutants reduce the sensitivity of PKA for cAMP, thereby delaying PKA activation (Bruystens et al. 2016; Linglart et al. 2011). HDXMS analysis of two ACRDYS-causing mutations, the T207A mutation in the PBC of CNB-A in RIα and the T690P in the PDE8 catalytic domain (T594 in the PDE4D catalytic domain) revealed that these mutant signal termination complexes were unable to mediate processive hydrolysis of cAMP (Venkatakrishnan et al. 2023). While RIα T207A-PDE8 complexes demonstrated 15-fold slower rates of cAMP release across both CNB sites, RIα:PDE8 T690P complexes were unaffected. However, when excess cAMP was present, both mutant complexes were unresponsive, indicating an impairment of cAMP processivity in ACRDYS. By stabilizing intermediate conformations of the dynamic signal termination complex, these mutations likely arrest cAMP processivity by RIα:PDE8 complexes. Targeting R:PDE complexes is a potential therapeutic strategy in treating diseases where PKA is dysregulated (Tulsian et al. 2021).

4.5.5 Feedback by PKA C-Subunit Accelerates PDE8-Mediated Termination of cAMP-PKA Signaling

While in vitro experiments revealed the half-lives ($t_{1/2}$) of cAMP release from RIα by PDE8 to be ~3–8 min, contrastingly in vivo FRET experiments revealed the intracellular timing of signal termination to span ~2–5 min across numerous cell types (Anton et al. 2022; Borner et al. 2011). This discrepancy represents a major gap in our understanding of the temporal regulation of signal termination. Termination can be divided broadly into two steps: (i) release of cAMP-bound to R, followed by (ii) PDE-mediated hydrolysis. Pulse-labeling and continuous labeling HDXMS experiments together demonstrated that PDE8 on its own was inefficient at dissociating cAMP from RIα, and this translated to a six-fold slower rate of hydrolysis than that for free cAMP (Tulsian et al. 2017; Venkatakrishnan et al. 2025). However, addition of C-subunit (Cα) greatly enhanced that rates of cAMP release from RIα resulting in half-lives ($t_{1/2}$) < 1 min. This was attributed to the allosteric communication between the Cα-binding and cAMP-binding sites on RIα where Cα binding dissociates cAMP from RIα but this cAMP can reassociate back thereby dissociating Cα in an oscillating process (Lu et al. 2019; Anand et al. 2002). Addition of PDE8 shifted the equilibrium to favor stable complexation with Cα, by hydrolyzing all dissociated cAMP before it reassociated with RIα (Venkatakrishnan et al. 2025). This results in rapid removal of all four bound cAMP molecules while simultaneously regenerating the PKA holoenzyme complex without the formation of a cAMP-free RIα intermediate. Holoenzyme regeneration termed "reset" was complete within 4 min which consolidates the intracellular timing of signal termination. This reveals a novel indirect role for the C-subunit in potentiating PDEs for rapid reset of PKA.

4.5.6 Product Feedback by 5′AMP in PKA Signaling

PDE-mediated cAMP hydrolysis generates 5′adenosine monophosphate (5′AMP), which is considered a passive end-product of PKA signaling. HDXMS combined with fluorescence polarization spectroscopy revealed that 5′AMP binds to RIα utilizing similar binding determinants at the CNBs as cAMP does due to the structural similarities between both adenine nucleotides (Tulsian et al. 2020). This alters the dynamics of RIα in a similar manner to cAMP, leading to global conformational changes across the protein, including the αB/C helix whose conformational dynamics is sensitive to changes in nucleotide occupancy at the CNBs. 5′AMP binding also increased the dynamics of the disordered linkers, indicating allostery from 5′AMP alters interchain dynamics in RIα. 5′AMP was also released from RIα in RIα:PDE8 complexes similar to cAMP. However, this 5′AMP is sequestered by PDE8 in its catalytic pocket as revealed by deuterium exchange protection at multiple ligand binding sites on PDE8, such as phosphate binding site (542–569), catalytic site (590–607) and substrate binding site (715–746). Catalytic pocket rearrangements in PDE8 upon RIα interaction provides an explanation for the binding of 5'AMP. This is representative of product feedback by 5'AMP, which has only been demonstrated before by deactivating the PDE (Huai et al. 2003b). Accumulation of 5'AMP is a hallmark of metabolic stress, resulting in reduced ATP:5'AMP ratios as observed in metabolic syndromes such as diabetes (Sakamoto et al. 2005; Scott et al. 2014; Steinberg and Carling 2019). This will have profound implications for PDE-mediated termination and isoform-specific interaction between R- and C-subunits of PKA.

4.6 Dynamics of AKAP:PKA Interactions

While PDE:R-subunit interactions primarily drive temporal regulation of PKA signaling, AKAPs are essential for spatial control of PKA. HDXMS analysis on AKAP:R-subunit interactions reveal R-isoform-specific interactions (Burns-Hamuro et al. 2005). D-AKAP2 is a dual-specific AKAP that interacts with both types of R-subunits. The 14-mer helix from D-AKAP2 binds tighter to D/D domain of RIα than RIIα. This is attributable to conformational differences in the D/D domains, where the RIIα D/D has a preformed AKAP-binding interface while RIα D/D domain undergoes a large conformational change to facilitate AKAP interactions (Banky et al. 2003). This conformational change is necessitated by the extended N-terminal helix and the disulfide bridge which locks the D/D of both RIα chains (Burns et al. 2003). This study provides structural and dynamic insights it AKAP specificity across R-isoforms and provides a molecular basis for spatial compartmentalization of PKA.

4.7 Conclusions

Although PKA was among the first protein kinases to be discovered in 1968, the molecular mechanisms that drive cAMP-PKA signaling have only been uncovered in recent years (Walsh et al. 1968). Compartmentalization into signalosomes exerts spatiotemporal regulation on PKA signaling. This has resulted in a paradigm shift from PKA functioning as a binary system of active and inactive states to a highly dynamic and complex system that operates by several intermediate steps. These mechanistic and structural details have greatly advanced our understanding and interpretation of allosteric regulation in PKA, and this approach can be applied to study numerous biomolecular assemblies. These studies were undisputedly facilitated by the advent of techniques probing protein dynamics such as HDXMS. Nevertheless, the complementation of dynamic in-solution approaches (HDXMS, NMR) with static high-resolution structural techniques (X-ray crystallography, cryo-electron microscopy) has been critical for mapping the activation and termination phases of PKA signaling.

While much is known about the structural and molecular mechanisms of PKA regulation, the broader question as to how the structural and functional diversity of R, C, PDEs, and AKAPs contributes to highly precise spatiotemporal regulation of PKA signaling across the various permutations and combinations of PKA signalosomes remains poorly understood. Furthermore, the pathophysiology of disease states impairing PKA regulation, in particular the allosteric effects of mutations on cAMP-mediated activation and termination, requires further investigation. Answering these questions will vastly improve our understanding of numerous intracellular signaling processes and enable us to develop novel therapeutic strategies against aberrant PKA signaling.

Conflicts of Interest The authors have no conflicts of interest to declare that are relevant to the content of this chapter.

References

Adams SR, Harootunian AT, Buechler YJ, Taylor SS, Tsien RY. Fluorescence ratio imaging of cyclic AMP in single cells. Nature. 1991;349:694–7.

Amieux PS, McKnight GS. The essential role of RI alpha in the maintenance of regulated PKA activity. Ann N Y Acad Sci. 2002;968:75–95.

Amieux PS, Cummings DE, Motamed K, Brandon EP, Wailes LA, Le K, Idzerda RL, McKnight GS. Compensatory regulation of RIalpha protein levels in protein kinase A mutant mice. J Biol Chem. 1997;272:3993–8.

Anand GS, Hughes CA, Jones JM, Taylor SS, Komives EA. Amide H/2H exchange reveals communication between the cAMP and catalytic subunit-binding sites in the R(I)alpha subunit of protein kinase A. J Mol Biol. 2002;323:377–86.

Anand GS, Law D, Mandell JG, Snead AN, Tsigelny I, Taylor SS, Ten Eyck LF, Komives EA. Identification of the protein kinase A regulatory RIalpha-catalytic subunit interface by amide H/2H exchange and protein docking. Proc Natl Acad Sci USA. 2003;100:13264–9.

Anand GS, Hotchko M, Brown SH, Ten Eyck LF, Komives EA, Taylor SS. R-subunit isoform specificity in protein kinase A: distinct features of protein interfaces in PKA types I and II by amide H/2H exchange mass spectrometry. J Mol Biol. 2007;374:487–99.

Anton SE, Kayser C, Maiellaro I, Nemec K, Moller J, Koschinski A, Zaccolo M, Annibale P, Falcke M, Lohse MJ, Bock A. Receptor-associated independent cAMP nanodomains mediate spatiotemporal specificity of GPCR signaling. Cell. 2022;185:1130–1142.e1111.

Arveseth CD, Happ JT, Hedeen DS, Zhu JF, Capener JL, Klatt Shaw D, Deshpande I, Liang J, Xu J, Stubben SL, Nelson IB, Walker MF, Kawakami K, Inoue A, Krogan NJ, Grunwald DJ, Huttenhain R, Manglik A, Myers BR. Smoothened transduces Hedgehog signals via activity-dependent sequestration of PKA catalytic subunits. PLoS Biol. 2021;19:e3001191.

Asirvatham AL, Galligan SG, Schillace RV, Davey MP, Vasta V, Beavo JA, Carr DW. A-kinase anchoring proteins interact with phosphodiesterases in T lymphocyte cell lines. J Immunol. 2004;173:4806–14.

Bachmann VA, Mayrhofer JE, Ilouz R, Tschaikner P, Raffeiner P, Rock R, Courcelles M, Apelt F, Lu TW, Baillie GS, Thibault P, Aanstad P, Stelzl U, Taylor SS, Stefan E. Gpr161 anchoring of PKA consolidates GPCR and cAMP signaling. Proc Natl Acad Sci USA. 2016;113:7786–91.

Bai Y, Milne JS, Mayne L, Englander SW. Primary structure effects on peptide group hydrogen exchange. Proteins. 1993;17:75–86.

Baillie GS, Scott JD, Houslay MD. Compartmentalisation of phosphodiesterases and protein kinase A: opposites attract. FEBS Lett. 2005;579:3264–70.

Banky P, Roy M, Newlon MG, Morikis D, Haste NM, Taylor SS, Jennings PA. Related protein-protein interaction modules present drastically different surface topographies despite a conserved helical platform. J Mol Biol. 2003;330:1117–29.

Beard MB, Olsen AE, Jones RE, Erdogan S, Houslay MD, Bolger GB. UCR1 and UCR2 domains unique to the cAMP-specific phosphodiesterase family form a discrete module via electrostatic interactions. J Biol Chem. 2000;275:10349–58.

Beca S, Ahmad F, Shen W, Liu J, Makary S, Polidovitch N, Sun J, Hockman S, Chung YW, Movsesian M, Murphy E, Manganiello V, Backx PH. Phosphodiesterase type 3A regulates basal myocardial contractility through interacting with sarcoplasmic reticulum calcium ATPase type 2a signaling complexes in mouse heart. Circ Res. 2013;112:289–97.

Bedioune I, Lefebvre F, Lechene P, Varin A, Domergue V, Kapiloff MS, Fischmeister R, Vandecasteele G. PDE4 and mAKAPbeta are nodal organizers of beta2-ARs nuclear PKA signalling in cardiac myocytes. Cardiovasc Res. 2018;114:1499–511.

Berman HM, Ten Eyck LF, Goodsell DS, Haste NM, Kornev A, Taylor SS. The cAMP binding domain: an ancient signaling module. Proc Natl Acad Sci USA. 2005;102:45–50.

Bolger GB, Dunlop AJ, Meng D, Day JP, Klussmann E, Baillie GS, Adams DR, Houslay MD. Dimerization of cAMP phosphodiesterase-4 (PDE4) in living cells requires interfaces located in both the UCR1 and catalytic unit domains. Cell Signal. 2015;27:756–69.

Borner S, Schwede F, Schlipp A, Berisha F, Calebiro D, Lohse MJ, Nikolaev VO. FRET measurements of intracellular cAMP concentrations and cAMP analog permeability in intact cells. Nat Protoc. 2011;6:427–38.

Brown KM, Lee LC, Findlay JE, Day JP, Baillie GS. Cyclic AMP-specific phosphodiesterase, PDE8A1, is activated by protein kinase A-mediated phosphorylation. FEBS Lett. 2012;586:1631–7.

Brown KM, Day JP, Huston E, Zimmermann B, Hampel K, Christian F, Romano D, Terhzaz S, Lee LC, Willis MJ, Morton DB, Beavo JA, Shimizu-Albergine M, Davies SA, Kolch W, Houslay MD, Baillie GS. Phosphodiesterase-8A binds to and regulates Raf-1 kinase. Proc Natl Acad Sci USA. 2013;110:E1533–42.

Bruystens JG, Wu J, Fortezzo A, Kornev AP, Blumenthal DK, Taylor SS. PKA RIalpha homodimer structure reveals an intermolecular interface with implications for cooperative cAMP binding and Carney complex disease. Structure. 2014;22:59–69.

Bruystens JG, Wu J, Fortezzo A, Del Rio J, Nielsen C, Blumenthal DK, Rock R, Stefan E, Taylor SS. Structure of a PKA RIalpha recurrent acrodysostosis mutant explains defective cAMP-dependent activation. J Mol Biol. 2016;428:4890–904.

Burns LL, Canaves JM, Pennypacker JK, Blumenthal DK, Taylor SS. Isoform specific differences in binding of a dual-specificity A-kinase anchoring protein to type I and type II regulatory subunits of PKA. Biochemistry. 2003;42:5754–63.

Burns-Hamuro LL, Hamuro Y, Kim JS, Sigala P, Fayos R, Stranz DD, Jennings PA, Taylor SS, Woods VL Jr. Distinct interaction modes of an AKAP bound to two regulatory subunit isoforms of protein kinase A revealed by amide hydrogen/deuterium exchange. Protein Sci. 2005;14:2982–92.

Buxton IL, Brunton LL. Compartments of cyclic AMP and protein kinase in mammalian cardiomyocytes. J Biol Chem. 1983;258:10233–9.

Carlisle Michel JJ, Dodge KL, Wong W, Mayer NC, Langeberg LK, Scott JD. PKA-phosphorylation of PDE4D3 facilitates recruitment of the mAKAP signalling complex. Biochem J. 2004;381:587–92.

Carr DW, Stofko-Hahn RE, Fraser ID, Bishop SM, Acott TS, Brennan RG, Scott JD. Interaction of the regulatory subunit (RII) of cAMP-dependent protein kinase with RII-anchoring proteins occurs through an amphipathic helix binding motif. J Biol Chem. 1991;266:14188–92.

Chandramohan A, Tulsian NK, Anand GS. Dissecting orthosteric contacts for a reverse-fragment-based ligand design. Anal Chem. 2017;89:7876–85.

Cho-Chung YS, Pepe S, Clair T, Budillon A, Nesterova M. cAMP-dependent protein kinase: role in normal and malignant growth. Crit Rev Oncol Hematol. 1995;21:33–61.

Conti M, Beavo J. Biochemistry and physiology of cyclic nucleotide phosphodiesterases: essential components in cyclic nucleotide signaling. Annu Rev Biochem. 2007;76:481–511.

Das R, Esposito V, Abu-Abed M, Anand GS, Taylor SS, Melacini G. cAMP activation of PKA defines an ancient signaling mechanism. Proc Natl Acad Sci USA. 2007;104:93–8.

Diskar M, Zenn HM, Kaupisch A, Prinz A, Herberg FW. Molecular basis for isoform-specific autoregulation of protein kinase A. Cell Signal. 2007;19:2024–34.

Doskeland SO, Ogreid D. Characterization of the interchain and intrachain interactions between the binding sites of the free regulatory moiety of protein kinase I. J Biol Chem. 1984;259:2291–301.

Elli FM, Bordogna P, de Sanctis L, Giachero F, Verrua E, Segni M, Mazzanti L, Boldrin V, Toromanovic A, Spada A, Mantovani G. Screening of PRKAR1A and PDE4D in a large Italian series of patients clinically diagnosed with Albright hereditary osteodystrophy and/or pseudo-hypoparathyroidism. J Bone Miner Res. 2016;31:1215–24.

Englander SW, Kallenbach NR. Hydrogen exchange and structural dynamics of proteins and nucleic acids. Q Rev Biophys. 1983;16:521–655.

Forlino A, Vetro A, Garavelli L, Ciccone R, London E, Stratakis CA, Zuffardi O. PRKACB and Carney complex. N Engl J Med. 2014;370:1065–7.

Francis SH, Turko IV, Corbin JD. Cyclic nucleotide phosphodiesterases: relating structure and function. Prog Nucleic Acid Res Mol Biol. 2001;65:1–52.

Gangal M, Clifford T, Deich J, Cheng X, Taylor SS, Johnson DA. Mobilization of the A-kinase N-myristate through an isoform-specific intermolecular switch. Proc Natl Acad Sci USA. 1999;96:12394–9.

Gonzalez GA, Montminy MR. Cyclic AMP stimulates somatostatin gene transcription by phosphorylation of CREB at serine 133. Cell. 1989;59:675–80.

Hamuro Y, Anand GS, Kim JS, Juliano C, Stranz DD, Taylor SS, Woods VL Jr. Mapping inter-subunit interactions of the regulatory subunit (RIalpha) in the type I holoenzyme of protein kinase A by amide hydrogen/deuterium exchange mass spectrometry (DXMS). J Mol Biol. 2004;340:1185–96.

Han P, Sonati P, Rubin C, Michaeli T. PDE7A1, a cAMP-specific phosphodiesterase, inhibits cAMP-dependent protein kinase by a direct interaction with C. J Biol Chem. 2006;281:15050–7.

Happ JT, Arveseth CD, Bruystens J, Bertinetti D, Nelson IB, Olivieri C, Zhang J, Hedeen DS, Zhu JF, Capener JL, Brockel JW, Vu L, King CC, Ruiz-Perez VL, Ge X, Veglia G, Herberg FW, Taylor SS, Myers BR. A PKA inhibitor motif within SMOOTHENED controls Hedgehog signal transduction. Nat Struct Mol Biol. 2022;29:990–9.

Hayes JS, Brunton LL. Functional compartments in cyclic nucleotide action. J Cyclic Nucleotide Res. 1982;8:1–16.

Herberg FW, Dostmann WR, Zorn M, Davis SJ, Taylor SS. Crosstalk between domains in the regulatory subunit of cAMP-dependent protein kinase: influence of amino terminus on cAMP binding and holoenzyme formation. Biochemistry. 1994;33:7485–94.

Herberg FW, Taylor SS, Dostmann WR. Active site mutations define the pathway for the cooperative activation of cAMP-dependent protein kinase. Biochemistry. 1996;35:2934–42.

Herberg FW, Doyle ML, Cox S, Taylor SS. Dissection of the nucleotide and metal-phosphate binding sites in cAMP-dependent protein kinase. Biochemistry. 1999;38:6352–60.

Hoofnagle AN, Resing KA, Ahn NG. Protein analysis by hydrogen exchange mass spectrometry. Annu Rev Biophys Biomol Struct. 2003;32:1–25.

Hoshi N, Langeberg LK, Scott JD. Distinct enzyme combinations in AKAP signalling complexes permit functional diversity. Nat Cell Biol. 2005;7:1066–73.

Huai Q, Wang H, Sun Y, Kim HY, Liu Y, Ke H. Three-dimensional structures of PDE4D in complex with roliprams and implication on inhibitor selectivity. Structure. 2003a;11:865–73.

Huai Q, Colicelli J, Ke H. The crystal structure of AMP-bound PDE4 suggests a mechanism for phosphodiesterase catalysis. Biochemistry. 2003b;42:13220–6.

Huai Q, Liu Y, Francis SH, Corbin JD, Ke H. Crystal structures of phosphodiesterases 4 and 5 in complex with inhibitor 3-isobutyl-1-methylxanthine suggest a conformation determinant of inhibitor selectivity. J Biol Chem. 2004a;279:13095–101.

Huai Q, Wang H, Zhang W, Colman RW, Robinson H, Ke H. Crystal structure of phosphodiesterase 9 shows orientation variation of inhibitor 3-isobutyl-1-methylxanthine binding. Proc Natl Acad Sci USA. 2004b;101:9624–9.

Hunter T. Why nature chose phosphate to modify proteins. Philos Trans R Soc Lond Ser B Biol Sci. 2012;367:2513–6.

Iffland A, Kohls D, Low S, Luan J, Zhang Y, Kothe M, Cao Q, Kamath AV, Ding YH, Ellenberger T. Structural determinants for inhibitor specificity and selectivity in PDE2A using the wheat germ in vitro translation system. Biochemistry. 2005;44:8312–25.

Kammer GM, Khan IU, Kammer JA, Olorenshaw I, Mathis D. Deficient type I protein kinase A isozyme activity in systemic lupus erythematosus T lymphocytes: II. Abnormal isozyme kinetics. J Immunol. 1996;157:2690–8.

Kannan N, Wu J, Anand GS, Yooseph S, Neuwald AF, Venter JC, Taylor SS. Evolution of allostery in the cyclic nucleotide binding module. Genome Biol. 2007;8:R264.

Katta V, Chait BT. Conformational changes in proteins probed by hydrogen-exchange electrospray-ionization mass spectrometry. Rapid Commun Mass Spectrom. 1991;5:214–7.

Kim C, Xuong NH, Taylor SS. Crystal structure of a complex between the catalytic and regulatory (RIalpha) subunits of PKA. Science. 2005;307:690–6.

Kim C, Cheng CY, Saldanha SA, Taylor SS. PKA-I holoenzyme structure reveals a mechanism for cAMP-dependent activation. Cell. 2007;130:1032–43.

Kirschner LS, Carney JA, Pack SD, Taymans SE, Giatzakis C, Cho YS, Cho-Chung YS, Stratakis CA. Mutations of the gene encoding the protein kinase A type I-alpha regulatory subunit in patients with the Carney complex. Nat Genet. 2000;26:89–92.

Kole HK, Abdel-Ghany M, Racker E. Specific dephosphorylation of phosphoproteins by protein-serine and -tyrosine kinases. Proc Natl Acad Sci USA. 1988;85:5849–53.

Krishnamurthy S, Moorthy BS, Liqin L, Anand GS. Dynamics of phosphodiesterase-induced cAMP dissociation from protein kinase A: capturing transient ternary complexes by HDXMS. Biochim Biophys Acta. 2013;1834:1215–21.

Krishnamurthy S, Moorthy BS, Xin Xiang L, Xin Shan L, Bharatham K, Tulsian NK, Mihalek I, Anand GS. Active site coupling in PDE:PKA complexes promotes resetting of mammalian cAMP signaling. Biophys J. 2014;107:1426–40.

Krishnamurthy S, Tulsian NK, Chandramohan A, Anand GS. Parallel allostery by cAMP and PDE coordinates activation and termination phases in cAMP signaling. Biophys J. 2015;109:1251–63.

Lehnart SE, Wehrens XH, Reiken S, Warrier S, Belevych AE, Harvey RD, Richter W, Jin SL, Conti M, Marks AR. Phosphodiesterase 4D deficiency in the ryanodine-receptor complex promotes heart failure and arrhythmias. Cell. 2005;123:25–35.

Leon DA, Herberg FW, Banky P, Taylor SS. A stable alpha-helical domain at the N terminus of the RIalpha subunits of cAMP-dependent protein kinase is a novel dimerization/docking motif. J Biol Chem. 1997;272:28431–7.

Li X, Li HP, Amsler K, Hyink D, Wilson PD, Burrow CR. PRKX, a phylogenetically and functionally distinct cAMP-dependent protein kinase, activates renal epithelial cell migration and morphogenesis. Proc Natl Acad Sci USA. 2002;99:9260–5.

Li X, Wilmanns M, Thornton J, Kohn M. Elucidating human phosphatase-substrate networks. Sci Signal. 2013;6:rs10.

Lignitto L, Carlucci A, Sepe M, Stefan E, Cuomo O, Nistico R, Scorziello A, Savoia C, Garbi C, Annunziato L, Feliciello A. Control of PKA stability and signalling by the RING ligase praja2. Nat Cell Biol. 2011;13:412–22.

Linglart A, Menguy C, Couvineau A, Auzan C, Gunes Y, Cancel M, Motte E, Pinto G, Chanson P, Bougneres P, Clauser E, Silve C. Recurrent PRKAR1A mutation in acrodysostosis with hormone resistance. N Engl J Med. 2011;364:2218–26.

Linglart A, Fryssira H, Hiort O, Holterhus PM, Perez de Nanclares G, Argente J, Heinrichs C, Kuechler A, Mantovani G, Leheup B, Wicart P, Chassot V, Schmidt D, Rubio-Cabezas O, Richter-Unruh A, Berrade S, Pereda A, Boros E, Munoz-Calvo MT, Castori M, Gunes Y, Bertrand G, Bougneres P, Clauser E, Silve C. PRKAR1A and PDE4D mutations cause acrodysostosis but two distinct syndromes with or without GPCR-signaling hormone resistance. J Clin Endocrinol Metab. 2012;97:E2328–38.

Lohmann SM, DeCamilli P, Einig I, Walter U. High-affinity binding of the regulatory subunit (RII) of cAMP-dependent protein kinase to microtubule-associated and other cellular proteins. Proc Natl Acad Sci USA. 1984;81:6723–7.

Lu TW, Wu J, Aoto PC, Weng JH, Ahuja LG, Sun N, Cheng CY, Zhang P, Taylor SS. Two PKA RIalpha holoenzyme states define ATP as an isoform-specific orthosteric inhibitor that competes with the allosteric activator, cAMP. Proc Natl Acad Sci USA. 2019;116:16347–56.

Lynch MJ, Baillie GS, Mohamed A, Li X, Maisonneuve C, Klussmann E, van Heeke G, Houslay MD. RNA silencing identifies PDE4D5 as the functionally relevant cAMP phosphodiesterase interacting with beta arrestin to control the protein kinase A/AKAP79-mediated switching of the beta2-adrenergic receptor to activation of ERK in HEK293B2 cells. J Biol Chem. 2005;280:33178–89.

MacKenzie SJ, Baillie GS, McPhee I, MacKenzie C, Seamons R, McSorley T, Millen J, Beard MB, van Heeke G, Houslay MD. Long PDE4 cAMP specific phosphodiesterases are activated by protein kinase A-mediated phosphorylation of a single serine residue in Upstream Conserved Region 1 (UCR1). Br J Pharmacol. 2002;136:421–33.

Manning G, Whyte DB, Martinez R, Hunter T, Sudarsanam S. The protein kinase complement of the human genome. Science. 2002;298:1912–34.

Mantovani G, Bondioni S, Alberti L, Gilardini L, Invitti C, Corbetta S, Zappa MA, Ferrero S, Lania AG, Bosari S, Beck-Peccoz P, Spada A. Protein kinase A regulatory subunits in human adipose tissue: decreased R2B expression and activity in adipocytes from obese subjects. Diabetes. 2009;58:620–6.

Marbach F, Stoyanov G, Erger F, Stratakis CA, Settas N, London E, Rosenfeld JA, Torti E, Haldeman-Englert C, Sklirou E, Kessler E, Ceulemans S, Nelson SF, Martinez-Agosto JA, Palmer CGS, Signer RH, Undiagnosed Diseases Network, Andrews MV, Grange DK, Willaert R, Person R, Telegrafi A, Sievers A, Laugsch M, Theiss S, Cheng Y, Lichtarge O, Katsonis P, Stocco A, Schaaf CP. Variants in PRKAR1B cause a neurodevelopmental disorder with autism spectrum disorder, apraxia, and insensitivity to pain. Genet Med. 2021;23:1465–73.

Marchmont RJ, Houslay MD. A peripheral and an intrinsic enzyme constitute the cyclic AMP phosphodiesterase activity of rat liver plasma membranes. Biochem J. 1980;187:381–92.

Martin BR, Deerinck TJ, Ellisman MH, Taylor SS, Tsien RY. Isoform-specific PKA dynamics revealed by dye-triggered aggregation and DAKAP1alpha-mediated localization in living cells. Chem Biol. 2007;14:1031–42.

Masson GR, Burke JE, Ahn NG, Anand GS, Borchers C, Brier S, Bou-Assaf GM, Engen JR, Englander SW, Faber J, Garlish R, Griffin PR, Gross ML, Guttman M, Hamuro Y, Heck AJR, Houde D, Iacob RE, Jorgensen TJD, Kaltashov IA, Klinman JP, Konermann L, Man P, Mayne L, Pascal BD, Reichmann D, Skehel M, Snijder J, Strutzenberg TS, Underbakke ES, Wagner C, Wales TE, Walters BT, Weis DD, Wilson DJ, Wintrode PL, Zhang Z, Zheng J, Schriemer DC, Rand KD. Recommendations for performing, interpreting and reporting hydrogen deuterium exchange mass spectrometry (HDX-MS) experiments. Nat Methods. 2019;16:595–602.

McClendon CL, Kornev AP, Gilson MK, Taylor SS. Dynamic architecture of a protein kinase. Proc Natl Acad Sci USA. 2014;111:E4623–31.

McKnight GS, Cummings DE, Amieux PS, Sikorski MA, Brandon EP, Planas JV, Motamed K, Idzerda RL. Cyclic AMP, PKA, and the physiological regulation of adiposity. Recent Prog Horm Res. 1998;53:139–59; discussion 160–131.

Mongillo M, McSorley T, Evellin S, Sood A, Lissandron V, Terrin A, Huston E, Hannawacker A, Lohse MJ, Pozzan T, Houslay MD, Zaccolo M. Fluorescence resonance energy transfer-based analysis of cAMP dynamics in live neonatal rat cardiac myocytes reveals distinct functions of compartmentalized phosphodiesterases. Circ Res. 2004;95:67–75.

Moorthy BS, Gao Y, Anand GS. Phosphodiesterases catalyze hydrolysis of cAMP-bound to regulatory subunit of protein kinase A and mediate signal termination. Mol Cell Proteomics. 2011a;10:M110.002295.

Moorthy BS, Badireddy S, Anand GS. Cooperativity and allostery in cAMP-dependent activation of protein kinase A: monitoring conformations of intermediates by amide hydrogen/deuterium exchange. Int J Mass Spectrom. 2011b;302:157–66.

Movsesian MA, Smith CJ, Krall J, Bristow MR, Manganiello VC. Sarcoplasmic reticulum-associated cyclic adenosine 5′-monophosphate phosphodiesterase activity in normal and failing human hearts. J Clin Invest. 1991;88:15–9.

Neves-Zaph SR. Phosphodiesterase diversity and signal processing within cAMP signaling networks. Adv Neurobiol. 2017;17:3–14.

Newlon MG, Roy M, Hausken ZE, Scott JD, Jennings PA. The A-kinase anchoring domain of type IIalpha cAMP-dependent protein kinase is highly helical. J Biol Chem. 1997;272:23637–44.

Newlon MG, Roy M, Morikis D, Hausken ZE, Coghlan V, Scott JD, Jennings PA. The molecular basis for protein kinase A anchoring revealed by solution NMR. Nat Struct Biol. 1999;6:222–7.

Omori K, Kotera J. Overview of PDEs and their regulation. Circ Res. 2007;100:309–27.

Pare GC, Bauman AL, McHenry M, Michel JJ, Dodge-Kafka KL, Kapiloff MS. The mAKAP complex participates in the induction of cardiac myocyte hypertrophy by adrenergic receptor signaling. J Cell Sci. 2005;118:5637–46.

Pizzoni A, Zhang X, Altschuler DL. From membrane to nucleus: a three-wave hypothesis of cAMP signaling. J Biol Chem. 2024;300:105497.

Qasim H, Rajaei M, Xu Y, Reyes-Alcaraz A, Abdelnasser HY, Stewart MD, Lahiri SK, Wehrens XHT, McConnell BK. AKAP12 upregulation associates with PDE8A to accelerate cardiac dysfunction. Circ Res. 2024;134:1006–22.

Redden JM, Dodge-Kafka KL. AKAP phosphatase complexes in the heart. J Cardiovasc Pharmacol. 2011;58:354–62.

Reikhardt BA, Shabanov PD. Catalytic subunit of PKA as a prototype of the eukaryotic protein kinase family. Biochemistry (Mosc). 2020;85:409–24.

Rhayem Y, Le Stunff C, Abdel Khalek W, Auzan C, Bertherat J, Linglart A, Couvineau A, Silve C, Clauser E. Functional characterization of PRKAR1A mutations reveals a unique molecular mechanism causing acrodysostosis but multiple mechanisms causing Carney complex. J Biol Chem. 2015;290:27816–28.

Sakamoto K, McCarthy A, Smith D, Green KA, Grahame Hardie D, Ashworth A, Alessi DR. Deficiency of LKB1 in skeletal muscle prevents AMPK activation and glucose uptake during contraction. EMBO J. 2005;24:1810–20.

Sassone-Corsi P. The cyclic AMP pathway. Cold Spring Harb Perspect Biol. 2012;4:a011148.

Sastri M, Barraclough DM, Carmichael PT, Taylor SS. A-kinase-interacting protein localizes protein kinase A in the nucleus. Proc Natl Acad Sci USA. 2005;102:349–54.

Scapin G, Patel SB, Chung C, Varnerin JP, Edmondson SD, Mastracchio A, Parmee ER, Singh SB, Becker JW, Van der Ploeg LH, Tota MR. Crystal structure of human phosphodiesterase 3B: atomic basis for substrate and inhibitor specificity. Biochemistry. 2004;43:6091–100.

Scott JW, Ling N, Issa SM, Dite TA, O'Brien MT, Chen ZP, Galic S, Langendorf CG, Steinberg GR, Kemp BE, Oakhill JS. Small molecule drug A-769662 and AMP synergistically activate naive AMPK independent of upstream kinase signaling. Chem Biol. 2014;21:619–27.

Shabb JB. Physiological substrates of cAMP-dependent protein kinase. Chem Rev. 2001;101:2381–411.

Shakur Y, Pryde JG, Houslay MD. Engineered deletion of the unique N-terminal domain of the cyclic AMP-specific phosphodiesterase RD1 prevents plasma membrane association and the attainment of enhanced thermostability without altering its sensitivity to inhibition by rolipram. Biochem J. 1993;292(Pt 3):677–86.

Shoji S, Parmelee DC, Wade RD, Kumar S, Ericsson LH, Walsh KA, Neurath H, Long GL, Demaille JG, Fischer EH, Titani K. Complete amino acid sequence of the catalytic subunit of bovine cardiac muscle cyclic AMP-dependent protein kinase. Proc Natl Acad Sci USA. 1981;78:848–51.

Skalhegg BS, Tasken K. Specificity in the cAMP/PKA signaling pathway. Differential expression, regulation, and subcellular localization of subunits of PKA. Front Biosci. 2000;5:D678–93.

Smith FD, Reichow SL, Esseltine JL, Shi D, Langeberg LK, Scott JD, Gonen T. Intrinsic disorder within an AKAP-protein kinase A complex guides local substrate phosphorylation. elife. 2013;2:e01319.

Smith FD, Esseltine JL, Nygren PJ, Veesler D, Byrne DP, Vonderach M, Strashnov I, Eyers CE, Eyers PA, Langeberg LK, Scott JD. Local protein kinase A action proceeds through intact holoenzymes. Science. 2017;356:1288–93.

Soberg K, Jahnsen T, Rognes T, Skalhegg BS, Laerdahl JK. Evolutionary paths of the cAMP-dependent protein kinase (PKA) catalytic subunits. PLoS One. 2013;8:e60935.

Spivey HO, Ovadi J. Substrate channeling. Methods. 1999;19:306–21.

Steichen JM, Kuchinskas M, Keshwani MM, Yang J, Adams JA, Taylor SS. Structural basis for the regulation of protein kinase A by activation loop phosphorylation. J Biol Chem. 2012;287:14672–80.

Steinberg SF, Brunton LL. Compartmentation of G protein-coupled signaling pathways in cardiac myocytes. Annu Rev Pharmacol Toxicol. 2001;41:751–73.

Steinberg GR, Carling D. AMP-activated protein kinase: the current landscape for drug development. Nat Rev Drug Discov. 2019;18:527–51.

Su Y, Dostmann WR, Herberg FW, Durick K, Xuong NH, Ten Eyck L, Taylor SS, Varughese KI. Regulatory subunit of protein kinase A: structure of deletion mutant with cAMP binding domains. Science. 1995;269:807–13.

Sung BJ, Hwang KY, Jeon YH, Lee JI, Heo YS, Kim JH, Moon J, Yoon JM, Hyun YL, Kim E, Eum SJ, Park SY, Lee JO, Lee TG, Ro S, Cho JM. Structure of the catalytic domain of human phosphodiesterase 5 with bound drug molecules. Nature. 2003;425:98–102.

Surdo NC, Berrera M, Koschinski A, Brescia M, Machado MR, Carr C, Wright P, Gorelik J, Morotti S, Grandi E, Bers DM, Pantano S, Zaccolo M. FRET biosensor uncovers cAMP nano-domains at beta-adrenergic targets that dictate precise tuning of cardiac contractility. Nat Commun. 2017;8:15031.

Taylor SS, Ilouz R, Zhang P, Kornev AP. Assembly of allosteric macromolecular switches: lessons from PKA. Nat Rev Mol Cell Biol. 2012;13:646–58.

Theurkauf WE, Vallee RB. Molecular characterization of the cAMP-dependent protein kinase bound to microtubule-associated protein 2. J Biol Chem. 1982;257:3284–90.

Tortora G, Damiano V, Bianco C, Baldassarre G, Bianco AR, Lanfrancone L, Pelicci PG, Ciardiello F. The RIalpha subunit of protein kinase A (PKA) binds to Grb2 and allows PKA interaction with the activated EGF-receptor. Oncogene. 1997;14:923–8.

Tsigelny I, Greenberg JP, Cox S, Nichols WL, Taylor SS, Ten Eyck LF. 600 ps molecular dynamics reveals stable substructures and flexible hinge points in cAMP dependent protein kinase. Biopolymers. 1999;50:513–24.

Tulsian NK, Krishnamurthy S, Anand GS. Channeling of cAMP in PDE-PKA complexes promotes signal adaptation. Biophys J. 2017;112:2552–66.

Tulsian NK, Ghode A, Anand GS. Adenylate control in cAMP signaling: implications for adaptation in signalosomes. Biochem J. 2020;477:2981–98.

Tulsian NK, Sin VJ, Koh HL, Anand GS. Development of phosphodiesterase-protein-kinase complexes as novel targets for discovery of inhibitors with enhanced specificity. Int J Mol Sci. 2021;22:5242.

Venkatakrishnan V, Ghode A, Tulsian NK, Anand GS. Impaired cAMP processivity by phosphodiesterase-protein kinase A complexes in acrodysostosis. Front Mol Biosci. 2023;10:1202268.

Venkatakrishnan V, Laremore TN, Buckley TSC, Armache JP, Anand GS. Multiplicity of regulatory subunit conformations defines structural ensemble of reset protein kinase A holoenzyme. J Am Chem Soc. 2025;147:14174–90.

Walker C, Wang Y, Olivieri C, Karamafrooz A, Casby J, Bathon K, Calebiro D, Gao J, Bernlohr DA, Taylor SS, Veglia G. Cushing's syndrome driver mutation disrupts protein kinase A allosteric network, altering both regulation and substrate specificity. Sci Adv. 2019;5:eaaw9298.

Walsh DA, Van Patten SM. Multiple pathway signal transduction by the cAMP-dependent protein kinase. FASEB J. 1994;8:1227–36.

Walsh DA, Perkins JP, Krebs EG. An adenosine 3′,5′-monophosphate-dependant protein kinase from rabbit skeletal muscle. J Biol Chem. 1968;243:3763–5.

Wang H, Liu Y, Chen Y, Robinson H, Ke H. Multiple elements jointly determine inhibitor selectivity of cyclic nucleotide phosphodiesterases 4 and 7. J Biol Chem. 2005;280:30949–55.

Wang H, Yan Z, Yang S, Cai J, Robinson H, Ke H. Kinetic and structural studies of phosphodiesterase-8A and implication on the inhibitor selectivity. Biochemistry. 2008;47:12760–8.

Weber IT, Takio K, Titani K, Steitz TA. The cAMP-binding domains of the regulatory subunit of cAMP-dependent protein kinase and the catabolite gene activator protein are homologous. Proc Natl Acad Sci USA. 1982;79:7679–83.

Wen W, Meinkoth JL, Tsien RY, Taylor SS. Identification of a signal for rapid export of proteins from the nucleus. Cell. 1995;82:463–73.

Wong W, Scott JD. AKAP signalling complexes: focal points in space and time. Nat Rev Mol Cell Biol. 2004;5:959–70.

Wu J, Jones JM, Nguyen-Huu X, Ten Eyck LF, Taylor SS. Crystal structures of RIalpha subunit of cyclic adenosine 5′-monophosphate (cAMP)-dependent protein kinase complexed with (Rp)-adenosine 3′,5′-cyclic monophosphothioate and (Sp)-adenosine 3′,5′-cyclic monophosphothioate, the phosphothioate analogues of cAMP. Biochemistry. 2004;43:6620–9.

Wu J, Brown SH, von Daake S, Taylor SS. PKA type IIalpha holoenzyme reveals a combinatorial strategy for isoform diversity. Science. 2007;318:274–9.

Yan Z, Wang H, Cai J, Ke H. Refolding and kinetic characterization of the phosphodiesterase-8A catalytic domain. Protein Expr Purif. 2009;64:82–8.

Yarwood SJ, Steele MR, Scotland G, Houslay MD, Bolger GB. The RACK1 signaling scaffold protein selectively interacts with the cAMP-specific phosphodiesterase PDE4D5 isoform. J Biol Chem. 1999;274:14909–17.

Zhang KY, Card GL, Suzuki Y, Artis DR, Fong D, Gillette S, Hsieh D, Neiman J, West BL, Zhang C, Milburn MV, Kim SH, Schlessinger J, Bollag G. A glutamine switch mechanism for nucleotide selectivity by phosphodiesterases. Mol Cell. 2004;15:279–86.

Zhang P, Smith-Nguyen EV, Keshwani MM, Deal MS, Kornev AP, Taylor SS. Structure and allostery of the PKA RIIbeta tetrameric holoenzyme. Science. 2012;335:712–6.

Zhang JZ, Lu TW, Stolerman LM, Tenner B, Yang JR, Zhang JF, Falcke M, Rangamani P, Taylor SS, Mehta S, Zhang J. Phase separation of a PKA regulatory subunit controls cAMP compartmentation and oncogenic signaling. Cell. 2020;182:1531–1544.e1515.

Zhang H, Cao X, Tang M, Zhong G, Si Y, Li H, Zhu F, Liao Q, Li L, Zhao J, Feng J, Li S, Wang C, Kaulich M, Wang F, Chen L, Li L, Xia Z, Liang T, Lu H, Feng XH, Zhao B. A subcellular map of the human kinome. elife. 2021;10:e64943.

Zheng J, Trafny EA, Knighton DR, Xuong NH, Taylor SS, Ten Eyck LF, Sowadski JM. 2.2 A refined crystal structure of the catalytic subunit of cAMP-dependent protein kinase complexed with MnATP and a peptide inhibitor. Acta Crystallogr D Biol Crystallogr. 1993;49:362–5.

Chapter 5
The TOM Complex in the Outer Membrane of Mitochondria: A Supramolecular Assembly for Protein Import

Stephan Nussberger, Robin Ghosh, and Shuo Wang

Abstract Currently, there is no comprehensive physical model explaining how unfolded polypeptide chains with diverse characteristics are transported into the mitochondria. On a molecular scale, the kinetics of how transit polypeptides approach, are captured by the protein translocation machinery at the outer mito- chondrial membrane, and cross the protein translocation pore to enter the intermem- brane space remain unclear. This knowledge gap is primarily due to the lack of dynamic single-molecule data on the "protein-conducting channels" involved in mitochondrial protein translocation. In this chapter, we explore the recently resolved sub-nanometer cryo-EM structures, which are a prerequisite for a fundamental understanding of the translocation mechanism, and our existing knowledge of the mitochondrial two-pore outer membrane protein translocation machinery (TOM complex). Particularly intriguing are recent findings from single-molecule TIRF microscopy indicating that the TOM core complex can function as a mechanosen- sor, with the pores closing upon interaction with nearby membrane structures. We emphasize novel and unexpected correlations between the structural components of the TOM complexes and their dynamic behavior within the membrane environment.

S. Nussberger (✉)
Department of Biophysics, Institute of Biomaterials and Biomolecular Systems, University of Stuttgart, Stuttgart, Germany
e-mail: nussberger@bio.uni-stuttgart.de

R. Ghosh
Department of Bioenergetics, Institute of Biomaterials and Biomolecular Systems, University of Stuttgart, Stuttgart, Germany

Institute for Energy Efficiency in Production, University of Stuttgart, Stuttgart, Germany

S. Wang
Department of Biophysics, Institute of Biomaterials and Biomolecular Systems, University of Stuttgart, Stuttgart, Germany

Department of Bionanoscience, Kavli Institute of Nanoscience Delft, Delft University of Technology, Delft, The Netherlands

© The Author(s), under exclusive license to Springer Nature Switzerland AG 2026
A. M. Pedley (ed.), *Supramolecular Protein Assemblies In Cells*, Advances in Experimental Medicine and Biology 1514,
https://doi.org/10.1007/978-3-032-26629-3_5

Keywords Mitochondria · Protein import · TOM complex · Electron cryo-microscopy · Single-molecule fluorescence microscopy · Droplet interface bilayer membranes · Native mass spectrometry · Mechanosensitivity

5.1 Introduction

Mitochondria contain over ~1300 distinct proteins in humans and more than ~1000 proteins in the baker's yeast *Saccharomyces cerevisiae* (Calvo et al. 2016; Smith and Robinson 2016; Morgenstern et al. 2017; Rath et al. 2021). As the mitochondrial genome encodes only a limited number of these proteins (13 in humans (Hällberg and Larsson 2014; Ott et al. 2016) and 8 in yeast (Foury et al. 1998)), the majority of nuclear-encoded proteins are imported from the cytosol (Reid and Schatz 1982; Wienhues et al. 1991; Neupert 1997; Pfanner and Geissler 2001). Although it is well known that this process occurs primarily post-translationally, there is also some biochemical and electron cryo-microscopy (cryo-EM) evidence suggesting co-translational import from cytosolic ribosomes localized at the outer mitochondrial membrane surface (Fujiki and Verner 1993; Garcia et al. 2007; Lesnik et al. 2015; Gold et al. 2017).

Over the last three decades, significant progress has been made in unravelling the biochemical pathways of post-translational protein import into the mitochondria (Wiedemann and Pfanner 2017; Hansen and Herrmann 2019; Eaglesfield and Tokatlidis 2021; Araiso et al. 2022; den Brave et al. 2024). Nuclear-encoded proteins targeted to the mitochondrial intermembrane space (IMS), the inner mitochondrial membrane (IMM), or the mitochondrial matrix typically have canonical mitochondrial targeting sequences. They are synthesized on cytosolic ribosomes with a cleavable, positively charged N-terminal targeting sequence that forms a basic, amphipathic α-helix (Schatz and Butow 1983; von Heijne 1986; Schatz 1987). Moving these precursors into the mitochondria is then facilitated by specific protein translocases (Fig. 5.1) located in the outer and inner membranes. While the energetics of protein translocation across the outer mitochondrial membrane (OMM) is still unclear, it is well established that translocation across the IMM is energized by the mitochondrial inner membrane electrochemical potential and the activity of heat shock protein 70 (mtHsp70) located in the mitochondrial matrix. Proteins with N-terminal transit sequences that reach the mitochondrial matrix are matured by removal of their targeting sequences via enzymatic cleavage (Arretz et al. 1991). Folding of the incoming protein into its mature conformation is then mediated by the interaction with mitochondrial chaperones. Precursor proteins that do not contain typical N-terminal transit sequences but carry targeting information in their sequences (Busch et al. 2023), are generally not targeted to the matrix but to other mitochondrial compartments. These include, for example, all proteins of the OMM.

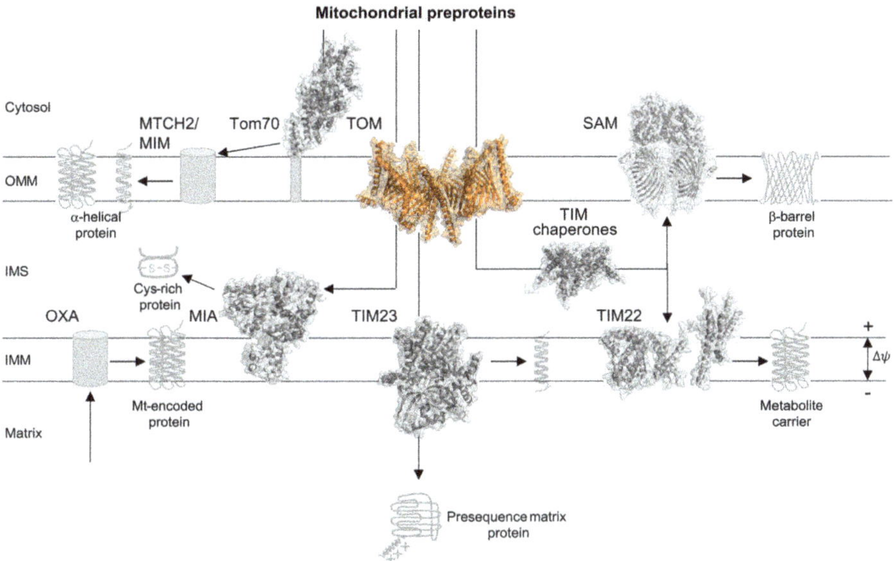

Fig. 5.1 Schematic of the major protein import machineries and import pathways in mitochondria. High-resolution structural information is currently available for five transmembrane mitochondrial protein translocases: TOM, SAM, TIM23, TIM22, MIA, and partially OXA (for review, see Busch et al. (2023)). Each assembly is a multi-protein complex with a central translocation subunit accompanied by accessory subunits (e.g., TIM chaperones Tim8/13 and Tim9/10) involved in precursor recognition or translocase scaffold stabilization. Outer membrane β-barrel proteins destined for transfer into the outer mitochondrial membrane are first transferred by TOM (PDB: 8B4I) across the OMM into the IMS and then sorted in cooperation with small TIM chaperones (PDB: 2BSK) into the outer membrane by the sorting and assembly machinery SAM (PDB: 2BTW). Outer membrane proteins with α-helical transmembrane domains (TMDs) bypass the IMS and are directly incorporated into the OMM from the cytosol via the mitochondrial import machinery complex MIM in yeast and via the mitochondrial insertase MTCH2, a member of the mitochondrial organic solute carrier family 25 (SLC25), and the subunit Tom70, which is only weakly associated with TOM, in mammals. It is important to note that proteins with only one transmembrane domain that are imported via MTCH2 do not require the help of Tom70 (Ruprecht and Kunji 2020). Cysteine-rich precursor proteins destined for the IMS are imported into the IMS via TOM in concert with the IMS/IMM protein complex MIA. Proteins containing presequences and primed for import into the mitochondrial matrix are first recognized by the translocase of the outer membrane TOM complex, then transported across the OMM into the IMS, and finally transferred across the IMM by the mitochondrial inner membrane complexes TIM23 (PDB: 8E1M). TIM23 operates with an inner membrane potential gradient, facilitating complete translocation into the matrix with the help of a presequence translocase-associated motor (PAM), which consists of mitochondrial Hsp70, J-domain containing cochaperones (Pam16 and Pam18), and associated proteins (Tim15 and Mge1) (Neupert and Herrmann 2007). Mitochondrial inner membrane carrier proteins are primed for import by cytosolic chaperones and then translocated across the outer membrane via the TOM complex. Upon entry into the IMS, carrier precursors are escorted by small TIM chaperones (PDB: 2BSK) to the TIM22 complex for insertion into the inner membrane. OMM, mitochondrial outer membrane; IMM, mitochondrial inner membrane; IMS, intermembrane space

The first step in mitochondrial protein import occurs at the OMM and is mediated by the Translocase of the Outer Membrane TOM (Fig. 5.1) (Pfaller et al. 1988; Kiebler et al. 1990; Künkele et al. 1998; Dekker et al. 1998; Saeki et al. 2000; Suzuki et al. 2000; Meisinger et al. 2001; Johnston et al. 2002; Kato and Mihara 2008). TOM facilitates the recognition, binding, and translocation of almost all mitochondrial proteins across the OMM and thus serves as the main gateway for nuclear-encoded precursors synthesized in the cytosol destined for the mitochondria.

The import of β-barrel precursor proteins into the OMM is facilitated by the concerted action of TOM and the Sorting and Assembly Machinery SAM (Fig. 5.1). The β-barrel proteins are first imported into the IMS by TOM and then targeted and finally inserted into the OMM via small TIM chaperones and SAM. The core component of the SAM machinery is Sam50, which, in conjunction with partner proteins, ensures the proper sorting and insertion (Wiedemann et al. 2003; Takeda et al. 2021). Immunoprecipitation and chemical cross-linking studies using detergent-solubilized mitochondrial membranes revealed the transient formation of TOM-SAM supramolecular complexes in the IMM, supporting this functional cooperation (Qiu et al. 2013).

In contrast to the import pathway of β-barrel precursors, nuclear-encoded α-helical tail-anchored, signal-anchored, and multi-pass proteins have been shown to insert into the OMM from the cytosol without the need for complete TOM machinery. In mammalian mitochondria, this process has recently been shown to be mediated by the TOM preprotein receptor subunit Tom70 and the newly identified import protein insertase MTCH2 (Fig. 5.1) or its paralog MTCH1 (Guna et al. 2022; Muthukumar et al. 2024). MTCH1 and MTCH2 are multitopic membrane proteins located in the OMM with sequence similarity to members of the organic solute carrier family SLC25 located in the IMM, which includes the ADP/ATP exchange translocator in the IMM. Research using functional and mutational analysis suggests that MTCH2 has evolved from an ancestral solute carrier (Guna et al. 2022; Muthukumar et al. 2024). Interestingly, in yeast, the insertion of α-helical proteins into the OMM is facilitated by a protein complex that is unrelated to the MTCH carrier proteins. In this organism, protein import is facilitated by a protein machinery called the mitochondrial import complex MIM (Fig. 5.1), which consists of two subunits: Mim1 (Becker et al. 2008, 2011), which can form an ion channel when reconstituted in a lipid bilayer, and Mim2 modifying the Mim1 channel activity (Dimmer et al. 2012; Krüger et al. 2017; Doan et al. 2020; Mazur et al. 2021). MIM associated with TOM has been shown to mediate Tom70-dependent import, particularly of multi-spanning and signal-anchored proteins. Free MIM complexes also appeared to import single-spanning proteins through a Tom70-independent mechanism. A third group of MIM complexes interacting with SAM has been proposed to deliver the small TOM subunits Tom5, Tom6, and Tom7 to the OMM for TOM complex assembly.

The majority of precursor proteins destined for the IMS, IMM, or mitochondrial matrix require further sorting and assembly. The translocase of the inner membrane TIM23 (Fig. 5.1) is responsible for the import of proteins with matrix target signals and certain membrane proteins of the IMM (Sim et al. 2023). The core components of the TIM23 complex, Tim23 and Tim17, function in conjunction with subunit Tim50 and other auxiliary proteins to guarantee effective and selective import of mitochondrial proteins (Popov-Čeleketić et al. 2008; Mokranjac and Neupert 2010; Callegari et al. 2020; Waegemann et al. 2015; Gomkale et al. 2021; Fielden et al. 2023). The protein-conducting channel of TIM23 is expected to be tightly regulated during polypeptide translocation in order to prevent loss of mitochondrial inner membrane electrochemical potential, which is necessary for ATP synthesis. Consistent with this, high-resolution cryo-EM data recently provided evidence that TIM23-mediated protein translocation is unlikely to require an aqueous channel (Sim et al. 2023). Recent low-resolution cryo-EM studies have suggested that TIM23 interacts with TOM during protein translocation via the subunit Tim17 (Zhou et al. 2023; Wang et al. 2024).

The IMM also harbors the protein translocase TIM22 (Fig. 5.1) (Qi et al. 2021; Zhang et al. 2021). TIM22 is responsible for the transfer of membrane proteins with internal targeting signals from the IMS to the IMM. Typical substrates are members of the mitochondrial solute carrier protein family (Endres et al. 1999; Kovermann et al. 2002; Rehling et al. 2003). Thus, unlike TIM23, TIM22 specifically mediates the insertion of precursor proteins with multiple transmembrane segments. The core component of the TIM22 complex is Tim22, which has been suggested to form a voltage-activated ion channel in the IMM through which precursor proteins are laterally released into the lipid bilayer (Kovermann et al. 2002). However, it is not clear as to how previous single-channel electrophysiology and low-resolution EM-based data of TIM22, which suggested a two-pore complex similar to TOM (Rehling et al. 2003), can be reconciled with the recent cryo-EM map of this complex at 3.7 Å resolution (Qi et al. 2021; Zhang et al. 2021), since the latter shows no visible pores. In recent years, supramolecular assemblies such as TOM-TIM22 have been identified, suggesting functional cooperation between the two protein import machineries (Kang et al. 2016).

Cysteine-rich IMS preproteins require specialized machineries for import and oxidative folding (Chacinska et al. 2004; Stojanovski et al. 2008), TOM and the Mitochondrial Intermembrane-space Assembly complex MIA (Fig. 5.1). MIA fulfills this role by facilitating the import and oxidative folding of IMS proteins, mediated by the IMS oxidoreductase Mia40 (Chacinska et al. 2008).

In summary, the import of nuclear-encoded proteins from the cytosol into the various mitochondrial compartments is a highly orchestrated process involving multiple translocases and auxiliary proteins that ensure the precise targeting, translocation, sorting, and assembly of precursor proteins. Their coordinated function is critical for maintaining mitochondrial biogenesis and functionality.

In the following, we will focus on the main protein entry point for almost all mitochondrial proteins. We will present the data from biochemical and cell biological studies in the context of new insights obtained from the recent elucidation of the cryo-EM structures of TOM core complexes (Fig. 5.2) from *Neurospora crassa*, *Saccharomyces cerevisiae*, and humans, respectively (for a comparative overview, see Nussberger et al. 2024), as well as new insights into the mechanistic dynamics

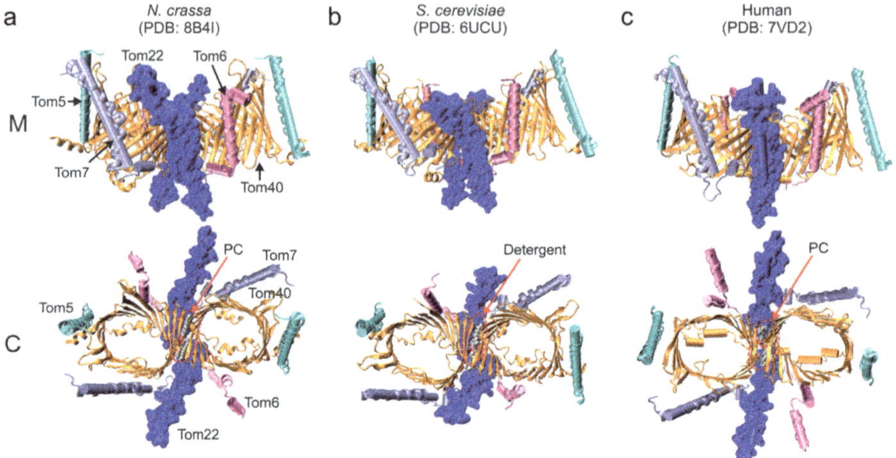

Fig. 5.2 Structural comparison of cryo-EM structures of the TOM core complexes from different species. Comparison of the structures of TOM core complexes from (**a**) *N. crassa* (PDB: 8B4I (Ornelas et al. 2023)), (**b**) *S. cerevisiae* (PDB: 6UCU (Tucker and Park 2019)), and (**c**) human (PDB: 7VD2 (Su et al. 2022)). The Tom subunits have been uniformly color-coded (see labels in (**a**) for assignments). Tom22 is shown in surface representation, and all other subunits are depicted in "cartoon" form. In the upper panels, the structures are viewed from the membrane plane (M, dotted lines), and in the lower panels, the view from the cytosol (C) is shown. The central lipid (*N. crassa*, human) or detergent (*S. cerevisiae*) is shown in VdW representation and highlighted (red dashed line). The Tom20 subunit of the *N. crassa*, which is adjacent to and associated with the TOM core complex (Ornelas et al. 2023), has been omitted for clarity

of the translocase obtained by recent studies using sophisticated single-molecule microscopy techniques (Wang et al. 2022; Wang and Nussberger 2023, 2024).

5.2 Composition and Function of the TOM Complex Subunits

In mammals and fungi, the TOM holo-complex is comprised of seven components (Pfanner et al. 2019; Nussberger et al. 2024). Two decades of work have shown that Tom40 is a β-barrel protein that forms the protein translocation channel, which is intimately associated with Tom22 and three low molecular weight proteins, Tom5, Tom6, and Tom7. Tom22 was initially assigned a role as a preprotein acceptor protein, but the recent structural data (Bausewein et al. 2017; Araiso et al. 2019; Tucker and Park 2019; Su et al. 2022; Ornelas et al. 2023), together with single-molecule dynamic measurements (Wang et al. 2022) suggest that Tom22 may have several roles in the translocation process. With the exception of Tom40, all of the TOM complex proteins are anchored to the OMM by a single transmembrane domain. The proteins Tom70 and Tom20 possess hydrophilic receptor domains extending into the cytosol. Tom22 is a membrane-spanning protein exhibiting, in addition, a large hydrophilic domain in the IMS (Court et al. 1996; Moczko et al. 1997; Rodriguez-Cousiño et al. 1998; van Wilpe et al. 1999). Tom22 acts as a facilitator of complex assembly: its cytosolic domain serves as a docking site for receptors Tom20 and Tom70 (Dekker et al. 1998; van Wilpe et al. 1999; Su et al. 2022; Ornelas et al. 2023) and directly interacts with Tom40 (van Wilpe et al. 1999; Ahting et al. 1999; Shiota et al. 2015; Bausewein et al. 2017; Su et al. 2022; Ornelas et al. 2023).

The TOM protein assembly has been examined extensively by biochemical, cell biological, and genetic methods. However, without detailed structural information, many of these studies presented only a blurry and often ambiguous picture. The recent structural data (Fig. 5.2) (Bausewein et al. 2017; Araiso et al. 2019; Tucker and Park 2019; Su et al. 2022; Ornelas et al. 2023) now allows us to consider some of the older observations (Künkele et al. 1998; Ahting et al. 1999, 2001) in a more discerning light than was available previously, so we feel it pertinent to begin with a very brief overview of the TOM complex assembly, and its core functional unit, the TOM core complex composed of the subunits Tom40, Tom22, Tom7, Tom6, and Tom5.

Thus, we now know that the fundamental functional unit of the TOM translocase is the TOM core complex, which is a dimer of two amphipathic 19-stranded Tom40 β-barrels (Gessmann et al. 2011; Lackey et al. 2014), which are associated via a Tom22 dimeric assembly. The structural fold of the Tom40 barrel shows a strong similarity to that of the mitochondrial voltage-dependent anion channel VDAC (Hiller et al. 2008; Bayrhuber et al. 2008; Ujwal et al. 2008). The Tom40-Tom22 units are arranged about a twofold symmetry axis in the outer membrane, and each Tom40-Tom22 monomer binds a single Tom5, Tom6, and Tom7 subunit, which are also embedded in the membrane. The TOM core complex can be isolated by standard biochemical methods and is a stable entity (for more details, see Sect. 5.4). The additional subunit Tom20 appears to be more loosely associated with the TOM core complex (Ahting et al. 1999; Ornelas et al. 2023), and some evidence exists that Tom20 interacts dynamically with the TOM core complex (Bhagawati et al. 2021).

The TOM receptor subunits Tom22 and Tom20 have been shown to recognize a diverse range of pre-proteins with N-terminal matrix targeting signals and proteins intended for insertion into the outer membrane (van Wilpe et al. 1999; Abe et al. 2000). In plant mitochondria, Tom20 is anchored into the outer mitochondrial membrane at the C-terminus, while in mammals and fungi, the N-terminus serves as the membrane anchor (Lister and Whelan 2006; Perry et al. 2008; Rimmer et al. 2011). Both Tom20 and Tom70, which were proposed to primarily bind precursor proteins with non-cleavable internal targeting signals (Brix et al. 1997; Melin et al. 2015; Backes et al. 2018), associate with the TOM core complex to form the TOM holo--complex (Su et al. 2022; Ornelas et al. 2023). Tom70 recognizes preproteins of the inner membrane metabolite carrier family and transfers them to Tom22 for further protein import (Pfanner and Neupert 1987; Ryan et al. 1999). It is noteworthy that a more recent study has challenged this previously held functional view, suggesting that Tom70 functions more in the recruitment of chaperones to the mitochondrial surface, as the previously described function of Tom70 to directly bind precursor proteins appeared to be largely dispensable (Backes et al. 2021). In this context, the question arises as to why Tom70 co-purifies with the TOM core complex and is present in TOM holo-complex preparations (Künkele et al. 1998; Ornelas et al. 2023).

Tom70 is anchored to the membrane at the N-terminus and is present only in animals and fungi (Dolezal et al. 2006; Endo and Yamano 2010). In plants, the function of Tom70 is presumed to be fulfilled by mtOM64, although the role of the latter remains to be elucidated (Chew et al. 2004; Lister et al. 2007). In yeast, a paralog of Tom70, known as Tom71 or Tom72, has been identified (Schlossmann et al. 1996). Upregulation of Tom71 can partially compensate for Tom70 deletion in knock-out mutants of *S. cerevisiae* (Koh et al. 2001). Although it is possible to isolate the TOM holo-complex in its entirety from mitochondrial outer membranes using mild nonionic detergents, such as digitonin and glyco-diosgenin (GDN) (Künkele et al. 1998; Ornelas et al. 2023), this is not feasible in the presence of nonionic detergents, such as *n*-dodecyl-β-D-maltoside (DDM) (Ahting et al. 1999). Here, the holo-complex dissociates into Tom70, Tom20, and the TOM core complex, consisting of the subunits Tom40, Tom22, Tom5, Tom6, and Tom7 as mentioned before. The initial work on the structure of the TOM protein translation machinery, therefore, already assumed that both Tom70 and Tom20 are peripherally connected to the TOM core complex (Ahting et al. 1999).

Since most translocated proteins must pass through the Tom40 protein pore, it is now unsurprising that Tom40 was previously shown to be an essential protein for cell viability (Baker et al. 1990). More intriguing are the observations that the key dimeric Tom22 structure is not absolutely essential, for instance, whereas simultaneous deletion of Tom20 and Tom70 is lethal (Ramage et al. 1993). It has been reported that yeast cells can survive without Tom22 (van Wilpe et al. 1999), although the electrophysiological behavior of the TOM core complex/Tom40 channels is modified to be mainly in the "open" conformation, and cell growth is affected, though possible. This would be consistent with early findings in microsporidian mitosomes, which were reported to retain only a subset of the typical mitochondrial protein import system components (Burri et al. 2006) and a TOM machinery consisting only of the two subunits Tom70 and Tom40 (Waller et al. 2009; Heinz and Lithgow 2013). In *Arabidopsis thaliana*, proteins Tom9-1 and Tom9-2, approximately 10 kDa in size, correspond to partially N-terminally truncated forms of yeast Tom22 (Parvin et al. 2017). Interestingly, the N-terminal cytosolic domain of yeast Tom22, when fused with the *A. thaliana* transmembrane domain but lacking the yeast C-terminal region, can restore cell growth in Tom22-depleted yeast strains (Maćasev et al. 2004). The abundance of negatively charged amino acid residues in the cytosolic domain of Tom22, which is conserved across fungal and animal species, initially led to the "acid-chain hypothesis," where the positively charged transit sequence is targeted to Tom22 (Kiebler et al. 1993; Bolliger et al. 1995; Maćasev et al. 2004). However, a subsequent study demonstrated that negatively charged residues in the cytosolic domain of Tom22 were not essential for binding and import (Nargang et al. 1998). In addition, sequence conservation of Tom22s in the IMS domain is relatively low; whereas fungi possess a negatively charged domain, mammals exhibit a neutral glutamine-rich domain instead (Yano et al. 2000).

The small TOM core complex subunits, Tom5, Tom6, and Tom7, participate in the assembly and contribute to the stability of the TOM core complex (Alconada et al. 1995; Hönlinger et al. 1996; Dietmeier et al. 1997; Allen et al. 2002; Schmitt et al. 2005; Sherman et al. 2005). Biochemical studies have indicated that Tom5 may be crucial for the import of Tom40 into the outer membrane (Becker et al. 2010) and for transferring preproteins from Tom22 to Tom40 at the cytosolic side of the complex (Dietmeier et al. 1997). Tom6 and Tom7 were proposed to play opposing roles, with Tom6 stabilizing the complex and Tom7 promoting its dissociation (Sherman et al. 2005). Taken in the context of our present structural picture of the TOM core complex (Fig. 5.2), it seems likely that these "roles" may be due to the structural interactions of the small Tom proteins with the Tom40-Tom22 units, thereby influencing their function.

5.3 Architecture of TOM: Insights from Biochemical Studies, Native Mass Spectrometry and Cryo-EM

All static methods for structural determination, such as X-ray crystallography or cryo-EM, can be subject to misleading interpretations, usually due to a lack of resolution, which may not accurately reflect the true physiological situation. Thus, it is essential that the structural data is supported by other techniques obtained from preparations that are in a more native state. For the TOM holo-complex and TOM core complex, this is particularly relevant, as different compositional assembly states have been reported previously (Künkele et al. 1998; Model et al. 2002, 2008; Shiota et al. 2015; Araiso et al. 2019; Tucker and Park 2019; Guan et al. 2021).

Early biochemical investigations suggested a total molecular mass of the TOM core and holo-complex in detergent solution ranging between 350 and 500 kDa (Künkele et al. 1998; Ahting et al. 1999, 2001; Model et al. 2002, 2008), with Tom proteins forming oligomers and Tom40 dimers as the fundamental structure (Dembowski et al. 2001). Chemical cross-linking studies analyzing the molecular organization of yeast Tom40 revealed that the α-helical transmembrane segments of the other Tom subunits interact with the external wall of the Tom40 β-barrel (Shiota et al. 2015). The analysis of the cross-linking data in the context of a Tom40-β-barrel model (which used the VDAC β-barrel as a starting point) led to the suggestion that the TOM complex was a trimer of three Tom40 β-barrel structures which sandwich a central Tom22 trimer. This arrangement was suggested to be consistent with early low-resolution structures (Künkele et al. 1998; Model et al. 2002). Further, the trimeric Tom40-Tom22 complexes were suggested to exchange with a dimeric isoform (Shiota et al. 2015). However, the model suggested by Shiota et al. seems to be inconsistent with more recent structural cryo-EM data obtained for both the Tom40 core as well as holo complexes at higher resolution.

The first structural views of the TOM architecture were initially provided by electron microscopy and single particle analysis at low resolution (Künkele et al. 1998; Ahting et al. 1999, 2001; Model et al. 2002, 2008). Purified TOM holo complexes from *S. cerevisiae* and *N. crassa* mitochondria displayed particles predominantly featuring two, occasionally three pores, each approximately 20 Å in diameter. These preparations were shown to be capable of accommodating unfolded or partially folded mitochondrial preproteins. However, the total molecular mass and subunit stoichiometry were largely reliant upon values deduced by gel filtration in detergent or the analysis of blue native gels, both of which can show significant deviations from the true values.

The most unambiguous determination of the stoichiometry of the TOM core complex was initially provided by native laser-induced liquid bead ion desorption (LILBID-MS) mass spectrometry of the purified detergent-solubilized complex in the absence of chemical cross-linking of individual Tom subunits (Fig. 5.3) (Mager et al. 2010). LILBID-MS allowed controlled dissociation of the oligomeric assembly (dissolved in the detergent *n*-dodecyl-β-maltoside (DDM)) into stable subcomplexes and individual subunits and yielded precise mass spectral data. At low laser intensity ("soft" dissociating conditions), the complete core complex yielded a broad m/z peak (where $z = -1$) with a maximum value corresponding to a total molecular mass of 148.0 (±1) kDa, which is in excellent correspondence with the predicted value for the $[\text{Tom40-Tom22-Tom5-Tom6-Tom7}]_2$ subunit assembly (147.580 kDa (Mager et al. 2010)). This early data has been confirmed in the recent work for the *N. crassa* TOM core complex (Ornelas et al. 2023). At higher laser intensity ("hard" dissociating conditions), complex assemblies are expected to dissociate into various combinations of interacting species (Fig. 5.3). In retrospect, these data provided a benchmark for the cryo-EM structural data presented recently (Bausewein et al. 2017; Ornelas et al. 2023).

A more detailed analysis of the LLBID-MS data of Mager et al. (2010) now performed in the light of the recent cryo-EM data of the *N. crassa* TOM core complex (PDB:8B4I (Ornelas et al. 2023)) provides some interesting insights (Fig. 5.3). As a preamble, we note that the medium resolution of the cryo-EM data (3.3 Å) prevents accurate predictions of amino acid packing and thus prevents energetic calculations from being performed reliably, so that often the assignment of subunit interactions must be supported by biochemical data. The LLBID data of Mager et al. (2010) provides several good examples here. First, nearly all the subunit combinations of the Tom40-Tom22-small-Tom "monomer" [1 × 40, 1 × 22 ± smT] can be observed (Fig. 5.3a), with the exception of the 3xsmT species, which cannot be detected reliably at low laser intensity. At high laser intensities, however, all smTs were detected. In addition, the Tom40-$[\text{Tom22}]_2$ species can also be observed, thus confirming the subunit organization deduced from the cryo-EM data. For the $[\text{Tom40-Tom22}]_2$ species, most of the possible combinations of small Tom subunits were assigned. Particularly interesting, however, is the observation of the Tom22-Tom (non-assigned) 1xsmT as well as the Tom22-2xsmT ionic species (Fig. 5.3a, highlighted yellow, Fig. 5.3c), which appears tenuous in the cryo-EM data (see Fig. 5.4). The LILBID MS data support the cryo-EM data in that Tom6 and Tom7

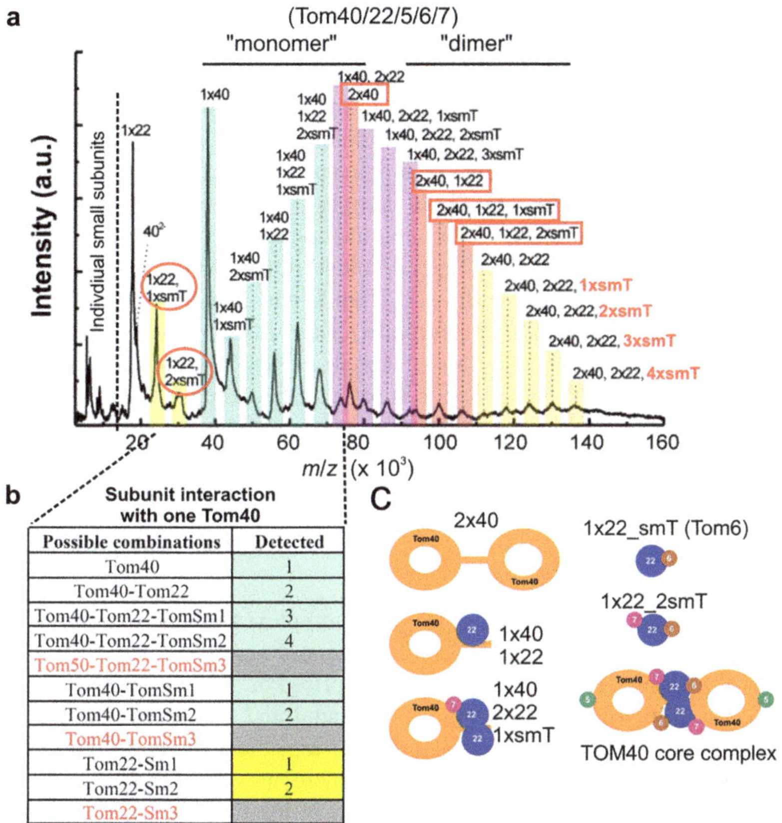

Possible combinations	Detected
Tom40	1
Tom40-Tom22	2
Tom40-Tom22-TomSm1	3
Tom40-Tom22-TomSm2	4
Tom50-Tom22-TomSm3	
Tom40-TomSm1	1
Tom40-TomSm2	2
Tom40-TomSm3	
Tom22-Sm1	1
Tom22-Sm2	2
Tom22-Sm3	

Fig. 5.3 Native mass spectrometry of the TOM core complex. (**a**) LILIBID mass spectrum of the TOM core complex obtained at low laser intensity. (Adapted from Mager et al. (2010)). Various molecular classes have been color-coded (note that the Tom5, Tom6, and Tom7 subunits cannot be resolved here and are summarized as smTs): yellow, single Tom40 (1 × 40) with one (1xsmT) or two (2xsmT) small Tom subunits bound; green, single Tom40-Tom22 (1 × 40,1 × 22) species with one or two small Tom subunits bound; magenta, Tom40 (1 × 40) or (Tom40)₂ dimer (2xT40) bound to a Tom22 dimer (2xT22) small Tom subunits; red-brown, [(Tom40)₂ dimer-Tom22] (2 × 40, 1 × 22) in combination with small Tom subunits; light brown, [Tom40-Tom22]₂ dimer in combination with small Tom subunits. (**b**) Comparison of the number of species expected from the cryo-- EM-derived single [Tom40-Tom22-Tom5-Tom6-Tom7] "monomer" or "dimer" compared to those detected by LLBID-MS. (**c**) Key molecular-ion species (peaks highlighted with red circles or boxes) which support the cryo-EM data (see main text). Individual subunits are color-coded

"link" each Tom22 subunit to *both* "monomeric" Tom40 subunits (Fig. 5.3c) and thus contribute to Tom40-dimer stability. The LILBID MS data also shows clearly that a Tom40-Tom40 interaction site (Fig. 5.3a, 2 × 40; Fig. 5.3c) exists in the *absence* of the Tom22 dimer, a fact which is not clearly visible in the cryo-EM data (Fig. 5.2). Thus, Tom22 stabilizes the Tom40 dimer but is *not* essential for Tom40 dimer formation. Finally, it is interesting that the observed LLBID MS-detected species tend to be of slightly lower molecular weight (about 500 Da) than those predicted from the amino acid sequences alone. It is tempting to suggest that the presence of the bound phospholipid, which is sandwiched between the two chains of the Tom22 dimer (Fig. 5.2) (not known in 2010), containing two fatty acyl chains with at least two double bonds, might account for the missing mass. Although a phosphatidyl choline molecule was fitted to the electron densities in the cryo-EM data, the mass difference would favor a phosphatidyl ethanolamine, which is another major lipid of the OMM (Hallermayer and Neupert 1974).

Native MS analyses of the TOM holo-complex have revealed clear peaks indicating individual Tom70 molecules forming subcomplexes with the small subunits Tom5, Tom6, Tom7, Tom20, Tom22, and Tom40 (Ornelas et al. 2023). Tom20 was shown to form subcomplexes with itself, Tom22 and Tom40, suggesting the presence of two copies of Tom20 per complex. This was consistent with a previous dimeric crystal structure of the cytosolic domain of Tom20 tethered through an integrated disulfide bond to a mitochondrial presequence peptide (Saitoh et al. 2007, 2011; PDB 2V1T).

5.4   Key Structural Aspects of the TOM Core Complex

The cryo-EM structures for the TOM core complexes obtained from *N. crassa*, yeast, and human OMMs (Fig. 5.2) show a two-pore complex with overall dimensions of ~130 Å by ~100 Å. Each pore is comprised of a stoichiometric complex of five subunits, Tom40, Tom22, Tom5, Tom6, and Tom7, arranged around a twofold axis of symmetry (Fig. 5.4). Viewed from the top, the 19-strand β-barrel of each pore shows a pronounced prolate ellipticity, where one end of the long axis is associated with the Tom22 subunit. The other end of the long axis is bounded by the Tom5 subunit, and the two opposing ends of the short axis are associated with Tom6 and Tom7, respectively. Thus, the small Tom subunits seem to be molecular "splints," serving to stabilize the Tom40 β-barrels against structural deformation. This hypothesis, however, has still to be confirmed experimentally. As mentioned previously, Tom40 exhibits a folding pattern that closely corresponds to that of the OMM voltage-gated anion channel VDAC (Hiller et al. 2008; Bayrhuber et al. 2008; Ujwal et al. 2008; Jaremko et al. 2016) and the mitochondrial division and morphology protein Mdm10 (Ellenrieder et al. 2016; Takeda et al. 2021). The odd number of β-strands distinguishes them from porins found in Gram-negative bacteria, suggesting a close evolutionary relationship between the VDAC and TOM complexes. This is further supported by the fact that Tom40 dimerization occurs via β-strands β1 and β19, mirroring dimer-formation observed for VDAC (Bayrhuber et al. 2008; Schredelseker et al. 2014).

Although the Tom40 subunits do exhibit a tenuous structural interface, thereby confirming the 2 × 40 species observed in the MS data above, the major source of Tom40 dimerization is via the transmembrane Tom22 subunits, which also possess an extensive dimerization interface with each other. Thus, we now recognize that the TOM core complex is a dimer of two [Tom40-Tom22-Tom5-Tom6-Tom7] "monomers." A striking aspect is that the Tom40 barrels of the "dimer" are strongly

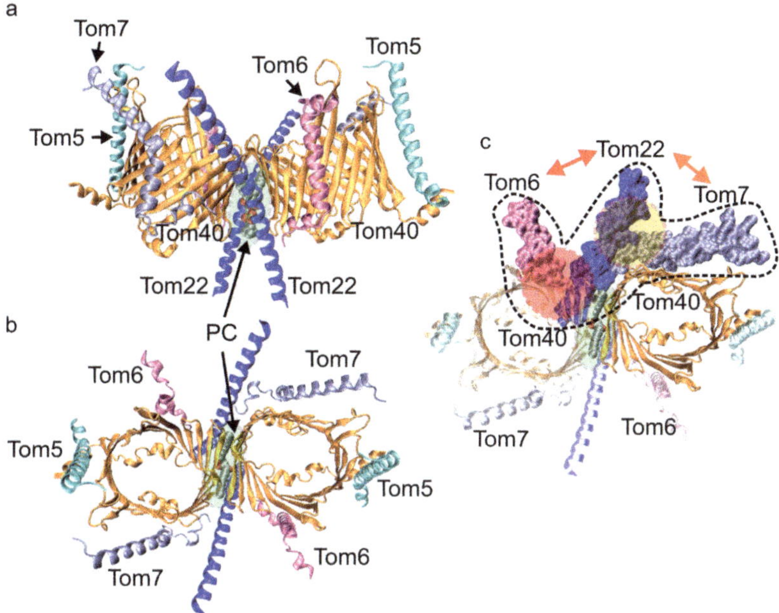

Fig. 5.4 Structural overview of the *N. crassa* Tom40 core complex as derived from cryo-EM data (Ornelas et al. 2023). (**a**) "Side" and (**b**) "top" views of the [Tom40-Tom22-Tom5-Tom6-Tom7]$_2$ dimer are shown, where the individual subunits have been color-coded. The "top" view corresponds to the view from the cytosol. The single-bound phospholipid is shown in VdW representation and highlighted by green shading. The dotted line in (**c**) indicates the subcomplex species probably observable as 2xT22 [Tom6-Tom22-Tom7] or 1xT22 ([Tom6-Tom22] + [Tom22-Tom7]) observed in the LLBID-MS data (see main text). The interacial domains between subunits are indicated by yellow and red shading

tilted (by about $20°$ from the bilayer normal) out of the bilayer plane, such that the combined cytosolic entrance forms a "funnel"-type structure. The tilt of the Tom40 subunits seems to be largely mediated via the twisted conformation of the Tom22-Tom22 dimer, where a highly conserved proline residue (for the *N. crassa* Tom22, this corresponds to Pro99) serves to stabilize an intramembrane bent conformation of the Tom22 α-helices (Fig. 5.6a).

Whereas the transmembrane and IMS domains of Tom22 seem to be rigid rods, part of the cytoplasmic domains of Tom22 may be flexible, as ordered electron densities for part of these domains are absent in the cryo-EM maps. However, it is thought that these domains may be critical for preprotein recognition at the cytoplasmic surface (Rodriguez-Cousiño et al. 1998).

The cryo-EM structure of *N. crassa* provides an explanation for the observation of the 1xT22 and 2xT22 species observed in the MS data presented above. The 2xT22 species is highly likely to be the [Tom6-Tom22-Tom7] species (Fig. 5.3c), and the "1xT22-1xsmT" species is probably [Tom6-Tom22] (Fig. 5.3c), as already shown by the copurification of Tom22 and Tom6 from *n*-octyl-ß-D-glucoside solubilized mitochondria (Ahting et al. 2001). The [Tom6-Tom22-Tom7] species would nicely explain the somewhat broad peak observed in the MS data (Fig. 5.3a).

Perhaps the most striking aspect of the cryo-EM-derived structure for all three species mentioned above is the observation of electron density corresponding to a single phospholipid (originally assigned to phosphatidyl choline) at the Tom22 dimer interface within the membrane bilayer. The fatty acyl chains of the phospholipid lie along the same axes as the Tom22 helices proceeding from the Pro99-mediated kink, and the phospholipid head group faces the IMS and is probably bound by polar interactions with the Tom22 subunits. Previously, it was suggested that the phospholipid molecule may act as a "molecular glue" for the Tom22 subunits (Su et al. 2022). However, our energetic considerations and dynamic measurements of TOM function (see below) suggest an alternative, though still key function of this moiety. We consider this, focusing on the *N. crassa* TOM core complex, in more detail below.

Overwhelmingly, the bulk of the biochemical and structural data so far would suggest that the physiological form of the Tom40 core complex is a [Tom40-Tom22-Tom5-Tom6-Tom7] dimer. However, higher orders of even symmetry may also be possible, which may explain the report of a tetrameric assembly of the complex (Tucker and Park 2019).

The analysis of single-particle datasets of *N. crassa* TOM core complex, combined with a model fusion protein containing the N-terminal transit sequence peptide of aldehyde dehydrogenase, yielded a 4 Å map revealing elongated densities traversing each translocation pore within the Tom40 barrel (Ornelas et al. 2023). Fourier difference maps indicated the possible involvement of Tyr60 and Leu335 in transit sequence recognition and translocation.

Previous studies have reported that Tom40 exists in a dynamic equilibrium between a dimeric assembly intermediate and a mature three-pore TOM complex (Künkele et al. 1998; Rapaport et al. 1998; Model et al. 2001, 2008; Shiota et al. 2015; Araiso et al. 2021, 2022; Guan et al. 2021). More recent high-resolution cryo-EM studies of the holo-complex suggest that the peripherally associated preprotein receptor subunit Tom20 (Su et al. 2022; Ornelas et al. 2023) may correspond to the previously assigned third pore. However, a definitive answer can only be attained through the elucidation of the high-resolution structure of the entire TOM holo complex, encompassing all subunits (Tom70, Tom40, Tom22, Tom20, Tom7, Tom6, and Tom5), under the most native conditions feasible.

5.5   Function and Mechanosensitivity of TOM Core Complex

Inspired by the success of nanopore sequencing of DNA and polypeptides (Kasianowicz et al. 1996; Brinkerhoff et al. 2021), the study of the current and voltage characteristics of TOM complexes and Tom40 channels in planar lipid bilayers has provided insight into their channel kinetics (Poynor et al. 2008; Romero-Ruiz et al. 2010; Mager et al. 2011) and responsiveness to mitochondrial transit peptides

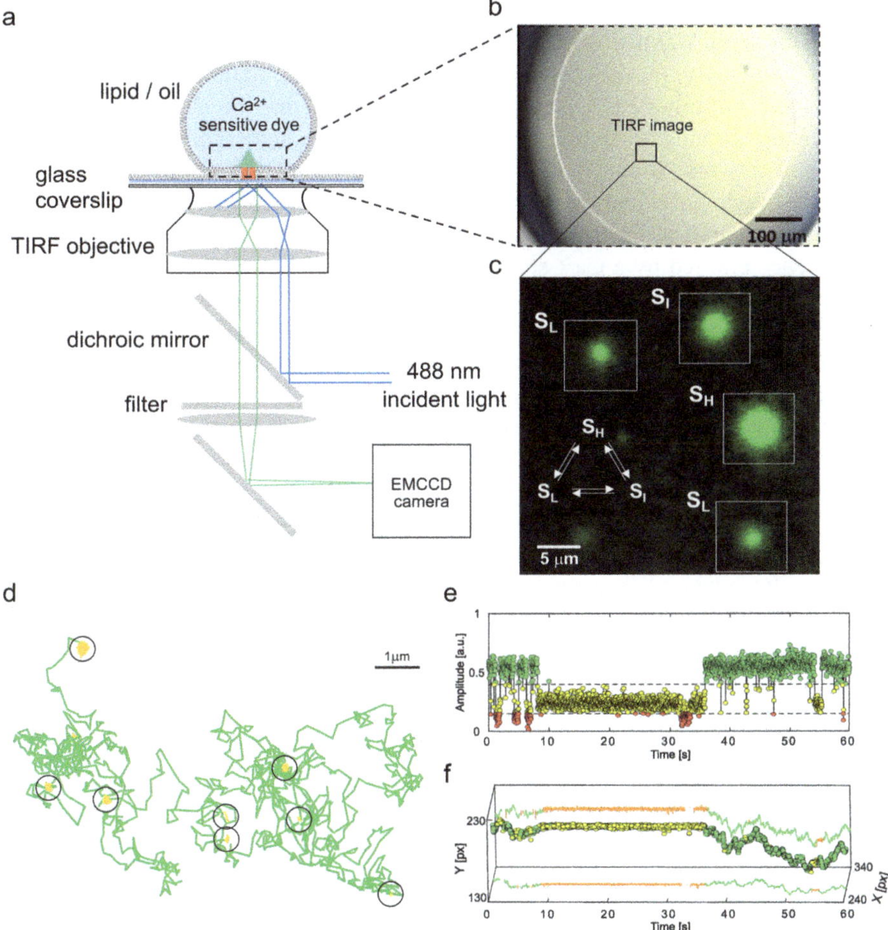

Fig. 5.5 Mechanosensitivity of the TOM core complex in lipid membranes. (**a**) Visualization of mitochondrial TOM channel activity in the absence of external electric fields using TIRF microscopy. An aqueous droplet with a lipid monolayer at an oil/water interface is applied to a lipid monolayer on an ultrathin agarose hydrogel, forming a lipid bilayer membrane. Single-channel activities are monitored by imaging Ca^{2+} flux through membrane-integrated ion channels using a Ca^{2+}-sensitive fluorescent dye. (**b**) Lipid bilayer membrane imaged with Hoffmann modulation contrast microscopy. (**c**) Membrane with individual TOM core complex molecules imaged by TIRF microscopy, showing fluorescent spots of high (S_H), medium (S_I), and low (S_L) intensity. (**d**) Trajectory of a TOM core complex molecule during a period of 30 s. The fluorescence intensities (**e**) and corresponding trajectories (**f**) of a moving and nonmoving TOM core complex molecule. The fluorescence intensities and trajectories of TOM in the two-pore open (S_H), single-pore (S_L), and two-pore closed (S_L) states are marked in green, yellow, and red, respectively

(Mahendran et al. 2012). These experiments revealed significant conformational changes between defined structural states induced by varying artificial electric fields. Interestingly, in vitro biophysical single-molecule optical studies (Fig. 5.5) (Wang and Nussberger 2023, 2024) have demonstrated that the TOM core complex, when freely moving in supported lipid membranes, undergoes conformational changes in response to mechanical cues (Wang et al. 2022). In these studies, Ca^{2+} ion flux through individual two-pore TOM core complexes and single Tom40 channels was measured in the absence of any artificial electric field by observing the fluorescence emitted by a Ca^{2+}-sensitive dye in an evanescent field near the membrane resting on ultrathin hydrogel films using total internal reflection fluorescence (TIRF) microscopy (Fig. 5.5a, b). Unlike traditional single-molecule tracking techniques, this approach did not require fluorescent fusion proteins or labels, which can affect lateral TOM complex movement in the membrane and channel function. The only factors influencing changes in ion flux related to protein conformational changes were the thermal motions of the protein and potential interactions with the hydrogel beneath the membrane.

As shown in Fig. 5.5c–e, a number of interesting observations were be made by monitoring the lateral mobility and single-channel activity of single TOM assemblies simultaneously: When the IMS domains of the two Tom22 receptors interact with structures adjacent to the membrane, lateral diffusion of TOM is arrested (the TOM complex is "stalled") (Fig. 5.5d) and, simultaneously, the two Tom40 pores are closed reversibly (Figs. 5.5e, f) (Wang et al. 2022). The "stalling" was shown to coincide with a reversible transition from active to less active or inactive states of the TOM channel, suggesting that its gating is sensitive to local membrane confinement. By restricting the lateral mobility of the complex through physical interactions in the membrane, TOM appeared to respond to mechanical interactions. The two Tom40 channels of the TOM core complex reversibly switched from an active (fully bright and two pores open) to a weakly active (moderately bright one pore open) and inactive (dark and two pores closed) state. This led to the intriguing hypothesis that TOM may be the first mechanosensitive β-barrel protein complex, in addition to the typical and well-known mechanosensitive α-helical protein channels (Jin et al. 2020; Goodman et al. 2023).

Considering that the TOM complex bends its membrane (Fig. 5.2) with an average radius of ~140 Å (Lomize et al. 2022), it has been concluded that it also bends the OMM towards the IMS so that the IMS domains Tom22 can dock to proteins of the IMM, leading to closure of the Tom40 channel. Comparison of the cryo-EM structures of the fungal and human TOM core complexes revealed two significantly different twists of the IMS domains of Tom22 (Nussberger et al. 2024). Assuming that these two conformations are not due to endogenous properties of the individual TOM species, it was hypothesized that these twists are responsible for the opening and closing of one of the two Tom40 channels. This is supported by all-atom molecular dynamics simulations indicating that movements of Tom22 are coupled to structural rearrangements within the Tom40 pore (Acharya et al. 2025).

Mitochondria are dynamic organelles that adapt their shape, network organization, and functions in response to mechanical, physical, and metabolic signals (Giacomello et al. 2020; Romani et al. 2022; Phuyal et al. 2023). It is thus quite possible that the spatial remodeling of the mitochondrial outer and inner membranes affects TOM complex function. Such dynamic changes in membrane structure could influence the positioning and channel activity of the TOM complex, potentially altering the efficiency and regulation of protein import into the mitochondria. Fusion and fission events involve significant changes in membrane architecture, which could affect the spatial organization and functionality of the TOM complex if the Tom40 pores undergo mechanosensitive closure during interactions with other proteins in the IMM.

5.6   The Lipid Localized at the Tom22-Tom22 Dimer Interface May Facilitate Rigid Body Conformational Change Required for TOM Mechanosensitivity

The positioning of a phospholipid between two interacting transmembrane α-helices (Fig. 5.6a), which is observed for all of the TOM complex structures available so far (Bausewein et al. 2017; Araiso et al. 2019; Tucker and Park 2019; Su et al. 2022; Ornelas et al. 2023) is, to our knowledge, a unique structural biological feature not observed previously. In contrast to a previous suggestion (Su et al. 2022), we consider it unlikely that the phospholipid is functioning as a molecular glue between Tom22 α-helices, since the tight interaction of α-helices without any "cofactor" is one of the oldest observations in structural biology. However, it may be significant that the strength of typical helix-helix interaction is usually about 1.5–2 kcal.mole^{-1}, i.e., significantly above the thermal energy limit (~0.6 kcal.mole^{-1}), which usually corresponds to a very rigid structure. So far, there is no biochemical evidence that the phospholipid is tightly bound to the Tom22, rather the cryo-EM data shows it to be completely embedded between the two α-helices and not accessible to the external membrane environment. The phospholipid seems to be loosely positioned at a region close to the Pro99-mediated kink by primarily polar interactions (in the *N. crassa* structure several key residues (Glu106, Tyr110) may participate in H-bonding with the phosphocholine head group (Fig. 5.6a)). The fatty acyl chains of the phospholipid seem to be aligned along the tilted axis of the Tom22 α-helix essentially by weak Van der Waals packing forces. Another interesting aspect is that the head-group is probably held in a high-energy configuration by the protein environment (Fig. 5.6b, upper panel), since in both single crystals (see Hauser et al. 1981) as well as lipid membranes (see Seelig 1978), the phosphocholine head group of all phospholipid species has been shown to be positioned parallel to the membrane plane (see Fig. 5.6b, lower panel).

In the context of the single-molecule studies, described above, which strongly suggest that conformational change of the Tom22 α-helices is a key aspect of the translocation mechanism, we consider it likely that the weakly bound and configurationally flexible phospholipid molecule may actually facilitate rigid body motion of the Tom22 α-helices, thus acting as a kind of "oily lubricant." The high-energy

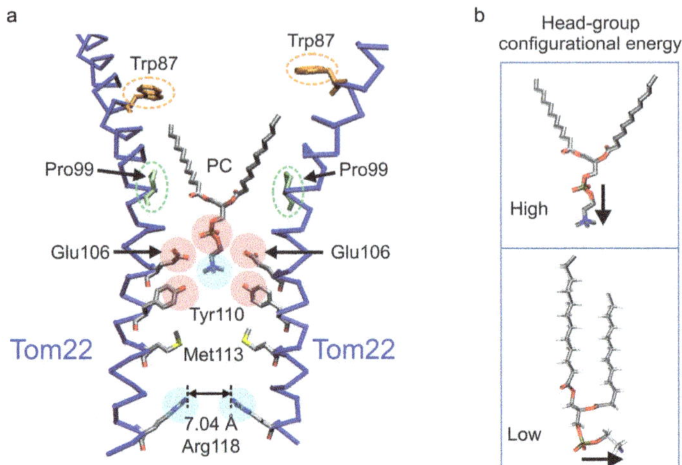

Fig. 5.6 Interactions and energetics of the phospholipid bound to the Tom22 dimer interface. (**a**) Putative binding interaction of the phosphatidylcholine sandwiched (also completely entrapped by the Tom22-Tom40 dimer (α-helical backbones shown in blue)) between the Tom22-Tom22 α-helical dimer interface of the *N. crassa* TOM core complex (PDB: 8B4I). In our interpretation of the cryo-EM structure, the polar phosphocholine moiety is positioned, though weakly bound, by two key residues (Glu106 and Tyr110) at the "kink" region mediated by Pro99. The fatty acyl chains are aligned along the α-helical vector directed to the cytosol, and appear to be arrested by a conserved Trp87 residue. The C-termini of the Tom22 α-helices appear to be in a strongly repulsive configuration due to the close proximity of the positively charged, respective Arg118 residues. (**b**) Comparison of the head group conformation deduced from fitting the electron densities of the cryo-EM data (PDB: 8B4I). The protein-bound head group conformation (upper panel) appears to be in a high-energy configuration when compared to the usual minimal energy configuration observed for phospholipid crystals (lower panel) (PDB: LAPETM10) (CSD database, Cambridge, U.K., see Hauser et al. (1981)) as well as in natural membranes (see Seelig 1978). The arrows indicate the positions of the head group vector with respect to the C_2–C_3 axis of the glycerol backbone. The latter is oriented parallel to the membrane plane in bilayer membranes.

configuration of the head group may even contribute to this event, since no external energy source is available at the OMM. This highly speculative hypothesis deserves rigorous testing in future studies.

5.7 Future Perspectives, Outlook, Concluding Remarks

The physics of the process of protein import into the mitochondria is still far from clear. Whereas the biochemical literature provides detailed insights into this process, only recently have systematic experiments on single-molecule translocation under in vitro conditions begun to uncover the physical mechanism by observing individual polypeptides on the inside and threading them through a proteinaceous nanopore (Movileanu 2009). The translocation event is mediated by the supramolecular TOM assembly complex, whereby an interplay between three components allows diverse polypeptide chains to be translocated through a nanopore. However, the origin of the driving force(s) is still unclear. The physical description of all events of the translocation process (partitioning, threading, and pore exit) is still challenging (Muthukumar 2014). All preproteins are topologically connected units maintaining electrical charges and hydrophobic groups in an aqueous electrolyte solution and are presented to the nanopore in a semi-folded state. Fundamental principles of statistical thermodynamics tell us that small systems (here, proteins) will have a very high degree of (energetic) fluctuation (Cooper 1976), so it seems likely that the low energy barrier for unfolding (in order to enter the translocation pore) is overcome by interaction with components of the TOM complex (presumably Tom20, Tom70, and/or Tom22). Within the pore, the conformational fluctuations will also be influenced by those of the protein-conducting channel.

Current cryo-EM work suggests that TIM23 and the IMS-located subunit Tim17 play a pivotal role in exerting a "pulling" force required for translocating matrix-targeted preproteins through TOM across the OMM. However, this hypothesis begs the question as to how a transit polypeptide can engage with a "pulling protein" in the IMS or IMM without initially passing through the TOM complex pore in a uni-directional manner. This initial "push" is particularly tantalizing, since all steps of the translocation mechanism must operate close to the thermal energy limit. Identifying these initial steps of protein translocation through Tom40 will be a significant challenge in the coming decade, requiring a unique marriage of structural biology, molecular biology, and single-molecule spectroscopy.

Acknowledgments The authors would like to thank all colleagues who contributed to the research and shared their experience in the journey toward the elucidation of the TOM structure. In particular, we would like to thank Pamela Ornelas, Thomas Bausewein, and Werner Kühlbrandt (Max-Planck-Institute of Biophysics, Frankfurt). This work was partly funded by the German Cluster4Future initiative "nanodiag BW" (BMBF 03ZU1208AM to S.N.).

References

Abe Y, Shodai T, Muto T, et al. Structural basis of presequence recognition by the mitochondrial protein import receptor Tom20. Cell. 2000;100:551–60. https://doi.org/10.1016/S0092-8674(00)80691-1.

Ahting U, Thun C, Hegerl R, et al. The Tom core complex: the general protein import pore of the outer membrane of mitochondria. J Cell Biol. 1999;147:959–68. https://doi.org/10.1083/jcb.147.5.959.

Ahting U, Thieffry M, Engelhardt H, et al. Tom40, the pore-forming component of the protein-conducting TOM channel in the outer membrane of mitochondria. J Cell Biol. 2001;153:1151–60. https://doi.org/10.1083/jcb.153.6.1151.

Alconada A, Kübrich M, Moczko M, et al. The mitochondrial receptor complex: the small subunit Mom8b/Isp6 supports association of receptors with the general insertion pore and transfer of preproteins. Mol Cell Biol. 1995;15:6196–205. https://doi.org/10.1128/MCB.15.11.6196.

Allen R, Egan B, Gabriel K, et al. A conserved proline residue is present in the transmembrane-spanning domain of Tom7 and other tail-anchored protein subunits of the TOM translocase. FEBS Lett. 2002;514:347–50. https://doi.org/10.1016/s0014-5793(02)02433-x.

Araiso Y, Tsutsumi A, Qiu J, et al. Structure of the mitochondrial import gate reveals distinct pre-protein paths. Nature. 2019;575:395–401. https://doi.org/10.1038/s41586-019-1680-7.

Araiso Y, Imai K, Endo T. Structural snapshot of the mitochondrial protein import gate. FEBS J. 2021;288:5300–10. https://doi.org/10.1111/febs.15661.

Araiso Y, Imai K, Endo T. Role of the TOM complex in protein import into mitochondria: structural views. Annu Rev Biochem. 2022;91:679–703. https://doi.org/10.1146/annurev-biochem-032620-104527.

Arretz M, Schneider H, Wienhues U, et al. Processing of mitochondrial precursor proteins. Biomed Biochim Acta. 1991;50:403–12.

Acharya A, Nussberger S, Wang S, et al. Dynamic coupling between Tom22 motions and Tom40 pore dynamics modulates ion transport in the mitochondrial TOM complex. J Chem Inf Model. 2025;65:12475–88. https://doi.org/10.1021/acs.jcim.5c01761

Backes S, Hess S, Boos F, et al. Tom70 enhances mitochondrial preprotein import efficiency by binding to internal targeting sequences. J Cell Biol. 2018;217:1369–82. https://doi.org/10.1083/jcb.201708044.

Backes S, Bykov YS, Flohr T, et al. The chaperone-binding activity of the mitochondrial surface receptor Tom70 protects the cytosol against mitoprotein-induced stress. Cell Rep. 2021;35:108936. https://doi.org/10.1016/j.celrep.2021.108936.

Baker KP, Schaniel A, Vestweber D, et al. A yeast mitochondrial outer membrane protein essential for protein import and cell viability. Nature. 1990;348:605–9. https://doi.org/10.1038/348605a0.

Bausewein T, Mills DJ, Langer JD, et al. Cryo-EM structure of the TOM core complex from *Neurospora crassa*. Cell. 2017;170:693–700. https://doi.org/10.1016/j.cell.2017.07.012.

Bayrhuber M, Meins T, Habeck M, et al. Structure of the human voltage-dependent anion channel. Proc Natl Acad Sci USA. 2008;105:15370–5. https://doi.org/10.1073/pnas.0808115105.

Becker T, Pfannschmidt S, Guiard B, et al. Biogenesis of the mitochondrial TOM complex: Mim1 promotes insertion and assembly of signal-anchored receptors. J Biol Chem. 2008;283:120–7. https://doi.org/10.1074/jbc.M706997200.

Becker T, Guiard B, Thornton N, et al. Assembly of the mitochondrial protein import channel. Mol Biol Cell. 2010;21:3106–13. https://doi.org/10.1091/mbc.e10-06-0518.

Becker T, Wenz L-S, Krüger V, et al. The mitochondrial import protein Mim1 promotes biogenesis of multispanning outer membrane proteins. J Cell Biol. 2011;194:387–95. https://doi.org/10.1083/jcb.201102044.

Bhagawati M, Arroum T, Webeling N, et al. The receptor subunit Tom20 is dynamically associated with the TOM complex in mitochondria of human cells. Mol Biol Cell. 2021;32:br1. https://doi.org/10.1091/mbc.E21-01-0042.

Bolliger L, Junne T, Schatz G, et al. Acidic receptor domains on both sides of the outer membrane mediate translocation of precursor proteins into yeast mitochondria. EMBO J. 1995;14:6318–26. https://doi.org/10.1002/j.1460-2075.1995.tb00322.x.

Brinkerhoff H, Kang ASW, Liu J, et al. Multiple rereads of single proteins at single–amino acid resolution using nanopores. Science. 2021;374:1509–13. https://doi.org/10.1126/science.abl4381.

Brix J, Dietmeier K, Pfanner N. Differential recognition of preproteins by the purified cytosolic domains of the mitochondrial import receptors Tom20, Tom22, and Tom70. J Biol Chem. 1997;272:20730–5. https://doi.org/10.1074/jbc.272.33.20730.

Burri L, Williams BAP, Bursac D, et al. Microsporidian mitosomes retain elements of the general mitochondrial targeting system. Proc Natl Acad Sci USA. 2006;103:15916–20. https://doi.org/10.1073/pnas.0604109103.

Busch JD, Fielden LF, Pfanner N, et al. Mitochondrial protein transport: versatility of translocases and mechanisms. Mol Cell. 2023;83:890–910. https://doi.org/10.1016/j.molcel.2023.02.020.

Callegari S, Cruz-Zaragoza LD, Rehling P. From TOM to the TIM23 complex – handing over of a precursor. Biol Chem. 2020;401:709–21. https://doi.org/10.1515/hsz-2020-0101.

Calvo SE, Clauser KR, Mootha VK. MitoCarta2.0: an updated inventory of mammalian mitochondrial proteins. Nucleic Acids Res. 2016;44:D1251–7. https://doi.org/10.1093/nar/gkv1003.

Chacinska A, Pfannschmidt S, Wiedemann N, et al. Essential role of Mia40 in import and assembly of mitochondrial intermembrane space proteins. EMBO J. 2004;23:3735–46. https://doi.org/10.1038/sj.emboj.7600389.

Chacinska A, Guiard B, Müller JM, et al. Mitochondrial biogenesis, switching the sorting pathway of the intermembrane space receptor Mia40. J Biol Chem. 2008;283:29723–9. https://doi.org/10.1074/jbc.M805356200.

Chew O, Lister R, Qbadou S, et al. A plant outer mitochondrial membrane protein with high amino acid sequence identity to a chloroplast protein import receptor. FEBS Lett. 2004;557:109–14. https://doi.org/10.1016/s0014-5793(03)01457-1.

Cooper A. Thermodynamic fluctuations in protein molecules. Proc Natl Acad Sci USA. 1976;73:2740–1. https://doi.org/10.1073/pnas.73.8.2740.

Court DA, Nargang FE, Steiner H, et al. Role of the intermembrane-space domain of the preprotein receptor Tom22 in protein import into mitochondria. Mol Cell Biol. 1996;16:4035–42. https://doi.org/10.1128/MCB.16.8.4035.

den Brave F, Schulte U, Fakler B, et al. Mitochondrial complexome and import network. Trends Cell Biol. 2024;34:578–94. https://doi.org/10.1016/j.tcb.2023.10.004.

Dekker PJ, Ryan MT, Brix J, et al. Preprotein translocase of the outer mitochondrial membrane: molecular dissection and assembly of the general import pore complex. Mol Cell Biol. 1998;18:6515–24. https://doi.org/10.1128/MCB.18.11.6515.

Dembowski M, Kunkele KP, Nargang FE, et al. Assembly of Tom6 and Tom7 into the TOM core complex of *Neurospora crassa*. J Biol Chem. 2001;276:17679–85. https://doi.org/10.1074/jbc.M009653200.

Dietmeier K, Hönlinger A, Bömer U, et al. Tom5 functionally links mitochondrial preprotein receptors to the general import pore. Nature. 1997;388:195–200. https://doi.org/10.1038/40663.

Dimmer KS, Papić D, Schumann B, et al. A crucial role for Mim2 in the biogenesis of mitochondrial outer membrane proteins. J Cell Sci. 2012;125:3464–73. https://doi.org/10.1242/jcs.103804.

Doan KN, Grevel A, Mårtensson CU, et al. The mitochondrial import complex MIM functions as main translocase for α-helical outer membrane proteins. Cell Rep. 2020;31:107567. https://doi.org/10.1016/j.celrep.2020.107567.

Dolezal P, Likic V, Tachezy J, et al. Evolution of the molecular machines for protein import into mitochondria. Science. 2006;313:314–8. https://doi.org/10.1126/science.1127895.

Eaglesfield R, Tokatlidis K. Targeting and insertion of membrane proteins in mitochondria. Front Cell Dev Biol. 2021;9:803205. https://doi.org/10.3389/fcell.2021.803205.

Ellenrieder L, Opaliński Ł, Becker L, et al. Separating mitochondrial protein assembly and endoplasmic reticulum tethering by selective coupling of Mdm10. Nat Commun. 2016;7:13021. https://doi.org/10.1038/ncomms13021.

Endo T, Yamano K. Transport of proteins across or into the mitochondrial outer membrane. Biochim Biophys Acta. 2010;1803:706–14. https://doi.org/10.1016/j.bbamcr.2009.11.007.

Endres M, Neupert W, Brunner M. Transport of the ADP/ATP carrier of mitochondria from the TOM complex to the TIM22·54 complex. EMBO J. 1999;18:3214–21. https://doi.org/10.1093/emboj/18.12.3214.

Fielden LF, Busch JD, Merkt SG, et al. Central role of Tim17 in mitochondrial presequence protein translocation. Nature. 2023;621:627–34. https://doi.org/10.1038/s41586-023-06477-8.

Foury F, Roganti T, Lecrenier N, et al. The complete sequence of the mitochondrial genome of *Saccharomyces cerevisiae*. FEBS Lett. 1998;440:325–31. https://doi.org/10.1016/s0014-5793(98)01467-7.

Fujiki M, Verner K. Coupling of cytosolic protein synthesis and mitochondrial protein import in yeast. Evidence for cotranslational import in vivo. J Biol Chem. 1993;268:1914–20. https://doi.org/10.1016/S0021-9258(18)53941-7.

Garcia M, Darzacq X, Delaveau T, et al. Mitochondria-associated yeast mRNAs and the biogenesis of molecular complexes. Mol Biol Cell. 2007;18:362–8. https://doi.org/10.1091/mbc.E06-09-0827.

Gessmann D, Flinner N, Pfannstiel J, et al. Structural elements of the mitochondrial preprotein-conducting channel Tom40 dissolved by bioinformatics and mass spectrometry. Biochim Biophys Acta. 2011;1807:1647–57. https://doi.org/10.1016/j.bbabio.2011.08.006.

Giacomello M, Pyakurel A, Glytsou C, et al. The cell biology of mitochondrial membrane dynamics. Nat Rev Mol Cell Biol. 2020;21:204–24. https://doi.org/10.1038/s41580-020-0210-7.

Gold VA, Chroscicki P, Bragoszewski P, et al. Visualization of cytosolic ribosomes on the surface of mitochondria by electron cryo-tomography. EMBO Rep. 2017;18:1786–800. https://doi.org/10.15252/embr.201744261.

Gomkale R, Linden A, Neumann P, et al. Mapping protein interactions in the active TOM-TIM23 supercomplex. Nat Commun. 2021;12:5715. https://doi.org/10.1038/s41467-021-26016-1.

Goodman MB, Haswell ES, Vásquez V. Mechanosensitive membrane proteins: usual and unusual suspects in mediating mechanotransduction. J Gen Physiol. 2023;155:e202213248. https://doi.org/10.1085/jgp.202213248.

Guan Z, Yan L, Wang Q, et al. Structural insights into assembly of human mitochondrial translocase TOM complex. Cell Discov. 2021;7:1–5. https://doi.org/10.1038/s41421-021-00252-7.

Guna A, Stevens TA, Inglis AJ, et al. MTCH2 is a mitochondrial outer membrane protein insertase. Science. 2022;378:317–22. https://doi.org/10.1126/science.add1856.

Hällberg BM, Larsson N-G. Making proteins in the powerhouse. Cell Metab. 2014;20:226–40. https://doi.org/10.1016/j.cmet.2014.07.001.

Hallermayer G, Neupert W. Lipid composition of mitochondrial outer and inner membranes of *Neurospora crassa*. Biol Chem. 1974;355:279–88. https://doi.org/10.1515/bchm2.1974.355.1.279.

Hansen KG, Herrmann JM. Transport of proteins into mitochondria. Protein J. 2019;38:330–42. https://doi.org/10.1007/s10930-019-09819-6.

Hauser H, Pascher I, Pearson RH, et al. Preferred conformation and molecular packing of phosphatidylethanolamine and phosphatidylcholine. Biochim Biophys Acta Biomembr. 1981;650:21–51. https://doi.org/10.1016/0304-4157(81)90007-1.

Heinz E, Lithgow T. Back to basics: a revealing secondary reduction of the mitochondrial protein import pathway in diverse intracellular parasites. Biochim Biophys Acta. 2013;1833:295–303. https://doi.org/10.1016/j.bbamcr.2012.02.006.

Hiller S, Garces RG, Malia TJ, et al. Solution structure of the integral human membrane protein VDAC-1 in detergent micelles. Science. 2008;321:1206–10. https://doi.org/10.1126/science.1161302.

Hönlinger A, Bömer U, Alconada A, et al. Tom7 modulates the dynamics of the mitochondrial outer membrane translocase and plays a pathway-related role in protein import. EMBO J. 1996;15:2125–37.

Jaremko M, Jaremko Ł, Villinger S, et al. High-resolution NMR determination of the dynamic structure of membrane proteins. Angew Chem Int Ed. 2016;55:10518–21. https://doi.org/10.1002/anie.201602639.

Jin P, Jan LY, Jan Y-N. Mechanosensitive ion channels: structural features relevant to mechanotransduction mechanisms. Annu Rev Neurosci. 2020;43:207–29. https://doi.org/10.1146/annurev-neuro-070918-050509.

Johnston AJ, Hoogenraad J, Dougan DA, et al. Insertion and assembly of human tom7 into the preprotein translocase complex of the outer mitochondrial membrane. J Biol Chem. 2002;277:42197–204. https://doi.org/10.1074/jbc.M205613200.

Kang Y, Baker MJ, Liem M, et al. Tim29 is a novel subunit of the human TIM22 translocase and is involved in complex assembly and stability. eLife. 2016;5:e17463. https://doi.org/10.7554/eLife.17463.

Kasianowicz JJ, Brandin E, Branton D, et al. Characterization of individual polynucleotide molecules using a membrane channel. Proc Natl Acad Sci USA. 1996;93:13770–3. https://doi.org/10.1073/pnas.93.24.13770.

Kato H, Mihara K. Identification of Tom5 and Tom6 in the preprotein translocase complex of human mitochondrial outer membrane. Biochem Biophys Res Commun. 2008;369:958–63. https://doi.org/10.1016/j.bbrc.2008.02.150.

Kiebler M, Pfaller R, Söllner T, et al. Identification of a mitochondrial receptor complex required for recognition and membrane insertion of precursor proteins. Nature. 1990;348:610–6. https://doi.org/10.1038/348610a0.

Kiebler M, Keil P, Schneider H, et al. The mitochondrial receptor complex: a central role of MOM22 in mediating preprotein transfer from receptors to the general insertion pore. Cell. 1993;74:483–92. https://doi.org/10.1016/0092-8674(93)80050-o.

Koh JY, Hájek P, Bedwell DM. Overproduction of PDR3 suppresses mitochondrial import defects associated with a TOM70 null mutation by increasing the expression of TOM72 in *Saccharomyces cerevisiae*. Mol Cell Biol. 2001;21:7576–86. https://doi.org/10.1128/MCB.2 1.22.7576-7586.2001.

Kovermann P, Truscott KN, Guiard B, et al. Tim22, the essential core of the mitochondrial protein insertion complex, forms a voltage-activated and signal-gated channel. Mol Cell. 2002;9:363–73. https://doi.org/10.1016/S1097-2765(02)00446-X.

Krüger V, Becker T, Becker L, et al. Identification of new channels by systematic analysis of the mitochondrial outer membrane. J Cell Biol. 2017;216:3485–95. https://doi.org/10.1083/jcb.201706043.

Künkele K-P, Heins S, Dembowski M, et al. The preprotein translocation channel of the outer membrane of mitochondria. Cell. 1998;93:1009–19. https://doi.org/10.1016/S0092-8674(00)81206-4.

Lackey SWK, Taylor RD, Go NE, et al. Evidence supporting the 19 β-strand model for Tom40 from cysteine scanning and protease site accessibility studies. J Biol Chem. 2014;289:21640–50. https://doi.org/10.1074/jbc.M114.578765.

Lesnik C, Golani-Armon A, Arava Y. Localized translation near the mitochondrial outer membrane: an update. RNA Biol. 2015;12:801–9. https://doi.org/10.1080/15476286.2015.1058686.

Lister R, Whelan J. Mitochondrial protein import: convergent solutions for receptor structure. Curr Biol. 2006;16:R197–9. https://doi.org/10.1016/j.cub.2006.2.24.

Lister R, Carrie C, Duncan O, et al. Functional definition of outer membrane proteins involved in preprotein import into mitochondria. Plant Cell. 2007;19:3739–59. https://doi.org/10.1105/tpc.107.50534.

Lomize AL, Todd SC, Pogozheva ID. Spatial arrangement of proteins in planar and curved membranes by PPM 3.0. Protein Sci. 2022;31:209–20. https://doi.org/10.1002/pro.4219.

Maćasev D, Whelan J, Newbigin E, et al. Tom22′, an 8-kDa trans-site receptor in plants and protozoans, is a conserved feature of the TOM complex that appeared early in the evolution of eukaryotes. Mol Biol Evol. 2004;21:1557–64. https://doi.org/10.1093/molbev/msh166.

Mager F, Sokolova L, Lintzel J, et al. LILBID-mass spectrometry of the mitochondrial preprotein translocase TOM. J Phys Condens Matter. 2010;22:454132. https://doi.org/10.1088/0953-8984/22/45/454132.

Mager F, Gessmann D, Nussberger S, et al. Functional refolding and characterization of two Tom40 isoforms from human mitochondria. J Membr Biol. 2011;242:11–21. https://doi.org/10.1007/s00232-011-9372-8.

Mahendran KR, Romero-Ruiz M, Schlösinger A, et al. Protein translocation through Tom40: kinetics of peptide release. Biophys J. 2012;102:39–47. https://doi.org/10.1016/j.bpj.2011.11.4003.

Mazur M, Kmita H, Wojtkowska M. The diversity of the mitochondrial outer membrane protein import channels: emerging targets for modulation. Molecules. 2021;26:4087. https://doi.org/10.3390/molecules26134087.

Meisinger C, Ryan MT, Hill K, et al. Protein import channel of the outer mitochondrial membrane: a highly stable Tom40-Tom22 core structure differentially interacts with preproteins, small tom proteins, and import receptors. Mol Cell Biol. 2001;21:2337–48. https://doi.org/10.1128/MCB. 21.7.2337-2348.2001.

Melin J, Kilisch M, Neumann P, et al. A presequence-binding groove in Tom70 supports import of Mdl1 into mitochondria. Biochim Biophys Acta. 2015;1853:1850–9. https://doi.org/10.1016/j.bbamcr.2015.04.021.

Moczko M, Bömer U, Kübrich M, et al. The intermembrane space domain of mitochondrial Tom22 functions as a trans binding site for preproteins with N-terminal targeting sequences. Mol Cell Biol. 1997;17:6574–84. https://doi.org/10.1128/mcb.17.11.6574.

Model K, Meisinger C, Prinz T, et al. Multistep assembly of the protein import channel of the mitochondrial outer membrane. Nat Struct Biol. 2001;8:361–70. https://doi.org/10.1038/86253.

Model K, Prinz T, Ruiz T, et al. Protein translocase of the outer mitochondrial membrane: role of import receptors in the structural organization of the TOM complex. J Mol Biol. 2002;316:657–66. https://doi.org/10.1006/jmbi.2001.5365.

Model K, Meisinger C, Kühlbrandt W. Cryo-electron microscopy structure of a yeast mitochondrial preprotein translocase. J Mol Biol. 2008;383:1049–57. https://doi.org/10.1016/j.jmb.2008.07.087.

Mokranjac D, Neupert W. The many faces of the mitochondrial TIM23 complex. Biochim Biophys Acta. 2010;1797:1045–54. https://doi.org/10.1016/j.bbabio.2010.01.026.

Morgenstern M, Stiller SB, Lübbert P, et al. Definition of a high-confidence mitochondrial proteome at quantitative scale. Cell Rep. 2017;19:2836–52. https://doi.org/10.1016/j.celrep.2017.06.014.

Movileanu L. Interrogating single proteins through nanopores: challenges and opportunities. Trends Biotechnol. 2009;27:333–41. https://doi.org/10.1016/j.tibtech.2009.02.008.

Muthukumar M. Macromolecular mechanisms of protein translocation. Protein Pept Lett. 2014;21:209–16. https://doi.org/10.2174/09298665113209990079.

Muthukumar G, Stevens TA, Inglis AJ, et al. Triaging of α-helical proteins to the mitochondrial outer membrane by distinct chaperone machinery based on substrate topology. Mol Cell. 2024;84:1101–1119.e9. https://doi.org/10.1016/j.molcel.2024.01.028.

Nargang FE, Rapaport D, Ritzel RG, et al. Role of the negative charges in the cytosolic domain of TOM22 in the import of precursor proteins into mitochondria. Mol Cell Biol. 1998;18:3173–81. https://doi.org/10.1128/MCB.18.6.3173.

Neupert W. Protein import into mitochondria. Annu Rev Biochem. 1997;66:863–917. https://doi.org/10.1146/annurev.biochem.66.1.863.

Neupert W, Herrmann JM. Translocation of proteins into mitochondria. Annu Rev Biochem. 2007;76:723–49. https://doi.org/10.1146/annurev.biochem.76.052705.163409.

Nussberger S, Ghosh R, Wang S. New insights into the structure and dynamics of the TOM complex in mitochondria. Biochem Soc Trans. 2024;52:911–22. https://doi.org/10.1042/BST20231236.

Ornelas P, Bausewein T, Martin J, et al. Two conformations of the Tom20 preprotein receptor in the TOM holo complex. Proc Natl Acad Sci USA. 2023;120:e2301447120. https://doi.org/10.1073/pnas.2301447120.

Ott M, Amunts A, Brown A. Organization and regulation of mitochondrial protein synthesis. Annu Rev Biochem. 2016;85:77–101. https://doi.org/10.1146/annurev-biochem-060815-014334.

Parvin N, Carrie C, Pabst I, et al. TOM9.2 is a calmodulin-binding protein critical for TOM complex assembly but not for mitochondrial protein import in *Arabidopsis thaliana*. Mol Plant. 2017;10:575–89. https://doi.org/10.1016/j.molp.2016.12.012.

Perry AJ, Rimmer KA, Mertens HDT, et al. Structure, topology and function of the translocase of the outer membrane of mitochondria. Plant Physiol Biochem. 2008;46:265–74. https://doi.org/10.1016/j.plaphy.2007.12.012.

Pfaller R, Steger HF, Rassow J, et al. Import pathways of precursor proteins into mitochondria: multiple receptor sites are followed by a common membrane insertion site. J Cell Biol. 1988;107:2483–90. https://doi.org/10.1083/jcb.107.6.2483.

Pfanner N, Geissler A. Versatility of the mitochondrial protein import machinery. Nat Rev Mol Cell Biol. 2001;2:339–49. https://doi.org/10.1038/35073006.

Pfanner N, Neupert W. Distinct steps in the import of ADP/ATP carrier into mitochondria. J Biol Chem. 1987;262:7528–36. https://doi.org/10.1016/S0021-9258(18)47598-9.

Pfanner N, Warscheid B, Wiedemann N. Mitochondrial protein organization: from biogenesis to networks and function. Nat Rev Mol Cell Biol. 2019;20:267–84. https://doi.org/10.1038/s41580-018-0092-0.

Phuyal S, Romani P, Dupont S, et al. Mechanobiology of organelles: illuminating their roles in mechanosensing and mechanotransduction. Trends Cell Biol. 2023;33:1049–61. https://doi.org/10.1016/j.tcb.2023.05.001.

Popov-Čeleketić D, Mapa K, Neupert W, et al. Active remodelling of the TIM23 complex during translocation of preproteins into mitochondria. EMBO J. 2008;27:1469–80. https://doi.org/10.1038/emboj.2008.79.

Poynor M, Eckert R, Nussberger S. Dynamics of the preprotein translocation channel of the outer membrane of mitochondria. Biophys J. 2008;95:1511–22. https://doi.org/10.1529/biophysj.108.131003.

Qi L, Wang Q, Guan Z, et al. Cryo-EM structure of the human mitochondrial translocase TIM22 complex. Cell Res. 2021;31:369–72. https://doi.org/10.1038/s41422-020-00400-w.

Qiu J, Wenz L-S, Zerbes RM, et al. Coupling of mitochondrial import and export translocases by receptor-mediated supercomplex formation. Cell. 2013;154:596–608. https://doi.org/10.1016/j.cell.2013.06.033.

Ramage L, Junne T, Hahne K, et al. Functional cooperation of mitochondrial protein import receptors in yeast. EMBO J. 1993;12:4115–23. https://doi.org/10.1002/j.1460-2075.1993.tb06095.x.

Rapaport D, Künkele K-P, Dembowski M, et al. Dynamics of the TOM complex of mitochondria during binding and translocation of preproteins. Mol Cell Biol. 1998;18:5256–62. https://doi.org/10.1128/mcb.18.9.5256.

Rath S, Sharma R, Gupta R, et al. MitoCarta3.0: an updated mitochondrial proteome now with sub-organelle localization and pathway annotations. Nucleic Acids Res. 2021;49:D1541–7. https://doi.org/10.1093/nar/gkaa1011.

Rehling P, Model K, Brandner K, et al. Protein insertion into the mitochondrial inner membrane by a twin-pore translocase. Science. 2003;299:1747–51. https://doi.org/10.1126/science.1080945.

Reid GA, Schatz G. Import of proteins into mitochondria. Extramitochondrial pools and post-translational import of mitochondrial protein precursors in vivo. J Biol Chem. 1982;257:13062–7. https://doi.org/10.1016/S0021-9258(18)33622-6.

Rimmer KA, Foo JH, Ng A, et al. Recognition of mitochondrial targeting sequences by the import receptors Tom20 and Tom22. J Mol Biol. 2011;405:804–18. https://doi.org/10.1016/j.jmb.2010.11.017.

Rodriguez-Cousiño N, Nargang FE, Baardman R, et al. An import signal in the cytosolic domain of the *Neurospora* mitochondrial outer membrane protein TOM22. J Biol Chem. 1998;273:11527–32. https://doi.org/10.1074/jbc.273.19.11527.

Romani P, Nirchio N, Arboit M, et al. Mitochondrial fission links ECM mechanotransduction to metabolic redox homeostasis and metastatic chemotherapy resistance. Nat Cell Biol. 2022;24:168–80. https://doi.org/10.1038/s41556-022-00843-w.

Romero-Ruiz M, Mahendran KR, Eckert R, et al. Interactions of mitochondrial presequence peptides with the mitochondrial outer membrane preprotein translocase TOM. Biophys J. 2010;99:774–81. https://doi.org/10.1016/j.bpj.2010.05.010.

Ruprecht JJ, Kunji ERS. The SLC25 mitochondrial carrier family: structure and mechanism. Trends Biochem Sci. 2020;45:244–58. https://doi.org/10.1016/j.tibs.2019.11.001.

Ryan MT, Müller H, Pfanner N. Functional staging of ADP/ATP carrier translocation across the outer mitochondrial membrane. J Biol Chem. 1999;274:20619–27. https://doi.org/10.1074/jbc.274.29.20619.

Saeki K, Suzuki H, Tsuneoka M, et al. Identification of mammalian TOM22 as a subunit of the preprotein translocase of the mitochondrial outer membrane. J Biol Chem. 2000;275:31996–2002. https://doi.org/10.1074/jbc.M004794200.

Saitoh T, Igura M, Obita T, et al. Tom20 recognizes mitochondrial presequences through dynamic equilibrium among multiple bound states. EMBO J. 2007;26:4777–87. https://doi.org/10.1038/sj.emboj.7601888.

Saitoh T, Igura M, Miyazaki Y, et al. Crystallographic snapshots of Tom20–mitochondrial presequence interactions with disulfide-stabilized peptides. Biochemistry. 2011;50:5487–96. https://doi.org/10.1021/bi200470x.

Schatz G. Signals guiding proteins to their correct locations in mitochondria. Eur J Biochem. 1987;165:1–6. https://doi.org/10.1111/j.1432-1033.1987.tb11186.x.

Schatz G, Butow RA. How are proteins imported into mitochondria? Cell. 1983;32:316–8. https://doi.org/10.1016/0092-8674(83)90450-6.

Schlossmann J, Lill R, Neupert W, et al. Tom71, a novel homologue of the mitochondrial preprotein receptor Tom70. J Biol Chem. 1996;271:17890–5. https://doi.org/10.1074/jbc.271.30.17890.

Schmitt S, Ahting U, Eichacker L, et al. Role of Tom5 in maintaining the structural stability of the TOM complex of mitochondria. J Biol Chem. 2005;280:14499–506. https://doi.org/10.1074/jbc.M413667200.

Schredelseker J, Paz A, López CJ, et al. High resolution structure and double electron-electron resonance of the zebrafish voltage-dependent anion channel 2 reveal an oligomeric population. J Biol Chem. 2014;289:12566–77. https://doi.org/10.1074/jbc.M113.497438.

Seelig J. 31P nuclear magnetic resonance and the head group structure of phospholipids in membranes. Biochim Biophys Acta. 1978;515:105–40. https://doi.org/10.1016/0304-4157(78)90001-1.

Sherman EL, Go NE, Nargang FE. Functions of the small proteins in the TOM complex of *Neurospora crasssa*. Mol Biol Cell. 2005;16:4172–82. https://doi.org/10.1091/mbc.E05-03-0187.

Shiota T, Imai K, Qiu J, et al. Molecular architecture of the active mitochondrial protein gate. Science. 2015;349:1544–8. https://doi.org/10.1126/science.aac6428.

Sim SI, Chen Y, Lynch DL, et al. Structural basis of mitochondrial protein import by the TIM23 complex. Nature. 2023;621:620–6. https://doi.org/10.1038/s41586-023-06239-6.

Smith AC, Robinson AJ. MitoMiner v3.1, an update on the mitochondrial proteomics database. Nucleic Acids Res. 2016;44:D1258–61. https://doi.org/10.1093/nar/gkv1001.

Stojanovski D, Müller JM, Milenkovic D, et al. The MIA system for protein import into the mitochondrial intermembrane space. Biochim Biophys Acta. 2008;1783:610–7. https://doi.org/10.1016/j.bbamcr.2007.10.004.

Su J, Liu D, Yang F, et al. Structural basis of Tom20 and Tom22 cytosolic domains as the human TOM complex receptors. Proc Natl Acad Sci. 2022;119:e2200158119. https://doi.org/10.1073/pnas.2200158119.

Suzuki H, Okazawa Y, Komiya T, et al. Characterization of rat TOM40, a central component of the preprotein translocase of the mitochondrial outer membrane. J Biol Chem. 2000;275:37930–6. https://doi.org/10.1074/jbc.M006558200.

Takeda H, Tsutsumi A, Nishizawa T, et al. Mitochondrial sorting and assembly machinery operates by β-barrel switching. Nature. 2021;590:163–9. https://doi.org/10.1038/s41586-020-03113-7.

Tucker K, Park E. Cryo-EM structure of the mitochondrial protein-import channel TOM complex at near-atomic resolution. Nat Struct Mol Biol. 2019;26:1158–66. https://doi.org/10.1038/s41594-019-0339-2.

Ujwal R, Cascio D, Colletier J-P, et al. The crystal structure of mouse VDAC1 at 2.3 Å resolution reveals mechanistic insights into metabolite gating. Proc Natl Acad Sci USA. 2008;105:17742–7. https://doi.org/10.1073/pnas.0809634105.

van Wilpe S, Ryan MT, Hill K, et al. Tom22 is a multifunctional organizer of the mitochondrial preprotein translocase. Nature. 1999;401:485–9. https://doi.org/10.1038/46802.

von Heijne G. Mitochondrial targeting sequences may form amphiphilic helices. EMBO J. 1986;5:1335–42. https://doi.org/10.1002/j.1460-2075.1986.tb04364.x.

Waegemann K, Popov-Čeleketić D, Neupert W, et al. Cooperation of TOM and TIM23 complexes during translocation of proteins into mitochondria. J Mol Biol. 2015;427:1075–84. https://doi.org/10.1016/j.jmb.2014.7.015.

Waller RF, Jabbour C, Chan NC, et al. Evidence of a reduced and modified mitochondrial protein import apparatus in microsporidian mitosomes. Eukaryot Cell. 2009;8:19–26. https://doi.org/10.1128/ec.00313-08.

Wang S, Nussberger S. Single-molecule imaging of lateral mobility and ion channel activity in lipid bilayers using total internal reflection fluorescence (TIRF). J Vis Exp. 2023;192:e64970. https://doi.org/10.3791/64970.

Wang S, Nussberger S. Tracking the activity and position of mitochondrial β-barrel proteins. Methods Mol Biol. 2024;2778:221–36. https://doi.org/10.1007/978-1-0716-3734-0_14.

Wang S, Findeisen L, Leptihn S, et al. Spatiotemporal stop-and-go dynamics of the mitochondrial TOM core complex correlates with channel activity. Commun Biol. 2022;5:1–11. https://doi.org/10.1038/s42003-022-03419-4.

Wang Q, Zhuang J, Huang R, et al. The architecture of substrate-engaged TOM–TIM23 supercomplex reveals preprotein proximity sites for mitochondrial protein translocation. Cell Discov. 2024;10:1–4. https://doi.org/10.1038/s41421-023-00643-y.

Wiedemann N, Pfanner N. Mitochondrial machineries for protein import and assembly. Annu Rev Biochem. 2017;86:685–714. https://doi.org/10.1146/annurev-biochem-060815-014352.

Wiedemann N, Kozjak V, Chacinska A, et al. Machinery for protein sorting and assembly in the mitochondrial outer membrane. Nature. 2003;424:565–71. https://doi.org/10.1038/nature01753.

Wienhues U, Becker K, Schleyer M, et al. Protein folding causes an arrest of preprotein translocation into mitochondria in vivo. J Cell Biol. 1991;115:1601–9. https://doi.org/10.1083/jcb.115.6.1601.

Yano M, Hoogenraad N, Terada K, et al. Identification and functional analysis of human Tom22 for protein import into mitochondria. Mol Cell Biol. 2000;20:7205–13. https://doi.org/10.1128/MCB.20.19.7205-7213.2000.

Zhang Y, Ou X, Wang X, et al. Structure of the mitochondrial TIM22 complex from yeast. Cell Res. 2021;31:366–8. https://doi.org/10.1038/s41422-020-00399-0.

Zhou X, Yang Y, Wang G, et al. Molecular pathway of mitochondrial preprotein import through the TOM–TIM23 supercomplex. Nat Struct Mol Biol. 2023;30:1996–2008. https://doi.org/10.1038/s41594-023-01103-7.

Chapter 6
Dynamic Assemblies in Genome Maintenance

Paras Gaur and Maria Spies

Abstract The integrity of the human genome is continuously challenged by diverse endogenous and exogenous threats that damage DNA and disrupt its replication. When the replication machinery encounters such obstacles, including lesions or non-canonical DNA structures, it may stall, initiate repair, or activate specialized pathways to bypass the impediment. Maintaining replication progression requires a coordinated and dynamic assembly of numerous nucleoprotein complexes that recognize, process, and resolve DNA damage and replication stalling structures. This chapter highlights how proliferating cell nuclear antigen (PCNA), poly(ADP-ribose) polymerase 1 (PARP1), and non-canonical DNA structures are integrated into higher-order supramolecular complexes that stabilize, remodel, or resolve stalled or damaged replication forks. The molecular events carried out by these supramolecular complexes are essential for preserving genomic integrity in human cells. Moreover, many of the factors involved emerge as attractive therapeutic targets for diseases driven by genome instability, including cancer.

Keywords DNA replication · DNA damage repair · Non-conical DNA structure · Proliferating cell nuclear antigen · PARP1 · G-quadruplex H-DNA · Condensates

P. Gaur (✉) · M. Spies (✉)
Department of Biochemistry and Molecular Biology, University of Iowa Carver College of Medicine, Iowa City, IA, USA

Holden Comprehensive Cancer Center, University of Iowa, Iowa City, IA, USA
e-mail: paras-gaur@uiowa.edu; maria-spies@uiowa.edu

A. M. Pedley (ed.), *Supramolecular Protein Assemblies In Cells*, Advances in Experimental Medicine and Biology 1514,
https://doi.org/10.1007/978-3-032-26629-3_6

113

6.1 Introduction: Dynamic Assemblies in Genome Maintenance

Genome integrity sustains life, while its instability drives a range of human diseases, including developmental disorders, bone marrow failure syndromes, and cancer. Each cell division demands near-perfect DNA replication, a feat that cannot be accomplished by isolated enzymes alone. Instead, replication and repair rely on dynamic, multi-protein complexes that coordinate DNA replication, remodel chromatin, recognize and signal the presence of DNA damage, and select the most appropriate pathway for its repair. These assemblies range from precisely coordinated complexes, such as the replisome, to dynamic arrangements involving partner exchange on the PCNA (proliferating cell nuclear antigen) clamp, to nucleoprotein condensates formed through liquid-liquid phase separation (LLPS). Together, these structures create an adaptable framework that safeguards genome stability under both normal and stress conditions.

Historically, replication and repair were viewed as processes driven by rigid macromolecular machines like the replisome, characterized by ordered recruitment and stable architecture to ensure processivity and fidelity. Recent studies reveal more dynamic organization and biomolecular condensates forming phase-separated compartments that concentrate proteins and nucleic acids. Biomolecular condensates exhibit tunable material properties, enabling rapid assembly and dissolution in response to cellular cues, providing spatiotemporal control over repair and replication, modulating enzyme concentrations, and creating microenvironments that accelerate repair (Spegg and Altmeyer 2021; Dall'Agnese et al. 2023; Wang et al. 2023; Kilic et al. 2019; Chin Sang et al. 2024; Chappidi et al. 2024; Wei et al. 2024; Levone et al. 2021).

Key players in DNA repair and replication, such as PARP1 (poly(ADP-ribose) polymerase 1), RPA (replication protein A), and PCNA, serve as protein-protein interaction hubs supporting the dynamic recruitment of various proteins that drive DNA metabolism in human cells. This chapter focuses on PARP1, PCNA, and non-canonical DNA structures that attract diverse repair factors, orchestrating replication and repair. PARP1, for example, promotes condensate formation by synthesizing poly(ADP-ribose) (PAR) chains at the sites of DNA damage, DNA breaks, and non-canonical DNA structures, creating an anionic scaffold that recruits DNA- and RNA-binding proteins and repair factors (e.g., FANCJ, REV1, FUS, EWS) into transient condensates that tether to target DNA and organize chromatin for repair (Chin Sang et al. 2024; Chappidi et al. 2024). Canonical repair foci, such as γH2AX and 53BP1, also exhibit liquid-like properties (Dall'Agnese et al. 2023; Wang et al. 2023; Kilic et al. 2019; Levone et al. 2021), highlighting the interplay between structured complexes and dynamic condensates (Dall'Agnese et al. 2023).

Certain repetitive DNA sequences can adopt noncanonical structures such as G-quadruplexes (G4s) (Ma et al. 2020), i-motifs (Abou Assi et al. 2018), hairpins formed by trinucleotide repeats (Pan et al. 2024; Qiu et al. 2015), and H-DNA (Hisey et al. 2024a). These structures regulate transcription, replication timing, and telomere maintenance, but when unresolved, they impede DNA and RNA metabolic processes (Tateishi-Karimata and Sugimoto 2021; Bansal et al. 2022). Dynamic assemblies of nucleoprotein complexes actively regulate pathway choice, replication restart, and chromatin signaling (Chin Sang et al. 2024; Chappidi et al. 2024; Wei et al. 2024; Ray Chaudhuri and Nussenzweig 2017; Moldovan et al. 2007). Their dysregulation drives mutagenesis, chemotherapy resistance, and neurodegeneration, while their unique properties offer therapeutic opportunities (Bansal et al. 2022). For example, cancer cells thrive under conditions that stress normal cells. One of their survival tricks is dependency on a narrow set of robust DNA repair mechanisms that compensate for accurate processes lost during tumorigenesis. For example, cancer cells displaying so-called BRCAness (i.e., deficiency in activities of tumor suppressors BRCA1, BRCA2, or PALB2) become addicted to DNA repair pathways that involve PARP1. This vulnerability is the basis of synthetic lethality: a situation where simultaneous inactivation of two genes results in catastrophic cell death, even though the loss of either one alone is tolerated (Chheda et al. 2021; Zinovyev et al. 2013; Ashworth and Lord 2018; Chandramouly et al. 2015; Hengel et al. 2016). When PARP inhibitors (PARPi) are administered in cancer cells displaying BRCAness, this combined defect becomes lethal, a so-called pharmacologically induced synthetic lethality (Zinovyev et al. 2013; Ashworth and Lord 2018; Helleday 2011; Jelinic and Levine 2014). This strategy has led to the FDA approval of drugs like olaparib (Lynparza), talazoparib (Talzenna), and rucaparib (Rubraca) for BRCA-mutated cancers (Ashworth and Lord 2018; Helleday 2011; Boussios et al. 2020; Kim et al. 2015; Balasubramaniam et al. 2017; Hoy 2018).

This chapter examines the molecular principles of dynamic nucleoprotein assemblies maintaining genome integrity, beginning with DNA replication and sources of damage, and then exploring how PCNA and PARP1 orchestrate repair and damage tolerance.

6.2 DNA Replication and Sources of Damage

Living organisms store their information in the form of double-stranded B-form DNA (Watson and Crick 1953, 1995). This structure provides stability while allowing readout and duplication of information. Complete, accurate, and timely duplication of the genetic material through DNA replication is essential for cell division, development, and organismal viability. Errors in DNA replication threaten genome integrity and can lead to diseases (Tateishi-Karimata and Sugimoto 2021; Bansal et al. 2022; Jensen and Rothenberg 2020; Satoh and Lindahl 1992; Maizels 2008). DNA replication is a highly organized, tightly regulated, and evolutionarily conserved process, carried out by a complex molecular machinery composed of many proteins that assemble into dynamic supramolecular structures collectively referred to as the replisome.

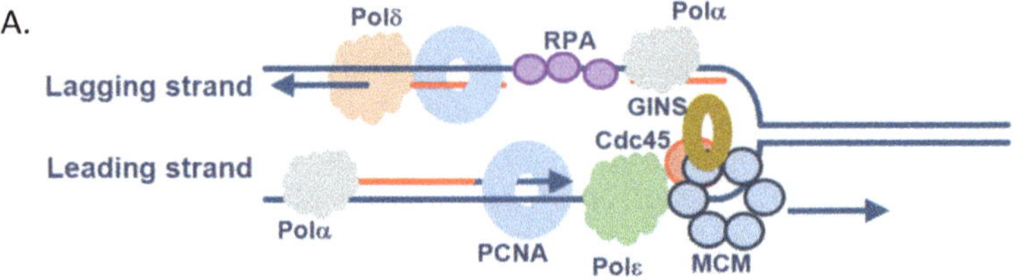

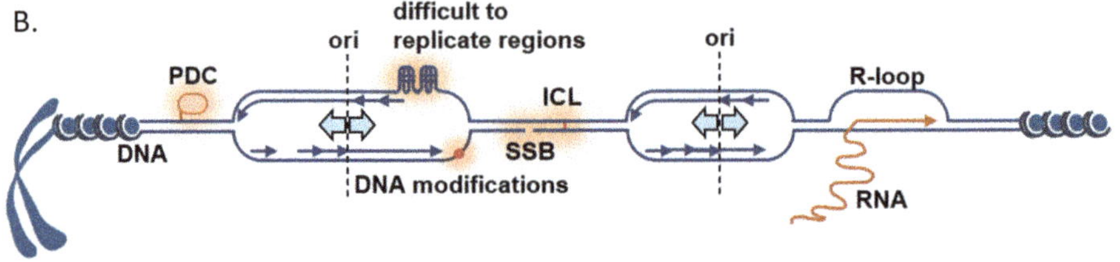

Fig. 6.1 Schematic diagram depicting. (**a**) The replication fork with leading and lagging strands, involving enzymes like Pol δ, Pol ε, and Pol α, along with proteins such as RPA, GINS, Cdc45, PCNA, and MCM. (**b**) The replication fork with difficult-to-replicate regions, including interstrand cross-links (ICL), single-strand breaks (SSB), and R-loops, marked by DNA modifications. Key sites like ori (origin of replication) and associated proteins are labeled

Access to genetic information is further shaped by chromatin organization and higher-order genome architecture. DNA is packaged with histones into chromatin, which regulates the spatial and temporal availability of replication origins and associated factors (Yunis and Bahr 1979; Abuelo and Moore 1969; Verschure et al. 2002). Coordinated control of DNA structure, chromatin state, and replisome activity ensures faithful genome duplication once and only once per cell cycle. DNA replication is initiated at defined regions called replication origins (see Fig. 6.1), where replication firing occurs. Efficient genome duplication requires multiple origins, with replication lasting from several minutes in yeast to hours in higher eukaryotes (Fig. 6.1b). It is important to note that there are 30,000–50,000 DNA replication origins in human cells that are nonuniformly distributed throughout the genome (Symeonidou et al. 2012).

Replisomes in all organisms assemble around ring-shaped hexameric helicases (Fig. 6.1). In eukaryotes, the central replicative helicase is the hetero-hexameric MCM2–7 complex, organized with an additional regulatory protein (Yeeles et al. 2017; Henrikus et al. 2024; Xu et al. 2023). However, how the MCM2–7 hexamer is assembled in living cells remains unknown. Recent studies reveal that MCM-binding protein (MCMBP) plays a critical role in its assembly in human cells (Saito et al. 2022). During S-phase transition in yeast, origin melting and assembly of the double hexamer poised for duplex unwinding occur in two steps. First, Cdc45, go-ichi-ni-san (GINS), and Polε engage MCM to form a double CMGE structure with two partially separated hexamers that initiate DNA melting (Henrikus et al. 2024). Second, Mcm10 triggers complete separation of the CMGEs, lagging-strand ejection, and helicase path crossing (Henrikus et al. 2024). Using biochemical reconstitution with cryo-EM, Henrikus and colleagues showed that Mcm10 splits double CMGE by engaging MCM's N-terminal homodimerization interface. DNA unwinding initiates from the N-terminal side, while the hexamer channel constricts, becoming too narrow for duplex DNA and thereby ejecting the lagging strand (Henrikus et al. 2024). CMG helicase translocates along the leading-strand template using a set of DNA-gripping elements in its C-terminal motor domains known as the Presensor-1 (PS1) and helix-2 Insert (H2I) loops (Iyer et al. 2004). These loops project into the helicase's central channel and act as coordinated molecular clamps that bind, stabilize, and pull the single-stranded DNA during forward movement. As CMG progresses on the leading strand in 3′ to 5′ direction, the PS1 and H2I loops from different MCM subunits collectively shift between planar and spiral configurations, producing a helical inchworm-like cycle that advances DNA in multi-nucleotide steps. This inchworm mechanism reflects CMG's intrinsic mode of DNA translocation (Eickhoff et al. 2019; Rzechorzek et al. 2020; Batra et al. 2025).

DNA replication on both leading and lagging strands begins when the DNA polymerase α-primase complex (Pol α-primase) synthesizes a short hybrid primer consisting of 8–12 RNA nucleotides, followed by 10–20 DNA nucleotides (see Fig. 6.1) (Jones et al. 2023). Using this primer, DNA polymerase ε extends the leading strand continuously in the 5′ → 3′ direction (Zhou et al. 2019). In contrast, the lagging strand is synthesized discontinuously as short Okazaki fragments (~150–200 nucleotides long in eukaryotes) (Pellegrini 2012; Pellegrini and Costa 2016; Okazaki and Okazaki 1969; Sugimoto et al. 1969; Sugimoto et al. 1968). A hetero-pentameric protein complex called replication factor C (RFC) binds to the primer-template junction and catalyzes the loading of PCNA in an ATP-dependent manner, which encircles DNA (see Fig. 6.1) (Majka and Burgers 2004; Bowman et al. 2004; Miyata et al. 2005; Mossi and Hubscher 1998). According to another model, the RFC complex assembles with PCNA in the presence of ATP; the opened PCNA ring twists like a spring washer and closes after ATP hydrolysis, followed by RFC dissociation from DNA-loaded PCNA. Once PCNA is loaded onto the primer strand, pol δ is recruited to synthesize the remainder of the Okazaki fragment and pol ε for synthesis at the leading strand (Zhou et al. 2019; Johnson et al. 1985, 2015; Pursell et al. 2007; McElhinny et al. 2008). When the replicative polymerase reaches the preceding Okazaki fragment, it partially displaces this fragment by the ongoing DNA synthesis, forming a flap structure that is then removed through the activity of FEN1 nuclease (Chapados et al. 2004); the resulting nick is sealed by DNA ligase I (Pascal et al. 2004). The recruitment of both FEN1 and DNA ligase is coordinated by PCNA, and PCNA also increases the processivity of the DNA replicative polymerase, which replicates the bulk of the DNA (Boehm et al. 2016a).

The processes enabling replication initiation and progression illustrate the remarkable coordination of supramolecular assemblies, from the multi-subunit CMG helicase to the dynamic interplay of primase, polymerases, PCNA, and ligases. This tightly regulated, evolutionarily conserved machinery ensures high-fidelity genome duplication, with each component performing a precise, interconnected role. Despite its precision, the progression of the replication forks is vulnerable to disruption. Replication forks frequently stall due to intrinsic or extrinsic challenges, including DNA damage (e.g., lesions, breaks), non-canonical DNA structures formed in difficult-to-replicate regions of the genome (e.g., some DNA sequences have a tendency to form G4s, hairpins, or triplexes), or conflicts with other molecular processes such as DNA repair or RNA transcription. Unresolved stalling can lead to fork breakage and genomic instability (Bouwman and Jonkers 2012; Ghosal and Chen 2013; Wolters and Schumacher 2013).

Defects in DNA replication, repair, tolerance, and damage response pathways are fundamental drivers of mutagenesis, genomic instability, and cancer progression (Bouwman and Jonkers 2012; Ghosal and Chen 2013; Wolters and Schumacher 2013; Tiwari and Wilson 3rd. 2019). These failures also contribute to aging and neurodegenerative disorders (Tiwari and Wilson 3rd. 2019). DNA damage arises from two primary sources: endogenous damage resulting from internal chemical reactions and exogenous damage caused by environmental agents.

Endogenous Damage Despite the high fidelity of DNA polymerases δ and ε and the presence of mismatch repair systems (Kunkel 2004, 2009, 2011), replication errors persist at rates of 10^{-6} to 10^{-8} per cell per generation (Kunkel 2004, 2009), arising from multiple mechanisms, including strand slippage that causes insertions and deletions (Lujan et al. 2025) and misincorporation of uracil along with altered dNTP pools (Saxena et al. 2024; Andersen et al. 2005; Kumar et al. 2011). Beyond replication errors, endogenous DNA damage can occur through various pathways. Base deamination is another major source of endogenous damage; cytosine spontaneously deaminates to uracil, and 5-methylcytosine to thymine. These transitions account for roughly one-third of hereditary mutations at CpG sites, while activation-induced deaminase (AID) and apolipoprotein B mRNA editing cytosine deaminases (APOBEC) enzymes drive physiological hypermutation processes (Duncan and Miller 1980; Pecori et al. 2022). AID catalyzes the deamination of 5-hydroxymethylcytosine (5hmC), a product generated through 5-methylcytosine (5mC) hydroxylation, into 5-hydroxymethyluracil (5hmU) (Franchini et al. 2012; Nawy 2013; Bhutani et al. 2011), while the base excision repair (BER) mechanism replaces 5hmC with unmodified cytosine. AID serves as a critical molecular orchestrator in the germinal center (GC) reaction within secondary lymphoid organs (SLOs), driving the production of high-affinity antibodies through somatic hypermutation; the dysregulation of AID leads to oncogenic mutations and promotes cancer malignancy (Lv et al. 2025). APOBEC deaminases are also widely associated with immune cell-derived cancer and virus-associated cancers (Cervantes-Gracia et al. 2021). The signatures of APOBECs, which feature in many cancer genomes, are considered to be potent mediators of tumorigenesis (Cervantes-Gracia et al. 2021; Law et al. 2020). The genome also suffers from abasic sites forming at approximately an estimated 50,000–200,000 AP sites, which are present in a mammalian cell (Nakamura and Swenberg 1999), unstable lesions that can lead to single-strand breaks (SSBs) and are addressed by BER or bypassed through translesion synthesis (TLS) (Lindahl and Barnes 2000; Cai et al. 2022). Additionally, reactive oxygen species (ROS) generated during metabolism produce more than 100 different types of base lesions, including 8-oxoguanine, thymine glycol, and formamidopyrimidine-adenine, while also inducing SSBs at approximately 2300 per cell per hour (Srinivas et al. 2019; Kasai and Nishimura 1983; Miller et al. 2004; Dizdaroglu et al. 2008). Finally, DNA methylation by S-adenosylmethionine generates several methylated bases, including N7-methylguanine, N3-methyladenine, and the particularly mutagenic O6-methylguanine (De Bont and van Larebeke 2004; Holliday and Ho 1998), which is repaired by the O^6-Methylguanine-DNA methyltransferase (MGMT) enzyme or through BER pathways.

Exogenous Damage Ionizing radiation produces various base lesions such as 8-oxoguanine and thymine glycol, along with DNA breaks featuring modified ends. Double-strand breaks (DSBs) resulting from ionizing radiation are primarily repaired through the nonhomologous end joining repair pathway (NHEJ) (Mavragani et al. 2019; Mahaney et al. 2009). Topoisomerase I (Top1), which normally relaxes DNA supercoils, can form stabilized transient cleavage complexes, and these complexes can be trapped by compounds like camptothecin, leading to a covalent protein-DNA cross-link and a break in the DNA strand (Pommier et al. 2022; Zhang et al. 2022). UV radiation forms cyclobutane pyrimidine dimers and 6–4 photoproducts (Chan et al. 1985; Kai-Feng et al. 2020). These lesions are addressed by nucleotide excision repair (NER), TLS, and homologous recombination (HR) (Paul et al. 2019). Alkylating agents include methyl methanesulfonate, ethyl methanesulfonate, and N-methyl-N′-nitro-N-nitrosoguanidine, which produce N7-methylguanine and O6-methylguanine adducts (Kondo et al. 2010). Aldehydes produced both endogenously (via metabolism and lipid peroxidation) and exogenously (cigarette smoke, chemotherapy, environmental pollutants) can produce DNA-protein cross-links, base adducts, and inter-/intra-strand cross-links (Blouin and Saini 2024). Cross-linking agents such as cisplatin block DNA replication and are exploited in cancer chemotherapy (Hu et al. 2016; Shoulkamy et al. 2025). Aromatic amines and polycyclic aromatic hydrocarbons, found in cigarette smoke, form bulky DNA adducts (Hang 2010). Benzo[a]pyrene, a particularly potent carcinogen, generates benzo[a]pyrene diol epoxide adducts (Gerhards et al. 2023; Pommier et al. 2000). Other electrophilic compounds, including N-nitrosamines, 4-nitroquinoline 1-oxide, and estrogen metabolites, cause various DNA adducts and strand breaks (Jones et al. 1989). Aflatoxin B1, a mycotoxin, forms N7-guanine adducts that are highly mutagenic (Smela et al. 2002).

6.3 PCNA-Driven DNA Damage Tolerance/Bypass

Cells employ an extensive DNA damage response (DDR) network to counteract genomic insults. Because DNA damage arises from diverse sources and varies in severity, multiple repair mechanisms have evolved to restore integrity (Friedberg 2005). These mechanisms include direct reversal by photolyase (Sancar 1996), HR (Helleday et al. 2007), NHEJ (Weterings and Van Gent 2004), microhomology-mediated end joining (MMEJ) (Sfeir and Symington 2015; Sfeir et al. 2024), single-strand annealing (SSA) (Bhargava et al. 2016), nucleotide excision repair (NER) (Fousteri and Mullenders 2008; Xiang et al. 2025), BER (Sancar 1996; Parikh et al. 1998), and mismatch repair (MMR) (Hsieh and Yamane 2008; Longley et al. 1997; Jiricny 2006).

Unrepaired DNA damage entering S phase obstructs replication, causing fork stalling. Stalled forks are unstable and prone to breakage, generating double-strand breaks (DSBs), leading to chromosomal aberrations and genomic rearrangements that threaten cell viability (Negrini et al. 2010; Shen 2011; Pikor et al. 2013; Macheret and Halazonetis 2015; Maslowska et al. 2025). DNA damage tolerance (DTT) pathways specialize in the rescue of the stalled replication fork (see Fig. 6.2). There are two major DTT pathways that exist: the first is error-prone TLS, and the other is template switching (see Fig. 6.2). The TLS is characterized by the switching of classical DNA polymerases to low-fidelity TLS polymerases (Minocha et al. 2026). In contrast, template switching is proposed to use a recombination-like mechanism by which the nascent DNA of the sister chromatid is utilized as a temporary template for replication (Branzei 2011). Most eukaryotic TLS polymerases, Pol η (eta), Pol ι (iota), Pol κ (kappa), and Rev1 (deoxycytidyl transferase), belong to the so-called Y-family of DNA polymerases (Minocha et al. 2026; Hoitsma et al. 2020). These polymerases have a wider active center than replicative polymerases; thus, they can incorporate nucleotides across the damaged ones (Bienko et al. 2005; Lehmann et al. 2007; Sale et al. 2012). Additionally, B-family TLS polymerase Pol ζ (zeta) can extend a primer-template junction containing a mismatch at the 3′ (Northam et al. 2014). TLS polymerases often display reduced fidelity, misincorporation of nucleotides, or inaccurate lesion bypass, which can promote mutagenesis and carcinogenesis (Bienko et al. 2005; Lehmann et al. 2007; Sale et al. 2012).

PCNA is a member of the DNA sliding clamp (β-clamp) family, forming a homotrimeric ring that encircles DNA and slides. Each PCNA subunit consists of two globular domains connected by an inter-domain loop (IDCL), together forming a stable trimeric clamp (Dieckman et al. 2012). Replication stress triggers extensive modification of PCNA at lysine 164 (K164), a key regulatory site that accepts ubiquitin and multiple ubiquitin-like modifiers. Monoubiquitylation of PCNA at K164 is carried out through the canonical E1-E2-E3 cascade: the E1 enzyme ubiquitin-like modifier activating enzyme 1 (Uba1) activates ubiquitin in an ATP-dependent reaction via formation of a ubiquitin-adenylate intermediate, which then forms a high-energy Uba1~ubiquitin thioester (Fenteany et al. 2019; Groen and Gillingwater 2015). Ubiquitin is subsequently transferred to the active-site cysteine of the E2 conjugating enzyme Rad6, generating a Rad6~ubiquitin thioester. Finally, the E3 ubiquitin ligase Rad18, in complex with Rad6, directs the transfer of ubiquitin specifically to the K164 side chain of PCNA, forming a PCNA-ubiquitin isopeptide (see Fig. 6.2) (Moldovan et al. 2007; Fenteany et al. 2019, 2020, 2022; Davies et al. 2008; Kanao and Masutani 2017; Leung et al. 2019; Gaur 2020; Gaur et al. 2021; Gaur and Tyagi 2023). Crystal structures of monoubiquitylated PCNA illuminate how ubiquitylation regulates polymerase exchange (Freudenthal et al. 2011). In the structure, ubiquitin is positioned on the back face of PCNA and engages PCNA through its canonical hydrophobic patch, without altering the overall PCNA

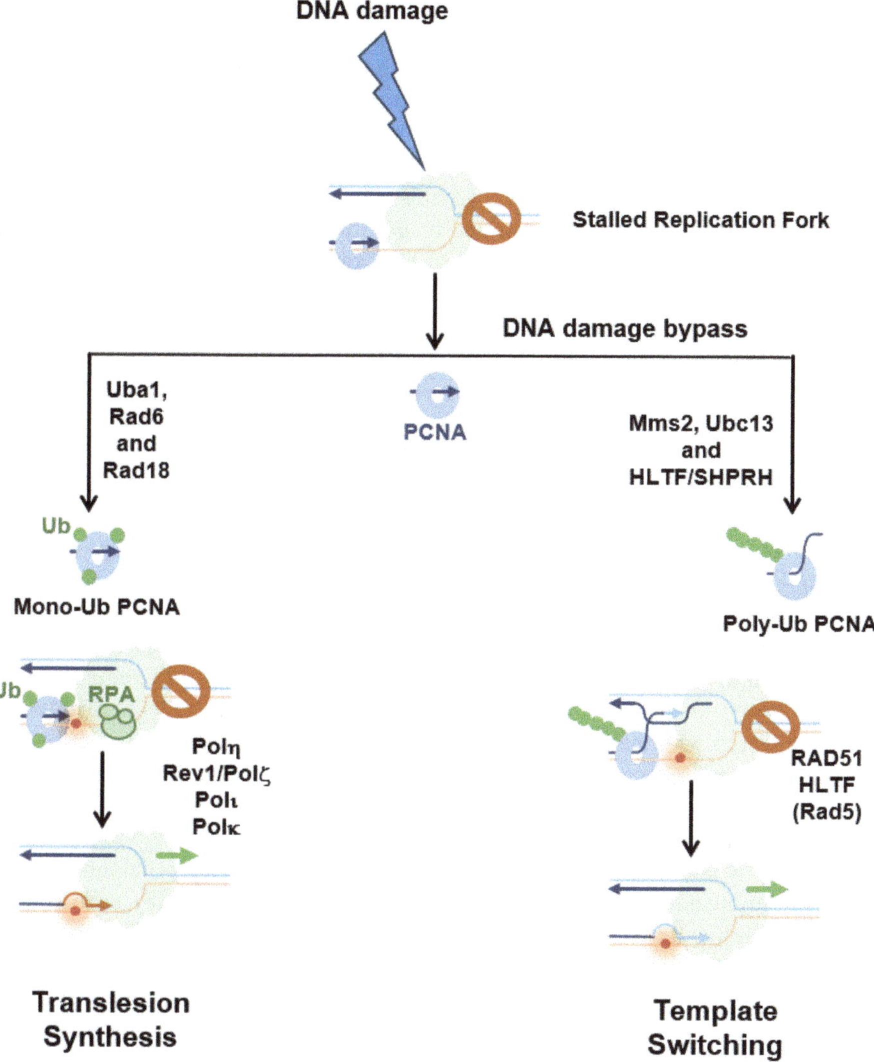

Fig. 6.2 Schematic illustrating the process of DNA damage bypass pathway regulated by PCNA ubiquitylation. The chart begins with DNA damage leading to a stalled replication fork. It shows two pathways for DNA damage bypass: monoubiquitylation of PCNA involving Uba1, Rad6, and Rad18, and polyubiquitylation involving Mms2, Ubc13, and HLTF/SHPRH. The monoubiquitylation pathway leads to translesion synthesis by recruiting Polη, Rev1/Polζ, Polι, and Polκ. The polyubiquitylation pathway leads to template switching, involving RAD51, HLTF, and Rad5. Arrows and symbols indicate the flow and interactions between components

conformation. This modification creates a new interaction surface that promotes recruitment of nonclassical polymerases, consistent with a "tool belt" model of polymerase exchange (Boehm et al. 2016b). Monoubiquitylation also reduces the binding affinity of canonical replication factors, shifting the balance toward TLS polymerases at stalled forks (Kanao and Masutani 2017; Leung et al. 2019; Gallo and Brown 2019; Bellí et al. 2022). These findings tie directly into a broader view of PCNA as a highly dynamic hub capable of supporting multiple simultaneous polymerase contacts, forming what can be described as a "Dyson sphere" of TLS options. Rev1 and Y-family polymerases, for example, can assemble with PCNA into large, flexible multi-protein complexes that include both PCNA tool belts and Rev1-mediated polymerase bridges. Single-molecule total internal reflection (smTIRF) microscopy experiments revealed that these assemblies are not static; instead, they interconvert between tool belt and bridge architectures without disassembly (Boehm et al. 2016b). This architectural plasticity provides a mechanism for rapid polymerase selection, allowing the replication machinery to sample multiple TLS polymerases and choose one best suited for the encountered lesion (Boehm et al. 2016b; Powers and Washington 2018). By hosting a repertoire of polymerases in tool belts and Rev1 bridges, PCNA enables dynamically controlled polymerase switching, ensuring that the cell maintains enough enzymatic "options" to navigate diverse DNA lesions efficiently (Moldovan et al. 2007; Boehm et al. 2016b).

Notably, TLS polymerases themselves (e.g., Pol η, Pol ι, and Rev1) are subject to monoubiquitylation (Sale et al. 2012). This additional layer of regulation has been proposed to support an ordered, sequential action of multiple TLS polymerases at a lesion. One model posits that monoubiquitylated PCNA can recruit TLS polymerases either because of their intrinsic ability to accommodate damaged bases or via specific ubiquitin-binding motifs within these enzymes (Minocha et al. 2026; Branzei 2011). An alternative scenario suggests that monoubiquitylation of a TLS polymerase would occur only when it fails to bypass the lesion efficiently. In that case, intramolecular interaction between its conjugated ubiquitin and its own ubiquitin-binding domain could signal its displacement from PCNA, thereby allowing the recruitment of another TLS polymerase. If none of the TLS polymerases can successfully bypass the lesion, prolonged stalling of the fork is thought to promote extension of the ubiquitin signal on PCNA, shifting the pathway toward damage tolerance (Moldovan et al. 2007; Bienko et al. 2005).

TLS is inherently mutagenic, as most TLS polymerases lack 3′-5′ exonucleolytic proofreading activity. While some TLS polymerases can accurately replicate specific lesions, many bypass events are error-prone and contribute to mutagenesis (Livneh et al. 2010; Sale 2013; Zhao and Washington 2017; Powers and Washington 2017). While template switching is generally considered an error-free bypass pathway, it can still lead to genomic instability (Branzei 2011). This process is associated with polyubiquitylation of PCNA at K164. In this pathway, an E2 enzyme complex composed of Mms2 and Ubc13, together with the E3 ligase Rad5 in yeast or its human orthologs HLTF and SHPRH, extends the monoubiquitin on PCNA into a K63-linked polyubiquitin chain (Lin et al. 2011; Unk et al. 2010; Gangavarapu et al. 2006; Parker and Ulrich 2009). These K63-linked chains differ from the

K48-linked polyubiquitin modifications that typically target substrates for proteasomal degradation and instead promote template switching. In template switching, the newly synthesized sister chromatid serves as an alternative, undamaged template, allowing the replication machinery to bypass the lesion without introducing mutations. Although the precise molecular steps of this pathway remain unclear, PCNA polyubiquitylation clearly marks a functional transition from mutagenic TLS to recombination-based lesion bypass (Moldovan et al. 2007).

PCNA ubiquitylation is dynamically reversible. The deubiquitylating enzyme ubiquitin-specific protease 1 (USP1) removes ubiquitin from PCNA (Mazloumi Aboukheili and Walden 2025), thereby modulating the duration and extent of TLS and template switching and preventing inappropriate or prolonged engagement of these pathways. How cells decide between monoubiquitylation and polyubiquitylation of PCNA at K164 is not yet clear, but this choice appears to be central to the orchestration of diverse PCNA-centered repair mechanisms.

In addition to ubiquitin, K164 of PCNA can be modified by several ubiquitin-like proteins (UBLs), including SUMO, NEDD8, and ISG15, each installed by distinct E1–E2–E3 enzyme cascades, with some overlap in components (Freudenthal et al. 2011; Bellí et al. 2022; Hoege et al. 2002; Stelter and Ulrich 2003; Haracska et al. 2004; Gali et al. 2012; Tsutakawa et al. 2015). These UBL modifications are functionally antagonistic to ubiquitylation in some contexts and add further complexity to PCNA regulation. SUMOylation of PCNA has been shown to modulate HR, limiting unscheduled recombination events at replication forks (Hoege et al. 2002; Stelter and Ulrich 2003). NEDDylation of PCNA restricts the recruitment of Pol η under conditions of oxidative stress, thereby fine-tuning TLS activity (Guan et al. 2018). ISGylation, in turn, facilitates the release of TLS polymerases from PCNA complexes, contributing to the timely termination of TLS (Park et al. 2014). Although the full range of functions of these UBL modifications remains to be elucidated, they collectively underscore the role of PCNA as a central molecular scaffold whose modification status governs the assembly, composition, and dynamics of DNA repair and tolerance complexes.

Importantly, ubiquitylation and SUMOylation both target the same lysine residue on PCNA (K164) (Bellí et al. 2022), yet a single PCNA trimer may still carry different modifications on individual subunits. This raises the possibility of "hybrid" clamps with composite functions. Such heterogeneously modified trimers could engage distinct interaction partners simultaneously, integrate multiple damage cues, and coordinate the stepwise deployment of TLS, template switching, and recombination pathways.

It is important to note that PCNA is highly abundant in the nucleus and participates extensively in DNA metabolism and the formation of dynamic multi-protein assemblies that support these processes. Zessin and colleagues demonstrated that PCNA exists in two distinct mobility populations in live mammalian cells (Zessin et al. 2016). The slow-diffusing population represents replication-engaged PCNA localized at replication foci and shows mild spatial confinement that likely reflects fork dynamics and local recycling within these foci. In contrast, the fast-diffusing population corresponds to freely mobile PCNA that is only partially constrained by the surrounding nuclear architecture, such as chromosome territories (Zessin et al. 2016). The ratio between these two populations remains constant throughout S-phase. It suggests that cells maintain a stable balance of active and inactive PCNA to support replication while minimizing replicative stress. Although the study does not provide direct evidence for PCNA participating in liquid-liquid phase separation, the behavior of the fast-diffusing pool raises interesting questions about whether PCNA might interact with or be influenced by phase-separated nuclear environments. At present, this remains speculative, but it represents a potentially intriguing area for future investigation.

6.4 PARP and Supramolecular Assemblies in DNA Repair

PARP1 is a founding member of the PARP family, which comprises 18 distinct members (Spiegel et al. 2021). One of the earliest cellular responses to DNA damage is the rapid synthesis of PAR chains by PARP family enzymes, primarily PARP1, at sites of DNA breaks (see Fig. 6.3) (Ame et al. 2004; Poltronieri and Miwa 2016; Ortega et al. 2025). The polymerization of ADP-ribose units derived from the ADP donor Nicotinamide adenine dinucleotide (NAD^+), resulting in the attachment of either linear or branched PAR polymers to itself or other target proteins (Chambon et al. 1963; Fujimura et al. 1967). These PAR polymers can reach ~200 units in length with branches every 2050 units, are highly negatively charged, and function as transient, multivalent docking platforms (D'Amours et al. 1999; Hassa and Hottiger 2008). They recruit a wide array of DNA damage response (DDR) and DNA metabolism proteins through non-covalent, electrostatic interactions with specialized PAR-binding motifs (PBMs), PAR-binding zinc fingers (PBZs), macrodomains, WWE domains, and others (Krietsch et al. 2013; Teloni and Altmeyer 2016). In many cases, these PAR-binding modules overlap with critical functional domains, enabling PAR to allosterically activate DNA binding, protein-protein interactions, nuclear localization, or other activities essential for repair (Gagne et al. 2008; Althaus et al. 1999; Malanga et al. 1998; Pleschke et al. 2000). To maintain genomic integrity, cells rely on excision repair mechanisms where PARP plays a central role, including in single-strand break repair (SSBR), BER, DSB repair, and NER (Ray Chaudhuri and Nussenzweig 2017).

6.4.1 A Central Function in Single-Strand Break Repair (SSBR)

PARP1 is a primary sensor of DNA SSBs (see Fig. 6.3b), which can arise spontaneously or from base modifications like oxidation and abasic site formation (Wei et al. 2024; Caldecott 2024; Gaur et al. 2025; Luedeman et al. 2022; Nosella et al. 2024;

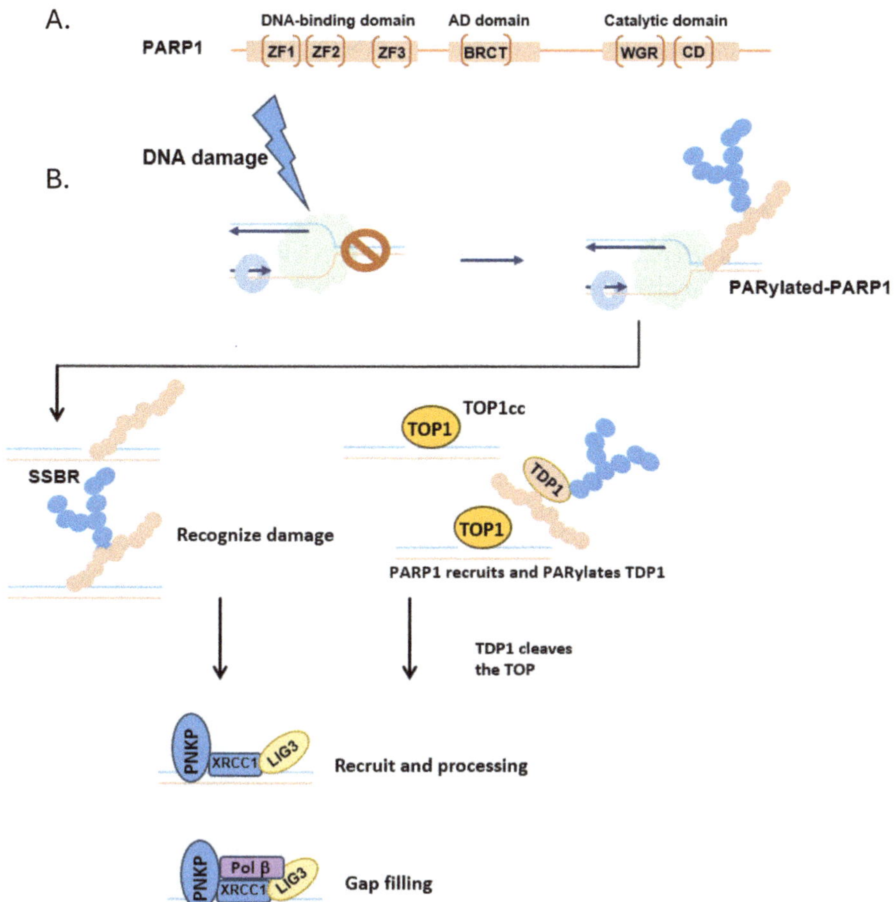

Fig. 6.3 Diagram showing (**a**) PARP1 structure and (**b**) the role of PARP1 in resolving TOP1 cleavage complexes (TOP1cc) and repairing single-strand breaks (SSBs)

Alemasova and Lavrik 2019). SSBs are typically resolved through a rapid global SSB repair pathway, although they can also be processed through HR-mediated mechanisms when converted to DSBs (Caldecott 2024; Davis and Maizels 2014). Historically, PARP1's central role in SSB detection and repair served as the conceptual foundation for developing PARPi (Ashworth and Lord 2018; Helleday 2011; Boussios et al. 2020).

Upon sensing an SSB, PARP1 rapidly binds the damaged site and catalyzes the addition of PAR chains onto itself and nearby proteins. This PARylation event activates PARP1 and provides a recruitment platform for downstream repair factors, enabling efficient initiation of the repair process.

A key component recruited by PARP1 is the scaffold protein X-ray repair cross-complementing protein 1 (XRCC1). XRCC1 assembles a repair complex of enzymes, including DNA ligase 3 (LIG3), DNA polymerase β, and polynucleotide kinase 3′-phosphatase (PNKP), to process and seal the break. The critical nature of this pathway is evident in human and mouse models, where XRCC1 mutations lead to defective SSBR and neurodegeneration (Hoch et al. 2017). This is caused by the futile hyperactivity of PARP1 at unrepaired SSBs, which depletes cellular NAD⁺ pools and triggers cell death. Beyond recruitment, PARP1 may also facilitate the gap-filling and final ligation steps of the repair process.

Facilitating the Repair of TOP1-Induced DNA Nicks PARP1 plays a critical role in repairing single-stranded DNA (ssDNA) nicks generated when TOP1 activity is interrupted (Pommier et al. 2022). Under normal conditions, TOP1 alleviates DNA supercoiling by introducing and re-ligating transient nicks (Pommier et al. 2016). When this process fails, TOP1 becomes covalently trapped on DNA, forming a TOP1 cleavage complex (TOP1cc) (see Fig. 6.3b) (Pommier et al. 2016). TOP1cc removal is mediated by tyrosyl-DNA phosphodiesterase 1 (TDP1), which hydrolyzes the phosphodiester bond linking the DNA 3′ end to the catalytic tyrosine of TOP1 (Zhang et al. 2022; Pouliot et al. 1999). PARP1 facilitates this process by PARylating TDP1, thereby stabilizing the enzyme and promoting its recruitment to the lesion. Following TOP1 removal, the exposed nick enters the canonical SSBR pathway, with XRCC1 recruitment again driven by PARP1 activity. This functional interplay between PARP1 and TDP1 explains the hypersensitivity of cells lacking either protein to TOP1 inhibitors.

Unlike its well-established role in SSBR, PARP1's contribution to base excision repair (BER), which addresses oxidative and alkylation-induced base damage, remains less defined. BER initiates by recognizing damaged bases, generating SSB intermediates, and employing factors shared with SSBR, including XRCC1. Some studies propose that PARP1 accelerates BER and senses SSB intermediates generated during repair (Dantzer et al. 2000; Demin et al. 2021). However, conflicting evidence indicates that PARP1 loss or inhibition does not consistently increase sensitivity to base-damaging agents (Dantzer et al. 1999; deMurcia et al. 1997; Pachkowski et al. 2009; Vodenicharov et al. 2000; Wang et al. 1997).

Promoting Chromatin Remodeling in Nucleotide Excision Repair PARP1 functions as a coordinator and scaffold in global genome NER (GG-NER) (see Fig. 6.4), leveraging its lesion-recognition capacity and dynamic chromatin interactions to facilitate recruitment and activity of multiple NER factors. PARP1 directly interacts with key NER components, including damage-specific DNA-binding protein 2 (DDB2), xeroderma pigmentosum group C (XPC), and Cockayne syndrome protein B (CSB), enhancing their recruitment, retention, and activity at damage sites. PARP1 interacts directly with DDB2 on chromatin after UV irradiation, independent of DNA damage binding protein 1 (DDB1) (Pines et al. 2012). Following UV irradiation, PARP1 binds DDB2 on chromatin independently of DDB1 (Lehmann et al. 2007). This interaction triggers localized PARylation, prolonging DDB2 retention and shielding it from ubiquitin-mediated degradation (Luijsterburg et al. 2012). In vitro approaches showed that PARP1 and DDB2 can simultaneously

Fig. 6.4 Schematic illustrating the process of global genome nucleotide excision repair (GG-NER). The sequence begins with histone PARylation, causing chromatin de-condensation, followed by ALC1 remodeling chromatin and recruiting XPC. Lesion verification is performed by XPB and XPD. Excision is carried out by ERCC1/XPF, with gap filling by Pol δ, Pol ε, and Pol κ, and ligation by LIG1 or LIG3

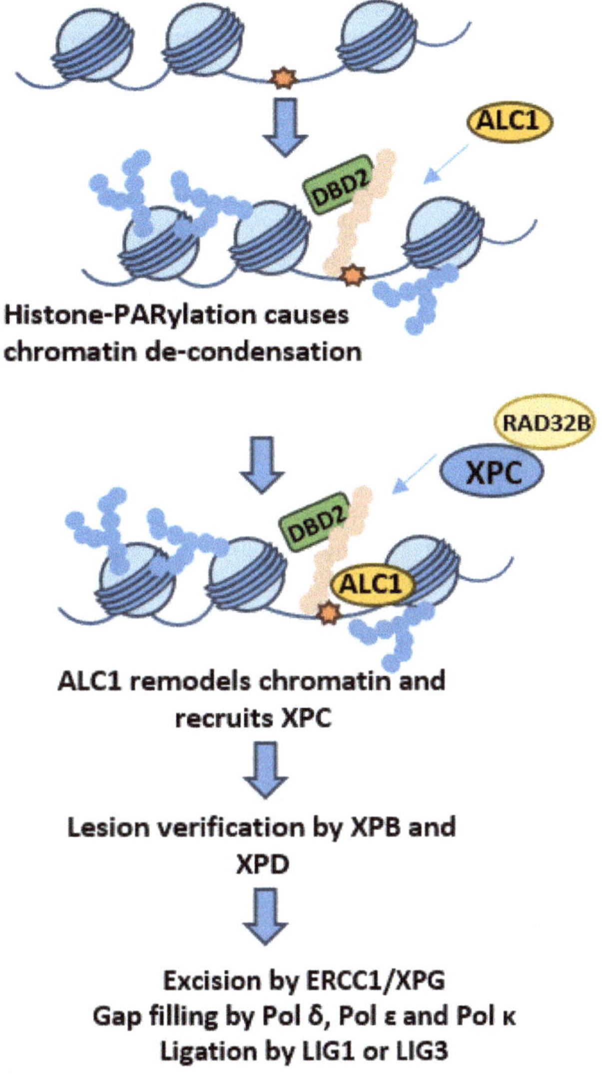

bind to a UV-induced cyclobutane pyrimidine dimer CPD (Purohit et al. 2016). This interaction also leads to the PARylation of histones, causing chromatin de-condensation and making the lesion more accessible (Ray Chaudhuri and Nussenzweig 2017; Pines et al. 2012). This PAR signal also recruits the chromatin-remodeling helicase amplified in liver cancer 1 (ALC1), which further opens up the DNA structure (Pines et al. 2012). This results in the recruitment of XPC and RAD23B (chromatin de-condensation), and lesion verification, which involves recruiting xeroderma pigmentosum group B and D (XPB and XPD) through the interaction of their binding partner xeroderma pigmentosum group A (XPA) with

PARP1. Following lesion verification, the bulky adduct is excised by the concerted action of two nucleases, excision repair cross-complementing group 1 protein (ERCC1) and xeroderma pigmentosum group B (XPG), and the resulting gap is filled by Pol δ, Pol ε, and Pol κ and ligated by ligase (LIG1 or LIG3) (Ray Chaudhuri and Nussenzweig 2017; Blessing et al. 2022).

PARP in Double-Strand Break Repair DSBs are among the most cytotoxic DNA lesions and are repaired through multiple pathways (mentioned earlier in the chapter). DSBs occur in two forms: double-ended DSBs (deDSBs), typically induced by exogenous agents, such as ionizing radiation or nucleases, and single-ended DSBs (seDSBs), which commonly arise during replication (Mehta and Haber 2014; Han and Huang 2020). deDSBs breaks can be repaired by HR, NHEJ, and, less frequently, MMEJ, also known as alternative end joining (Sfeir and Symington 2015; Sfeir et al. 2024; Truong et al. 2013); however, most mammalian cells favor NHEJ. While seDSBs are primarily repaired by the HR pathway (Mehta and Haber 2014; Lieber 2010) with MMEJ as a possible alternative (Wang et al. 2019). MMEJ has been shown to repair seDSBs that are mediated by Cas9^{D10A} or those arising from replication fork breakage (Wang et al. 2019). It has also implicated Polθ in post-replicative single-stranded DNA (ssDNA) gap filling, which is essential in HR-deficient (HRD) cells or following PARP inhibitor (PARPi) treatment (Belan et al. 2022; Mann et al. 2022; Schrempf et al. 2022).

6.4.2 *PARP-Driven Chromatin Remodeling at Double-Strand Breaks*

PARP1 responds early at DSBs, but it requires extensive chromatin remodeling (see Fig. 6.5a). PARP contributes to chromatin remodeling through histone PARylation, which is achieved by cofactor histone PARylation factor (HPF1) (Smith et al. 2023). In addition to PARP1 autoPARylation, together they redirect PARylation to the serine residue of histones (Leidecker et al. 2016). This serine ADP-ribosylation of histones promotes chromatin decompaction and recruitment of DDR proteins (Leidecker et al. 2016; Palazzo et al. 2018). Cells lacking HPF1 have reduced recruitment of XRCC4, APLF, and BRCA1 and impaired activity of both HR and NHEJ (Smith et al. 2023). PARP1 also transiently regulates the binding of linker histone H1 (Azad et al. 2018) and facilitates the eviction of nucleosomes around DSBs (Yang et al. 2020).

Importantly, chromatin relaxation at sites of DNA damage is initiated and tightly coordinated by PARP1 activity, which drives dynamic cycles of expansion and subsequent condensation through regulated PARylation to recruit repair factors and increase DNA accessibility (Burgess et al. 2014). While the subsequent chromatin compaction signal is executed via ATM, this phase is critically dependent on the prior PARP1-driven expansion. This compaction involves recruiting macroH2A1.2

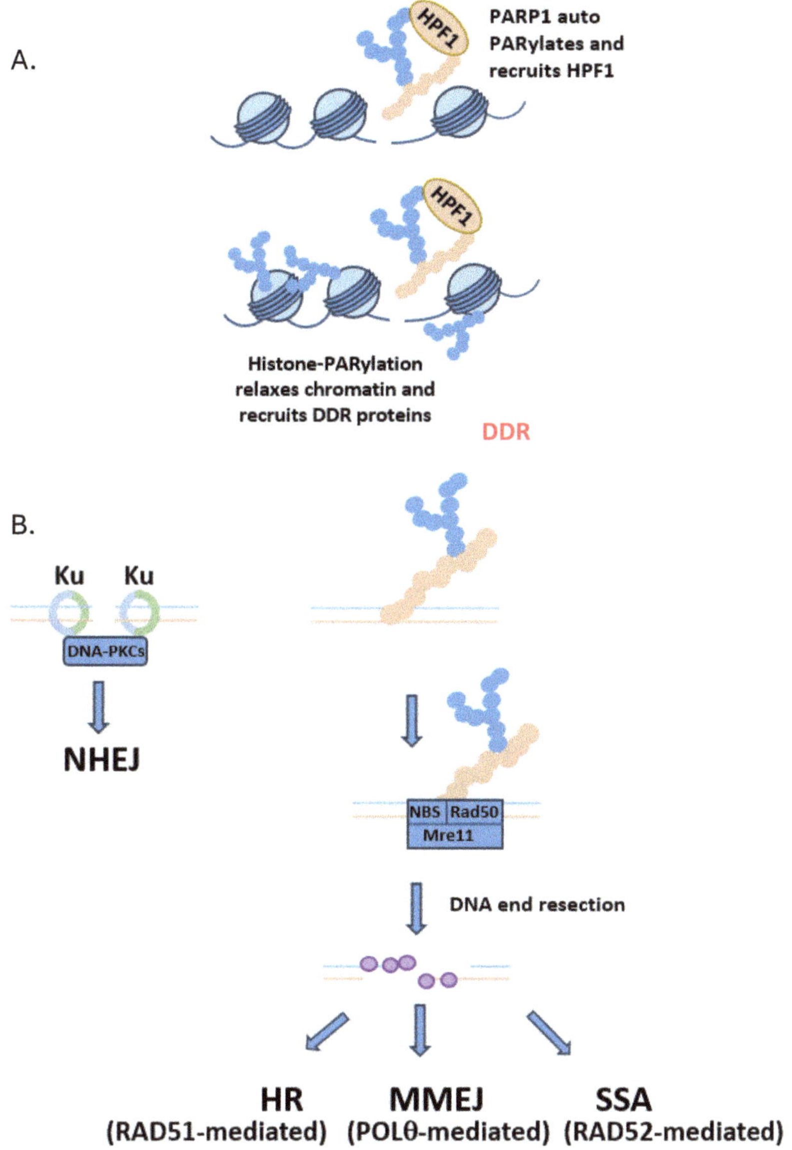

Fig. 6.5 Schematic showing (**a**) PARP-dependent chromatin remodeling at double-strand breaks and (**b**) how PARP regulates DNA end resection to guide repair through NHEJ, MMEJ, homologous recombination (HR), or single-strand annealing (SSA)

(which lacks the PAR-binding motif to support HR), while the distinct macroH2A1.1 isoform actively utilizes PAR binding to promote NHEJ/MMEJ pathways (Khurana et al. 2014; Giallongo et al. 2022; Sebastian et al. 2020). Moreover, the critical HR regulator KDM5A is also recruited via PAR binding, underscoring that PARP1 dictates downstream pathway decision-making regardless of whether the chromatin is expanding or compacting (Kumbhar et al. 2021).

6.4.3 PARP Regulation of End Resection

The decision between NHEJ, MMEJ, SSA, and HR at DSBs is governed by PARP1-mediated regulation of end resection, alongside cell cycle stage, chromatin state, and DNA sequence (see Fig. 6.5b) (Ghosal and Chen 2013; Helleday et al. 2007; Mehta and Haber 2014). Resection serves as the critical determinant: minimal resection favors NHEJ (driven by 53BP1, DNA-PK, and shieldin) (Weterings and Van Gent 2004; Xiang et al. 2025), while short-range resection enables MMEJ (via MRN-CtIP) (Truong et al. 2013), and extensive resection commits to homology-directed repair by HR (via EXO1 or DNA2-BLM/WRN) (Helleday et al. 2007) or SSA mediated by RAD52. Resection proceeds $5' \rightarrow 3'$, initiated by the MRN complex (Mre11-Rad50-NBS1) and CtIP, aided by EXO1, DNA2, and RecQ helicases such as BLM and WRN (Dantzer et al. 2000), to produce $3'$ OH ssDNA overhangs. Short-range resection exposes small microhomologies (on average ~3–20 bp) that flank the break site (Daley and Wilson 2005; Chang et al. 2017), a step shared between MMEJ and HR (Truong et al. 2013; Sfeir et al. 2024), enabling MMEJ pathway engagement.

PARP1 acts as a pro-resection factor that directly opposes NHEJ by competing with Ku70/Ku80 (Ku) and DNA-PKcs for DNA end binding in a cell-cycle-regulated manner (see Fig. 6.5b) (Ceccaldi and Cejka 2025; Mansour et al. 2010, 2013; Wang et al. 2006; Audebert et al. 2004; Yang et al. 2018). Beyond competition, PARP1 actively promotes short-range resection by recruiting MRE11, NBS1, RAD50, and BRCA1 to DSBs (Haince et al. 2008). During long-range resection, PARP1 PARylation exhibits dual regulatory effects: PAR chains recruit key HR proteins EXO1 and BRCA2 to initiate extensive processing via EXO1 (Zhang et al. 2015), yet PARylation of BRCA1 itself inhibits resection by blocking BRCA2 and EXO1 recruitment to DSBs (Lodovichi et al. 2023), likely preventing excessive processing. Supporting this regulatory balance, PARP inhibition increases RPA and RAD51 foci (Caron et al. 2019), while a BRCA1 PARylation mutant shows hyper-resection with elevated gene conversion and chromosomal rearrangements (Lodovichi et al. 2023; Hu et al. 2014). Thus, while PARylation initiates long-range resection, its primary function appears to be preventing excessive resection and suppressing recombinogenic HR. PARP2 also facilitates DSB resection but surprisingly does so without requiring its PARylation activity. Instead, it restricts 53BP1 from accumulating at break sites (Fouquin et al. 2017). PARP3's role in resection appears more complex and varies by cellular context. Research in mouse cells demonstrated that PARP3 enhances resection during chromosomal rearrangements (Layer et al. 2018). However, contrasting evidence indicates that PARP3 inhibits resection through two mechanisms: by adding mono-ADP-ribose (MAR) groups to the Ku70/Ku80 complex and by facilitating Ku recruitment to the laser-damaged site (Beck et al. 2014).

In NHEJ, this involves limited resection of 5′ or 3′ overhangs by exonuclease or endonuclease activity, exposing or generating short microhomologies (≤ 4 nucleotides) that align the ends for ligation (Vogt and He 2023). NHEJ initiates when the Ku70/80 heterodimer binds DSBs with high affinity for duplex DNA ends, followed by the recruitment of DNA-PKcs to form the DNA-PK complex (Vogt and He 2023; Blier et al. 1993). Activation of this complex requires a transition from an end-protecting state to a catalytically active form through major conformational changes in DNA-PKcs, with kinase activity dependent on DNA binding and enhanced by Ku-DNA interaction (Vogt and He 2023; Carter et al. 1990). These rearrangements appear dynamic throughout the repair. The DNA-PK complex mediates end-bridging with other core factors, including LigIV, XRCC4, XLF, and PAXX (Vogt and He 2023).

6.4.4 Downstream Homologous Recombination Pathway and PARP Activity

HR pathway for DSB repair requires the generation of long 3′ single-stranded DNA (ssDNA) tails through end resection, a step shared with MMEJ (Truong et al. 2013; Sfeir et al. 2024). Resection proceeds 5′ → 3′ to produce 3′-OH ssDNA overhangs. Briefly, resection is initiated by the MRN complex (Mre11-Rad50-NBS1) and CtIP, aided by EXO1, DNA2, and RecQ helicases such as BLM and WRN (Dantzer et al. 2000).

The resulting ssDNA is coated by RPA, which protects it but competes with binding by the RAD51 recombinase. This inhibition is removed by mediator proteins such as BRCA2, which loads RAD51 and promotes the formation of the RAD51 nucleoprotein filament active in the central steps of HR. BRCA2 is assisted by RAD51 paralogs (RAD51B, RAD51C, RAD51D, XRCC2, XRCC3) (Wright et al. 2018; Ranjha et al. 2018; Wiese et al. 2007). If appropriate homology is not found, resected DSBs may be repaired by SSA mediated by RAD52 or by Pol-theta-mediated end joining (Bhargava et al. 2016; Bhat et al. 2022, 2023; Kelso et al. 2019). Additional factors, including RAD54, PALB2, RAD51AP1, and HOP2-MND1, stabilize the synaptic complex and promote D-loop formation (Dray et al. 2010). DNA synthesis at the D-loop requires PCNA, RFC, and replicative polymerases, while helicases like Srs2, Mph1, and BLM, along with Top3-Rmi1, dismantle extended D-loops to favor synthesis-dependent strand annealing (SDSA) and avoid crossovers (Liu et al. 2017; Heyer et al. 2010). Alternatively, double Holliday junctions may form and are resolved by structure-selective nucleases (MUS81-EME1, SLX1-SLX4, GEN1) or dissolved by the BLM-TopIIIα-RMI1 complex to maintain genome stability (Wright et al. 2018; Ranjha et al. 2018; Yang et al. 2010).

6.4.5   PARP1 Role in Single-Ended DSBs

seDSBs, likely the predominant endogenous break type (Han and Huang 2020; Ranjha et al. 2018), are primarily repaired by HR with MMEJ as backup (Sfeir et al. 2024), while NHEJ is suppressed (Chanut et al. 2016; Vriend et al. 2016; Britton et al. 2020). HR repair of seDSBs differs mechanistically from double-ended DSBs (blunt ends); for instance, BRCA1's resection function is dispensable at seDSBs (Pavani et al. 2024). PARP's role in seDSB repair remains poorly defined, partly because seDSBs can arise from unrepaired SSBs or fork breakage (Caldecott 2024), making it difficult to separate PARP's HR function from its SSB repair role. XRCC1 deletion increases HR repair (Vriend et al. 2016; Fan et al. 2007), and since PARP1 functions in SSB repair and at forks, its inhibition likely affects HR by increasing seDSB prevalence (Caldecott 2024; Zeman and Cimprich 2014; Bryant et al. 2009). HR reporter assays with nickase-induced SSBs show no effect or increased HR upon PARP inhibition, suggesting PARP1 isn't essential for nick-induced HR (Vriend et al. 2016; Metzger et al. 2013). However, PARP inhibition after hydroxyurea-induced replication stress reduced Mre11 and RPA foci and decreased HR-mediated restarting of stalled forks (Bryant et al. 2009).

6.4.6   PARP in Microhomology-Mediated End Joining

MMEJ, also known as alternative end-joining, is characterized by annealing of short microhomologies near a DNA break, followed by fill-in DNA synthesis to complete repair (Sfeir et al. 2024). These small microhomologies (on average ~3–20 bp) that flank the break site (Daley and Wilson 2005; Chang et al. 2017) and that are exposed by short-range resection of the DSB, a step that is shared with HR (Truong et al. 2013). The central enzyme in this pathway is DNA polymerase theta (Polθ), encoded by POLQ (Sfeir et al. 2024; Wyatt et al. 2016; Li et al. 2025; Ito et al. 2025). Originally described as a backup to NHEJ (Han and Huang 2020; Sfeir et al. 2024; Wang et al. 2006), MMEJ is now recognized as a distinct repair pathway that functions in both NHEJ-proficient and NHEJ-deficient cells (Han and Huang 2020; Bennardo et al. 2008). HR and MMEJ can be synthetically lethal, likely due to their roles at seDSBs, where NHEJ is suppressed, and MMEJ becomes essential when HR is lost. Supporting evidence shows MMEJ repairs seDSBs from Cas9D10A cleavage or replication collapse. Recent studies also implicate Polθ in filling post-replicative ssDNA gaps, critical in HR-deficient cells and after PARP inhibitor treatment.

PARP1, together with XRCC1 and ligase 3, was among the first set of proteins identified as essential for MMEJ in mammalian cells (Wang et al. 2006; Audebert et al. 2004; Robert et al. 2009; Jia et al. 2013; Sharma et al. 2015; Liang et al. 2008; Sfeir and de Lange 2012). MMEJ is intrinsically error-prone, as microhomology annealing results in deletions, while Polθ is an inaccurate polymerase that frequently introduces mutations and can produce templated insertions (Schimmel et al. 2017). Early work using HeLa nuclear extracts showed that PARP1 can mediate DNA synapsis at overhangs independently of Ku70/80 and DNA-PKcs (Audebert et al. 2004). Ku competes with PARP1 for DNA binding, and PARP inhibition compromises NHEJ-independent repair in plasmid end-joining assays (Wang et al. 2006; Sfeir and de Lange 2012; Fleury et al. 2023). PARP1-dependent plasmid end-joining is observed in the absence of Lig4 or Ku80, but not DNA-PKcs, indicating that PARP1's contribution is most evident when specific NHEJ components are disrupted.

PARP1 also enhances Polθ accumulation at DNA damage sites (Luedeman et al. 2022; Mateos-Gomez et al. 2015; Kais et al. 2016). Polθ itself is subject to PARylation, which inhibits its DNA-binding and polymerase activity. Following PARP1-dependent recruitment of Polθ, its activation is thought to require de-PARylation by PARG (Vekariya et al. 2024; Hottiger 2015). Consistent with this model, inhibition of either PARP or PARG diminishes MMEJ activity (Luedeman et al. 2022; Vekariya et al. 2024). Although PARP3's role in MMEJ has not been directly tested, indirect evidence suggests it may function as a negative regulator. PARP3 has been shown to limit end resection, a step essential for MMEJ, and PARP3-depleted cells display larger deletions at I-SceI-induced breaks, a characteristic feature of MMEJ-mediated repair (Beck et al. 2014).

6.5 Phase-Separation and Condensate Dynamics

Beyond its catalytic functions, *PARP1* plays a crucial structural role through biomolecular condensate formation, representing a fundamental organizational principle in DNA damage response. PAR, a nucleic acid-like polymer, drives LLPS to form membraneless compartments that selectively enrich repair factors (Chin Sang et al. 2024; Chappidi et al. 2024). PAR condensate formation varies by cellular context and PARP family member. Under proteasomal stress, PAR undergoes PARP2-dependent (not PARP1) LLPS, co-condensing with proteasome and K6-linked ubiquitin chains in the nucleus. PAR directly interacts with ubiquitin chains, revealing crosstalk between ADP-ribosylation and ubiquitin signaling. These condensates co-localize with stalled replication forks, attenuating replication stress and stabilizing forks to sustain genomic integrity (Chappidi et al. 2024; Wei et al. 2024; Caron et al. 2019). As a multivalent modification, PAR triggers phase separation through high-valency interaction networks.

DSB sites assemble through co-condensation of PARP1 multimers with DNA, creating structures that exert mechanical forces to tether DNA ends while becoming enzymatically active for PAR synthesis (Chappidi et al. 2024; Wei et al. 2024; Caron et al. 2019). PARylation triggers a hierarchical cascade: PARP1 release from DNA ends enables effector recruitment, particularly FET family proteins (FUS, EWS, TAF15) containing positively charged RG/RGG-rich C-termini that bind anionic PAR (Altmeyer et al. 2015; Lee et al. 2020; Mastrocola et al. 2013). FUS is a highly abundant protein with nuclear concentrations of 4–8 µM (Patel et al. 2015) and undergoes PAR-dependent phase separation to nucleate liquid-like condensates that organize DNA repair and γH2AX nanofoci clustering (Levone et al. 2021; Singatulina et al. 2019).

EWS, TAF15, and hnRNP family members similarly co-assemble into PAR-seeded compartments (Altmeyer et al. 2015; Lee et al. 2020; Mastrocola et al. 2013; Matveeva et al. 2016; Klaric et al. 2021). SAFB, a heterochromatin-associated RNA-binding protein, recruits to PAR-rich damage sites via its R/G-rich domain, promoting γH2AX spreading and signaling (Altmeyer et al. 2013). However, PAR may compete with satellite RNAs for SAFB binding, raising questions about heterochromatin reorganization during repair (Huo et al. 2020).

RPA is an eukaryotic ssDNA-binding protein with critical roles in DNA replication, recombination, and repair (Dueva and Iliakis 2020). It is a heterotrimer consisting of three subunits of 70, 32, and 14 kDa (RPA1, 2, 3) (Wold 1997; Caldwell and Spies 2020; Chen and Wold 2014; Fan and Pavletich 2012). Four DNA-binding OB-folds enable dynamic, high-affinity interaction between RPA and ssDNA, as well as RPA's capacity to melt secondary DNA structures, including G4s (Caldwell and Spies 2020; Ray et al. 2013; Granger et al. 2024; Olson et al. 2023). RPA binds ssDNA independent of sequence but has a slight preference for G-rich sequences. A recent study by Spegg et al. demonstrates that RPA readily forms dynamic condensates, with sub-stoichiometric amounts of ssDNA acting as a potent trigger for phase separation, in contrast to RNA or double-stranded DNA, which have no such effect (Spegg et al. 2023). The RPA2 subunit is essential for condensation, and multi-site phosphorylation within its N-terminal intrinsically disordered region fine-tunes RPA self-interaction and condensate behavior. Using quantitative proximity proteomics, the study further links RPA condensation to telomere clustering and the maintenance of telomere integrity in cancer cells. Together, these findings suggest a model in which RPA-coated ssDNA is sequestered into highly dynamic condensates that influence nuclear architecture and promote genome stability (Spegg et al. 2023).

Human SSA protein *RAD52* (Bhat et al. 2023; Honda et al. 2025), comprising an undecameric ring enveloped by eleven ~200-residue disordered regions, undergoes homotypic phase separation at submicromolar concentrations (250 nM with molecular crowding) (Alshareedah et al. 2026). These RAD52 condensates function as DNA repair hubs, selectively recruiting key homologous recombination components such as ssDNA, TERRA RNA, RPA, and RAD51 (Alshareedah et al. 2026).

6.6 Non-Canonical DNA Structures Interfere with Replication

In addition to DNA damage, non-canonical DNA structures can stall the fork and cause replication stress (Hisey et al. 2024a; Bansal et al. 2022; Castillo Bosch et al. 2014). The human genome contains diverse classes of repetitive DNA

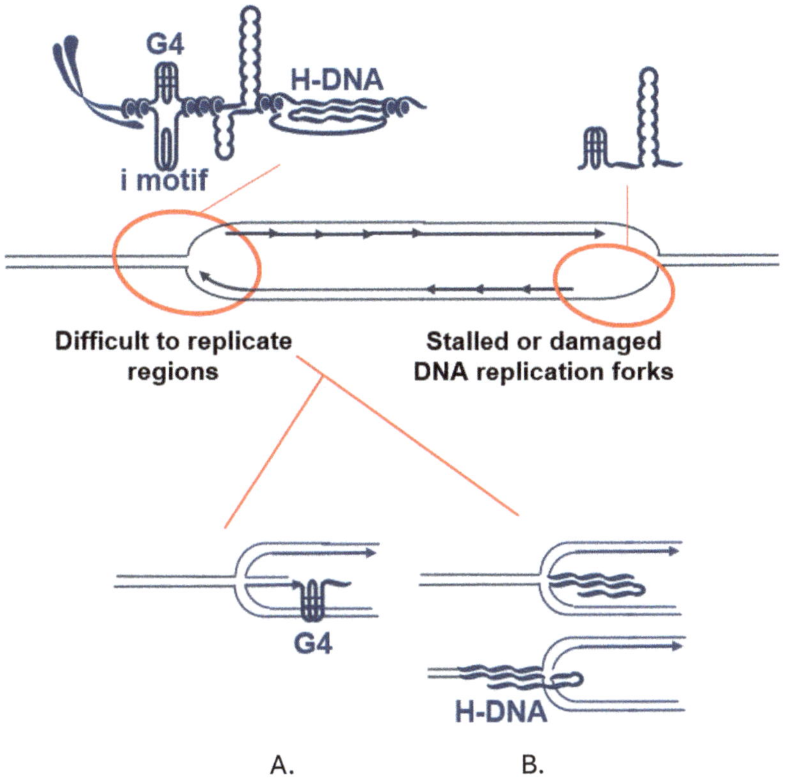

Fig. 6.6 Schematic showing DNA replication across non-canonical DNA structures, including (**a**) G-quadruplexes (G4) and (**b**) H-DNA (triplex DNA)

sequences. These include guanine-rich regions that can fold into G4 structures (see Fig. 6.6). G4s are stabilized by Hoogsteen hydrogen bonding and monovalent cations and can adopt parallel, antiparallel, or hybrid topologies; these structures are associated with cancers (Rhodes and Lipps 2015; Hänsel-Hertsch et al. 2017). The genome also harbors trinucleotide repeat (TNR) sequences, including (CGG) n and (CAG)n expansions. These sequences are prone to strand slippage during replication, forming TNR-induced hairpin structures that have been implicated in neurological disorders such as Huntington's disease and fragile X syndrome (Cohen et al. 2025). Additionally, purine-pyrimidine mirror repeats, exemplified by (GAA)n/(TTC)n sequences, can slip out of the canonical double helix and adopt alternative non-B DNA conformations, including H-DNA (see Fig. 6.6) (Hisey et al. 2024a). Such structural transitions have been associated with conditions ranging from Burkitt lymphoma and polycystic kidney disease to various repeat expansion diseases (Hisey et al. 2024a; Cohen et al. 2025). Despite growing recognition of non-canonical DNA structures as contributors to replication stress, the molecular mechanisms by which cells coordinate their resolution remain poorly defined.

6.6.1 G-Quadruplex DNA: Formation, Consequence, and Resolution

G4s form naturally during DNA replication on both the leading and lagging strand templates. The human genome contains over 10,000 sequences with G4-forming potential, and these structures arise transiently when DNA becomes single-stranded during replication fork progression. On the lagging strand, G4s form in the template during Okazaki fragment synthesis, while on the leading strand, they can form in the displaced parental strand or when the template becomes temporarily exposed due to uncoupling between the replicative helicase and polymerase. During replication, CMG helicase is the central eukaryotic replicative helicase that moves $3'$ to $5'$ along the leading-strand template, unwinding parental DNA while coordinating polymerase engagement at the fork. It can stall when it encounters G4 DNA structures. A study by Batra et al. shows that when CMG encounters a preformed G4 on the leading-strand template, it pulls the folded structure into its N-terminal entry chamber but cannot thread it through the narrow C-terminal motor pore formed by the PS1 and H2I translocation loops (Batra et al. 2025). This creates a precise steric block inside the helicase, arresting forward movement even though the upstream fork architecture and downstream DNA contacts remain intact. Thus, G4s do not alter the helicase mechanism but instead freeze CMG mid-stride, providing direct structural evidence for how G4s on the leading strand impede replication fork progression and contribute to replication stress.

To overcome these stalled forks, cells employ specialized helicases and polymerases that collaborate to resolve G4 structures. DNA replication through noncanonical structures: G4s can be resolved during replication through the collaboration of specialized polymerases and G4-specific helicases, such as FANCJ, BLM, WRN, PIF1, and RHAU (Tippana et al. 2016; Duan et al. 2015; Wu et al. 2015, 2017). FANCJ (also known as BACH1 or BRIP1) is a superfamily-2 helicase implicated in Fanconi anemia (FA) and classified as a tumor suppressor, with mutations linked to breast and ovarian cancers (Cantor and Calvo 2017). FANCJ is important for replication progression, as it protects cells from replication stress (Cantor and Calvo 2017). FANCJ interacts with G4s, unfolding and refolding them to facilitate replication (Wu and Spies 2016). It co-localizes with RPA in cells exposed to replication stress or DNA damage, supporting its role in maintaining genome stability (Gupta et al. 2007). Incomplete processing of G4s can lead to replication stress and double-stranded breaks (Lee et al. 2021; Odermatt et al. 2020; Brosh and Wu 2021; Wu et al. 2008). FANCJ's ability to resolve stable G4s enables RPA to bind ssDNA, promoting smooth replication past these structures (Lee et al. 2021; Odermatt et al. 2020; Brosh and Wu 2021; Wu et al. 2008). Other helicases, such as BLM, WRN, RTEL1, PIF1, and ChlR1/DDX11 (Fry and Loeb 1999; Sanders 2010; Sun et al. 1998; Vannier et al. 2013; Wu et al. 2012), are capable of unwinding G4s, but they cannot compensate for FANCJ deficiency under G4-induced replication stress (Castillo Bosch et al. 2014; Wu et al. 2008; Schwab et al. 2013), leading to G4 accumulation and genome destabilizing consequences (Barthelemy et al. 2016;

Matsuzaki et al. 2015). RPA is the main eukaryotic ssDNA-binding protein with critical roles in DNA replication, recombination, and repair (Wold 1997; Caldwell and Spies 2020; Chen and Wold 2014; Fan and Pavletich 2012). Four DNA-binding OB-folds enable dynamic, high-affinity interaction between RPA and ssDNA, as well as RPA's capacity to melt secondary DNA structures, including G4s (Caldwell and Spies 2020; Ray et al. 2013; Olson et al. 2023; Granger et al. 2024; Prakash et al. 2011a, b; Popuri et al. 2008; Salas et al. 2006). Rev1 is a translesion DNA synthesis polymerase that specifically binds to and unfolds G4s, preventing their refolding (Lin et al. 1999; Nair et al. 2005). Among FANCJ's key interactors at G4s are Rev1 and RPA (Gupta et al. 2007; Sarkies et al. 2012). FANCJ binds Rev1 via its Rev1-interaction region (RIR) motif, enabling Rev1's exclusive incorporation of cytosines during synthesis (Boehm and Washington 2016). The loss of Rev1 increased mutation frequency, which was further exacerbated by the G4 stabilizer pyridostatin (PDS). Complementation with WT partially suppressed this effect, whereas G4-binding-deficient mutants did not underscore Rev1's critical role in G4 resolution during DNA replication (Schiavone et al. 2014). The final protein player is PARP1 (Soldatenkov et al. 2008; Edwards et al. 2021). PARP1 performs poly(ADP-ribosyl)ation (PARylation) by synthesizing negatively charged poly (ADP-ribose) (PAR) polymers using NAD^+ as a substrate (Hassa and Hottiger 2008). PAR chains synthesized by PARP1 can form long structures that branch several times and regulate DNA repair (Munoz-Gamez et al. 2009; Pascal 2018). PARP1 binds specifically to folded G4s (Soldatenkov et al. 2008). G4 binding by PARP1 triggers its catalytic activity and leads to PAR synthesis. PARP1 binds folded G4s with high affinity and is catalytically activated upon G4 recognition (Soldatenkov et al. 2008).

6.6.2 *G-Quadruplex-Associated Phase Separation*

Many proteins that undergo LLPS have been shown to interact with G4s, and they shape nuclear and cytoplasmic organization (Pavlova et al. 2023; Zhang et al. 2019; Robinson et al. 2021). In the nucleus, G4s associate with phase-separated compartments that regulate transcription, chromatin architecture, and ribonucleoprotein assembly (Pavlova et al. 2023; Zhang et al. 2019). Proteins involved in infection sensing and innate immune signaling include several DEAD box helicases, such as DDX1, DDX2, DDX5, Dbp1, and Dbp2. These proteins shuttle between the nucleus and cytoplasm (Taschuk and Cherry 2020; Ali 2021); in the nucleus, they regulate transcription, and in the cytoplasm, they sense viral nucleic acids and promote antiviral responses, in part by driving stress granule assembly through their IDRs (Taschuk and Cherry 2020; Ali 2021).

Nucleophosmin (NPM1) supports ribosome biogenesis and recognizes G4s through its aromatic C-terminal domain and a neighboring IDR (Mitrea et al. 2018). This region interacts with arginine-rich motifs of nucleolar proteins and likely contributes to nucleolar assembly around ribosomal RNA (Gallo et al. 2012). SERBP1 functions in transcription, chromatin remodeling, and the assembly of promyelo-cytic leukemia bodies in the nucleus (Liu et al. 2021; Martini et al. 2021). It contains arginine-glycine/arginine-glycine-glycine (RG/RGG) repeat domains that recognize G4s and promote phase separation (Baudin et al. 2021).

The FUS protein also binds G4s through its RGG domains and is enriched at endogenous DNA G4 sites. These domains support G4 recognition, nonspecific nucleic acid binding through electrostatic interactions, and FUS self-association (Ji et al. 2022). RGG domains, together with aromatic-rich N-terminal regions, play a central role in FUS phase separation (Ji et al. 2022; Murthy et al. 2021; Murray et al. 2017). Additional contributors include nucleolin (NCL) (Dempsey et al. 1999), heterogeneous ribonucleoprotein A1 (hnRNP A1) (Ghosh and Singh 2018; Paramasivam et al. 2009), and H1-driven phase separation (Mimura et al. 2021).

6.6.3 *H-DNA: Formation, Consequences, and Resolution*

H-DNA is an intramolecular triplex structure formed by homopurine/homopyrimidine mirror repeats that fold back and bind via Hoogsteen or reverse Hoogsteen hydrogen bonds (Hisey et al. 2024a). Like G4s, these structures arise transiently during replication, transcription, or repair when DNA becomes unwound or negatively supercoiled. During replication, H-DNA acts as a strong barrier to DNA polymerases, causing fork stalling and sometimes fork reversal, particularly when the purine-rich strand serves as the lagging strand template. This stalling can lead to double-strand breaks (DSBs) (Patel et al. 2004), checkpoint activation, and mutagenesis (Rossetti et al. 2001; Liu et al. 2012; Hisey et al. 2024b). Multiple protein players interact with H-DNA: replication proteins such as DNA polymerases α and δ exhibit pausing at H-motifs; helicases like FANCJ, DDX3X, ChlR1 (DDX11), and RPA help unwind triplexes (Barthelemy et al. 2016; Vasquez et al. 2002); and repair proteins, including nucleotide excision repair factors XPA, XPF, and XPG, as well as flap endonuclease FEN1, bind and cleave H-DNA either to resolve the structure or inadvertently generate DSBs (Vasquez et al. 2002; Zhao et al. 2018; Xu et al. 2024; Guo et al. 2015). Mismatch repair proteins MutSα, MutSβ, and MutLα also contribute to instability at H-motifs (Neil et al. 2021; Kim et al. 2008).

These interactions underline the pathogenic roles of H-DNA in diseases such as autosomal dominant polycystic kidney disease (ADPKD), where a large polypyrimidine repeat in intron 21 of PKD1 stalls replication (Hisey et al. 2024a; Cordido et al. 2017; Schroth and Ho 1995). Friedreich's ataxia (FRDA) is caused by expanded GAA repeats in intron 1 of the *FXN* gene, forming H-DNA under physiological conditions and causing stalling of replication forks (Campuzano et al. 1996). Cerebellar ataxia, neuropathy, and vestibular areflexia syndrome (CANVAS), caused by AAGGG repeat expansions in *RFC1*, shows in vitro evidence of H-DNA formation and replication stalling (Hisey et al. 2024b). Cancers, including follicular lymphoma, Burkitt lymphoma, and diffuse large B-cell lymphoma, where H-DNA motifs at BCL2, c-MYC, and BCL6 loci promote fragility and translocation. Collectively, H-DNA's ability to impede replication and recruit repair proteins makes it a potent driver of genome instability and disease (Hisey et al. 2024a).

6.7 Summary

Genome maintenance depends on coordinated assemblies that integrate enzymatic activity, signaling, and structural organization to protect replication accuracy and repair DNA damage. This chapter outlined the shift from classical complexes such as the replisome and the PCNA clamp to dynamic condensates formed through liquid-liquid phase separation. These structures provide spatial and temporal control over repair pathways and support rapid responses to replication stress and DNA lesions.

The discussion covered the core principles of DNA replication, major sources of endogenous and exogenous damage, and the mechanisms that drive damage tolerance. PCNA functions as a central hub whose modifications, including ubiquitylation, SUMOylation, NEDDylation, and ISGylation, control the choice between translesion synthesis and template switching. PARP1 acts as a sensor and scaffold that promotes chromatin remodeling, factor recruitment, and phase separation at damage sites. These assemblies support genome stability and influence therapeutic responses seen with PARPi and with new strategies that target PCNA ubiquitylation.

Non-canonical DNA structures, such as G4 and H-DNA, add complexity by creating replication barriers that need specialized helicases and accessory proteins. Failure to resolve these structures drives genomic instability and disease, which supports their value as drug targets.

Supramolecular assemblies represent a major shift in how genome maintenance is understood. Their dynamic behavior, regulatory range, and clinical impact make them strong targets for next-generation cancer therapies.

Acknowledgments We acknowledge the support by NIH/NIGMS R35GM131704, NIH/NCI R01CA232425, and NIH/NCI P30 CA086862 awards.

Competing Interests The authors declare no competing financial interests.

References

Abou Assi H, Garavís M, González C, Damha MJ. i-Motif DNA: structural features and significance to cell biology. Nucleic Acids Res. 2018;46:8038–56.
Abuelo JG, Moore DE. The human chromosome. Electron microscopic observations on chromatin fiber organization. J Cell Biol. 1969;41:73–90.

Alemasova EE, Lavrik OI. Poly(ADP-ribosyl)ation by PARP1: reaction mechanism and regulatory proteins. Nucleic Acids Res. 2019;47:3811–27.

Ali MAM. DEAD-box RNA helicases: the driving forces behind RNA metabolism at the crossroad of viral replication and antiviral innate immunity. Virus Res. 2021;296:198352.

Alshareedah I, Pangeni S, Dewan PA Jr, Honda M, Liao TW, Spies M, Ha T. The human RAD52 complex undergoes phase separation and facilitates bundling and end-to-end tethering of RAD51 presynaptic filaments. Nucleic Acids Res. 2026;54:gkag043.

Althaus FR, Kleczkowska HE, Malanga M, Muntener CR, Pleschke JM, Ebner M, Auer B. Poly ADP-ribosylation: a DNA break signal mechanism. Mol Cell Biochem. 1999;193:5–11.

Altmeyer M, Toledo L, Gudjonsson T, Grofte M, Rask MB, Lukas C, Akimov V, Blagoev B, Bartek J, Lukas J. The chromatin scaffold protein SAFB1 renders chromatin permissive for DNA damage signaling. Mol Cell. 2013;52:206–20.

Altmeyer M, Neelsen KJ, Teloni F, Pozdnyakova I, Pellegrino S, Grofte M, Rask MBD, Streicher W, Jungmichel S, Nielsen ML, et al. Liquid demixing of intrinsically disordered proteins is seeded by poly(ADP-ribose). Nat Commun. 2015;6:8088.

Ame JC, Spenlehauer C, de Murcia G. The PARP superfamily. Bioessays. 2004;26:882–93.

Andersen S, Heine T, Sneve R, König I, Krokan HE, Epe B, Nilsen H. Incorporation of dUMP into DNA is a major source of spontaneous DNA damage, while excision of uracil is not required for cytotoxicity of fluoropyrimidines in mouse embryonic fibroblasts. Carcinogenesis. 2005;26:547–55.

Ashworth A, Lord CJ. Synthetic lethal therapies for cancer: what's next after PARP inhibitors? Nat Rev Clin Oncol. 2018;15:564–76.

Audebert M, Salles B, Calsou P. Involvement of poly(ADP-ribose) polymerase-1 and XRCC1/ DNA ligase III in an alternative route for DNA double-strand breaks rejoining. J Biol Chem. 2004;279:55117–26.

Azad GK, Ito K, Sailaja BS, Biran A, Nissim-Rafinia M, Yamada Y, Brown DT, Takizawa T, Meshorer E. PARP1-dependent eviction of the linker histone H1 mediates immediate early gene expression during neuronal activation. J Cell Biol. 2018;217:473–81.

Balasubramaniam S, Beaver JA, Horton S, Fernandes LL, Tang S, Horne HN, Liu J, Liu C, Schrieber SJ, Yu J, et al. FDA approval summary: rucaparib for the treatment of patients with deleterious BRCA mutation-associated advanced ovarian cancer. Clin Cancer Res. 2017;23:7165–70.

Bansal A, Kaushik S, Kukreti S. Non-canonical DNA structures: diversity and disease association. Front Genet. 2022;13:959258.

Barthelemy J, Hanenberg H, Leffak M. FANCJ is essential to maintain microsatellite structure genome-wide during replication stress. Nucleic Acids Res. 2016;44:6803–16.

Batra S, Allwein B, Kumar C, Devbhandari S, Brüning JG, Bahng S, Lee CM, Marians KJ, Hite RK, Remus D. G-quadruplex-stalled eukaryotic replisome structure reveals helical inchworm DNA translocation. Science. 2025;387:eadt1978.

Baudin A, Moreno-Romero AK, Xu X, Selig EE, Penalva LOF, Libich DS. Structural characterization of the RNA-binding protein SERBP1 reveals intrinsic disorder and atypical RNA binding modes. Front Mol Biosci. 2021;8:744707.

Beck C, Boehler C, Barbat JG, Bonnet ME, Illuzzi G, Ronde P, Gauthier LR, Magroun N, Rajendran A, Lopez BS, et al. PARP3 affects the relative contribution of homologous recombination and nonhomologous end-joining pathways. Nucleic Acids Res. 2014;42:5616–32.

Belan O, Sebald M, Adamowicz M, Anand R, Vancevska A, Neves J, Grinkevich V, Hewitt G, Segura-Bayona S, Bellelli R, et al. POLQ seals post-replicative ssDNA gaps to maintain genome stability in BRCA-deficient cancer cells. Mol Cell. 2022;82:4664–4680.e9.

Bellí G, Colomina N, Castells-Roca L, Lorite NP. Post-translational modifications of PCNA: guiding for the best DNA damage tolerance choice. J Fungi. 2022;8:621.

Bennardo N, Cheng A, Huang N, Stark JM. Alternative-NHEJ is a mechanistically distinct pathway of mammalian chromosome break repair. PLoS Genet. 2008;4:e1000110.

Bhargava R, Onyango DO, Stark JM. Regulation of single-strand annealing and its role in genome maintenance. Trends Genet. 2016;32:566–75.

Bhat DS, Spies MA, Spies M. A moving target for drug discovery: structure activity relationship and many genome (de)stabilizing functions of the RAD52 protein. DNA Repair. 2022;120:103421.

Bhat DS, Malacaria E, Di Biagi L, Razzaghi M, Honda M, Hobbs KF, Hengel SR, Pichierri P, Spies MA, Spies M. Therapeutic disruption of RAD52-ssDNA complexation via novel drug-like inhibitors. NAR Cancer. 2023;5:zcad018.

Bhutani N, Burns DM, Blau HM. DNA demethylation dynamics. Cell. 2011;146:866–72.

Bienko M, Green CM, Crosetto N, Rudolf F, Zapart G, Coull B, Kannouche P, Wider G, Peter M, Lehmann AR, et al. Biochemistry: ubiquitin-binding domains in Y-family polymerases regulate translesion synthesis. Science. 2005;310:1821–4.

Blessing C, Apelt K, Van den Heuvel D, Gonzalez-Leal C, Rother MB, Van der Woude M, González-Prieto R, Yifrach A, Parnas A, Shah RG, et al. XPC-PARP complexes engage the chromatin remodeler ALC1 to catalyze global genome DNA damage repair. Nat Commun. 2022;13:4762.

Blier PR, Griffith AJ, Craft J, Hardin JA. Binding of Ku protein to DNA – measurement of affinity for ends and demonstration of binding to nicks. J Biol Chem. 1993;268:7594–601.

Blouin T, Saini N. Aldehyde-induced DNA-protein crosslinks- DNA damage, repair and mutagenesis. Front Oncol. 2024;14:1478373.

Boehm EM, Washington MT. RIP to the PIP: PCNA-binding motif no longer considered specific: PIP motifs and other related sequences are not distinct entities and can bind multiple proteins involved in genome maintenance. Bioessays. 2016;38:1117–22.

Boehm EM, Gildenberg MS, Washington MT. The many roles of PCNA in eukaryotic DNA replication. Enzymes. 2016a;39:231–54.

Boehm EM, Spies M, Washington MT. PCNA tool belts and polymerase bridges form during translesion synthesis. Nucleic Acids Res. 2016b;44:8250–60.

Boussios S, Karihtala P, Moschetta M, Abson C, Karathanasi A, Zakynthinakis-Kyriakou N, Ryan JE, Sheriff M, Rassy E, Pavlidis N. Veliparib in ovarian cancer: a new synthetically lethal therapeutic approach. Investig New Drugs. 2020;38:181–93.

Bouwman P, Jonkers J. The effects of deregulated DNA damage signalling on cancer chemotherapy response and resistance. Nat Rev Cancer. 2012;12:587–98.

Bowman GD, O'Donnell M, Kuriyan J. Structural analysis of a eukaryotic sliding DNA clamp-clamp loader complex. Nature. 2004;429:724–30.

Branzei D. Ubiquitin family modifications and template switching. FEBS Lett. 2011;585:2810–7.

Britton S, Chanut P, Delteil C, Barboule N, Frit P, Calsou P. ATM antagonizes NHEJ proteins assembly and DNA-ends synapsis at single-ended DNA double strand breaks. Nucleic Acids Res. 2020;48:9710–23.

Brosh RM, Wu Y-L. An emerging picture of FANCJ's role in G4 resolution to facilitate DNA replication. NAR Cancer. 2021;3:zcab034.

Bryant HE, Petermann E, Schultz N, Jemth AS, Loseva O, Issaeva N, Johansson F, Fernandez S, McGlynn P, Helleday T. PARP is activated at stalled forks to mediate Mre11-dependent replication restart and recombination. EMBO J. 2009;28:2601–15.

Burgess RC, Burman B, Kruhlak MJ, Misteli T. Activation of DNA damage response signaling by condensed chromatin. Cell Rep. 2014;9:1703–17.

Cai Y, Cao HF, Wang F, Zhang YF, Kapranov P. Complex genomic patterns of abasic sites in mammalian DNA revealed by a high-resolution SSiNGLe-AP method. Nat Commun. 2022;13:5868.

Caldecott KW. Causes and consequences of DNA single-strand breaks. Trends Biochem Sci. 2024;49:68–78.

Caldwell CC, Spies M. Dynamic elements of replication protein A at the crossroads of DNA replication, recombination, and repair. Crit Rev Biochem Mol Biol. 2020;55:482–507.

Campuzano V, Montermini L, Molto MD, Pianese L, Cossee M, Cavalcanti F, Monros E, Rodius F, Duclos F, Monticelli A, et al. Friedreich's ataxia: autosomal recessive disease caused by an intronic GAA triplet repeat expansion. Science. 1996;271:1423–7.

Cantor SB, Calvo JA. Fork protection and therapy resistance in hereditary breast cancer. Cold Spring Harb Symp Quant Biol. 2017;82:339–48.

Caron M-C, Sharma AK, O'Sullivan J, Myler LR, Ferreira MT, Rodrigue A, Coulombe Y, Ethier C, Gagné J-P, Langelier M-F, et al. Poly(ADP-ribose) polymerase-1 antagonizes DNA resection at double-strand breaks. Nat Commun. 2019;10:2954.

Carter T, Vancurova I, Sun I, Lou W, Deleon S. A DNA-activated protein-kinase from HeLa-cell nuclei. Mol Cell Biol. 1990;10:6460–71.

Castillo Bosch P, Segura-Bayona S, Koole W, van Heteren JT, Dewar JM, Tijsterman M, Knipscheer P. FANCJ promotes DNA synthesis through G-quadruplex structures. EMBO J. 2014;33:2521–33.

Ceccaldi R, Cejka P. Mechanisms and regulation of DNA end resection in the maintenance of genome stability. Nat Rev Mol Cell Biol. 2025;26:586–99.

Cervantes-Gracia K, Gramalla-Schmitz A, Weischedel J, Chahwan R. APOBECs orchestrate genomic and epigenomic editing across health and disease. Trends Genet. 2021;37:1028–43.

Chambon P, Weill JD, Mandel P. Nicotinamide mononucleotide activation of new DNA-dependent polyadenylic acid synthesizing nuclear enzyme. Biochem Biophys Res Commun. 1963;11:39–43.

Chan GL, Doetsch PW, Haseltine WA. Cyclobutane pyrimidine dimers and (6-4) photoproducts block polymerization by DNA polymerase I. Biochemistry. 1985;24:5723–8.

Chandramouly G, McDevitt S, Sullivan K, Kent T, Luz A, Glickman JF, Andrake M, Skorski T, Pomerantz RT. Small-molecule disruption of RAD52 rings as a mechanism for precision medicine in BRCA-deficient cancers. Chem Biol. 2015;22:1491–504.

Chang HHY, Pannunzio NR, Adachi N, Lieber MR. Non-homologous DNA end joining and alternative pathways to double-strand break repair. Nat Rev Mol Cell Biol. 2017;18:495–506.

Chanut P, Britton S, Coates J, Jackson SP, Calsou P. Coordinated nuclease activities counteract Ku at single-ended DNA double-strand breaks. Nat Commun. 2016;7:12889.

Chapados BR, Hosfield DJ, Han S, Qiu J, Yelent B, Shen B, Tainer JA. Structural basis for FEN-1 substrate specificity and PCNA-mediated activation in DNA replication and repair. Cell. 2004;116:39–50.

Chappidi N, Quail T, Doll S, Vogel LT, Aleksandrov R, Felekyan S, Kuhnemuth R, Stoynov S, Seidel CAM, Brugues J, et al. PARP1-DNA co-condensation drives DNA repair site assembly to prevent disjunction of broken DNA ends. Cell. 2024;187:945–961.e918.

Chen R, Wold MS. Replication protein A: single-stranded DNA's first responder: dynamic DNA-interactions allow replication protein A to direct single-strand DNA intermediates into different pathways for synthesis or repair. Bioessays. 2014;36:1156–61.

Chheda P, Spies M, Spies MA. Small-molecule effectors of DNA repair proteins: applications for development of cancer therapeutics and research. In: Burger's medicinal chemistry and drug discovery. New York: Wiley; 2021. p. 1–39.

Chin Sang C, Moore G, Tereshchenko M, Zhang H, Nosella ML, Dasovich M, Alderson TR, Leung AKL, Finkelstein IJ, Forman-Kay JD, et al. PARP1 condensates differentially partition DNA repair proteins and enhance DNA ligation. EMBO Rep. 2024;25:5635–66.

Cohen CB, Coombes MC, Merlo CP, Kontor CA, Meah R, Whitaker AM. Base excision repair within structure-forming repeat sequences and its impact on cancer and other diseases. NAR Cancer. 2025;7:zcaf051.

Cordido A, Besada-Cerecedo L, García-González MA. The genetic and cellular basis of autosomal dominant polycystic kidney disease – a primer for clinicians. Front Pediatr. 2017;5:279.

D'Amours D, Desnoyers S, D'Silva I, Poirier GG. Poly(ADP-ribosyl)ation reactions in the regulation of nuclear functions. Biochem J. 1999;342(Pt 2):249–68.

Daley JM, Wilson TE. Rejoining of DNA double-strand breaks as a function of overhang length. Mol Cell Biol. 2005;25:896–906.

Dall'Agnese G, Dall'Agnese A, Banani SF, Codrich M, Malfatti MC, Antoniali G, Tell G. Role of condensates in modulating DNA repair pathways and its implication for chemoresistance. J Biol Chem. 2023;299:104800.

Dantzer F, Schreiber V, Niedergang C, Trucco C, Flatter E, De la Rubia G, Oliver J, Rolli V, Ménissier-de Murcia J, de Murcia G. Involvement of poly(ADP-ribose) polymerase in base excision repair. Biochimie. 1999;81:69–75.

Dantzer F, de la Rubia G, Murcia JMD, Hostomsky Z, de Murcia G, Schreiber V. Base excision repair is impaired in mammalian cells lacking poly(ADP-ribose) polymerase-1. Biochemistry. 2000;39:7559–69.

Davies AA, Huttner D, Daigaku Y, Chen S, Ulrich HD. Activation of ubiquitin-dependent DNA damage bypass is mediated by replication protein A. Mol Cell. 2008;29:625–36.

Davis L, Maizels N. Homology-directed repair of DNA nicks via pathways distinct from canonical double-strand break repair. Proc Natl Acad Sci USA. 2014;111:E924–32.

De Bont R, van Larebeke N. Endogenous DNA damage in humans: a review of quantitative data. Mutagenesis. 2004;19:169–85.

Demin AA, Hirota K, Tsuda M, Adamowicz M, Hailstone R, Brazina J, Gittens W, Kalasova I, Shao Z, Zha S, et al. XRCC1 prevents toxic PARP1 trapping during DNA base excision repair. Mol Cell. 2021;81:3018–3030.e3015.

Dempsey LA, Sun H, Hanakahi LA, Maizels N. G4 DNA binding by LR1 and its subunits, nucleolin and hnRNP D, A role for G-G pairing in immunoglobulin switch recombination. J Biol Chem. 1999;274:1066–71.

deMurcia JM, Niedergang C, Trucco C, Ricoul M, Dutrillaux B, Mark M, Oliver FJ, Masson M, Dierich A, LeMeur M, et al. Requirement of poly(ADP-ribose) polymerase in recovery from DNA damage in mice and in cells. Proc Natl Acad Sci USA. 1997;94:7303–7.

Dieckman LM, Freudenthal BD, Washington MT. PCNA structure and function: insights from structures of PCNA complexes and post-translationally modified PCNA. Subcell Biochem. 2012;62:281–99.

Dizdaroglu M, Kirkali G, Jaruga P. Formamidopyrimidines in DNA: mechanisms of formation, repair, and biological effects. Free Radic Biol Med. 2008;45:1610–21.

Dray E, Etchin J, Wiese C, Saro D, Williams GJ, Hammel M, Yu X, Galkin VE, Liu DQ, Tsai MS, et al. Enhancement of RAD51 recombinase activity by the tumor suppressor PALB2. Nat Struct Mol Biol. 2010;17:1255–9.

Duan XL, Liu NN, Yang YT, Li HH, Li M, Dou SX, Xi XG. G-quadruplexes significantly stimulate Pif1 helicase-catalyzed duplex DNA unwinding. J Biol Chem. 2015;290:7722–35.

Dueva R, Iliakis G. Replication protein A: a multifunctional protein with roles in DNA replication, repair and beyond. NAR Cancer. 2020;2:zcaa022.

Duncan BK, Miller JH. Mutagenic deamination of cytosine residues in DNA. Nature. 1980;287:560–1.

Edwards A, Marecki JC, Byrd AK, Gao J, Raney KD. G-quadruplex loops regulate PARP-1 enzymatic activation. Nucleic Acids Res. 2021;49:416–31.

Eickhoff P, Kose HB, Martino F, Petojevic T, Ali FA, Locke J, Tamberg N, Nans A, Berger JM, Botchan MR, et al. Molecular basis for ATP-hydrolysis-driven DNA translocation by the CMG helicase of the eukaryotic replisome. Cell Rep. 2019;28:2673–2688.e8.

Fan J, Pavletich NP. Structure and conformational change of a replication protein A heterotrimer bound to ssDNA. Genes Dev. 2012;26:2337–47.

Fan JH, Wilson PF, Wong HK, Urbin SS, Thompson LH, Wilson DM. XRCC1 down-regulation in human cells leads to DNA-damaging agent hypersensitivity, elevated sister chromatid exchange, and reduced survival of mutant cells. Environ Mol Mutagen. 2007;48:491–500.

Fenteany G, Gaur P, Hegedűs L, Dudás K, Kiss E, Wéber E, Hackler L, Martinek T, Puskás LG, Haracska L. Multilevel structure-activity profiling reveals multiple green tea compound families that each modulate ubiquitin-activating enzyme and ubiquitination by a distinct mechanism. Sci Rep. 2019;9:12801.

Fenteany G, Gaur P, Sharma G, Pintér L, Kiss E, Haracska L. Robust high-throughput assays to assess discrete steps in ubiquitination and related cascades. BMC Mol Cell Biol. 2020;21:21.

Fenteany G, Sharma G, Gaur P, Borics A, Weber E, Kiss E, Haracska L. A series of xanthenes inhibiting Rad6 function and Rad6-Rad18 interaction in the PCNA ubiquitination cascade. iScience. 2022;25:104053.

Fleury H, MacEachern MK, Stiefel CM, Anand R, Sempeck C, Nebenfuehr B, Maurer-Alcalá K, Ball K, Proctor B, Belan O, et al. The APE2 nuclease is essential for DNA double-strand break repair by microhomology-mediated end joining. Mol Cell. 2023;83:1429–1445.e8.

Fouquin A, Guirouilh-Barbat J, Lopez B, Hall J, Amor-Guéret M, Pennaneach V. PARP2 controls double-strand break repair pathway choice by limiting 53BP1 accumulation at DNA damage sites and promoting end-resection. Nucleic Acids Res. 2017;45:12325–39.

Fousteri M, Mullenders LHF. Transcription-coupled nucleotide excision repair in mammalian cells: molecular mechanisms and biological effects. Cell Res. 2008;18:73–84.

Franchini DM, Schmitz KM, Petersen-Mahrt SK. 5-Methylcytosine DNA demethylation: more than losing a methyl group. Annu Rev Genet. 2012;46:419–41.

Freudenthal BD, Brogie JE, Gakhar L, Kondratick CM, Washington MT. Crystal structure of SUMO-modified proliferating cell nuclear antigen. J Mol Biol. 2011;406:9–17.

Friedberg EC. Suffering in silence: the tolerance of DNA damage. Nat Rev Mol Cell Biol. 2005;6:943–53.

Fry M, Loeb LA. Human werner syndrome DNA helicase unwinds tetrahelical structures of the fragile X syndrome repeat sequence d(CGG)n. J Biol Chem. 1999;274:12797–802.

Fujimura S, Hasegawa S, Shimizu Y, Sugimura T. Polymerization of the adenosine 5′-diphosphate-ribose moiety of nicotinamide-adenine dinucleotide by nuclear enzyme. I. Enzymatic reactions. Biochim Biophys Acta. 1967;145:247–59.

Gagne JP, Isabelle M, Lo KS, Bourassa S, Hendzel MJ, Dawson VL, Dawson TM, Poirier GG. Proteome-wide identification of poly(ADP-ribose) binding proteins and poly(ADP-ribose)-associated protein complexes. Nucleic Acids Res. 2008;36:6959–76.

Gali H, Juhasz S, Morocz M, Hajdu I, Fatyol K, Szukacsov V, Burkovics P, Haracska L. Role of SUMO modification of human PCNA at stalled replication fork. Nucleic Acids Res. 2012;40:6049–59.

Gallo D, Brown GW. Post-replication repair: Rad5/HLTF regulation, activity on undamaged templates, and relationship to cancer. Crit Rev Biochem Mol Biol. 2019;54:301–32.

Gallo A, Lo Sterzo C, Mori M, Di Matteo A, Bertini I, Banci L, Brunori M, Federici L. Structure of nucleophosmin DNA-binding domain and analysis of its complex with a G-quadruplex sequence from the c-MYC promoter. J Biol Chem. 2012;287:26539–48.

Gangavarapu V, Haracska L, Unk I, Johnson RE, Prakash S, Prakash L. Mms2-Ubc13-dependent and -independent roles of Rad5 ubiquitin ligase in postreplication repair and translesion DNA synthesis in Saccharomyces cerevisiae. Mol Cell Biol. 2006;26:7783–90.

Gaur P. Discovery of small-molecule inhibitors of Uba1 and development of step-specific assays for PCNA ubiquitination. PhD thesis, University of Szeged, Szeged, Hungary. 2020.

Gaur P, Tyagi C. Unraveling the mechanism of action of myricetin in the inhibition of hUba1 approximately ubiquitin thioester bond formation via in silico molecular modeling techniques. ACS Omega. 2023;8:30432–41.

Gaur P, Fenteany G, Tyagi C. Mode of inhibitory binding of epigallocatechin gallate to the ubiquitin-activating enzyme Uba1 viaaccelerated molecular dynamics. RSC Adv. 2021;11:8264–76.

Gaur P, Bain FE, Meah R, Spies M. Single-molecule analysis of PARP1-G-quadruplex interaction. bioRxiv. 2025. https://doi.org/10.1101/2025.01.06.631587.

Gerhards M, Bohme A, Schubert K, Kodritsch B, Ulrich N. DNA adducts as link between in vitro and in vivo carcinogenicity – a case study with benzo[a]pyrene. Curr Res Toxicol. 2023;4:100097.

Ghosal G, Chen J. DNA damage tolerance: a double-edged sword guarding the genome. Transl Cancer Res. 2013;2:107–29.

Ghosh M, Singh M. RGG-box in hnRNPA1 specifically recognizes the telomere G-quadruplex DNA and enhances the G-quadruplex unfolding ability of UP1 domain. Nucleic Acids Res. 2018;46:10246–61.

Giallongo S, Reháková D, Biagini T, Lo Re O, Raina P, Lochmanová G, Zdráhal Z, Resnick I, Pata P, Pata I, et al. Histone variant macroH2A1.1 enhances nonhomologous end joining-dependent DNA double-strand-break repair and reprogramming efficiency of human iPSCs. Stem Cells. 2022;40:35–48.

Granger SL, Sharma R, Kaushik V, Razzaghi M, Honda M, Gaur P, Bhat DS, Labenz SM, Heinen JE, Williams BA, et al. Human hnRNPA1 reorganizes telomere-bound replication protein A. Nucleic Acids Res. 2024;52:12422–37.

Groen EJN, Gillingwater TH. UBA1: at the crossroads of ubiquitin homeostasis and neurodegeneration. Trends Mol Med. 2015;21:622–32.

Guan J, Yu S, Zheng X. NEDDylation antagonizes ubiquitination of proliferating cell nuclear antigen and regulates the recruitment of polymerase η in response to oxidative DNA damage. Protein Cell. 2018;9:365–79.

Guo MH, Hundseth K, Ding H, Vidhyasagar V, Inoue A, Nguyen CH, Zain R, Lee JS, Wu YL. A distinct triplex DNA unwinding activity of ChlR1 helicase. J Biol Chem. 2015;290:5174–89.

Gupta R, Sharma S, Sommers JA, Kenny MK, Cantor SB, Brosh RM Jr. FANCJ (BACH1) helicase forms DNA damage inducible foci with replication protein A and interacts physically and functionally with the single-stranded DNA-binding protein. Blood. 2007;110:2390–8.

Haince JF, McDonald D, Rodrigue A, Déry U, Masson JY, Hendzel MJ, Poirier GG. PARP1-dependent kinetics of recruitment of MRE11 and NBS1 proteins to multiple DNA damage sites. J Biol Chem. 2008;283:1197–208.

Han J, Huang J. DNA double-strand break repair pathway choice: the fork in the road. Genome Instab Dis. 2020;1:10–9.

Hang B. Formation and repair of tobacco carcinogen-derived bulky DNA adducts. J Nucleic Acids. 2010;2010:709521.

Hänsel-Hertsch R, Di Antonio M, Balasubramanian S. DNA G-quadruplexes in the human genome: detection, functions and therapeutic potential. Nat Rev Mol Cell Biol. 2017;18:279.

Haracska L, Torres-Ramos CA, Johnson RE, Prakash S, Prakash L. Opposing effects of ubiquitin conjugation and SUMO modification of PCNA on replicational bypass of DNA lesions in Saccharomyces cerevisiae. Mol Cell Biol. 2004;24:4267–74.

Hassa PO, Hottiger MO. The diverse biological roles of mammalian PARPS, a small but powerful family of poly-ADP-ribose polymerases. Front Biosci. 2008;13:3046–82.

Helleday T. The underlying mechanism for the PARP and BRCA synthetic lethality: clearing up the misunderstandings. Mol Oncol. 2011;5:387–93.

Helleday T, Lo J, van Gent DC, Engelward BP. DNA double-strand break repair: from mechanistic understanding to cancer treatment. DNA Repair. 2007;6:923.

Hengel SR, Malacaria E, Constantino LFD, Bain FE, Diaz A, Koch BG, Yu LP, Wu M, Pichierri P, Spies MA, et al. Small-molecule inhibitors identify the RAD52-ssDNA interaction as critical for recovery from replication stress and for survival of BRCA2 deficient cells. elife. 2016;5:e14740.

Henrikus SS, Gross MH, Willhoft O, Pühringer T, Lewis JS, McClure AW, Greiwe JF, Palm G, Nans A, Diffley JFX, et al. Unwinding of a eukaryotic origin of replication visualized by cryo-EM (vol 31, 1265, 2024). Nat Struct Mol Biol. 2024;31:1808.

Heyer WD, Ehmsen KT, Liu J. Regulation of homologous recombination in eukaryotes. Annu Rev Genet. 2010;44:113–39.

Hisey JA, Masnovo C, Mirkin SM. Triplex H-DNA structure: the long and winding road from the discovery to its role in human disease. NAR Mol Med. 2024a;1:ugae024.

Hisey JA, Radchenko EA, Mandel NH, Mcginty RJ, Matos-Rodrigues G, Rastokina A, Masnovo C, Ceschi S, Hernandez A, Nussenzweig A, et al. Pathogenic CANVAS (AAGGG) repeats stall DNA replication due to the formation of alternative DNA structures. Nucleic Acids Res. 2024b;52:4361–74.

Hoch NC, Hanzlikova H, Rulten SL, Tetreault M, Komulainen E, Ju L, Hornyak P, Zeng Z, Gittens W, Rey SA, et al. XRCC1 mutation is associated with PARP1 hyperactivation and cerebellar ataxia. Nature. 2017;541:87–91.

Hoege C, Pfander B, Moldovan GL, Pyrowolakis G, Jentsch S. RAD6-dependent DNA repair is linked to modification of PCNA by ubiquitin and SUMO. Nature. 2002;419:135–41.

Hoitsma NM, Whitaker AM, Schaich MA, Smith MR, Fairlamb MS, Freudenthal BD. Structure and function relationships in mammalian DNA polymerases. Cell Mol Life Sci. 2020;77:35–59.

Holliday R, Ho T. Gene silencing and endogenous DNA methylation in mammalian cells. Mutat Res. 1998;400:361–8.

Honda M, Razzaghi M, Gaur P, Malacaria E, Marozzi G, Di Biagi L, Aiello FA, Paintsil EA, Stanfield AJ, Deppe BJ, et al. The RAD52 double-ring remodels replication forks restricting fork reversal. Nature. 2025;641:512–9.

Hottiger MO. Nuclear ADP-ribosylation and its role in chromatin plasticity, cell differentiation, and epigenetics. Annu Rev Biochem. 2015;84:227–63.

Hoy SM. Talazoparib: first global approval. Drugs. 2018;78:1939–46.

Hsieh P, Yamane K. DNA mismatch repair: molecular mechanism, cancer, and ageing. Mech Ageing Dev. 2008;129:391–407.

Hu YD, Petit SA, Ficarro SB, Toomire KJ, Xie AY, Lim E, Cao SLA, Park E, Eck MJ, Scully R, et al. PARP1-driven poly-ADP-ribosylation regulates BRCA1 function in homologous recombination-mediated DNA repair. Cancer Discov. 2014;4:1430–47.

Hu JC, Lieb JD, Sancar A, Adar S. Cisplatin DNA damage and repair maps of the human genome at single-nucleotide resolution. Proc Natl Acad Sci USA. 2016;113:11507–12.

Huo XR, Ji LZ, Zhang YW, Lv P, Cao X, Wang QF, Yan ZX, Dong SS, Du D, Zhang F, et al. The nuclear matrix protein SAFB cooperates with major satellite RNAs to stabilize heterochromatin architecture partially through phase separation. Mol Cell. 2020;77:368–383.e7.

Ito F, Li ZY, Minakhin L, Khant HA, Pomerantz RT, Chen XS. Structural basis for Polθ-helicase DNA binding and microhomology-mediated end-joining. Nat Commun. 2025;16:3725.

Iyer LM, Leipe DD, Koonin EV, Aravind L. Evolutionary history and higher order classification of AAA plus ATPases. J Struct Biol. 2004;146:11–31.

Jelinic P, Levine DA. New insights into PARP inhibitors' effect on cell cycle and homology-directed DNA damage repair. Mol Cancer Ther. 2014;13:1645–54.

Jensen RB, Rothenberg E. Preserving genome integrity in human cells via DNA double-strand break repair. Mol Biol Cell. 2020;31:859–65.

Ji Y, Li F, Qiao Y. Modulating liquid-liquid phase separation of FUS: mechanisms and strategies. J Mater Chem B. 2022;10:8616–28.

Jia Q, den Dulk-Ras A, Shen HX, Hooykaas PJJ, de Pater S. Poly(ADP-ribose)polymerases are involved in microhomology mediated back-up non-homologous end joining in Arabidopsis thaliana. Plant Mol Biol. 2013;82:339–51.

Jiricny J. The multifaceted mismatch-repair system. Nat Rev Mol Cell Biol. 2006;7:335–46.

Johnson LM, Snyder M, Chang LMS, Davis RW, Campbell JL. Isolation of the gene encoding yeast DNA polymerase I. Cell. 1985;43:369–77.

Johnson RE, Klassen R, Prakash L, Prakash S. A major role of DNA polymerase δ in replication of both the leading and lagging DNA strands. Mol Cell. 2015;59:163–75.

Jones CJ, Edwards SM, Waters R. The repair of identified large DNA adducts induced by 4-nitroquinoline-1-oxide in normal or xeroderma pigmentosum group A human fibroblasts, and the role of DNA polymerases alpha or delta. Carcinogenesis. 1989;10:1197–201.

Jones ML, Aria V, Baris Y, Yeeles JTP. How Pol α-primase is targeted to replisomes to prime eukaryotic DNA replication. Mol Cell. 2023;83:2911–2924.e16.

Kai-Feng H, Sidorova JM, Nghiem P, Kawasumi M. The 6-4 photoproduct is the trigger of UV-induced replication blockage and ATR activation. Proc Natl Acad Sci USA. 2020;117:12806–16.

Kais Z, Rondinelli B, Holmes A, O'Leary C, Kozono D, D'Andrea AD, Ceccaldi R. FANCD2 maintains fork stability in BRCA1/2-deficient tumors and promotes alternative end-joining DNA repair. Cell Rep. 2016;15:2488–99.

Kanao R, Masutani C. Regulation of DNA damage tolerance in mammalian cells by post-translational modifications of PCNA. In: Mutation research – fundamental and molecular mechanisms of mutagenesis, vol. 803–805. Elsevier; 2017. p. 82–8.

Kasai H, Nishimura S. Hydroxylation of the C-8 position of deoxyguanosine by reducing agents in the presence of oxygen. Nucleic Acids Symp Ser. 1983;(12):165–7.

Kelso AA, Lopezcolorado FW, Bhargava R, Stark JM. Distinct roles of RAD52 and POLQ in chromosomal break repair and replication stress response. PLoS Genet. 2019;15:e1008319.

Khurana S, Kruhlak MJ, Kim J, Tran AD, Liu JP, Nyswaner K, Shi L, Jailwala P, Sung MH, Hakim O, et al. A macrohistone variant links dynamic chromatin compaction to BRCA1-dependent genome maintenance. Cell Rep. 2014;8:1049–62.

Kilic S, Lezaja A, Gatti M, Bianco E, Michelena J, Imhof R, Altmeyer M. Phase separation of 53BP1 determines liquid-like behavior of DNA repair compartments. EMBO J. 2019;38:e101379.

Kim HM, Narayanan V, Mieczkowski PA, Petes TD, Krasilnikova MM, Mirkin SM, Lobachev KS. Chromosome fragility at GAA tracts in yeast depends on repeat orientation and requires mismatch repair. EMBO J. 2008;27:2896–906.

Kim G, Ison G, McKee AE, Zhang H, Tang S, Gwise T, Sridhara R, Lee E, Tzou A, Philip R, et al. FDA approval summary: olaparib monotherapy in patients with deleterious germline BRCA-mutated advanced ovarian cancer treated with three or more lines of chemotherapy. Clin Cancer Res. 2015;21:4257–61.

Klaric JA, Wust S, Panier S. New faces of old friends: emerging new roles of RNA-binding proteins in the DNA double-strand break response. Front Mol Biosci. 2021;8:668821.

Kondo N, Takahashi A, Ono K, Ohnishi T. DNA damage induced by alkylating agents and repair pathways. J Nucleic Acids. 2010;2010:543531.

Krietsch J, Rouleau M, Pic E, Ethier C, Dawson TM, Dawson VL, Masson JY, Poirier GG, Gagne JP. Reprogramming cellular events by poly(ADP-ribose)-binding proteins. Mol Asp Med. 2013;34:1066–87.

Kumar D, Abdulovic AL, Viberg J, Nilsson AK, Kunkel TA, Chabes A. Mechanisms of mutagenesis due to imbalanced dNTP pools. Nucleic Acids Res. 2011;39:1360–71.

Kumbhar R, Sanchez A, Perren J, Gong FD, Corujo D, Medina F, Devanathan SK, Xhemalce B, Matouschek A, Buschbeck M, et al. Poly(ADP-ribose) binding and macroH2A mediate recruitment and functions of KDM5A at DNA lesions. J Cell Biol. 2021;220:e202006149.

Kunkel TA. DNA replication fidelity. J Biol Chem. 2004;279:16895–8.

Kunkel TA. Evolving views of DNA replication (in)fidelity. Cold Spring Harb Symp Quant Biol. 2009;74:91–101.

Kunkel TA. Balancing eukaryotic replication asymmetry with replication fidelity. Curr Opin Chem Biol. 2011;15:620–6.

Law EK, Levin-Klein R, Jarvis MC, Kim H, Argyris PP, Carpenter MA, Starrett GJ, Temiz NA, Larson LK, Durfee C, et al. APOBEC3A catalyzes mutation and drives carcinogenesis in vivo. J Exp Med. 2020;217:e20200261.

Layer JV, Cleary JP, Brown AJ, Stevenson KE, Morrow SN, Van Scoyk A, Blasco RB, Karaca E, Meng FL, Frock RL, et al. Parp3 promotes long-range end joining in murine cells. Proc Natl Acad Sci USA. 2018;115:10076–81.

Lee SG, Kim N, Kim SM, Park IB, Kim H, Kim S, Kim BG, Hwang JM, Baek IJ, Gartner A, et al. Ewing sarcoma protein promotes dissociation of poly(ADP-ribose) polymerase 1 from chromatin. EMBO Rep. 2020;21:e48676.

Lee WTC, Yin Y, Morten MJ, Tonzi P, Gwo PP, Odermatt DC, Modesti M, Cantor SB, Gari K, Huang TT, et al. Single-molecule imaging reveals replication fork coupled formation of G-quadruplex structures hinders local replication stress signaling. Nat Commun. 2021;12:2525.

Lehmann AR, Niimi A, Ogi T, Brown S, Sabbioneda S, Wing JF, Kannouche PL, Green CM. Translesion synthesis: Y-family polymerases and the polymerase switch. DNA Repair. 2007;6:891–9.

Leidecker O, Bonfiglio JJ, Colby T, Zhang Q, Atanassov I, Zaja R, Palazzo L, Stockum A, Ahel I, Matic I. Serine is a new target residue for endogenous ADP-ribosylation on histones. Nat Chem Biol. 2016;12:998–1000.

Leung W, Baxley RM, Moldovan GL, Bielinsky AK. Mechanisms of DNA damage tolerance: post-translational regulation of PCNA. Genes. 2019;10:10.

Levone BR, Lenzken SC, Antonaci M, Maiser A, Rapp A, Conte F, Reber S, Mechtersheimer J, Ronchi AE, Mühlemann O, et al. FUS-dependent liquid-liquid phase separation is important for DNA repair initiation. J Cell Biol. 2021;220:e202008030.

Li YZ, Dang NK, He W, Returan M, Carvajal-Maldonado D, Guerin AT, Xu H, Liu B, Wood RD. Pol θ-mediated end-joining uses microhomologies containing mismatches. Nat Commun. 2025;16:6085.

Liang L, Deng L, Nguyen SC, Zhao X, Maulion CD, Shao C, Tischfield JA. Human DNA ligases I and III, but not ligase IV, are required for microhomology-mediated end joining of DNA double-strand breaks. Nucleic Acids Res. 2008;36:3297–310.

Lieber MR. The mechanism of double-strand DNA break repair by the nonhomologous DNA end-joining pathway. Annu Rev Biochem. 2010;79:181–211.

Lin W, Xin H, Zhang Y, Wu X, Yuan F, Wang Z. The human REV1 gene codes for a DNA template-dependent dCMP transferase. Nucleic Acids Res. 1999;27:4468–75.

Lin JR, Zeman MK, Chen JY, Yee MC, Cimprich KA. SHPRH and HLTF act in a damage-specific manner to coordinate different forms of postreplication repair and prevent mutagenesis. Mol Cell. 2011;42:237–49.

Lindahl T, Barnes DE. Repair of endogenous DNA damage. Cold Spring Harb Symp Quant Biol. 2000;65:127–33.

Liu GQ, Myers S, Chen XM, Bissler JJ, Sinden RR, Leffak M. Replication fork stalling and checkpoint activation by a locus mirror repeat polypurine-polypyrimidine (Pu-Py) tract. J Biol Chem. 2012;287:33412–23.

Liu J, Ede C, Wright WD, Gore SK, Jenkins SS, Freudenthal BD, Washington MT, Veaute X, Heyer WD. Srs2 promotes synthesis-dependent strand annealing by disrupting DNA polymerase δ-extending D-loops. elife. 2017;6:e22195.

Liu X, Xiong Y, Zhang C, Lai R, Liu H, Peng R, Fu T, Liu Q, Fang X, Mann S, et al. G-quadruplex-induced liquid-liquid phase separation in biomimetic protocells. J Am Chem Soc. 2021;143:11036–43.

Livneh Z, Ziv O, Shachar S. Multiple two-polymerase mechanisms in mammalian translesion DNA synthesis. Cell Cycle. 2010;9:729–35.

Lodovichi S, Quadri R, Sertic S, Pellicioli A. PARylation of BRCA1 limits DNA break resection through BRCA2 and EXO1. Cell Rep. 2023;42:112060.

Longley MJ, Pierce AJ, Modrich P. DNA polymerase δ is required for human mismatch repair in vitro. J Biol Chem. 1997;272:10917–21.

Luedeman ME, Stroik S, Feng WJ, Luthman AJ, Gupta GP, Ramsden DA. Poly(ADP) ribose polymerase promotes DNA polymerase theta-mediated end joining by activation of end resection. Nat Commun. 2022;13:4547.

Luijsterburg MS, Lindh M, Acs K, Vrouwe MG, Pines A, van Attikum H, Mullenders LH, Dantuma NP. DDB2 promotes chromatin decondensation at UV-induced DNA damage. J Cell Biol. 2012;197:267–81.

Lujan SA, Zhou ZX, Kunkel TA. Evidence that transient replication errors initiate nuclear genome mutations. Nucleic Acids Res. 2025;53:gkaf679.

Lv ZW, Jiao JN, Xue WY, Shi XY, Wang RH, Wu JH. Activation-induced cytidine deaminase in tertiary lymphoid structures: dual roles and implications in cancer prognosis. Front Oncol. 2025;15:1555491.

Ma Y, Iida K, Nagasawa K. Topologies of G-quadruplex: biological functions and regulation by ligands. Biochem Biophys Res Commun. 2020;531:3–17.

Macheret M, Halazonetis TD. DNA replication stress as a hallmark of cancer. Annu Rev Pathol. 2015;10:425–48.

Mahaney BL, Meek K, Lees-Miller SP. Repair of ionizing radiation-induced DNA double-strand breaks by non-homologous end-joining. Biochem J. 2009;417:639–50.

Maizels N. Genomic stability: FANCJ-dependent G4 DNA repair. Curr Biol. 2008;18:R613–4.

Majka J, Burgers PMJ. The PCNA-RFC families of DNA clamps and clamp loaders. Prog Nucleic Acid Res Mol Biol. 2004;78:227–60.

Malanga M, Pleschke JM, Kleczkowska HE, Althaus FR. Poly(ADP-ribose) binds to specific domains of p53 and alters its DNA binding functions. J Biol Chem. 1998;273:11839–43.

Mann A, Ramirez-Otero MA, De Antoni A, Hanthi YW, Sannino V, Baldi G, Falbo L, Schrempf A, Bernardo S, Loizou J, et al. POLtheta prevents MRE11-NBS1-CtIP-dependent fork breakage in the absence of BRCA2/RAD51 by filling lagging-strand gaps. Mol Cell. 2022;82:4218–4231.e4218.

Mansour WY, Rhein T, Dahm-Daphi J. The alternative end-joining pathway for repair of DNA double-strand breaks requires PARP1 but is not dependent upon microhomologies. Nucleic Acids Res. 2010;38:6065–77.

Mansour WY, Borgmann K, Petersen C, Dikomey E, Dahm-Daphi J. The absence of Ku but not defects in classical non-homologous end-joining is required to trigger PARP1-dependent end-joining. DNA Repair. 2013;12:1134–42.

Martini S, Davis K, Faraway R, Elze L, Lockwood N, Jones A, Xie X, McDonald NQ, Mann DJ, Armstrong A, et al. A genetically-encoded crosslinker screen identifies SERBP1 as a PKCepsilon substrate influencing translation and cell division. Nat Commun. 2021;12:6934.

Maslowska KH, Wong RP, Ulrich HD, Pages V. Post-replicative lesion processing limits DNA damage-induced mutagenesis. Nucleic Acids Res. 2025;53:gkaf198.

Mastrocola AS, Kim SH, Trinh AT, Rodenkirch LA, Tibbetts RS. The RNA-binding protein fused in sarcoma (FUS) functions downstream of poly(ADP-ribose) polymerase (PARP) in response to DNA damage. J Biol Chem. 2013;288:24731–41.

Mateos-Gomez PA, Gong FD, Nair N, Miller KM, Lazzerini-Denchi E, Sfeir A. Mammalian polymerase θ promotes alternative NHEJ and suppresses recombination. Nature. 2015;518:254–7.

Matsuzaki K, Borel V, Adelman CA, Schindler D, Boulton SJ. FANCJ suppresses microsatellite instability and lymphomagenesis independent of the Fanconi anemia pathway. Genes Dev. 2015;29:2532–46.

Matveeva E, Maiorano J, Zhang QY, Eteleeb AM, Convertini P, Chen J, Infantino V, Stamm S, Wang JP, Rouchka EC, et al. Involvement of PARP1 in the regulation of alternative splicing. Cell Discov. 2016;2:15046.

Mavragani IV, Nikitaki Z, Kalospyros SA, Georgakilas AG. Ionizing radiation and complex DNA damage: from prediction to detection challenges and biological significance. Cancers (Basel). 2019;11:1789.

Mazloumi Aboukheili AM, Walden H. USP1 in regulation of DNA repair pathways. DNA Repair (Amst). 2025;146:103807.

McElhinny SAN, Gordenin DA, Stith CM, Burgers PMJ, Kunkel TA. Division of labor at the eukaryotic replication fork. Mol Cell. 2008;30:137–44.

Mehta A, Haber JE. Sources of DNA double-strand breaks and models of recombinational DNA repair. Cold Spring Harb Perspect Biol. 2014;6:a016428.

Metzger MJ, Stoddard BL, Monnat RJ. PARP-mediated repair, homologous recombination, and back-up non-homologous end joining-like repair of single-strand nicks. DNA Repair. 2013;12:529–34.

Miller H, Fernandes AS, Zaika E, McTigue MM, Torres MC, Wente M, Iden CR, Grollman AP. Stereoselective excision of thymine glycol from oxidatively damaged DNA. Nucleic Acids Res. 2004;32:338–45.

Mimura M, Tomita S, Shinkai Y, Hosokai T, Kumeta H, Saio T, Shiraki K, Kurita R. Quadruplex folding promotes the condensation of linker histones and DNAs via liquid-liquid phase separation. J Am Chem Soc. 2021;143:9849–57.

Minocha S, Oliva-Santiago M, Gadi SA, Duxin JP. Just bypass it: mechanisms of DNA damage tolerance. Trends Biochem Sci. 2026;1:107–24.

Mitrea DM, Cika JA, Stanley CB, Nourse A, Onuchic PL, Banerjee PR, Phillips AH, Park CG, Deniz AA, Kriwacki RW. Self-interaction of NPM1 modulates multiple mechanisms of liquid-liquid phase separation. Nat Commun. 2018;9:842.

Miyata T, Suzuki H, Oyama T, Mayanagi K, Ishino Y, Morikawa K. Open clamp structure in the clamp-loading complex visualized by electron microscopic image analysis. Proc Natl Acad Sci USA. 2005;102:13795–800.

Moldovan GL, Pfander B, Jentsch S. PCNA, the maestro of the replication fork. Cell. 2007;129:665–79.

Mossi R, Hubscher U. Clamping down on clamps and clamp loaders--the eukaryotic replication factor C. Eur J Biochem. 1998;254:209–16.

Munoz-Gamez JA, Rodriguez-Vargas JM, Quiles-Perez R, Aguilar-Quesada R, Martin-Oliva D, de Murcia G, Menissier de Murcia J, Almendros A, Ruiz de Almodovar M, Oliver FJ. PARP-1 is involved in autophagy induced by DNA damage. Autophagy. 2009;5:61–74.

Murray DT, Kato M, Lin Y, Thurber KR, Hung I, McKnight SL, Tycko R. Structure of FUS protein fibrils and its relevance to self-assembly and phase separation of low-complexity domains. Cell. 2017;171:615–27. e616

Murthy AC, Tang WS, Jovic N, Janke AM, Seo DH, Perdikari TM, Mittal J, Fawzi NL. Molecular interactions contributing to FUS SYGQ LC-RGG phase separation and co-partitioning with RNA polymerase II heptads. Nat Struct Mol Biol. 2021;28:923–35.

Nair DT, Johnson RE, Prakash L, Prakash S, Aggarwal AK. Rev1 employs a novel mechanism of DNA synthesis using a protein template. Science. 2005;309:2219–22.

Nakamura J, Swenberg JA. Endogenous apurinic/apyrimidinic sites in genomic DNA of mammalian tissues. Cancer Res. 1999;59:2522–6.

Nawy T. Dynamics of DNA demethylation. Nat Methods. 2013;10:466.

Negrini S, Gorgoulis VG, Halazonetis TD. Genomic instability--an evolving hallmark of cancer. Nat Rev Mol Cell Biol. 2010;11:220–8.

Neil AJ, Hisey JA, Quasem I, McGinty RJ, Hitczenko M, Khristich AN, Mirkin SM. Replication-independent instability of Friedreich's ataxia GAA repeats during chronological aging. Proc Natl Acad Sci USA. 2021;118:e2013080118.

Northam MR, Moore EA, Mertz TM, Binz SK, Stith CM, Stepchenkova EI, Wendt KL, Burgers PMJ, Shcherbakova PV. DNA polymerases ζ and Rev1 mediate error-prone bypass of non-B DNA structures. Nucleic Acids Res. 2014;42:290–306.

Nosella ML, Kim TH, Huang SK, Harkness RW, Goncalves M, Pan A, Tereshchenko M, Vahidi S, Rubinstein JL, Lee HO, et al. Poly(ADP-ribosyl)ation enhances nucleosome dynamics and organizes DNA damage repair components within biomolecular condensates. Mol Cell. 2024;84:429–446.e417.

Odermatt DC, Lee WTC, Wild S, Jozwiakowski SK, Rothenberg E, Gari K. Cancer-associated mutations in the iron-sulfur domain of FANCJ affect G-quadruplex metabolism. PLoS Genet. 2020;16:e1008740.

Okazaki T, Okazaki R. Mechanism of DNA chain growth. IV. Direction of synthesis of T4 short DNA chains as revealed by exonucleolytic degradation. Proc Natl Acad Sci USA. 1969;64:1242–8.

Olson CL, Barbour AT, Wieser TA, Wuttke DS. RPA engages telomeric G-quadruplexes more effectively than CST. Nucleic Acids Res. 2023;51:5073–86.

Ortega P, Bournique E, Li JY, Sanchez A, Santiago G, Harris BR, Striepen J, Maciejowski J, Green AM, Buisson R. Mechanism of DNA replication fork breakage and PARP1 hyperactivation during replication catastrophe. Sci Adv. 2025;11:eadu0437.

Pachkowski BF, Tano K, Afonin V, Elder RH, Takeda S, Watanabe M, Swenberg JA, Nakamura J. Cells deficient in PARP-1 show an accelerated accumulation of DNA single strand breaks, but not AP sites, over the PARP-1-proficient cells exposed to MMS. Mutat Res. 2009;671:93–9.

Palazzo L, Leidecker O, Prokhorova E, Dauben H, Matic I, Ahel I. Serine is the major residue for ADP-ribosylation upon DNA damage. elife. 2018;7:e34334.

Pan F, Xu P, Roland C, Sagui C, Weninger K. Structural and dynamical properties of nucleic acid hairpins implicated in trinucleotide repeat expansion diseases. Biomolecules. 2024;14:1278.

Paramasivam M, Membrino A, Cogoi S, Fukuda H, Nakagama H, Xodo LE. Protein hnRNP A1 and its derivative Up1 unfold quadruplex DNA in the human KRAS promoter: implications for transcription. Nucleic Acids Res. 2009;37:2841–53.

Parikh SS, Mol CD, Slupphaug G, Bharati S, Krokan HE, Tainer JA. Base excision repair initiation revealed by crystal structures and binding kinetics of human uracil-DNA glycosylase with DNA. EMBO J. 1998;17:5214–26.

Park JM, Yang SW, Yu KR, Ka SH, Lee SW, Seol JH, Jeon YJ, Chung CH. Modification of PCNA by ISG15 plays a crucial role in termination of error-prone translesion DNA synthesis. Mol Cell. 2014;54:626–38.

Parker JL, Ulrich HD. Mechanistic analysis of PCNA poly-ubiquitylation by the ubiquitin protein ligases Rad18 and Rad5. EMBO J. 2009;28:3657–66.

Pascal JM. The comings and goings of PARP-1 in response to DNA damage. DNA Repair (Amst). 2018;71:177–82.

Pascal JM, O'Brien PJ, Tomkinson AE, Ellenberger T. Human DNA ligase I completely encircles and partially unwinds nicked DNA. Nature. 2004;432:473–8.

Patel HP, Lu L, Blaszak RT, Bissler JJ. PKD1 intron 21: triplex DNA formation and effect on replication. Nucleic Acids Res. 2004;32:1460–8.

Patel A, Lee HO, Jawerth L, Maharana S, Jahnel M, Hein MY, Stoynov S, Mahamid J, Saha S, Franzmann TM, et al. A liquid-to-solid phase transition of the ALS protein FUS accelerated by disease mutation. Cell. 2015;162:1066–77.

Paul D, Mu H, Zhao H, Ouerfelli O, Jeffrey PD, Broyde S, Min JH. Structure and mechanism of pyrimidine-pyrimidone (6-4) photoproduct recognition by the Rad4/XPC nucleotide excision repair complex. Nucleic Acids Res. 2019;47:6015–28.

Pavani R, Tripathi V, Vrtis KB, Zong DL, Chari R, Callen E, Pankajam AV, Zhen G, Matos-Rodrigues G, Yang JJ, et al. Structure and repair of replication-coupled DNA breaks. Science. 2024;385:eado3867.

Pavlova I, Iudin M, Surdina A, Severov V, Varizhuk A. G-quadruplexes in nuclear biomolecular condensates. Genes (Basel). 2023;14:1076.

Pecori R, Di Giorgio S, Paulo Lorenzo J, Nina Papavasiliou F. Functions and consequences of AID/APOBEC-mediated DNA and RNA deamination. Nat Rev Genet. 2022;23:505–18.

Pellegrini L. The Pol α-primase complex. In: The eukaryotic replisome: a guide to protein structure and function, vol. 62. Dordrecht: Springer; 2012. p. 157–70.

Pellegrini L, Costa A. New insights into the mechanism of DNA duplication by the eukaryotic replisome. Trends Biochem Sci. 2016;41:859–71.

Pikor L, Thu K, Vucic E, Lam W. The detection and implication of genome instability in cancer. Cancer Metastasis Rev. 2013;32:341–52.

Pines A, Vrouwe MG, Marteijn JA, Typas D, Luijsterburg MS, Cansoy M, Hensbergen P, Deelder A, de Groot A, Matsumoto S, et al. PARP1 promotes nucleotide excision repair through DDB2 stabilization and recruitment of ALC1. J Cell Biol. 2012;199:235–49.

Pleschke JM, Kleczkowska HE, Strohm M, Althaus FR. Poly(ADP-ribose) binds to specific domains in DNA damage checkpoint proteins. J Biol Chem. 2000;275:40974–80.

Poltronieri P, Miwa M. Editorial (thematic issue: overview on ADP ribosylation and PARP superfamily of proteins). Curr Protein Pept Sci. 2016;17:630–2.

Pommier Y, Kohlhagen G, Pourquier P, Sayer JM, Kroth H, Jerina DM. Benzo[a]pyrene diol epoxide adducts in DNA are potent suppressors of a normal topoisomerase I cleavage site and powerful inducers of other topoisomerase I cleavages. Proc Natl Acad Sci USA. 2000;97:2040–5.

Pommier Y, Sun Y, Huang SN, Nitiss JL. Roles of eukaryotic topoisomerases in transcription, replication and genomic stability. Nat Rev Mol Cell Biol. 2016;17:703–21.

Pommier Y, Nussenzweig A, Takeda S, Austin C. Human topoisomerases and their roles in genome stability and organization. Nat Rev Mol Cell Biol. 2022;23:407–27.

Popuri V, Bachrati CZ, Muzzolini L, Mosedale G, Costantini S, Giacomini E, Hickson ID, Vindigni A. The human RecQ helicases, BLM and RECQ1, display distinct DNA substrate specificities. J Biol Chem. 2008;283:17766–76.

Pouliot JJ, Yao KC, Robertson CA, Nash HA. Yeast gene for a Tyr-DNA phosphodiesterase that repairs topoisomerase I complexes. Science. 1999;286:552–5.

Powers KT, Washington MT. Analyzing the catalytic activities and interactions of eukaryotic translesion synthesis polymerases. Methods Enzymol. 2017;592:329–56.

Powers KT, Washington MT. Eukaryotic translesion synthesis: choosing the right tool for the job. DNA Repair. 2018;71:127–34.

Prakash A, Kieken F, Marky LA, Borgstahl GE. Stabilization of a G-quadruplex from unfolding by replication protein a using potassium and the porphyrin TMPyP4. J Nucleic Acids. 2011a;2011:529828.

Prakash A, Natarajan A, Marky LA, Ouellette MM, Borgstahl GE. Identification of the DNA-binding domains of human replication protein a that recognize G-quadruplex DNA. J Nucleic Acids. 2011b;2011:896947.

Purohit NK, Robu M, Shah RG, Geacintov NE, Shah GM. Characterization of the interactions of PARP-1 with UV-damaged DNA in vivo and in vitro. Sci Rep. 2016;6:19020.

Pursell ZF, Isoz I, Lundstrom EB, Johansson E, Kunkel TA. Yeast DNA polymerase epsilon participates in leading-strand DNA replication. Science. 2007;317:127–30.

Qiu Y, Niu H, Vukovic L, Sung P, Myong S. Molecular mechanism of resolving trinucleotide repeat hairpin by helicases. Structure. 2015;23:1018–27.

Ranjha L, Howard SM, Cejka P. Main steps in DNA double-strand break repair: an introduction to homologous recombination and related processes. Chromosoma. 2018;127:187–214.

Ray Chaudhuri A, Nussenzweig A. The multifaceted roles of PARP1 in DNA repair and chromatin remodelling. Nat Rev Mol Cell Biol. 2017;18:610–21.

Ray S, Qureshi MH, Malcolm DW, Budhathoki JB, Celik U, Balci H. RPA-mediated unfolding of systematically varying G-quadruplex structures. Biophys J. 2013;104:2235–45.

Rhodes D, Lipps HJ. G-quadruplexes and their regulatory roles in biology. Nucleic Acids Res. 2015;43:8627–37.

Robert I, Dantzer F, Reina-San-Martin B. Parp1 facilitates alternative NHEJ, whereas Parp2 suppresses IgH/c-myc translocations during immunoglobulin class switch recombination. J Exp Med. 2009;206:1047–56.

Robinson J, Raguseo F, Nuccio SP, Liano D, Di Antonio M. DNA G-quadruplex structures: more than simple roadblocks to transcription? Nucleic Acids Res. 2021;49:8419–31.

Rossetti S, Strmecki L, Gamble V, Burton S, Sneddon V, Peral B, Roy S, Bakkaloglu A, Komel R, Winearls CG, et al. Mutation analysis of the entire PKD1 gene: genetic and diagnostic implications. Am J Hum Genet. 2001;68:46–63.

Rzechorzek NJ, Hardwick SW, Jatikusumo VA, Chirgadze DY, Pellegrini L. CryoEM structures of human CMG-ATPγS-DNA and CMG-AND-1 complexes. Nucleic Acids Res. 2020;48:6980–95.

Saito Y, Santosa V, Ishiguro KI, Kanemaki MT. MCMBP promotes the assembly of the MCM2-7 hetero-hexamer to ensure robust DNA replication in human cells. elife. 2022;11:e77393.

Salas TR, Petruseva I, Lavrik O, Bourdoncle A, Mergny JL, Favre A, Saintome C. Human replication protein A unfolds telomeric G-quadruplexes. Nucleic Acids Res. 2006;34:4857–65.

Sale JE. Translesion DNA synthesis and mutagenesis in eukaryotes. Cold Spring Harb Perspect Biol. 2013;5:a012708.

Sale JE, Lehmann AR, Woodgate R. Y-family DNA polymerases and their role in tolerance of cellular DNA damage. Nat Rev Mol Cell Biol. 2012;13:141–52.

Sancar A. DNA excision repair. Annu Rev Biochem. 1996;65:43–81.

Sanders CM. Human Pif1 helicase is a G-quadruplex DNA-binding protein with G-quadruplex DNA-unwinding activity. Biochem J. 2010;430:119–28.

Sarkies P, Murat P, Phillips LG, Patel KJ, Balasubramanian S, Sale JE. FANCJ coordinates two pathways that maintain epigenetic stability at G-quadruplex DNA. Nucleic Acids Res. 2012;40:1485–98.

Satoh MS, Lindahl T. Role of poly(ADP-ribose) formation in DNA repair. Nature. 1992;356:356–8.

Saxena S, Nabel CS, Seay TW, Patel PS, Kawale AS, Crosby CR, Tigro H, Oh E, Vander Heiden MG, Hata AN, et al. Unprocessed genomic uracil as a source of DNA replication stress in cancer cells. Mol Cell. 2024;84:2036.

Schiavone D, Guilbaud G, Murat P, Papadopoulou C, Sarkies P, Prioleau MN, Balasubramanian S, Sale JE. Determinants of G quadruplex-induced epigenetic instability in REV1-deficient cells. EMBO J. 2014;33:2507–20.

Schimmel J, Kool H, van Schendel R, Tijsterman M. Mutational signatures of non-homologous and polymerase theta-mediated end-joining in embryonic stem cells. EMBO J. 2017;36:3634–49.

Schrempf A, Bernardo S, Arasa Verge EA, Ramirez Otero MA, Wilson J, Kirchhofer D, Timelthaler G, Ambros AM, Kaya A, Wieder M, et al. POLtheta processes ssDNA gaps and promotes replication fork progression in BRCA1-deficient cells. Cell Rep. 2022;41:111716.

Schroth GP, Ho PS. Occurrence of potential cruciform and H-DNA forming sequences in genomic DNA. Nucleic Acids Res. 1995;23:1977–83.

Schwab RA, Nieminuszczy J, Shin-ya K, Niedzwiedz W. FANCJ couples replication past natural fork barriers with maintenance of chromatin structure. J Cell Biol. 2013;201:33–48.

Sebastian R, Hosogane EK, Sun EG, Tran AD, Reinhold WC, Burkett S, Sturgill DM, Gudla PR, Pommier Y, Aladjem MI, et al. Epigenetic regulation of DNA repair pathway choice by macroH2A1 splice variants ensures genome stability. Mol Cell. 2020;79:836–845.e7.

Sfeir A, de Lange T. Removal of shelterin reveals the telomere end-protection problem. Science. 2012;336:593–7.

Sfeir A, Symington LS. Microhomology-mediated end joining: a back-up survival mechanism or dedicated pathway? Trends Biochem Sci. 2015;40:701–14.

Sfeir A, Tijsterman M, McVey M. Microhomology-mediated end-joining chronicles: tracing the evolutionary footprints of genome protection. Annu Rev Cell Dev Biol. 2024;40:195–218.

Sharma S, Javadekar SM, Pandey M, Srivastava M, Kumari R, Raghavan SC. Homology and enzymatic requirements of microhomology-dependent alternative end joining. Cell Death Dis. 2015;6:e1697.

Shen Z. Genomic instability and cancer: an introduction. J Mol Cell Biol. 2011;3:1–3.

Shoulkamy MI, Mohammed TO, Ide H, Nakano T. DNA-protein cross-links emerge as major contributors to chemotherapeutic cytotoxicity at physiological equitoxic doses. Sci Rep. 2025;15:23330.

Singatulina AS, Hamon L, Sukhanova MV, Desforges B, Joshi V, Bouhss A, Lavrik OI, Pastre D. PARP-1 activation directs FUS to DNA damage sites to form PARG-reversible compartments enriched in damaged DNA. Cell Rep. 2019;27:1809–1821.e1805.

Smela ME, Hamm ML, Henderson PT, Harris CM, Harris TM, Essigmann JM. The aflatoxin B(1) formamidopyrimidine adduct plays a major role in causing the types of mutations observed in human hepatocellular carcinoma. Proc Natl Acad Sci USA. 2002;99:6655–60.

Smith R, Zentout S, Rother M, Bigot N, Chapuis C, Mihut A, Zobel FF, Ahel I, van Attikum H, Timinszky G, et al. HPF1-dependent histone ADP-ribosylation triggers chromatin relaxation to promote the recruitment of repair factors at sites of DNA damage. Nat Struct Mol Biol. 2023;30:678–91.

Soldatenkov VA, Vetcher AA, Duka T, Ladame S. First evidence of a functional interaction between DNA quadruplexes and poly(ADP-ribose) polymerase-1. ACS Chem Biol. 2008;3:214.

Spegg V, Altmeyer M. Biomolecular condensates at sites of DNA damage: more than just a phase. DNA Repair. 2021;106:103179.

Spegg V, Panagopoulos A, Stout M, Krishnan A, Reginato G, Imhof R, Roschitzki B, Cejka P, Altmeyer M. Phase separation properties of RPA combine high-affinity ssDNA binding with dynamic condensate functions at telomeres. Nat Struct Mol Biol. 2023;30:451–62.

Spiegel JO, Van Houten B, Durrant JD. PARP1: structural insights and pharmacological targets for inhibition. DNA Repair (Amst). 2021;103:103125.

Srinivas US, Tan BWQ, Vellayappan BA, Jeyasekharan AD. ROS and the DNA damage response in cancer. Redox Biol. 2019;25:101084.

Stelter P, Ulrich HD. Control of spontaneous and damage-induced mutagenesis by SUMO and ubiquitin conjugation. Nature. 2003;425:188–91.

Sugimoto K, Okazaki T, Okazaki R. Mechanism of DNA chain growth, II. Accumulation of newly synthesized short chains in E. coli infected with ligase-defective T4 phages. Proc Natl Acad Sci USA. 1968;60:1356.

Sugimoto K, Okazaki T, Imae Y, Okazaki R. Mechanism of DNA chain growth. 3. Equal annealing of T4 nascent short DNA chains with the separated complementary strands of the phage DNA. Proc Natl Acad Sci USA. 1969;63:1343.

Sun H, Karow JK, Hickson ID, Maizels N. The Bloom's syndrome helicase unwinds G4 DNA. J Biol Chem. 1998;273:27587–92.

Symeonidou IE, Taraviras S, Lygerou Z. Control over DNA replication in time and space. FEBS Lett. 2012;586:2803–12.

Taschuk F, Cherry S. DEAD-box helicases: sensors, regulators, and effectors for antiviral defense. Viruses. 2020;12:181.

Tateishi-Karimata H, Sugimoto N. Roles of non-canonical structures of nucleic acids in cancer and neurodegenerative diseases. Nucleic Acids Res. 2021;49:7839–55.

Teloni F, Altmeyer M. Readers of poly(ADP-ribose): designed to be fit for purpose. Nucleic Acids Res. 2016;44:993–1006.

Tippana R, Hwang H, Opresko PL, Bohr VA, Myong S. Single-molecule imaging reveals a common mechanism shared by G-quadruplex-resolving helicases. Proc Natl Acad Sci USA. 2016;113:8448–53.

Tiwari V, Wilson DM 3rd. DNA damage and associated DNA repair defects in disease and premature aging. Am J Hum Genet. 2019;105:237–57.

Truong LN, Li YJ, Shi LDZ, Hwang PYH, He J, Wang HL, Razavian N, Berns MW, Wu XH. Microhomology-mediated End Joining and Homologous Recombination share the initial end resection step to repair DNA double-strand breaks in mammalian cells. Proc Natl Acad Sci USA. 2013;110:7720–5.

Tsutakawa SE, Yan CL, Xu XJ, Weinacht CP, Freudenthal BD, Yang K, Zhuang ZH, Washington MT, Tainer JA, Ivanov I. Structurally distinct ubiquitin- and sumo-modified PCNA: implications for their distinct roles in the DNA damage response. Structure. 2015;23:724–33.

Unk I, Hajdú I, Blastyák A, Haracska L. Role of yeast Rad5 and its human orthologs, HLTF and SHPRH in DNA damage tolerance. DNA Repair. 2010;9:257–67.

Vannier JB, Sandhu S, Petalcorin MI, Wu X, Nabi Z, Ding H, Boulton SJ. RTEL1 is a replisome-associated helicase that promotes telomere and genome-wide replication. Science. 2013;342:239–42.

Vasquez KM, Christensen J, Li L, Finch RA, Glazer PM. Human XPA and RPA DNA repair proteins participate in specific recognition of triplex-induced helical distortions. Proc Natl Acad Sci USA. 2002;99:5848–53.

Vekariya U, Minakhin L, Chandramouly G, Tyagi M, Kent T, Sullivan-Reed K, Atkins J, Ralph D, Nieborowska-Skorska M, Kukuyan AM, et al. PARG is essential for Polθ-mediated DNA end-joining by removing repressive poly-ADP-ribose marks. Nat Commun. 2024;15:5822.

Verschure PJ, Van Der Kraan I, Enserink JM, Mone MJ, Manders EM, Van Driel R. Large-scale chromatin organization and the localization of proteins involved in gene expression in human cells. J Histochem Cytochem. 2002;50:1303–12.

Vodenicharov MD, Sallmann FR, Satoh MS, Poirier GG. Base excision repair is efficient in cells lacking poly(ADP-ribose) polymerase 1. Nucleic Acids Res. 2000;28:3887–96.

Vogt A, He Y. Structure and mechanism in non-homologous end joining. DNA Repair. 2023;130:103547.

Vriend LEM, Prakash R, Chen CC, Vanoli F, Cavallo F, Zhang Y, Jasin M, Krawczyk PM. Distinct genetic control of homologous recombination repair of Cas9-induced double-strand breaks, nicks and paired nicks. Nucleic Acids Res. 2016;44:5204–17.

Wang ZQ, Stingl L, Morrison C, Jantsch M, Los M, SchulzeOsthoff K, Wagner EF. PARP is important for genomic stability but dispensable in apoptosis. Genes Dev. 1997;11:2347–58.

Wang ML, Wu WZ, Wu WQ, Rosidi B, Zhang LH, Wang HC, Iliakis G. PARP-1 and Ku compete for repair of DNA double strand breaks by distinct NHEJ pathways. Nucleic Acids Res. 2006;34:6170–82.

Wang Z, Song Y, Li S, Kurian S, Xiang R, Chiba T, Wu X. DNA polymerase theta (POLQ) is important for repair of DNA double-strand breaks caused by fork collapse. J Biol Chem. 2019;294:3909–19.

Wang YL, Zhao WW, Shi J, Wan XB, Zheng J, Fan XJ. Liquid-liquid phase separation in DNA double-strand breaks repair. Cell Death Dis. 2023;14:746.

Watson JD, Crick FHC. Molecular structure of nucleic acids – a structure for deoxyribose nucleic acid. Nature. 1953;171:737–8.

Watson JD, Crick FHC. Molecular-structure of nucleic-acids – a structure for deoxyribose nucleic-acid (Reprinted from Nature, Vol 171, Pg 737, 1953). Ann N Y Acad Sci. 1995;758:737–8.

Wei X, Zhou F, Zhang L. PARP1-DNA co-condensation: the driver of broken DNA repair. Signal Transduct Target Ther. 2024;9:135.

Weterings E, Van Gent DC. The mechanism of non-homologous end-joining: a synopsis of synapsis. DNA Repair. 2004;3:1425–35.

Wiese C, Dray E, Groesser T, Filippo JS, Shi I, Collins DW, Tsai MS, Williams GJ, Rydberg B, Sung P, et al. Promotion of homologous recombination and genomic stability by RAD51AP1 via RAD51 recombinase enhancement. Mol Cell. 2007;28:482–90.

Wold MS. Replication protein A: a heterotrimeric, single-stranded DNA-binding protein required for eukaryotic DNA metabolism. Annu Rev Biochem. 1997;66:61–92.

Wolters S, Schumacher B. Genome maintenance and transcription integrity in aging and disease. Front Genet. 2013;4:19.

Wright WD, Shah SS, Heyer WD. Homologous recombination and the repair of DNA double-strand breaks. J Biol Chem. 2018;293:10524–35.

Wu CG, Spies M. G-quadruplex recognition and remodeling by the FANCJ helicase. Nucleic Acids Res. 2016;44:8742–53.

Wu Y, Shin-ya K, Brosh RM Jr. FANCJ helicase defective in Fanconia anemia and breast cancer unwinds G-quadruplex DNA to defend genomic stability. Mol Cell Biol. 2008;28:4116–28.

Wu Y, Sommers JA, Khan I, de Winter JP, Brosh RM Jr. Biochemical characterization of Warsaw breakage syndrome helicase. J Biol Chem. 2012;287:1007–21.

Wu WQ, Hou XM, Li M, Dou SX, Xi XG. BLM unfolds G-quadruplexes in different structural environments through different mechanisms. Nucleic Acids Res. 2015;43:4614–26.

Wu WQ, Hou XM, Zhang B, Fosse P, Rene B, Mauffret O, Li M, Dou SX, Xi XG. Single-molecule studies reveal reciprocating of WRN helicase core along ssDNA during DNA unwinding. Sci Rep. 2017;7:43954.

Wyatt DW, Feng WJ, Conlin MP, Yousefzadeh MJ, Roberts SA, Mieczkowski P, Wood RD, Gupta GP, Ramsden DA. Essential roles for polymerase θ-mediated end joining in the repair of chromosome breaks. Mol Cell. 2016;63:662–73.

Xiang S, Huang Y, Xie J. Non-homologous end-joining (NHEJ): physiological function in Mycobacterium and application in gene editing. Sheng Wu Gong Cheng Xue Bao. 2025;41:1280–90.

Xu ZC, Feng JR, Yu DQ, Huo YJ, Ma XH, Lam WH, Liu Z, Li XD, Ishibashi T, Dang SY, et al. Synergism between CMG helicase and leading strand DNA polymerase at replication fork. Nat Commun. 2023;14:5849.

Xu HZ, Ye J, Zhang KX, Hu QX, Cui TX, Tong C, Wang MQ, Geng HC, Shui KM, Sun Y, et al. Chemoproteomic profiling unveils binding and functional diversity of endogenous proteins that interact with endogenous triplex DNA. Nat Chem. 2024;16:1811.

Yang J, Bachrati CZ, Ou JW, Hickson ID, Brown GW. Human topoisomerase IIIα is a single-stranded DNA decatenase that is stimulated by BLM and RMI1. J Biol Chem. 2010;285:21426–36.

Yang G, Liu C, Chen SH, Kassab MA, Hoff JD, Walter NG, Yu XC. Super-resolution imaging identifies PARP1 and the Ku complex acting as DNA double-strand break sensors. Nucleic Acids Res. 2018;46:3446–57.

Yang G, Chen YB, Wu JX, Chen SH, Liu XH, Singh AK, Yu XC. Poly(ADP-ribosyl)ation mediates early phase histone eviction at DNA lesions. Nucleic Acids Res. 2020;48:3001–13.

Yeeles JTP, Janska A, Early A, Diffley JFX. How the eukaryotic replisome achieves rapid and efficient DNA replication. Mol Cell. 2017;65:105–16.

Yunis JJ, Bahr GF. Chromatin fiber organization of human interphase and prophase chromosomes. Exp Cell Res. 1979;122:63–72.

Zeman MK, Cimprich KA. Causes and consequences of replication stress. Nat Cell Biol. 2014;16:2–9.

Zessin PJM, Sporbert A, Heilemann M. PCNA appears in two populations of slow and fast diffusion with a constant ratio throughout S-phase in replicating mammalian cells. Sci Rep. 2016;6:18779.

Zhang F, Shi J, Bian C, Yu X. Poly(ADP-ribose) mediates the BRCA2-dependent early DNA damage response. Cell Rep. 2015;13:678–89.

Zhang Y, Yang M, Duncan S, Yang X, Abdelhamid MAS, Huang L, Zhang H, Benfey PN, Waller ZAE, Ding Y. G-quadruplex structures trigger RNA phase separation. Nucleic Acids Res. 2019;47:11746–54.

Zhang H, Xiong Y, Su D, Wang C, Srivastava M, Tang M, Feng X, Huang M, Chen Z, Chen J. TDP1-independent pathways in the process and repair of TOP1-induced DNA damage. Nat Commun. 2022;13:4240.

Zhao L, Washington MT. Translesion synthesis: insights into the selection and switching of DNA polymerases. Genes. 2017;8:24.

Zhao JH, Wang GL, del Mundo IM, McKinney JA, Lu XL, Bacolla A, Boulware SB, Zhang CS, Zhang HH, Ren PY, et al. Distinct mechanisms of nuclease-directed DNA-structure-induced genetic instability in cancer genomes. Cell Rep. 2018;22:1200–10.

Zhou ZX, Lujan SA, Burkholder AB, Garbacz MA, Kunkel TA. Roles for DNA polymerase δ in initiating and terminating leading strand DNA replication. Nat Commun. 2019;10:3992.

Zinovyev A, Kuperstein I, Barillot E, Heyer WD. Synthetic lethality between gene defects affecting a single non-essential molecular pathway with reversible steps. PLoS Comput Biol. 2013;9:e1003016.

Chapter 7
PCNA Macromolecular Complexes: PCNA Serves as a Molecular Hub Regulating Multiple Cellular Processes Inside and Outside of the Nucleus

Hamsini Kala, Dana Abou Abbas, Linda H. Malkas, and Robert J. Hickey

Abstract With over 27,100 papers currently written about Proliferating Cell Nuclear Antigen (PCNA) and more than 11,700 papers describing a role for the protein and its relationship to cancer, this chapter will just touch upon some of the important role's PCNA has within cells and what types of proteins might interact with PCNA. This chapter is not meant to provide a thorough review of the multitude of protein complexes utilizing PCNA as a cofactor but rather highlight only some of those direct or indirect interactions between PCNA and its binding partners which together are involved in maintaining the genome, regulating cytoplasmic activities, and participating in immune surveillance. Additionally, this chapter will provide a high-level overview of how the caPCNA isoform is being used as a platform for a new class of anticancer therapeutic agent that selectively kills cancer cells, while leaving non-cancer cells unharmed.

Keywords Proliferating cell nuclear antigen · DNA replication and repair · Genome stability · Replication stress · Post-translational modifications · Protein-

H. Kala
Irell & Manella Graduate School of Biological Sciences, Department of Cancer Biology and Molecular Medicine, Beckman Research Institute of City of Hope, Duarte, CA, USA

D. A. Abbas
Irell & Manella Graduate School of Biological Sciences, Department of Molecular Diagnostics and Experimental Therapeutics, Beckman Research Institute of City of Hope, Duarte, CA, USA

L. H. Malkas
Department of Molecular Diagnostics and Experimental Therapeutics, Beckman Research Institute of City of Hope, Duarte, CA, USA
e-mail: lmalkas@coh.org

R. J. Hickey (✉)
Department of Cancer Biology and Molecular Medicine, Beckman Research Institute of City of Hope, Duarte, CA, USA
e-mail: rohickey@coh.org

© The Author(s), under exclusive license to Springer Nature Switzerland AG 2026
A. M. Pedley (ed.), *Supramolecular Protein Assemblies In Cells*, Advances in Experimental Medicine and Biology 1514,
https://doi.org/10.1007/978-3-032-26629-3_7

protein interactions · Immune surveillance · Cancer-associated PCNA isoform · Small-molecule inhibitors

7.1 Demographics of Proliferating Cell Nuclear Antigen

PCNA is encoded by a single gene that is highly conserved from human to archae-bacteria. It is located between human chromosome 20p12.3 and 20p13. Two variants exist: with one variant having an un-spliced length of 11,670 base pairs (bp), a spliced length of 1355 bp, and a protein length of 261 amino acids. The second variant has an un-spliced length of 5049 bp, a spliced length of 1319 bp, and a protein length of 261 amino acids. The differences in transcript length are derived from the transcription start and ending sites and the translation start or termination site within the gene (Webb et al. 1990; Stoimenov and Helleday 2011). Three possible pseudo-genes, (PCNAP, PCNAP1, and PCNAP2) have also been identified. PCNAP is located within human chromosome X-p11 (Webb et al. 1990; Ku et al. 1989); while PCNAP1 and PCNAP2 are located as tandem genes on human chromosome 4-q24 (Taniguchi et al. 1996). Additionally, there are at least two recognized possible pseudo-genes; one on human chromosome 11-p15.1 and the other on chromosome Xp11.3 (Webb et al. 1990). Knock-out models of PCNA prove to be embryonic lethal (Roa et al. 2008), and according to the OMIM and Human Locus Specific Mutation Databases (Human Locus Specific Mutation Databases 2011), there are no known diseases, caused by mutation or loss of function of the PCNA protein (Stoimenov and Helleday 2011). The mammalian protein itself is assembled from three molecules of PCNA protein arranged into a ring-like structure with a center opening wide enough to allow DNA to easily slide through the opening. The PCNA protein can undergo a series of different post-translational modifications such as, but not limited to acetylation, phosphorylation, methyl esterification, glycosylation, palmitoylation, ubiquitination, SUMOylation, and Neddylation (Zhang et al. 2021). As a group, these modifications occur at specific sites across the protein and likely contribute to the changes in conformation and protein-protein interactions required for PCNA to contribute to its varied roles within the cell.

7.2 Introduction

Proliferating cell nuclear antigen is a pivotal protein that acts as a sliding clamp and scaffold in multiple nuclear and cytoplasmic processes. First identified as an antigen in autoimmune disorders (i.e., Lupus) and later recognized as a "cyclin" due to its cell cycle correlated expression level, PCNA is now known to be central to DNA replication and repair (Moldovan et al. 2007) among other processes. It was originally thought to reside exclusively within the nucleus of the cell (hence its name), where it participated in a variety of biochemical pathways required for the duplication, repair, and maintenance of chromosomal DNA. More recently, PCNA has been found to participate in multiple other cellular processes (Essers et al. 2000) and can also be found in the cytoplasm and on the outer cellular membrane. As mentioned, PCNA is a homo-trimeric protein consisting of three identical subunits of the PCNA molecule. These subunits form a ring-shaped structure that encircles DNA, enabling it to tether and organize numerous enzymes and proteins at the replication fork and at sites of DNA damage (Webb et al. 1990; Kelman 1997). Through interactions with nearly 200+ binding partners, PCNA coordinates the functions of vital macromolecular complexes required within the nucleus for genome duplication, DNA damage tolerance, and the maintenance of genomic stability (Webb et al. 1990; Mailand et al. 2013). In addition, emerging research shows that PCNA's role extends beyond DNA metabolism into broader cellular signaling including cell cycle control, apoptosis regulation, and metabolic adaptation, all of which are highly relevant to cancer biology (Mailand et al. 2013). Additionally, PCNA has a role in immune surveillance via a p44 complex found on the outer membrane of the cell (Witko-Sarsat et al. 2010).

7.3   Structure and Function of PCNA

PCNA consists of three ~29-kDa monomers each of which is folded into two domains that are connected to one another by an inter-domain connecting loop (IDCL) (Moldovan et al. 2007). The three monomers assemble head-to-tail to form a closed ring with pseudo-sixfold symmetry, creating a doughnut-shaped clamp through which duplex DNA can slide (Moldovan et al. 2007). The inner face of the ring is positively charged and contacts DNA, while the outer face is negatively charged (Moldovan et al. 2007). This structure allows PCNA to topologically clamp around DNA, preventing it from dissociating and guiding the double helix to slide freely along, and within, a multiprotein polymerase complex mediating the duplication and potentially the repair of the DNA duplex template. PCNA's movement on DNA has been likened to a cogwheel or switch mechanism that ensures it can track the DNA backbone, while providing enough flexibility for the polymerase to proofread newly synthesized DNA or bypass lesions within the DNA (Moldovan et al. 2007).

PCNA itself has no enzymatic activity; rather, its primary function is to serve as a molecular platform or scaffold that recruits and coordinates multiple enzymes and factors (Mailand et al. 2013) involved in the replication and repair of DNA. Many proteins bind to PCNA via short chemically similar motifs involving one or more aromatic and/or hydrophobic amino acids within a shallow surface pocket within the IDCL that is referred to as the PCNA-interacting peptide (PIP) box, composed of an eight amino acid sequence (~aa126–133) that docks a variety of proteins into PCNA's IDCL region (Warbrick 1998). The Inter-Domain Connector Loop is itself reported as a 17-amino-acid sequence between PCNA amino acids 118–134 and joins the right (aa1–117) and left (aa135–261) domains of the protein. This PIP box interacts with a series of proteins involved in DNA replication and repair including Pol Δ and ε, Fen1, p15, DNMT1, p21, RNase H2, Pol i, Pol k, Pol η, PARG, CDT2, TRA18, etc. (Gonzalez-Magana and Blanco 2020). Another motif found within the PCNA protein interacts with certain stress response proteins at PCNA's APIM binding site (i.e., AlkB homolog 2 PCNA-Interacting Motif) (Ohayon et al. 2016). Through these interactions, PCNA acts as a tool-belt, holding and exchanging proteins at sites of DNA replication or repair as needed. Switching of PCNA partners may occur by competition (where a higher-affinity partner displaces another) or can be regulated by PCNA directed post-translational modifications that alter its binding preferences (Mailand et al. 2013). This dynamic assembly and disassembly of PCNA-protein complexes enable rapid responses to changing conditions during DNA synthesis or damage repair.

7.4 PCNA in DNA Replication

During S-phase, PCNA is indispensable for high-fidelity and processive DNA replication. Historically, Pol δ was believed to perform leading-strand synthesis, while Pol ε was thought to have an undefined but parallel role. Foundational studies (Kunkel and Burgers 2008) revised this model, showing that Pol ε is the primary leading-strand polymerase, whereas Pol δ predominantly extends Okazaki fragments on the lagging strand. Pol α, once thought to be the lagging-strand polymerase, is now known to initiate synthesis on both strands by generating short RNA–DNA primers. Although Pol α does not require PCNA for its primase-associated activity, PCNA coordinates the essential polymerase-switching events that follow primer synthesis. This refined understanding explains why Pol α can still interact functionally at the fork: PCNA does not recruit Pol α directly but orchestrates the transition from Pol α to the high-fidelity polymerases that complete strand elongation (Maga and Hübscher 2003).

Beyond polymerases, PCNA orchestrates the maturation of Okazaki fragments during lagging strand replication. After Pol δ synthesizes a new Okazaki fragment and dissociates, PCNA remains attached at the nick and recruits flap endonuclease-1 (FEN1) and DNA ligase I, two enzymes essential for processing and joining Okazaki fragments (Hingorani and O'Donnell 2000; Balakrishnan and Bambara 2013). FEN1, guided by PCNA, cleaves the 5′ flap of the RNA primer that was displaced by Pol δ, and DNA ligase I then seals the nicks between adjacent fragments (Hingorani and O'Donnell 2000; Balakrishnan and Bambara 2013). Disruption of the FEN1–PCNA interaction has been shown to stall replication forks and uncouple leading/lagging strand synthesis, underscoring the critical nature of this complex (Balakrishnan and Bambara 2013). PCNA also works with RNase H2 (to remove RNA primers) and coordinates the sequential hand-off from Pol δ to FEN1 to ligase in an orderly "tool switching" mechanism for efficient lagging strand completion (Moldovan et al. 2007; Balakrishnan and Bambara 2013).

In addition to core replication enzymes, PCNA interacts with factors for chromatin assembly and epigenetic maintenance during replication. For example, the histone chaperone CAF-1 binds PCNA to deposit new nucleosomes behind the replication fork (Moldovan et al. 2007; Mailand et al. 2013). PCNA also recruits DNA methyltransferase-1 (DNMT1) to newly replicated DNA, ensuring proper maintenance of DNA methylation patterns on the daughter strands (Moldovan et al. 2007). Furthermore, PCNA is implicated in establishing sister chromatid cohesion through a complex with cohesin-loading factors (such as Ctf18-RFC), linking replication to chromosome segregation (Moldovan et al. 2007; Mailand et al. 2013). In summary, the PCNA trimer at the fork acts as a maestro of replication, holding together a large multiprotein complex that synthesizes DNA, replaces primers, proofreads errors, and immediately packages the new DNA into chromatin.

7.5 PCNA in DNA Repair Pathways

PCNA's function as a scaffold extends into virtually every major DNA repair pathway, reflecting its role at the crossroads of DNA replication and repair (Moldovan et al. 2007). When DNA damage or mismatches are encountered, PCNA recruits repair enzymes to resolve these issues, often using the same sliding clamp mechanism to coordinate repair synthesis.

- *Mismatch Repair (MMR):* In post-replicative MMR, PCNA interacts with the MutSα (MSH2-MSH6) and MutLα (MLH1-PMS2) complexes to fix base mispairing or insertion/deletion loops. PCNA binding is thought to help identify the newly synthesized (error-containing) strand by interacting with MSH6, and it stimulates the endonuclease activity of PMS2 on that strand (Moldovan et al. 2007; Mailand et al. 2013). PCNA also recruits Exonuclease 1 (Exo1) to cut out the error-containing DNA segment and then helps DNA Pol δ to resynthesize the correct sequence (Moldovan et al. 2007). Thus, PCNA is required for the efficient removal and replacement of mismatched bases after replication.
- *Base Excision Repair (BER):* In long-patch BER (a sub-pathway of BER), PCNA coordinates a similar hand-off as in Okazaki fragment repair. After a DNA glycosylase and AP endonuclease create a single-stranded break at a damaged base, DNA Pol δ/ε performs repair synthesis, often displacing the downstream DNA into a flap. PCNA-bound FEN1 then cleaves the flap, and ligase I seals the strand (Moldovan et al. 2007; Shivji and Kenny 1992). PCNA dramatically stimulates the strand-displacement DNA synthesis by Pol δ in BER and organizes the Pol–FEN1–ligase sequence of events (Moldovan et al. 2007; Hingorani and O'Donnell 2000). BER can also occur via a short-patch mode (single nucleotide replacement) that may not require PCNA, but PCNA is essential for the more frequent long-patch BER (Moldovan et al. 2007).
- *Nucleotide Excision Repair (NER):* PCNA is recruited during the late stage of NER, after the damage-containing oligonucleotide is excised. It forms a complex

with DNA Pol δ/ε to fill in the single-stranded gap with new DNA (Moldovan et al. 2007). PCNA also binds the NER factor(s) XPG (which contains a PIP box binding motif) and possibly XPA, while helping to coordinate the dual incisions and subsequent DNA resynthesis (Mailand et al. 2013). The NER of UV-induced lesions was one of the first repair processes recognizing the involvement of PCNA in this type of repair, as UV-damaged DNA replication requires PCNA and Pol δ (Moldovan et al. 2007).

- *Double-Strand Break Repair:* Although PCNA is not a classical factor in homologous recombination (HR), it participates after a break is repaired in the later stages of the process. PCNA facilitates the DNA synthesis step of HR when the invading 3′ end primes the synthesis of the new DNA strand using the homologous DNA template (Moldovan et al. 2007; Zeman and Cimprich 2014). PCNA loading at these sites may help recruit Pol δ for repair synthesis in concert with factors like RAD51 (Moldovan et al. 2007; Essers et al. 2000). In some alternative end-joining pathways, PCNA might also assist Pol β or λ in gap filling (Moldovan et al. 2007; Zeman and Cimprich 2014). Moreover, PCNA's role in maintaining genome stability indirectly aids HR by preventing excessive recombination; for instance, PCNA modified by SUMO (see below) can restrain aberrant recombination events (Moldovan et al. 2007).

- *Translesion DNA Synthesis (TLS) and Damage Tolerance:* One of PCNA's most critical roles in DNA damage response is to enable DNA damage tolerance through TLS. When replication forks stall at DNA lesions, PCNA is mono ubiquitinated at lysine 164 by the E2/E3 enzymes RAD6/RAD18. This ubiquitination acts as a switch that attracts specialized TLS polymerases (such as Pol η, Pol ι, Pol κ, or Rev1) to PCNA (Garg and Burgers 2005; Hoege et al. 2002). The TLS polymerase can then replace the replicative polymerase and synthesize past the lesion in an error-prone but damage-tolerant manner (Webb et al. 1990). Alternatively, if PCNA becomes poly-ubiquitinated (via Lys63-linked chains by Ubc13-Mms2/Rad5), an error-free template switching pathway is promoted, using the undamaged sister chromatid as a template (Garg and Burgers 2005; Hoege et al. 2002). These PCNA modifications thereby channel the fork into either TLS or a recombinational bypass, respectively. In contrast, PCNA SUMOylation (attachment of SUMO at K164, mainly studied in yeast) helps to block unscheduled homologous recombination, for example, by recruiting the anti-recombinase Srs2 helicase (Moldovan et al. 2007; Maga and Hübscher 2003). In all, post-translational modifications of PCNA serve as molecular signals that guide the choice of the repair/tolerance pathway at stalled forks, ensuring cell survival while minimizing mutagenesis.

Through participation in these pathways, PCNA has earned the reputation of being a "major scaffold protein for DNA damage response" by interacting with dozens of repair proteins to maintain the integrity of the genome (Moldovan et al. 2007; Mailand et al. 2013). PCNA's central position in replication and repair helps cells cope with DNA lesions without catastrophic fork collapse, thus preventing genomic instability.

7.6 PCNA Interactions with Cell Cycle Regulators

Proper coordination between DNA replication, repair, and the cell division cycle is crucial for cell survival. PCNA lies at this interface by interacting with key cell cycle regulators, thereby linking DNA synthesis/repair status to cell cycle progression decisions. One prominent PCNA-binding protein is p21[Cip1/Waf1] (CDKN1A), a cyclin-dependent kinase (CDK) inhibitor whose expression is induced by p53 during DNA damage responses. Besides inhibiting CDKs to cause cell cycle arrest, p21 directly binds PCNA via a PIP-box binding domain in its C-terminus (Warbrick 1998; Abbas and Dutta 2009). The crystal structure of the p21–PCNA complex shows p21 occupies one of the PCNA monomer interfaces without breaking the trimer; effectively masking the binding site(s) on that PCNA to preclude its interaction with other proteins (Smith et al. 1994). By doing so, p21 blocks replication by preventing PCNA from recruiting DNA polymerases and other replication factors, halting DNA synthesis, and giving the cell time to repair DNA damage (Abbas and Dutta 2009). This PCNA-binding function of p21 is critical for its ability to enforce S-phase checkpoints independently of CDK inhibition (Abbas and Dutta 2009). In effect, p21 serves as a molecular "brake" at replication forks via PCNA, ensuring that DNA repair precedes replication of damaged templates. Notably, p21 can also protect cells from apoptosis in some contexts by allowing time for repair, by pausing the cell cycle through PCNA mediated CDK inhibition. By doing so, p21 helps cells avoid triggering programmed cell death in response to unrepaired DNA damage (Abbas and Dutta 2009). However, in terminally differentiated cells such as neutrophils, induction of p21 has the opposite effect; it disrupts the PCNA scaffold and prevents the induction of apoptosis (discussed below) and triggering cell death (Witko-Sarsat et al. 2010).

Another important DNA repair associated interaction is between PCNA and the Gadd45 protein (i.e., the growth arrest and DNA damage 45 protein), which is also a p53-responsive stress protein. Gadd45 was found to bind PCNA and compete with PCNA's normal partners, thereby inhibiting cell cycle progression and facilitating DNA repair (Smith et al. 1994). Specifically, Gadd45 binding to PCNA was shown to stimulate nucleotide excision repair and block entry into the S-phase of the cell cycle by displacing replication factors from PCNA (Smith et al. 1994). This supports a model where under genotoxic stress, p53-induced proteins like p21, and Gadd45 converge on PCNA to tilt the balance from proliferation toward DNA repair or the induction of apoptosis.

PCNA also interacts with the anti-apoptotic Bcl-2 family protein MCL1. MCL1 contains a PCNA-binding motif and was shown to bind PCNA in human cells (Fujise et al. 2000). Interestingly, while MCL1's main known function is to sequester pro-apoptotic factors, its interaction with PCNA serves a distinct role in regulating S-phase progression. Overexpression of MCL1 slows DNA synthesis through the S phase, whereas a mutant MCL1, which is unable to bind PCNA, fails to induce this delay (Fujise et al. 2000). Notably, the PCNA-binding mutant of MCL1 retains its anti-apoptotic activity (Fujise et al. 2000), indicating that MCL1–PCNA interaction specifically mediates a cell cycle checkpoint function independent of apoptosis regulation. Overall, through proteins such as p21, Gadd45, and MCL1, PCNA serves as a hub where signals for DNA damage, cell cycle arrest, and cell survival intersect. These interactions ensure that cells do not continue replicating a damaged genome and are "given the time" required to either repair the damage or, if the damage is irreparable, initiate apoptosis.

7.7 PCNA in Apoptosis Signaling and Cell Survival

Beyond its canonical nuclear roles, PCNA has emerged as a surprising player in apoptosis regulation. Recent studies reveal that PCNA can act in the cytoplasm as a platform to modulate apoptotic machinery and other stress response pathways, thereby influencing cell fate independently of DNA replication (Witko-Sarsat et al. 2010; Ohayon et al. 2016).

One remarkable example is found in terminally differentiated cells like neutrophils, which are short-lived immune cells that do not proliferate. In neutrophils, PCNA relocates to the cytoplasm during their maturation (Witko-Sarsat et al. 2010). Cytosolic PCNA was discovered to bind directly to multiple *procaspases*—specifically procaspase –3, –8, –9, and –10—which are the inactive precursors of caspases that execute apoptosis (Witko-Sarsat et al. 2010). By binding these procaspases, PCNA sequesters them and prevents their activation (Witko-Sarsat et al. 2010). In essence, PCNA forms an anti-apoptotic complex in neutrophils that delays or blocks the initiation of apoptosis, thus extending the cell's lifespan. This PCNA-mediated survival mechanism is so critical that if PCNA's interaction with procaspases is disrupted, neutrophils rapidly undergo programmed cell death (Witko-Sarsat et al. 2010). For instance, when pro-inflammatory conditions subside, the cyclin-dependent kinase inhibitor p21 is induced in neutrophils. This p21 binds to PCNA, displaces procaspases from the PCNA scaffold, and triggers the apoptotic cascade (Witko-Sarsat et al. 2010). This suggests a model in which pro-survival signals maintain PCNA in a caspase-bound state, whereas pro-apoptotic signals (like loss of survival factors or induction of p21) cause PCNA to release caspases, which in turn allows the cell death process to proceed (Witko-Sarsat et al. 2010). In support of this, experimental PCNA inhibitors have been shown to disrupt the PCNA–procaspase interaction and accelerate neutrophil apoptosis (Witko-Sarsat et al. 2010; Søgaard et al. 2018a).

The anti-apoptotic role of PCNA is not limited to neutrophils. In cancer cells and other contexts, PCNA's influence on survival pathways is an area of active research. Cytosolic PCNA forms complexes with multiple metabolic and stress-response proteins. PCNA has been found associated with cytosolic oncoproteins (e.g., PI3K/AKT, p47phox which regulated ROS production, S100A8/S100A9 which drives inflammation) and metabolic enzymes (i.e., alpha enolase, glyceraldehyde dehydrogenase, and 6-phosphoglucanate via PCNA's PIP box or APIM binding motifs), hinting that PCNA might integrate signals for cell survival under cells that are under stress (Ohayon et al. 2016). A striking finding in leukemia cells is that PCNA can drive metabolic changes that favor survival. Chemotherapy-resistant acute myeloid leukemia (AML) cells showed an unexpected accumulation of PCNA in the cytoplasm due to increased nuclear export (Ohayon et al. 2016). In these drug-resistant cells, cytosolic PCNA interacts with NAMPT, a key enzyme in NAD^+ biosynthesis, and this interaction led to enhanced NAD^+ production and glycolytic flux (Ohayon et al. 2016). The result is metabolic reprogramming, (i.e., higher glycolysis and energy output), which supports cell survival and the development of drug resistance (Ohayon et al. 2016). In fact, patient AML samples with high cytosolic PCNA showed similar metabolic advantages, identifying a novel PCNA-driven survival pathway distinct from its nuclear role (Mailand et al. 2013). Cytosolic PCNA has also been reported to be associated with nucleophosmin (NPM1) in leukemia models and with procaspase-9 in neuroblastoma cells under nitrosative stress, where S-nitrosylation weakens its inhibitory interaction and permits apoptosis. Together, these observations highlight that cytosolic PCNA may scaffold diverse metabolic and pro-survival signaling modules. These findings identify PCNA as a potential coordinator of the Warburg-like metabolic shifts in cancer cells, where increased glycolysis and NAD^+ levels help cancer cells cope with the cytotoxic stresses induced by chemotherapy (Ohayon et al. 2016; Søgaard et al. 2018b).

There is also evidence that post-translational modifications of PCNA can regulate its anti-apoptotic function. For example, in human neuroblastoma cells under nitric oxide stress, PCNA was found to be S-nitrosylated, a modification that reduces PCNA's binding to procaspase-9, thereby allowing caspase-9 activation and apoptosis to proceed (Søgaard et al. 2018b). In other words, S-nitrosylation serves as a switch that turns *off* PCNA's protective effect on caspase-9, tipping the balance toward cell death in response to nitrosative stress. This illustrates how cells might actively regulate PCNA's scaffolding function in the cytosol to permit induction of the apoptosis process.

In summary, PCNA can act as a molecular guardian against apoptosis by binding key apoptosis regulators (like caspases or apoptosis-inhibitory proteins) and keeping them in an inactive state. Cancer cells and other long-lived cells appear to exploit this function, by retaining or exporting PCNA into the cytoplasm, where these cells then gain an extra layer of protection against programmed cell death, which enables them to better survive life-threatening conditions (e.g., DNA damage, metabolic stress, etc.). This also implies that disrupting PCNA's interactions in these cellular contexts can *promote* apoptosis, a concept that has significant implications for cancer therapy.

7.8 PCNA in Cancer: Proliferation and Therapy Resistance

Given PCNA's fundamental role in DNA replication, it is not surprising that PCNA is highly expressed in rapidly dividing cells. In pathology and particularly for hematopoietic malignancies, PCNA expression levels have been used, along with Ki67, as a proliferation marker. Elevated levels of PCNA in tumor biopsies are generally correlated with high cell proliferation rates and have been used as a prognostic biomarker. While this use has proven to be generally useful for hematopoietic malignancies (Gu et al. 2018), it has proven to be somewhat unreliable as a clinical prognostic tool for various solid tumors (Warbrick 1998).

Given PCNA's fundamental role in DNA replication, it is not surprising that PCNA is highly expressed in rapidly dividing cells. However, large surveys across solid tumor studies found that PCNA immunostaining is highly variable and often less predictive of clinical outcome than Ki67, due in part to PCNA expression in non-cycling cells engaged in DNA repair and the development of resistance to anti-cancer therapeutic interventions. In contrast, PCNA proved more reliable as a proliferation metric in hematopoietic malignancies, where proliferative burden appears to correlate more directly with disease progression (Gu et al. 2018). Early work using PCNA-selective antibodies as a prognostic biomarker for breast cancer, along with more than 70 publications on this and other types of cancer, supported the conclusion that PCNA alone lacks strong prognostic power in solid tumors but retains utility in selected blood cancers. Despite these limitations, pan-cancer transcriptomic analyses confirm that PCNA is widely overexpressed across human tumors (Søgaard et al. 2018b).

Despite its shortcomings as a prognostic tool for solid tumor progression, pan-cancer analyses have confirmed that PCNA is frequently overexpressed in human cancers, and its expression level does correlate with the proliferative capacity of cells within multiple tumor types (Søgaard et al. 2018b). Cancer cells, which endure replicative stress and DNA damage, are especially reliant on PCNA's replication and repair functions to maintain the proliferating cancer cell genome. PCNA is essential for cancer cell growth and survival (Søgaard et al. 2018b), in-large-part because it enables robust DNA synthesis and efficient repair of DNA lesions that would otherwise kill the cell. Experimental depletion or inhibition of PCNA typically results in stalled DNA replication, accumulation of DNA damage, and induction of apoptosis in cancer cells (Søgaard et al. 2018b).

An unfortunate tenant of treating cancers with cytotoxic agents is that cancer cells often develop resistance to treatments such as chemotherapy and radiation. These treatments aim to induce DNA damage or disrupt metabolic functions such that the cancer cells succumb to disruption in distinct cellular processes or alterations to their micro-environment. PCNA sits at the nexus of many drug resistance mechanisms by virtue of its role in inducing/sustaining tolerance to DNA damage. For example, if a chemotherapeutic drug causes DNA adducts or crosslinks, cancer cells with upregulated PCNA can better recruit TLS polymerases or repair pathways to bypass or fix these lesions, thereby evading cell death. Overexpression of PCNA and its partner proteins enable tumors to augment their capacity for DNA repair. Moreover, as described above, PCNA's relocation to the cytoplasm in some cancer cells confers metabolic and anti-apoptotic advantages that help them survive treatment (Ohayon et al. 2016). The PCNA–NAMPT interaction in drug-resistant leukemia is a clear example: by boosting NAD$^+$ levels and glycolysis, the cells can better cope with energy demands and the stresses caused by chemotherapeutic agents, which in turn renders treatments less effective (Ohayon et al. 2016).

There is also evidence that cancer cells may harbor, or select for, alterations in PCNA or its regulatory pathways. Studies have identified a cancer-associated iso-form of PCNA (caPCNA) that is present in a broad range of tumor cells, but not in normal cells or benign proliferative disease (Gu et al. 2018; Li et al. 2021). This caPCNA may be a conformational variant or specifically modified isoform of PCNA that exposes certain epitopes (for instance, the PIP box region within PCNA's Inter-Domain Connector Loop (IDCL segment encompassing amino acids L126–Y133) and makes this binding domain more accessible in cancer cells to interaction with antibodies directed toward this short peptide region of the PCNA protein than in non-cancer cells) (Gu et al. 2018; Li et al. 2021). This region of PCNA contains the shallow docking site (i.e., the PIP box domain) of PCNA utilized by many of PCNA's binding partners. The existence of caPCNA suggests that tumor cells might utilize a form of PCNA with altered interaction surfaces, potentially to better engage certain proteins involved in crucial pathways needed to maintain the cancer cell or to escape regulation. Importantly, this difference offers an opportunity for selec-tively treating cancer cells without disrupting PCNA function in non-cancer cells. Thus, therapeutic agents have been designed to specifically inhibit caPCNA and enable selective disruption of PCNA function in the cancer cell, which leads to the selective killing of the cancer cell while sparing proliferating non-cancer cells that use the "normal cell" PCNA isoform.

In addition to the roles already described for PCNA within the cytoplasm, PCNA interacts with the γ-tubulin network and plays a role in microtubule nucleation and anchoring at the centrosome. Localization studies demonstrate that PCNA can be detected in proximity to spindle microtubules and centrosomal structures during mitosis, suggesting a spatial relationship with both structural and nucleating com-ponents of the microtubule network. Further disruption of PCNA: γ-tubulin interac-tions result in aberrant mitotic events, including mitotic slippage and death. The interaction between PCNA and γ-tubulin appears highly specific as neither α- or β-tubulin co-immunoprecipitated with PCNA (Wendel et al. 2026).

Overall, PCNA can be thought of as a central node in cancer cell physiology: it drives proliferation, enables DNA repair-based resistance, modulates energy pro-duction, and suppresses apoptosis, all of which contributes to cancer progression and treatment failure. Consequently, PCNA and its protein interactions have become attractive targets for the development of new cancer therapies focused on disrupting PCNA functions in cancer cells (Ohayon et al. 2016; Shivji and Kenny 1992; Zeman and Cimprich 2014; Søgaard et al. 2018b; Berti and Vindigni 2016).

7.9 PCNA as a Link Between Genome Stability and Immune Surveillance

To understand how PCNA plays an important role in the systemic, mucosal, and innate immune systems, it helps to first define what the innate immune system is and how/what the innate immune system is detecting. The innate immune system is the body's "first responder" layer of defense. Unlike the systemic or mucosal immune system which recognizes specific systemic pathogens which are recognized by circulating or compartmentalized antibodies or T-cells/receptors, the innate immune system senses patterns and contexts associated with dangerous pathogen-associated molecular patterns (PAMPs) and damage-associated molecular patterns (DAMPs) (Matzinger 2002). One of the strongest DAMPs is the presence of double-stranded DNA in the wrong cellular compartment, such as in the cytosol (the fluid in the space between the nuclear and outer cellular membrane) or extracellular space. Under normal conditions, DNA is tightly packaged in a highly organized fashion inside the nucleus and mitochondria, while DNA sensors (such as cGAS) sit in or near the cytosol. As long as this compartmentalization remains intact, our own DNA stays "invisible" to these sensors and does not trigger an inflammatory response (Janeway Jr and Medzhitov 2002).

Thinking about compartmentalization also helps clarify how different parts of the immune system respond when genome integrity is lost. The innate immune system is fundamentally different from the acquired (adaptive) immune system. Innate immunity acts first and does not rely upon prior exposure or immune memory. Instead, it responds to general signs of danger, such as DNA appearing in the wrong cellular compartment. The acquired immune system, in contrast, is slower and more specific, relying on B and T cells that recognize defined antigens to generate long-lasting immune memory. Importantly, innate immune sensing provides the initial signal that determines whether these downstream adaptive responses are engaged at all (Abbas et al. 2010; Murphy and Weaver 2016).

There are also important differences between the processes associated with systemic immunity and mucosal immunity. Systemic immune responses primarily operate in the bloodstream and internal tissues and are often measured by circulating antibodies. Mucosal immunity, by contrast, is located on barrier surfaces such as the epithelial "surface" layer of the respiratory and gastrointestinal tracts, where local immune cells and secretory antibodies provide protection from infection and micro-sized foreign bodies (e.g., pollen). Because these immune compartments are physically separated, immune responses generated in the blood do not always translate into effective protection at mucosal sites (Sato et al. 2012; Alqahtani 2024).

This difference became especially clear during the COVID-19 pandemic. Vaccination primarily induced systemic antibodies that circulated in the blood with the goal of reducing severe disease, while infection of the respiratory tract triggered a strong mucosal immune response. Highlighting how immune protection depends on where immune sensing is initiated (Lavelle and Ward 2022), circulating antibodies alone were generally insufficient to fully control viral replication at mucosal surfaces.

PCNA links genome stability to immune surveillance because it has a role in preventing the types of replication errors and chromosome damage that results in the movement of DNA out of the nucleus and into the cytosol or even the extracellular space. When PCNA is functioning properly, replication forks are stable, DNA lesions are repaired or bypassed in a controlled way, chromosomes segregate correctly, and nuclear DNA, from the immune system's perspective, remains "quiet." In contrast, when PCNA's functions are overwhelmed (e.g., by strong oncogene signaling) or disrupted (by mutation or inhibition), replication problems can escalate into double-strand breaks, under-replicated regions, and chromosomal fragments (Kotsantis et al. 2018). These types of structural damage, if not corrected properly, are likely to become packaged into micronuclei, where they can then be detected by innate immune sensors in the cytoplasm, or expelled into the extracellular space where antibodies and specialized immune cells can detect these particles.

7.10 How Replication Stress Triggers Immune Responses

DNA replication is not a smooth, error-free process. As the replication fork moves along the DNA, many obstacles can slow or stall it; this overall situation results in what is termed replication stress. Common causes of replication stress include:

- DNA lesions such as the formation of thymidine dimers or single-strand breaks in DNA that block the movement of DNA polymerases.
- Reduced nucleotide triphosphate levels, which can slow DNA synthesis when the building blocks of DNA are depleted.
- Collision between the movement of the DNA and RNA synthetic machinery along the same chromosome segment, especially within regions that are highly transcribed, setting up transcription-replication conflicts.
- Disruption of specific DNA repair mechanisms or aberrant cell signaling associated with the DNA repair processes.
- Strong oncogenic signaling, which can shorten DNA repair periods and force cells into S-phase before they are fully prepared to progress into the G2 phase of the cell cycle.

When replication stress is mild, cells can activate the ATR–Chk1 pathway, pause cell-cycle progression, and give themselves time to fix the problem (Cimprick and Cortez 2008). PCNA plays a major role here in recruiting damage-tolerance factors such as ubiquitin-regulated translesion polymerases and specific DNA repair proteins to the stalled replication forks (Moldovan et al. 2007). Through its modifications and binding partners, PCNA helps decide whether the cell repairs the lesion, bypasses it, or uses recombination-based mechanisms to repair the damage and restart fork movement (Branzei and Foiani 2010). In these situations, the ATR–Chk1 pathway acts as the cell's main way of detecting and stabilizing a stalled replication fork. When the fork slows down, single-stranded DNA builds up and becomes coated by RPA proteins. This signal activates ATR, which, in turn, activates Chk1. Chk1 then slows the cell cycle and helps keep the fork from breaking, while repairs are made (Zou and Elledge 2003). The specific response chosen at the stalled fork depends heavily on how PCNA is modified. Mono-ubiquitination of PCNA at lysine 164 by the Rad6–Rad18 complex promotes the recruitment of translesion synthesis polymerases—such as Pol η, Pol κ, Rev1, and Pol ζ—which allow synthesis to continue directly across damaged templates when conventional polymerases stall (Moldovan et al. 2007). For a more accurate bypass, PCNA can be tagged with K63-linked poly-ubiquitin chains, which can steer the stalled fork into template switching mode. Here, the cell temporarily copies information from the newly synthesized strand of the sister chromatid, often using fork reversal or recombination, to move past the lesion. Because the template is undamaged, this process avoids introducing mutations (Mirsanaye et al. 2021). Another important modification is SUMOylation of PCNA. SUMO-modified PCNA recruits proteins such as Srs2, which helps prevent unwanted or excessive homologous recombination during S-phase. This keeps repair activities under control and protects the stability of the replication fork (Papouli et al. 2005).

When replication stress is severe or long-lasting, these protective mechanisms are not enough, and the DNA replication fork can collapse. A collapsed fork often results in a double-stranded break in DNA, or leaves stretches of DNA under-replicated prior to the cell entering mitosis. Both outcomes result in chromosomal instability, resulting in either gains, losses, or rearrangements of chromosome segments that persist over future cell divisions. Chromosomal instability is not just a genetic issue; it also creates aberrant DNA structures and mis-segregated fragments that can leave the main nucleus and eventually end up in the cytosol, where they become visible to innate immune DNA sensors such as cGAS, which detects cytosolic double-stranded DNA and activates STING-dependent type I interferon signaling, and other DNA-sensing pathways such as AIM2. This molecular cascade can form inflammasomes in response to cytosolic DNA containing IFI-16, which senses mis-localized self-DNA and contributes to the inflammatory signaling process (Unterholzner et al. 2020).

In short:

- PCNA helps keep replication stress under tight control by stabilizing replication forks and coordinating repair processes.
- When replication stress escapes that control, broken or mis-segregated chromosomes are produced.
- These types of chromosomal errors promote the development of "immune-visible" DNA within the cytosol which activates the innate immune response.

7.11 Micronuclei: How Mis-segregated DNA Becomes Visible to the Immune System

One of the clearest bridges between chromosomal instability and innate immunity is the formation of micronuclei. A micronucleus is a tiny, extra nucleus that forms when a full chromosome or a chromosome fragment fails to separate properly with the other chromosomes during mitosis and is unable to be incorporated into either of the main daughter nuclei. Under a fluorescence microscope, micronuclei appear as small, DAPI-positive dots sitting next to the main nucleus. Initially, micronuclei are surrounded by their own nuclear envelope, so the DNA inside is still technically "nuclear." However, this envelope is often built incorrectly: it has patchy nuclear pore complexes, delayed or incomplete assembly, and poor integration with the rest of the nuclear membrane system. Landmark studies showed that these micronuclear envelopes are mechanically fragile and tend to rupture, sometimes well after cell division has finished (Kwon et al. 2020). When the micronuclear envelope breaks, its DNA content spills directly into the cytosol. At that moment, DNA that originally came from the chromosomes is now sharing the same space as cytosolic DNA sensors (Kwon et al. 2020). In addition, this cytoplasmic DNA can become packaged into small vesicles slated for transport into the extracellular space (Dou et al. 2017). These vesicles then appear to trigger an immune mediated response recruiting specific types of immune cells such as natural killer T cells, lymphocytes, macrophages, and neutrophils, which destroy or remove the vesicles and the cell producing this extracellular material (Tsering et al. 2024).

These discoveries provided crucial information for the field of Immunology because they answered a central question: how can self-DNA from the nucleus end up in the cytosol without the whole cell breaking open? Micronuclei provide the explanation. They are a kind of "nuclear waste" container which forms as a result of chromosomal instability. When they rupture, they produce high-concentration patches of DNA that are very effective at activating DNA-sensing pathways.

However, the question remains, how and when do these chromosomal DNA fragments develop? In a recent article by Krupina et al. (2025), a process known as chromothripsis (the process of fragmenting whole chromosomes) within cancer cells being mediated at least in part by endonuclease N4BP2 is described. This process is found within the cytoplasm of cancer cells. Micronuclei are known for being structurally fragile, and upon rupture of the micronucleus, the N4BP2 nuclease enters the micronucleus and promotes fragmentation of the contained DNA. This process fragments chromosomes at specific sites of DNA damage, which in turn creates double-stranded DNA breaks and enables these fragments to recombine with other chromosomal fragments to create extra-chromosomal DNA. This extra-chromosomal DNA then contributes to the evolving aggressiveness of these cancer cells.

PCNA's role at this step is upstream rather than by direct interaction. By stabilizing replication forks, promoting complete replication of difficult regions (like common fragile sites), and coordinating repair, PCNA reduces the chances that chromosomes will break or mis-segregate during mitosis. When PCNA function is compromised or overwhelmed, more chromosome fragments and lagging chromosomes are generated. This leads to more micronuclei, more ruptures, and more DNA becoming visible to the immune system in the cytosol (Zhang et al. 2015).

7.12 The cGAS–STING Pathway: The Main Sensor for Cytosolic DNA

Once DNA appears in the cytosol (e.g., after a micronucleus ruptures), it is detected primarily by a sensor called cGAS (cyclic GMP–AMP synthase). cGAS is an enzyme that binds double-stranded DNA in a sequence-independent manner (it responds to the DNA backbone and length rather than specific base sequences). When cGAS binds DNA for a sufficient length of time and with the right geometry, it becomes catalytically active and synthesizes a small cyclic dinucleotide second messenger called cGAMP (cyclic GMP–AMP) (Sun et al. 2013).

cGAMP then binds to a protein called STING (stimulator of interferon genes), which is located on the membranes of the endoplasmic reticulum in resting cells. Once activated by cGAMP, STING changes conformation and moves from the ER to the Golgi and then to perinuclear vesicles. Along this journey, STING recruits and activates the kinase TBK1, which in turn phosphorylates the transcription factor IRF3. Phosphorylated IRF3 dimerizes, enters the nucleus, and drives the expression of type I interferons (such as IFN-β) and a broad set of interferon-stimulated genes. STING can also engage NF-κB signaling, further boosting inflammatory cytokine production. Type I interferons are powerful cytokines that alert neighboring cells, upregulate antigen-presentation machinery (like MHC class I), and bridge innate and adaptive immunity (Hopfner and Hornung 2020).

At the level of the immune system, activation of the cGAS–STING pathway helps damaged or stressed cells get noticed and removed. Type I interferons produced downstream of STING increase how well cells display antigens on MHC class I molecules, making them easier for cytotoxic CD8[+] T-cells to recognize and kill. Interferon signaling also promotes the release of chemokines that attract natural killer (NK) T-cells and dendritic cells to the affected tissue. NK-cells can directly eliminate cells showing signs of stress or genomic damage, while dendritic cells pick up antigens from these cells and activate adaptive immune responses. In this way, elevated cGAS–STING signaling turns internal DNA damage into a clear immune signal that helps coordinate the detection and removal of potentially dangerous cells (Woo et al. 2014).

Based upon current evidence, PCNA does not directly bind or regulate cGAS or STING. Instead, it influences this pathway indirectly by controlling the abundance of DNA reaching the cytosol. When PCNA is functioning properly and the cell is not under replication stress, cells produce fewer micronuclei, their envelopes rupture less often, and relatively little DNA finds its way into the cytoplasm for cGAS binding. When replication stress is high, cytosolic DNA accumulates, cGAS–STING signaling is activated more strongly, and innate immune signaling intensifies. This is how a "pure replication factor" like PCNA ends up having a strong impact on immune outcomes.

7.13 Escape from Immune Surveillance

In addition to this indirect effect, PCNA has also been shown to play a direct role in immune escape, particularly through modulation of natural killer (NK) cell activity. NK-cells are critical components of the innate immune system that eliminate stressed, infected, or transformed cells through receptor-mediated recognition. Their responses are governed by a balance of activating and inhibitory signals transmitted through surface receptors, including NKp44 (also known as p44). Although NKp44 is often described as an activating receptor, NKp44's signaling output depends strongly on the nature of the ligand it encounters (Matzinger 2002).

Multiple studies have demonstrated that PCNA can function as an intracellular inhibitory ligand for NKp44. In stressed or transformed cells, including many cancer cells, PCNA is not strictly confined to the nucleus and can accumulate in the cytoplasm. During formation of the immune synapse between an NK-cell and a target cell, cytoplasmic PCNA becomes accessible and engages the NKp44 signaling complex. Rather than activating NK-cell cytotoxicity, this interaction delivers an inhibitory signal that suppresses NK-cell–mediated killing and reduces cytokine production, allowing the target cell to evade innate immune surveillance (Rosental et al. 2011).

Recent work has further strengthened this model by showing that PCNA can function as an immune checkpoint–like molecule. A recent study demonstrated that cytoplasmic and cell-associated PCNA directly inhibits NK-cell cytotoxicity through NKp44 engagement and that blocking this interaction with monoclonal antibodies restores NK-cell activity against tumor cells (Knaneh et al. 2023). These findings indicate that PCNA-containing protein complexes actively suppress NK-cell effector functions rather than merely escaping immune detection passively.

7.14 Keeping the Peace: PCNA and Immune Homeostasis in Dividing Tissues

At least three major organs within the body are constantly undergoing rather rapid cell division. These include:

- The hematopoietic system, where bone marrow stem and progenitor cells continuously produce red blood cells, various lineages of immune cells, platelets, and a variety of other factors.

- The intestinal epithelium, which is renewed every 3–7 days to maintain the gut barrier and promote at least 1/3 of the systemic immune system.
- The epidermis and other epithelia, which constantly replace damaged or lost cells.

In these tissues, a certain level of replication stress is essentially built into normal physiology. If every minor replication glitch or transient fork stall triggered a full innate immune response, these organs would be in a state of chronic inflammation. Over time, chronic inflammation would exhaust stem cell pools, damage tissue architecture, and impair organ function. Consistent with this notion, it is already known that chronic, low-grade type I interferon signaling contributes to autoimmune and inflammatory diseases (Yu et al. 2021).

PCNA helps prevent this from happening by making replication more robust and orderly. It supports smooth fork progression, coordinates multiple DNA repair pathways, and allows damage to be bypassed in a controlled way when necessary. Because of this, most replication problems are handled internally and quietly, without leaving behind broken chromosomes, micronuclei, or cytosolic DNA. In other words, PCNA contributes to immune homeostasis in proliferative tissues by stopping "false alarms" before they reach the level of innate immune activation. During normal DNA synthesis, PCNA tethers the main replicative polymerases, DNA polymerase δ and DNA polymerase ε, to DNA, ensuring continuous and processive fork progression (Moldovan et al. 2007). At the same time, PCNA serves as a platform for recruiting proteins involved in lagging-strand maturation, such as FEN1 and DNA ligase I, which process Okazaki fragments and prevent the accumulation of single-stranded DNA gaps that could destabilize the replication fork (Jónsson et al. 1998).

When forks encounter DNA damage or obstacles, PCNA coordinates controlled damage responses rather than allowing replication forks to collapse. Post-translational modifications of PCNA regulate which pathways are engaged. Monoubiquitination of PCNA by the Rad6–Rad18 complex promotes the recruitment of translesion synthesis polymerases, including Pol η, Pol κ, Rev1, and Pol ζ, which allow replication to continue past damaged templates and prevent prolonged fork stalling (Moldovan et al. 2007). For higher-fidelity damage bypass, polyubiquitinated PCNA recruits factors involved in template switching, a process that uses the newly synthesized sister chromatid as a temporary template and avoids introducing mutations (Strzalka and Ziemienowicz 2011).

PCNA also actively restrains inappropriate repair events that could destabilize chromosomes. SUMOylation of PCNA recruits the helicase Srs2, which suppresses excessive homologous recombination at replication forks and prevents toxic recombination intermediates from forming during S phase (Papouli et al. 2005). In addition, PCNA interacts with mismatch repair proteins such as MSH2–MSH6 and MLH1–PMS2, ensuring that replication errors are corrected promptly before they can be fixed into mutations or lead to replication-associated breaks (Li 2008).

This illustrates a broader concept: immune tolerance to self does not rely only on immune checkpoints or regulatory cells such as T-cells, NK-cells, macrophages, or neutrophils. It also depends on intracellular genome-maintenance pathways such as those organized by PCNA that ensure self-DNA stays properly compartmentalized and does not accidentally look like a danger signal.

7.15 PCNA, Chromosomal Instability, and Immune Evasion in Cancer

Cancer cells live in a very different environment from normal cells. Oncogenes push them to divide rapidly, often in the presence of ongoing DNA damage, hypoxia, and limited nutrients. This creates chronic replication stress and high levels of chromosomal instability (CIN). In principle, CIN and the resulting micronuclei should make tumor cells highly immunogenic, because they generate many abnormal DNA structures and mutations that could be recognized by the immune system. However, many tumors manage to avoid elimination by actively buffering replication stress rather than resolving it cleanly. One way they do this is by boosting pathways that help them tolerate replication stress.

PCNA contributes directly to this process by stabilizing replication forks and coordinating damage-bypass mechanisms that allow DNA synthesis to continue under conditions that would otherwise cause replication fork collapse. Elevated PCNA levels promote the recruitment of translesion synthesis polymerases and template-switching factors, enabling replication to proceed across damaged or under-replicated DNA, while limiting catastrophic chromosome breakage (Moldovan et al. 2007; Ulrich 2011).

PCNA is often overexpressed in cancers and has been used as a marker of proliferation in pathology. Rather than eliminating chromosomal instability, high PCNA activity allows tumor cells to maintain a "tolerable" level of instability—sufficient to drive mutation and clonal evolution but insufficient to produce the large amounts of cytosolic DNA that would strongly activate innate immune sensing pathways. High levels or hyper-usage of PCNA therefore helps tumor cells keep their replication forks moving and avoid outright replication catastrophe (Strzalka and Ziemienowicz 2011).

When PCNA and related repair/tolerance pathways are very active, tumors can maintain a "sweet spot" of genomic instability: there is enough instability to create genetic diversity, drive clonal evolution, and promote drug resistance, but not so much that huge amounts of DNA spill into the cytosol and strongly activate cGAS–STING. In this way, PCNA-supported replication buffering can contribute to immune evasion, helping tumors remain "immunologically cold," poorly infiltrated by cytotoxic Tcells, and poorly recognized by the immune system, even though they are genetically abnormal (Bakhoum and Cantley 2018).

Also, PCNA has been shown to play a direct role in immune escape, particularly through modulation of natural killer (NK) cell activity. NK-cells are critical components of the innate immune system that eliminate stressed, infected, or transformed cells through receptor-mediated recognition. Their responses are governed by a balance of activating and inhibitory signals transmitted through surface receptors, including NKp44 (also known as p44). Although NKp44 is often described as an activating receptor, its signaling outcome depends strongly on the nature of the ligand it encounters (Matzinger 2002).

Multiple studies have demonstrated that PCNA can function as an intracellular inhibitory ligand for NKp44. In stressed or transformed cells, including many cancer cells, PCNA is not strictly confined to the nucleus and can accumulate in the cytoplasm. During formation of the immune synapse between an NK-cell and a target cell, cytoplasmic PCNA becomes accessible and engages the NKp44 signaling complex. Rather than activating NK-cell cytotoxicity, this interaction delivers an inhibitory signal that suppresses NK-cell–mediated killing and reduces cytokine production, allowing the target cell to evade innate immune surveillance (Woo et al. 2014).

Recent work has further strengthened this model by showing that PCNA can function as an immune checkpoint–like molecule. A recent study demonstrated that cytoplasmic and cell-associated PCNA directly inhibits NK-cell cytotoxicity through NKp44 engagement and that blocking this interaction with monoclonal antibodies restores NK-cell activity against tumor cells (Rosental et al. 2011). These findings indicate that PCNA-containing protein complexes actively suppress NK-cell effector functions rather than merely escaping immune detection passively.

At a mechanistic level, this replication buffering effect reflects PCNA's role as a scaffold for large protein complexes that stabilize stressed replication forks and limit catastrophic genome fragmentation. Elevated PCNA levels promote sustained recruitment of translesion synthesis polymerases, fork remodeling factors, and template-switching machinery, allowing replication to continue through damaged or under-replicated regions without generating extensive double-strand breaks or chromosome shattering (Moldovan et al. 2007; Strzalka and Ziemienowicz 2011). In rapidly proliferating tumor cells, PCNA also contributes to the resolution of transcription–replication conflicts, a major source of genome instability, by coordinating replication fork progression through regions occupied by RNA polymerase II complexes and by limiting the persistence of stalled forks and RNA–DNA hybrid structures (García-Muse et al. 2016). By reducing replication fork collapse, chromosome fragmentation, and excessive micronuclei formation, PCNA-centered complexes limit the amount of self-DNA that escapes into the cytosol and would otherwise activate strong innate immune sensing pathways. In parallel, as discussed above, PCNA-containing complexes can directly suppress innate immune effector functions through interactions such as engagement of the NKp44 receptor on natural killer cells, providing a complementary, non–replication-based mechanism of immune evasion (Iraqi et al. 2022).

7.16   What Happens to Immunity When PCNA Is Disrupted?

If PCNA function is partially compromised, for example, by pharmacological inhibition or disruption of specific regulatory interactions rather than complete loss, the replication landscape changes dramatically. Replication forks stall and collapse more frequently, DNA breaks accumulate, and chromosomes are mis-segregated at higher rates. Under these conditions, reduced coordination between PCNA, the replicative polymerases (Pol δ and Pol ε), and fork-stabilizing factors such as RPA, Timeless–Tipin, and Claspin leads to prolonged fork stalling and increased fork collapse into double-stranded DNA breaks (Zou and Elledge 2003).

As DNA breaks accumulate, errors in replication completion and checkpoint control propagate into mitosis. Defective coordination between replication-associated repair proteins (such as BRCA1, BRCA2, and RAD51) and chromosome segregation machinery increases the frequency of lagging chromosomes and acentric fragments during anaphase (Burrell et al. 2013). These mis-segregated chromosome fragments are a primary source of micronuclei. Micronuclei form when whole chromosomes or chromosome fragments fail to be incorporated into the main daughter nuclei and instead become enclosed within their own, structurally defective nuclear envelope.

Micronuclear envelopes are inherently fragile due to abnormal assembly of nuclear Lamins and nuclear pore complexes, and they frequently rupture after mitosis. This rupture allows chromosomal DNA to spill directly into the cytosol, where it becomes accessible to innate immune sensors (Mackenzie et al. 2017). As noted previously, cytosolic double-stranded DNA is then detected by cGAS, in a sequence-independent manner, and synthesizes cGAMP upon activation (Sun et al. 2013), which engages STING, triggering recruitment of TBK1 and phosphorylation of IRF3, which inturn leads to the induction of type I interferons and a broad inflammatory transcriptional program (Hopfner and Hornung 2020). Together, these protein–protein interactions link replication fork failure to immune activation. Disruption of PCNA-centered replication and repair complexes increases fork collapse and chromosome mis-segregation, promotes micronuclei formation and rupture, and ultimately converts replication-associated genome instability into a potent innate immune signal through activation of the cGAS–STING pathway.

Importantly, this immune activation is cell-intrinsic: it occurs because the cell's own genome is damaged and mis-localized and is not the direct result of external factors such as viral infection or exposure to DNA damaging toxins. In cancer, such activation can increase antigen presentation (e.g., by upregulating MHC class I and other interferon-stimulated genes) and potentially make tumors more visible to the immune system, especially in combination with checkpoint blockade.

Recent work has explicitly linked defects in key DNA repair and replication-associated protein complexes such as BRCA2 loss, mismatch repair (MMR) deficiency, and RNase H2 deficiency to the formation of cGAS-positive micronuclei and activation of interferon responses; demonstrating that unresolved replication stress and genome instability can themselves act as triggers of immune surveillance. BRCA2 plays a central role in stabilizing stalled replication forks and promoting RAD51-mediated homologous recombination; when BRCA2 is lost, impaired RAD51 loading leads to fork degradation, accumulation of double-strand breaks, increased chromosome fragmentation, and promotion of micronuclei formation and cGAS activation (Burrell et al. 2013; Mackenzie et al. 2017).

Similarly, defects in the mismatch repair machinery particularly loss of components such as MSH2, MSH6, MLH1, or PMS2, which normally interact with PCNA at replication forks, allow replication errors and small insertion–deletion loops to persist. These unresolved lesions increase replication fork stalling and chromosome breakage, which then elevates the frequency of lagging chromosomes and the formation of micronuclei. Many of these micronuclei subsequently rupture, which deposits self-DNA into the cytosol, resulting in activation of the cGAS–STING pathway and type I interferon signaling (Unterholzner et al. 2020; Mackenzie et al. 2017).

RNase H2 deficiency provides a complementary example in which genome instability arises from defective processing of RNA–DNA hybrids. RNase H2 normally removes ribonucleotides from DNA-RNA hybrid DNA observed during the DNA replication process and during transcription-replication conflict resolution. Loss of RNase H2 leads to persistent RNA–DNA hybrids, increased replication stress, and replication fork collapse, resulting in chromosome fragmentation and micronuclei formation. As noted previously, these micronuclei are frequently cGAS-positive and drive chronic interferon responses, as demonstrated in both cellular and in vivo models of RNase H2 deficiency (Rosental et al. 2011; Mackenzie et al. 2017).

Together, these examples illustrate how disruption of distinct but interconnected macromolecular complexes involved in replication fork protection, repair pathway choice, and RNA–DNA hybrid resolution converge on a common outcome: chromosome mis-segregation, micronuclear rupture, and activation of innate immune DNA-sensing pathways. In this framework, immune surveillance does not require foreign DNA or infection but instead arises directly from failures in genome maintenance processes thus linking DNA replication stress and DNA repair defects to intrinsic inflammatory signaling.

7.17 PCNA as an Indirect Regulator of Immune Surveillance

Putting all these pieces together, PCNA can be viewed as an indirect immuno-regulator. It is not a classical immune protein like a cytokine, a pattern-recognition receptor, or a transcription factor. Instead, its influence comes from its central position in DNA replication and repair:

- In the absence of DNA replication stress, chromosomal instability is limited, and nuclear DNA remains compartmentalized away from cytosolic sensors such as cGAS, AIM2, and IFI-16. Innate immune pathways such as cGAS–STING stay mostly quiet.
- At times of high DNA replication, stress, mutations, deletions, insertions, and recombination errors accumulate, micronuclei and cytosolic DNA increases, and innate immune pathways become activated, triggering interferon production and inflammatory signaling.

These effects arise from disrupted PCNA-centered complexes at replication forks, including altered interactions with replicative polymerases (Pol δ/ε), damage-tolerance factors such as Rad6–Rad18 and translesion synthesis polymerases, and fork-protection and recombination proteins including RPA, BRCA1/2, and RAD51, which together increase fork collapse, chromosome breakage, and the generation of immune-visible cytosolic DNA.

Thus, PCNA helps decide whether genome instability stays as a "silent" internal problem or becomes an immunological signal that can recruit and activate immune cells.

7.18 Targeting PCNA Complexes for Cancer Therapy

Because PCNA is so crucial to cancer cell survival, strategies to disrupt PCNA function or its interactions have gained interest as potential broad-spectrum anticancer approaches. However, directly inhibiting PCNA's sliding clamp activity is challenging and would be toxic to cancer and non-cancer cells alike. Thus, recent efforts from several laboratories (Ohayon et al. 2016; Søgaard et al. 2018a; Søgaard et al. 2018b; Gu et al. 2018; Li et al. 2021) have focused on either selectively targeting the unique ways in which PCNA operates in cancer cells or blocking specific protein–PCNA interactions that are more critical for cancer cell proliferation and maintenance than in normal tissue.

One approach has been the development of peptides that mimic PCNA-binding motifs, thereby acting as competitors that prevent endogenous PCNA partners from docking with full-length PCNA. One example of this approach involves utilizing the APIM (AlkB homolog 2 PCNA-interacting motif) sequence as an inhibitor of PCNA associated activity since this amino acid sequence is present in many stress-response proteins that bind PCNA (Ohayon et al. 2016). Based on this observation, Müller and colleagues developed a cell-penetrating APIM-containing peptide, designated ATX-101, to selectively disrupt PCNA-dependent stress-response signaling in cancer cells (Ohayon et al. 2016). ATX-101 displaces APIM-containing proteins from the PCNA scaffold and induces robust, caspase-dependent apoptosis in multiple myeloma cells, independent of cell cycle status (Ohayon et al. 2016). Notably, ATX-101 exhibited cancer-selective toxicity, sparing normal cells, and significantly enhanced the efficacy of DNA-damaging chemotherapy such as melphalan in preclinical tumor models (Ohayon et al. 2016). These findings support the concept that disrupting PCNA's interaction network can selectively compromise cancer cell survival by simultaneously impairing replication, repair, and stress-response pathways (Ohayon et al. 2016).

Prior to the development of ATX-101, Gu and colleagues described a cancer-selective peptide, termed caPeptide, derived from regions uniquely exposed in the cancer-associated PCNA isoform (caPCNA) (Søgaard et al. 2018a). This peptide selectively disrupted caPCNA protein–protein interactions and induced apoptosis in cancer cells, while sparing nonmalignant cells (Søgaard et al. 2018a). Together, these studies provided early proof-of-principle that cancer-selective targeting of PCNA interactions is feasible.

7.19 Small-Molecule Inhibitors of PCNA-Binding Partner Interaction

Beyond peptides, small-molecule PCNA inhibitors have also been developed. The compound T2AA and related derivatives (including the T3 series) bind PCNA and disrupt PIP-box–mediated protein-PCNA interactions, leading to impaired DNA replication and repair in cancer cells (Gu et al. 2018). More recently, work published by Gu et al. (2023) demonstrated that small-molecule PCNA inhibitors can induce severe replication stress selectively in tumor cells without promoting cytotoxicity in non-cancer cells (Li et al. 2021; Gu et al. 2023). This occurs in part by disrupting PCNA-mediated resolution of transcription–replication conflicts (Gu et al. 2023), which leads to double-stranded breaks in the chromosomal DNA and subsequent induction of an apoptotic response by these cancer cells. Collectively, these peptide- and small-molecule–based approaches highlight the therapeutic potential of directly targeting PCNA's interaction network rather than its essential clamp function.

7.20 A Cancer-Selective Inhibitor of the caPCNA Isoform

Our initial effort to identify a cancer-selective small-molecule PCNA inhibitor led to the discovery of AOH1160 and its more metabolically stable analog AOH1996. These compounds were identified through a large-scale structure-based in silico docking screen of more than 6.5 million compounds targeting the PCNA PIP-box binding region encompassing PCNA residues L126–Y133 (Gu et al. 2018, 2023). AOH1160 selectively induced cytotoxicity across a broad spectrum of cancer cell types at nanomolar concentrations, while exhibiting minimal toxicity toward non-malignant cells (Gu et al. 2018). Mechanistically, AOH1160 interferes with DNA replication, blocks homologous recombination repair, induces S-phase arrest, and promotes accumulation of unrepaired DNA damage, ultimately triggering apoptosis (Gu et al. 2018). In vivo, AOH1160 demonstrated oral bioavailability and suppressed tumor growth in animal models without overt systemic toxicity (Gu et al. 2018). The subsequent development of an optimized analog (i.e., AOH1996) further established PCNA as a druggable target, and this small-molecule represents a first-in-class strategy for selectively disrupting the cancer-associated PCNA interactome (Gu et al. 2018) without also disrupting the utilization of the PCNA molecule in proliferating non-cancer cells.

The rationale for developing AOH1996 as a cancer-selective PCNA inhibitor is based on the ability to simultaneously undermine multiple survival pathways required for malignant cell proliferation. By blocking the replication and repair machinery utilizing PCNA, these agents drive cancer cells into a lethal level of replication stress, which is in-part characterized by stalled replication forks, irreparable DNA damage, and activation of apoptosis. Importantly, selective targeting of the cancer-associated PCNA isoform circumvents many limitations associated with traditional approaches that rely on reducing PCNA expression levels or nonselectively disrupting the PCNA interactome. This strategy provides a compelling framework for exploiting a previously "undruggable" non-enzymatic scaffold protein and offers a promising avenue for improving therapeutic outcomes for patients while minimizing or eliminating toxicity in the cells of normal/healthy proliferating tissues.

7.21 Exploiting PCNA Binding Interactions for Treating Cancer: Current Research and Unresolved Questions

Cancer-Specific PCNA Dynamics The discovery of a cancer-associated PCNA isoform (caPCNA) raises questions about its nature. The search leading to the discovery of the caPCNA protein was prompted by the observation that the isolated intact DNA synthetic apparatus (i.e., the DNA synthesome) from cancer cells was highly error-prone when copying a DNA template relative to the DNA synthesome from non-cancer cells (Sekowski et al. 1998). A structural biochemistry analysis of the DNA synthesome from both cancer and non-cancer cell lines led to the discovery of a PCNA isoform that had an apparent unique cancer associated acidic pI when electrophoresed through two-dimensional polyacrylamide gels (Bechtel et al. 1998). In contrast, non-cancer cells only expressed an isoform of PCNA having an apparent basic pI. Together, Sekowski et al. (1998) and Bechtel et al. (1998) determined that the expression of this cancer associated isoform of PCNA correlated precisely with the decrease in DNA replication fidelity exhibited by cancer cells, while the DNA synthesome of non-cancer cells maintained a consistently lower mutation frequency than its cancer cell counterpart and expressed only the basic isoform of PCNA. Our initial hypotheses suggested that caPCNA might arise from gene mutations or alternative RNA splicing. However, Bechtel et al. (66) performed PCNA gene sequencing and protein structural analyses on the PCNA expressed by cancer and non-cancer cells and determined that there were no changes in the coding region of the PCNA gene in either cell type, nor was there evidence for an alternatively spliced gene transcript, suggesting that the protein structural differences observed were likely due to differential post-translational modification of the two PCNA isoforms. Subsequent structural analyses determined that the caPCNA isoform is a product of a differential post-translational modification, which appears to open the PIP box binding domain and potentially leads to a particular conformational state of PCNA found only in cancer cells? This alteration exposes normally buried epitopes on the cancer isoform of PCNA, including those presumably within the PIP box binding domain (L126–Y133) of the PCNA protein (Hoelz et al. 2006). Additionally, polyclonal antibodies prepared against the PIP box binding domain of PCNA demonstrated that these antibodies selectively reacted with the nuclei of cancer cells with little or no binding to that of proliferating non-cancer cells or that of cells found in benign proliferative tissues (Malkas et al. 2006). This difference in the extent or type of post-translational modification distinguishing the caPCNA isoform from its non-cancer cell counterpart provided a rationale for exploring the potential therapeutic vulnerability expression of this protein confers onto the cancer cell.

PCNA Post-Translational Modification Code While Ubiquitination and SUMOylation of PCNA have been studied extensively in the context of DNA damage tolerance, other modifications (phosphorylation, acetylation, methyl esterification, glycosylation, nitration, etc.) are not as well understood. For instance, acetylation of PCNA was reported to promote PCNA's dissociation from chromatin and could signal completion of DNA repair (Shivji and Kenny 1992); the detailed mechanisms and regulatory enzymes involved remain to be clarified. How the interplay between multiple modifications within PCNA (a "PCNA code") dictate which protein complexes PCNA forms at a given time remains unanswered. While ubiquitination and SUMOylation of PCNA are well studied, the potential roles for additional PTMs such as methyl esterification of acidic amino acids (Hoelz et al. 2006), phosphorylation, acetylation (Shivji and Kenny 1992), glycosylation, and methylation of basic amino acids warrant additional scrutiny, and together their interplay may constitute a combinatorial "PCNA code." This code may in fact be a series of codes that are signatures for the functional ability of PCNA to act so quickly when switching modes from (e.g., DNA synthesis to DNA damage repair, apoptosis induction to apoptosis prevention, induction of mitotic catastrophe vs. preserving faithful chromosomal segregation during mitosis, and a variety of other cellular processes utilizing PCNA as a cofactor).

PCNA in Chromatin and Transcription PCNA's role in coordinating replication–transcription conflicts is an emerging area of study. As described previously, transcription of actively replicating DNA not uncommonly leads to the transcription fork colliding with the DNA replication fork. PCNA functions within both the DNA replication and transcription machinery and has a recognized but not well-understood role in resolving the spatial and temporal conflicts that arise as both multiprotein complexes, mediating these cellular processes, move in opposite directions toward one another along the same segment of double-stranded DNA. The collision stalls both DNA replication fork movement and transcript synthesis, and thereby induces DNA replication stress. If this conflict is not resolved relatively quickly, it results in double-stranded DNA breaks (Li et al. 2021). The resolution of transcription–replication conflict involves PCNA recruiting specialized factors to potentially restart replication and transiently suppress transcription (Moldovan et al. 2007), while, at the same time, coordinating the movement of the DNA synthetic complex past the collision point on the DNA strand. How PCNA helps maintain epigenetic information (through partners like DNMT1, or histone modifiers) and how it might influence gene expression directly are questions linking PCNA to the regulation and maintenance of the genome and extends well beyond its role in DNA replication and transcription, per se. PCNA's role(s) in resolving transcription–replication conflicts and helping maintain epigenetic continuity remains to be elucidated.

Cytosolic PCNA Functions The paradigm-shifting discovery that PCNA can act in the cytosol (in neutrophils and leukemia cells) opens many questions. Which other cell types or conditions promote PCNA nuclear export? What proteins interact with cytoplasmic PCNA in terminally differentiated vs. proliferating cells? What cellular signals promote these specific interactions (e.g., stress kinases, differentiation pathways), or control the partitioning of PCNA between the nuclear and cytoplasmic compartments? Moreover, PCNA's role in metabolism—such as scaffolding metabolic enzymes like NAMPT to adjust NAD^+ levels (Mailand et al. 2013)—suggests PCNA could be a nexus between DNA replication stress and metabolic reprogramming. Investigating PCNA's potential involvement in other metabolic or signaling pathways (e.g., does cytosolic PCNA affect mTOR signaling, how does it drive/inhibit production of reactive oxygen species (ROS), or what role does it play in regulating fatty acid metabolism and oxidative phosphorylation) remains an area for investigation. PCNA's export to the cytosol, seen in a variety of cell types such as neutrophils, specific forms of leukemia such as AML, and in nitrosative-stressed neuroblastoma cells, raises questions about its non-nuclear biology.

Therapeutic Window and Resistance Although PCNA is an essential protein whose complete functional loss is incompatible with cell viability (Moldovan et al. 2007), the development of cancer-selective PCNA inhibitors has demonstrated that lethality is not dictated simply by global PCNA inhibition. In the case of AOH1996, selectivity arises from structure-based targeting of a cancer associated PCNA conformation rather than differences in PCNA expression levels between normal and malignant cells (Abbas and Dutta 2009). Importantly, the favorable tolerability observed in preclinical studies for AOH1996 is not indicative of a traditional "therapeutic window"-based approach employing partial PCNA suppression in cancer cells while tolerating some lower level of inhibition in non-cancer cells. Instead, the favorable tolerability reflects selective inhibition of the cancer-associated isoform (Kelman 1997) while leaving intact the utilization of the non-cancer isoform of the protein. In this context, observed toxicities in vivo are more consistent with excipient-related effects rather than compound-intrinsic toxicity (Ku et al. 1989). Preliminary studies with ATX-101 (Kunkel and Burgers 2008) and AOH1996 demonstrate that disruption of PCNA function can markedly enhance the cytotoxic consequences of DNA damage (Essers et al. 2000; Ohayon et al. 2016). However, this effect should not be interpreted as classical pathway synergy, as PCNA is a component of multiple DNA damage response and repair mechanisms (Kunkel and Burgers 2008) as well as multiple cytoplasmic-mediated cellular processes. Rather, PCNA inhibition collapses the structural scaffold required for damage tolerance and repair, thereby preventing cancer cells from resolving lesions induced by other therapies

(Ohayon et al. 2016), and as previously described multiple cytoplasmic-based processes such as energy metabolism and apoptosis induction. Thus, combination efficacy reflects pathway convergence at PCNA rather than independent mechanistic synergy. AOH1996 and several of its structural analogs take advantage of the unique characteristics of the caPCNA protein isoform and the cellular physiology of individual cancer types to selectively inhibit the activity of the caPCNA protein utilized by these cancer cells, and in combination with novel and/or traditional standard-of-care therapeutics, AOH1996 has been shown in early clinical trial to improve the clinical management of patients with specific forms of cancer without exhibiting any dose-limiting toxicities.

Because PCNA function is central to multiple nuclear processes involved in DNA replication, DNA repair, and DNA fork stability (Ohayon et al. 2016), the potential development of resistance to most PCNA-directed therapies could traditionally be thought to potentially arise through compensatory mechanisms such as altered clamp-loader dynamics, changes in fork remodeling pathways, or increased reliance on damage bypass processes (Malkas et al. 2006; Moldovan et al. 2010). However, because of the critical nature of PCNA in so many cellular processes, unintended cytotoxicity associated with nonselective PCNA inhibitors is more likely to occur before resistance to AOH1996 develops. A fuller mechanistic understanding of the interaction of PCNA with its many binding partners both in the nucleus and cytoplasm is critical for optimizing PCNA-targeted agents and defining rational combination strategies that exploit the structural characteristics of the cancer-associated PCNA and the specific cellular physiological characteristics associated with specific types of cancer being treated.

7.22 Continuing Work

PCNA stands out as a multifunctional maestro in the cell, conducting the complex orchestration within the cell involved in the regulation of DNA replication and repair, RNA transcription, cellular energy metabolism, apoptosis induction, cell division, and immune surveillance. Through its ability to form diverse macromolecular complexes, PCNA ensures that cells faithfully duplicate their genome, correct errors, and adapt to various forms of DNA damage, while also being integral to all of the other cellular functions needed to maintain cellular viability, maturation, and/or propagation. Thus, in cancer, PCNA's multiple mechanistic roles not only allow tumor cells to proliferate rapidly and resist efforts to stop cancer growth, but it's involvement in so many cellular processes present's an Achilles' heel for treating cancer cells that researchers can exploit to improve the clinical management of the disease.

The study of PCNA complexes continues to yield fundamental insights into how cells maintain genomic stability and how this balance is subverted in diseases like cancer. From the replication fork to the cytoplasm, PCNA's presence is synonymous with a cell's proliferative capacity and its ability to survive challenges. By charting the network of PCNA interactions and discovering ways to perturb them, we edge closer to therapies that can disarm cancer cells of their resilience. Future research will undoubtedly reveal more about the fine-tuned regulation of PCNA's activities and answer outstanding questions about its noncanonical roles. What remains clear is that PCNA is a central node in the web of life-and-death decisions made by the cell. In many respects, it appears to be an ancient protein conserved in function over evolutionary time. PCNA is not the simple scaffold it was once thought to be, but a rather complex regulator of no less than nine distinct cellular processes, with at least 200 known binding partners. PCNA appears to tether multiple proteins in close proximity to one another so that they can facilitate complex biochemical reactions in a coordinated manner, which can simply be described as a biochemical pathway. In essence, PCNA is at the core of these processes by arranging the key proteins in these pathways within close proximity to one another and thereby creates complex molecular machines that efficiently carry out the biochemical reactions associated with each of these pathways. How PCNA coordinates these biochemical reactions is itself an intricate and fascinating subject deserving of thoughtful study. Additionally, PCNA's role in disease processes and the involvement of specific post-translational modifications as a potential driver for these disease processes suggests that these modified PCNA isoforms could be a promising target for therapeutic intervention (Mailand et al. 2013; Witko-Sarsat et al. 2010).

Acknowledgments The authors would like to gratefully acknowledge the Irell and Manella Graduate School of the Beckman Research Institute of the City of Hope National Comprehensive Cancer Center for providing financial support for HK and DA during the writing of this manuscript. We also wish to thank Dr. Nicholas Wallace (School of Health Sciences, Kansas State University) for providing, ahead of publication, information related to his findings associated with the selective inhibition of cervical cancer cell mitosis by the small-molecule inhibitor AOH1996. The writing of this manuscript was also supported in part within the RJH and LHM laboratories by sponsored research program grants from RLL, LLC (supporting the clinical development of the selective inhibitor of the cancer associated PCNA isoform (AOH1996)), directed philanthropic gifts to the City of Hope for the LHM and RJH laboratories, and research grants within the Beckman Research Institute of the City of Hope (CDMRP Breast Cancer Research Program grant W81XWH-19-1-0327) and the NIH Comprehensive Cancer Center Support Grant P30CA033572.

Disclosures Two of the authors, RJH and LHM, are among the co-founders of RLL, LLC biotech.

References

Abbas T, Dutta A. p21 in cancer: intricate networks and multiple activities. Nat Rev Cancer. 2009;9(6):400–14.

Abbas AK, Lichtman AH, Pillai S. Cellular and molecular immunology. 6th ed. Elsevier Saunders; 2010.

Alqahtani SAM. Mucosal immunity in COVID-19: a comprehensive review. Front Immunol. 2024;15:1433452. https://doi.org/10.3389/fimmu.2024.1433452.

Bakhoum SF, Cantley LC. The multifaceted role of chromosomal instability in cancer. Nat Rev Cancer. 2018;18(9):533–45. https://doi.org/10.1038/s41568-018-0022-7.

Balakrishnan L, Bambara RA. Flap endonuclease 1. Annu Rev Biochem. 2013;82:119–38.

Bechtel P, Hickey RJ, Schnaper L, Sekowski JW, Long BJ, Freund R, Liu N, Rodriguez-Valenzuela C, Malkas LH. A unique form of proliferating cell nuclear antigen is present in malignant breast cells. Cancer Res. 1998;58(15):3264–9. PMID: 9699653

Berti M, Vindigni A. Replication stress: getting back on track. Nat Struct Mol Biol. 2016;23(2):103–9.

Branzei D, Foiani M. Maintaining genome stability at the replication fork. Nat Rev Mol Cell Biol. 2010;11(3):208–19. https://doi.org/10.1038/nrm2852.

Burrell RA, et al. Replication stress links structural and numerical cancer chromosomal instability. Nature. 2013;494(7438):492–6. https://doi.org/10.1038/nature11935.

Cimprick KA, Cortez D. ATR: an essential regulator of genome integrity. Nat Rev Mol Cell Biol. 2008;9(8):616–27. https://doi.org/10.1038/nrm2450.

Dou Z, et al. Cytosolic chromatin triggers inflammation via cGAS. Nature. 2017;550(7674):402–6. https://doi.org/10.1038/nature24050.

Essers J, Theil AF, Baldeyron C, et al. Nuclear dynamics of PCNA in DNA replication and repair. Curr Biol. 2000;10(1):1–4.

Fujise K, Zhang D, Liu J, Yeh ET. Regulation of apoptosis and cell cycle progression by MCL1. Mol Cell Biol. 2000;20(11):3745–53.

García-Muse T, et al. Transcription-replication conflicts: how they occur and how they are resolved. Nat Rev Mol Cell Biol. 2016;17:553–63. https://doi.org/10.1038/nrm.2016.88.

Garg P, Burgers PM. Ubiquitinated proliferating cell nuclear antigen activates translesion DNA polymerases. J Biol Chem. 2005;280(45):36623–9.

Gonzalez-Magana A, Blanco FJ. Human PCNA structure, function, and interactions. Biomolecules. 2020;10(4):570. https://doi.org/10.3390/biom10040570. PMCID: PMC7225939 PMID7225939, PMID: 32276417

Gu L, Chu P, Lingeman R, et al. The anticancer activity of a first-in-class small molecule targeting PCNA. Oncotarget. 2018;9(25):17760–74.

Gu L, et al. Small molecule targeting of transcription-replication conflict for selective chemotherapy. Cell Chem Biol. 2023;30:1235–1247.e6. https://doi.org/10.1016/j.chembiol.2023.07.001.

Hingorani MM, O'Donnell M. Sliding clamps: a (tail)ored fit. Curr Biol. 2000;10(1):R25–9.

Hoege C, Pfander B, Moldovan GL, et al. RAD6-dependent DNA repair is linked to modification of PCNA by ubiquitin and SUMO. Nature. 2002;419(6903):135–41.

Hoelz D, Arnold RJ, Dobrolecki LE, Abdel-Aziz W, Loehrer AP, Novotny MV, Schnaper L, Hickey RJ, Malkas LH. The discovery of labile methylesters on proliferating cell nuclear antigen by MS/MS. Proteomics. 2006;6(17):4808–16. https://doi.org/10.1002/pmic.200600142.

Hopfner KP, Hornung V. Molecular mechanisms of cGAS–STING signaling. Nat Rev Mol Cell Biol. 2020;21(9):501–21. https://doi.org/10.1038/s41580-020-0244-x.

Human Locus Specific Mutation Databases. Houston: Human Genome Variation Society. 2011.

Iraqi M, et al. Blocking the PCNA/NKp44 checkpoint to stimulate NK cell responses to multiple myeloma. Int J Mol Sci. 2022;23:4717. https://doi.org/10.3390/ijms23094717.

Janeway CA Jr, Medzhitov R. Innate immune recognition. Annu Rev Immunol. 2002;20:197–216. https://doi.org/10.1146/annurev.immunol.20.083001.084359.

Jónsson ZO, et al. Regulation of DNA replication and repair proteins through interaction with the front side of proliferating cell nuclear antigen. EMBO J. 1998;17:2412–25. https://doi.org/10.1093/emboj/17.8.2412.

Kelman Z. PCNA: structure, functions and interactions. Oncogene. 1997;14(6):629–40.

Knaneh J, et al. mAb14, a monoclonal antibody against cell surface PCNA: a potential tool for Sezary syndrome diagnosis and targeted immunotherapy. Cancer. 2023;15:4421. https://doi.org/10.3390/cancers15174421.

Kotsantis P, et al. Mechanisms of oncogene-induced replication stress. Nat Rev Cancer. 2018;18(7):447–60. https://doi.org/10.1038/s41568-018-0021-8.

Krupina K, et al. Chromothripsis and ecDNA initiated by N4BP2 nuclease fragmentation of cytoplasm-exposed chromosome. Science. 2025;390:1156–63. https://doi.org/10.1126/science.ado0997.

Ku DH, Travali S, Calabretta B, Huebner K, Baserga R. Human gene for proliferating cell nuclear antigen has pseudogenes and localizes to chromosome 20. Somat Cell Mol Genet. 1989;15(4):297–307. https://doi.org/10.1007/BF01534969. PMID: 2569765

Kunkel T, Burgers PM. Dividing a workload at a replication fork. Trends Cell Biol. 2008;18(11):521–7. https://doi.org/10.1016/J.TCB.2008.08.005.

Kwon M, et al. Small but mighty: the causes and consequences of micronucleus rupture. Exp Mol Med. 2020;52:1777–86. https://doi.org/10.1038/s12276-020-00529-z.

Lavelle EC, Ward RW. Mucosal vaccines fortifying the frontiers. Nat Rev Immunol. 2022;22(4):236–50. https://doi.org/10.1038/s41577-021-00583-2.

Li GM. Mechanisms and functions of DNA mismatch repair. Cell Res. 2008;18(1):85–98. https://doi.org/10.1038/cr.2007.115.

Li CM, Haratipour P, Lingeman RG, Perry JJ, Gu L, Hickey RJ, Malkas LH. Novel peptide therapeutic approaches for cancer treatment. Cells. 2021;10(11):2908. https://doi.org/10.3390/cells10112908.

Mackenzie KJ, et al. cGAS surveillance of micronuclei links genome instability to innate immunity. Nature. 2017;548(7668):461–5. https://doi.org/10.1038/nature23449.

Maga G, Hübscher U. Proliferating cell nuclear antigen (PCNA): a dancer with many partners. J Cell Sci. 2003;116(Pt 15):3051–60.

Mailand N, Gibbs-Seymour I, Bekker-Jensen S. Regulation of PCNA–protein interactions for genome stability. Nat Rev Mol Cell Biol. 2013;14(5):269–82.

Malkas LH, Herbert BS, Abdel-Aziz W, et al. A cancer associated PCNA expressed in breast cancer cells is a potential Theapeutic target. Cancer Res. 2006;66(2):1003–10.

Matzinger P. The danger model: a renewed sense of self. Science. 2002;296(5566):301–5. https://doi.org/10.1126/science.1071059.

Mirsanaye AS, et al. Ubiquitylation at stressed replication forks: mechanisms and functions. Trends Cell Biol. 2021;31:584–97. https://doi.org/10.1016/j.tcb.2021.01.008.

Moldovan GL, Pfander B, Jentsch S. PCNA, the maestro of the replication fork. Cell. 2007;129(4):665–79. https://doi.org/10.1016/j.cell.2007.05.003.

Moldovan GL, et al. DNA polymerase POLN participates in cross-link repair and homologous recombination. Mol Cell Biol. 2010;30:1088–96. https://doi.org/10.1128/MCB.01124-09.

Murphy K, Weaver C. Janeway's immunobiology. 9th ed. Garland Science; 2016.

Ohayon D, De Chiara A, Chapuis N, et al. Cytoplasmic proliferating cell nuclear antigen coordinates glycolysis and promotes survival of acute myeloid leukemia cells. Sci Rep. 2016;6:35561.

Papouli E, et al. Crosstalk between SUMO and ubiquitin on PCNA regulates homologous recombination. Cell. 2005;123(1):39–51. https://doi.org/10.1016/j.cell.2005.08.035.

Roa S, Avdievich E, Peled JU, MacCarthy T, Werling U, Kuang F-L, Kan R, Zhao C, Bergman A, Cohen PE, Edelmann W, Scharff MD. Ubiquitylated PCNA plays a role in somatic hypermutation and class-switch recombination and is required for meiotic progression. Proc Natl Acad Sci USA. 2008;105(42):16248–53. https://doi.org/10.1073/pnas.0808182105. Epub. 2008 Oct. 142008

Rosental B, et al. Proliferating cell nuclear antigen is a novel inhibitory ligand for NKp44. J Immunol. 2011;187(11):5693–702. https://doi.org/10.4049/jimmunol.1101557.

Sato S, et al. The mucosal immune system of the respiratory tract. Curr Opin Virol. 2012;2:225–32. https://doi.org/10.1016/j.coviro.2012.03.009.

Sekowski J, Malkas LH, Schnaper L, Bechtel PE, Long BJ, Hickey RJ. Human breast cancer cells contain an error-prone DNA replication apparatus. Cancer Res. 1998;58(15):3259–63. PMID: 9699652

Shivji KK, Kenny MK. Wood RD proliferating cell nuclear antigen is required for DNA excision repair. Cell. 1992;69(2):367–74. https://doi.org/10.1016/0092-8674(92)90416-a.

Smith ML, Chen IT, Zhan Q, et al. Interaction of the p53-regulated protein Gadd45 with proliferating cell nuclear antigen. Science. 1994;266(5189):1376–80.

Søgaard CK, Moestue SA, Rye MB, Kim J, Nepal A, Liabakk N-B, Bachke S, Bathen TF, Otterlei M, Hill DK. APIM-peptide targeting PCNA improves the efficacy of docetaxel treatment in the tramp mouse model of prostate cancer. Oncotarget. 2018a;9(14):11752–66. https://doi.org/10.18632/oncotarget.24357.

Søgaard CK, Blindheim A, Røst LM, et al. APIM-peptide targeting of PCNA increases the efficacy of anticancer therapy. PLoS One. 2018b;13(6):e0199097.

Stoimenov I, Helleday T. PCNA (proliferating cell nuclear antigen). Atlas Genet Cytogenet Oncol Hematol. 2011; http://atlasgeneticsoncology.org/gene/41670/pcna-(proliferating-cell-nclear-antigen

Strzalka W, Ziemienowicz A. Proliferating cell nuclear antigen (PCNA): a key factor in DNA replication and cell cycle regulation. Ann Bot. 2011;107(7):1127–40. https://doi.org/10.1093/aob/mcq243.

Sun L, et al. Cyclic GMP–AMP synthase is a cytosolic DNA sensor. Science. 2013;339(6121):786–91. https://doi.org/10.1126/science.1232458.

Taniguchi Y, Katsumata Y, Koido S, Suemizu H, Yoshimura S, Moriuchi T, Okumura K, Kagotani K, Taguchi H, Imanishi T, Gojobori T, Inoko H. Cloning, sequencing, and chromosomal localization of two tandemly arranged human pseudogenes for the proliferating cell nuclear antigen (PCNA). Mamm Genome. 1996;7(12):906–8. https://doi.org/10.1007/s003359900266. PMID: 8995762

Tsering T, et al. Extracellular vesicle-associated DNA: ten years since its discovery in human blood. Cell Death Dis. 2024;15:668. https://doi.org/10.1038/s41419-024-07003-y.

Ulrich HD. Regulating post-replication repair by ubiquitin and SUMO. DNA Repair (Amst). 2011;10(5):461–9. https://doi.org/10.1016/j.dnarep.2011.03.002.

Unterholzner L, et al. IFI16 is an innate immune sensor for intracellular DNA. Genes (Basel). 2020;11(4):409. https://doi.org/10.3390/genes11040409.

Warbrick E. PCNA binding through a conserved motif. BioEssays. 1998;20(3):195–9.

Webb GC, Parsons PG, Chenevix-Trench G. Localization of the gene for human Proliferating Cell Nuclear Antigen/cyclin by in situ hybridization. Hu. Genet. 1990;86(1):84–6. https://doi.org/10.1007/00205180.

Wendel S, Brooke G, Hu C, Sandoval A, Haratipour P, Gu L, et al. Targeting cancer-associated PCNA with AOH1996 induces mitotic death and enhances cisplatin therapy in cervical cancer. Submitted for publication. 2026. (Personal communication from Dr. Nicholas Wallace).

Witko-Sarsat V, Mocek J, Bouayad D, et al. Proliferating cell nuclear antigen acts as an anti-apoptotic factor in mature neutrophils. J Exp Med. 2010;207(12):2633–45.

Woo SR, et al. STING-dependent cytosolic DNA sensing mediates innate immune recognition of cancer. Immunity. 2014;41(5):830–42. https://doi.org/10.1016/j.immuni.2014.10.017.

Yu L, et al. Cytosolic DNA sensing by cGAS: regulation, function, and human diseases. Sig Transduct Target Ther. 2021;6:170. https://doi.org/10.1038/s41392-021-00554-y.

Zeman MK, Cimprich KA. Causes and consequences of replication stress. Nat Cell Biol. 2014;16(1):2–9.

Zhang CZ, et al. Chromothripsis from DNA damage in micronuclei. Nature. 2015;522(7555):179–84. https://doi.org/10.1038/nature14493.

Zhang S, Zhou T, Wang Z, Yi F, Li C, Guo W, Xu H, Cui H, Dong X, Liu J, Song X, Cao L. Post-translational modifications of PCNA in control of DNA synthesis and DNA damage tolerance-the implications in carcinogenesis. Int J Biol Sci. 2021;17(14):4047–59. https://doi.org/10.7150/ijbs.64628. https://www.ijbs.com/v17p4047.htm.

Zou L, Elledge SJ. Sensing DNA damage through ATRIP recognition of RPA-ssDNA complexes. Science. 2003;300(5625):1542–8. https://doi.org/10.1126/science.1083430.

Chapter 8
Regulation of the Anti-termination RNA Transcription Complex by Lon-Mediated Lambda N Degradation

Marianita Castro, Sanghyuk Lee, and Irene Lee

Abstract The lambda N protein (λN) is an intrinsically unstructured protein functioning to mediate anti-termination transcription of lambda phage RNA during the infection of *Escherichia coli* (E. coli) host by interacting with the host's macromolecular transcription machinery. In *E. coli* host, λN is primarily turned over by the ATP-dependent protease Lon. While it has been demonstrated that *Escherichia coli* Lon (ELon) readily degrades purified λN in vitro, it is unclear how ELon degrades λN in the RNA anti-termination transcription complex. The N-dependent anti-termination transcription mechanism of *Escherichia coli* RNA polymerase (ERNAP) and the quality control functions of Lon have been extensively studied and reviewed in literature. By contrast, very little is known about Lon's function as a regulatory protease. Herein, we provide a survey of literature and newfound evidence showing how ELon regulates anti-termination RNA transcription assemblies by using λN as a substrate. Elucidating how a substrate interacts with the various components in the assembly (ERNAP versus ELon) can dictate whether anti-termination transcription or degradation occurs.

Keywords Anti-termination transcription complex · *Escherichia coli* RNA polymerase · Lambda N · ATP-dependent protease Lon · Protein degradation · Electron microscopy

M. Castro · S. Lee · I. Lee (✉)
Department of Chemistry, Case Western Reserve University, Cleveland, OH, USA
e-mail: ixl13@case.edu

A. M. Pedley (ed.), *Supramolecular Protein Assemblies In Cells*, Advances in Experimental Medicine and Biology 1514,
https://doi.org/10.1007/978-3-032-26629-3_8

8.1   N-Dependent Anti-termination Transcription

In cells, RNA transcription initiates by RNAP binding to the promoter of the DNA template. Incorporation of incoming nucleotide triphosphate based on the coding information of the DNA template follows. The transcription process ends when RNAP encounters a termination signal (terminator) in the primer template and falls off the transcript (Alberts et al. 2002). There are instances where RNA transcription proceeds past the standard termination signals. In the lambda phage infection of *E. coli*, the partition between standard transcription versus anti-termination transcription dictates the life cycle of lambda phage development in the host. The lambda phage gene product, lambda N (λN) protein, is crucial in direction anti-termination transcription (Goodson and Winkler 2018). λN, an intrinsically unstructured protein, adopts defined structures upon interaction with connate proteins or RNA partners. In vitro, λN can bind to RNAP and the NUT RNA sequence to form a minimal anti-termination transcription complex that can carry out run-off transcription (Rees et al. 1996; Van Gilst and von Hippel 1997, 2000; Das et al. 2003). The core ERNAP is a 400 kDa heterosubunit complex consisting of the subunits α, α, β', β, and ω. When saturated with the sigma factor 70, the core ERNAP becomes holo ERNAP. The sigma factor directs ERNAP to initiate transcription at specific bacterial and phage promoters.

λN alone can also mediate non-processive transcriptional anti-termination at high molar excess over ERNAP devoid of accessory proteins and the nut sequence (Rees et al. 1996). Study by Said et al. showed that the carboxyl terminal of λN, staring at residues 73, was important for interaction with ERNAP and rendering anti-termination transcription function (Said et al. 2017). In vivo, transcriptional

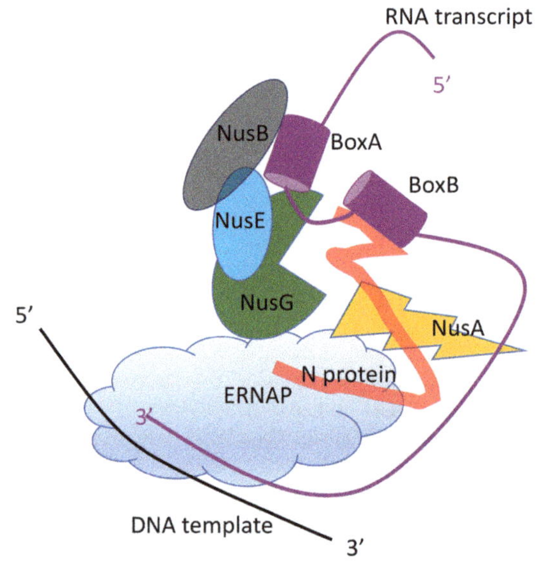

Fig. 8.1 The N-dependent anti-termination transcription complex that carries out processive anti-termination transcription is composed of the core ERNAP, the host NUS factors, the DNA-RNA hybrid containing the NUT sequences

anti-termination is processive (Goodson and Winkler 2018); accessory proteins and the NUT RNA bind different regions of λN to form an anti-termination complex under physiologically relevant protein concentrations (Fig. 8.1). A detailed mapping of the structure and functions of λN binding to different protein partners could be found in Said et al. and Krupp et al. (Said et al. 2017; Krupp et al. 2019). In this processive anti-termination transcription complex, λN interacts with RNAP, Nus A, and the Box B RNA hairpin. The *E. coli* elongation factors NusE, also known as S10, and NusB form a heterodimer that binds to the box A sequence of the NUT site in the RNA transcript. The NUT site is upstream of a standard transcription termination signal. The elongation factors NusA, NusB, NusG, and NusE stabilize the λN-RNAP DNA primer-RNA transcript complex and remodel the structure of RNAP to bypass the dissociation of RNAP from the primer template at the termination site. The cooperation among the different Nus proteins also prevents the formation of any hairpin structures in the RNA transcript to ensure uninterrupted transcription.

Nuclear magnetic resonance spectroscopic studies show that when not bound to accessory proteins or RNA, full-length and truncated forms of λN are intrinsically disordered (Legault et al. 1998; Mogridge et al. 1998; Van Gilst et al. 1997; Prasch et al. 2006). Krupp et al. utilized alanine substitution and deletion mutagenesis, along with high-resolution cryo-electron microscopy, to demonstrate that R89, R96, and K98/K100/K102 of λN interact with ERNAP in the transcription complex to confer anti-termination activity (Krupp et al. 2019). Although λN also interacts with ERNAP in other regions, the interactions between its carboxyl-terminal with different parts of ERNAP and nucleic acids contribute significantly to the effectiveness of the transcriptional anti-termination complex. The anti-termination activity of λN could be decreased stepwise by incrementally deleting the carboxyl terminal flanking residues 85–107 but rescued by a synthetic peptide constituting residues 88–107. Work pioneered by the von Hippel and the Greenblatt labs revealed that the carboxyl terminal of λN is important for interacting with *E. coli* ERNAP in the transcriptional anti-termination complex (Van Gilst and von Hippel 1997; Mogridge et al. 1998). Independently, NusA and BoxB RNA interact with residues 34 to 47 and 2 to 19 in λN, respectively. Rees et al. demonstrated that at high concentrations (~30-fold molar excess over transcription complexes), λN alone (without the NUT sequence to supply BoxB RNA and any accessory proteins such as NusA) could induce transcriptional anti-termination in vitro, with an estimated K_d of 5 μM for the transcription complex (Conant et al. 2008).

In vivo, λN is an endogenous substrate of Elon (Rees et al. 1996; Mogridge et al. 1998). The timing of λN expression and degradation dictates whether λ phage adopts a lytic or a lysogenic life cycle in the infected host. *E. coli* lacking Lon expression commits to lysis much faster than the wild-type cells (Gottesman et al. 1981a). The carboxyl terminal of λN, staring at residues 99, is important for initiating substrate interaction with ELon, which constitutes the rate-limiting step in the degradation of λN. In vitro, K_d of ELon for λN is 1.4 μM (Mikita et al. 2013). Given the similarity of the K_d of λN with ERNAP versus ELon, it is plausible that the competing interaction between ERNAP and Lon for the carboxyl terminal of λN dictates the fate and function of λN in vivo.

8.2 ATP-Dependent Proteases

ATP-dependent proteases are serine or threonine proteases containing a highly conserved ATPase module that couples the binding and hydrolysis of ATP to degrade abnormal or damaged proteins and short-lived regulatory proteins to maintain proper cellular function (Rep and Grivell 1996a; Langer and Neupert 1996; Savel'ev et al. 1998; Gottesman and Maurizi 1992; Maurizi 1987; Goff and Goldberg 1987). Based on the sequence and structural homology of the ATPase module, these proteases are classified as members of the AAA+ (*A*TPase-*A*ssociated with diverse cellular *A*ctivities) protein family (Neuwald et al. 1999; Hanson and Whiteheart 2005). Lon (named after *E. coli* bacteria that lack *lon* exhibits ELongated phenotype) and ClpA(X)P (*C*aseinoytic *P*roteinase complex partnered with subunit A or subunit X) bELong to this family. Based on the generally known functions of the ATPase module of these proteases and their propensity to degrade damaged cellular proteins, ATP-dependent proteases are considered quality-controlled protease machines. However, many of these protease machines also participate in the regulation of essential cellular processes such as nucleic acid metabolism by degrading certain short-lived timing proteins such as the λN anti-termination transcription factor (by ELon) or the sigma subunit of RNA polymerase (RNAP; by *E. coli* ClpXP) (Gottesman and Maurizi 1992; Gottesman et al. 1997; Gottesman 1996).

Lon, also known as protease La, is crucial in degrading certain damaged proteins and short-lived regulatory proteins in the cells (Rep and Grivell 1996b; Maurizi 1992a; Suzuki et al. 1997a; Goldberg 1992; Gottesman and Maurizi 1992; Kaser and Langer 2000; van Dyck and Langer 1999). This enzyme is a homo-hexameric ATP-dependent protease that resides in the cytosol of prokaryotes, lysosomes, and mitochondria of eukaryotes (Gottesman 2003; Maurizi 1992b; Goldberg et al. 1994; Maupin-Furlow et al. 2005; Rep and Grivell 1996a; van Dyck et al. 1998; Suzuki et al. 1997b). Mitochondrial Lon degrades oxidatively damaged proteins, thereby maintaining mitochondrial DNA integrity and function. This function is significant as it contributes to the overall health and functionality of the cell (Lu et al. 2007; van Dijl et al. 1998; Bulteau et al. 2006; Davies and Lin 1988). In *Saccharomyces cerevisiae*, Lon-deficient mutants suffer from large deletions in their mitochondrial genome and fail to process mitochondrial RNA transcripts (van Dyck et al. 1994, 1998; Suzuki et al. 1994). In some bacteria, Lon regulates the methylation of chromosomal DNA to enable proper cell differentiation (Wright et al. 1996) or transcription of bacteriophage RNA during host infection (Echols 1971).

8.3 Structure and Function of the Homo-oligomeric ATP-Dependent Protease ELon

ELon is a homohexamer with a molecular weight of 540 kDa (Goldberg et al. 1994; Botos et al. 2004a; Park et al. 2006). Oligomerization requires Mg^{2+} but not ATP (Rudyak et al. 2001). Each subunit contains three functional domains: the N-terminal domain (N-domain), the ATPase domain, and the protease domain (P-domain) at the carboxyl-terminal. The N-domain is proposed to mediate substrate recognition—the Walker Box A and B motifs within the ATPase domain that bind and catalyze ATP hydrolysis. The P-domain shows the highest evolutionary conservation surrounding the proteolytic active site serine residue (Rotanova et al. 2004). Analytical ultracentrifugation and electron microscopy studies reveal that bacterial Lon proteases are hexameric ring-shaped structures containing a central cavity where the proteolytic sites reside (Park et al. 2006; Rudyak et al. 2001). Limited proteolytic footprinting studies of ELon reveal that adenine nucleotide protects the enzyme from nonspecific proteolysis, indicating at least one conformational change generated by binding to MgATP (Patterson et al. 2004; Vasilyeva et al. 2002). Insights into the structure and function of Lon are obtained from crystal structures of various forms of bacterial and human Lon (Botos et al. 2004a, b; Cha et al. 2010; Duman and Lowe 2010; Rasulova et al. 1998; Shew et al. 1990; Garcia-Nafria et al. 2010). A Ser-Lys dyad constituting the proteolytic site is also observed in the crystal structures (Botos et al. 2004a; Cha et al. 2010; Duman and Lowe 2010). In full-length ELon, mutation of either S679A in the proteolytic site or K722A in the catalytic dyad abolishes proteolytic but not ATPase activity (Rotanova et al. 2004; Amerik et al. 1991; Starkova et al. 1998; Fisher and Glockshuber 1993).

8.4 Functional Mechanics of Substrate Recognition by Lon

Comparing the degradation profiles of several endogenous or heterologous substrates of ELon led to the postulate that solvent-exposed hydrophobic peptide patches in an unfolded protein function as recognition tags for the protease and initiation of substrate translocation often occurred at these tags (Choy et al. 2007; Gonzalez et al. 1998; Gur and Sauer 2008; Shah and Wolf Jr. 2006). Deleting such tags stabilizes proteins that are otherwise degraded by Lon (Gonzalez et al. 1998; Shah and Wolf Jr. 2006; Ishii and Amano 2001; Ishii et al. 2000). In some cases, protein degradation is initiated by incorporating a recognition tag into a stable heterologous reporter protein (Gonzalez et al. 1998; Gur and Sauer 2008; Shah and Wolf Jr. 2006). According to the literature, other signals or features of substrates that are not hydrophobic can also target proteins for degradation by bacterial Lon. Analysis of the peptide sequences within the transcriptional activator SoxS required for Lon-mediated proteolysis shows that hydrophobicity is not necessary for forming the substrate's recognition sequence of the substrate (Shah and Wolf Jr. 2006).

Studies examining the mechanism by which ELon degrades the DNA binding protein HUβ show that the recognition and binding sites within this substrate are independent of the initial cleavage site (Liao et al. 2010).

8.5   A Working Model for ATP-Dependent Proteolysis

Lon is a processive protease, with the unfolding and translocating of protein substrate to the proteolytic chamber formed by the oligomeric Lon subunits constituting the rate-limiting step (Wlodawer et al. 2022). Proteolytic cleavage of the scissile peptide bonds occurs once the entirely unfolded protein is delivered into the proteolytic chamber. As such the molecular interaction between Lon and the protein substrate determinant signal is crucial to commitment of protein degradation. A general mechanistic scheme accounting for the role of the ATPase in ATP-dependent proteolysis is illustrated in Fig. 8.2 (Licht and Lee 2008). The ATPase domain of the protease first binds/interacts with a specific recognition peptide within a protein substrate (step 1). The presence of MgATP induces a series of conformational changes within the enzyme subunits that leads to the unfolding of substrate (step 2), followed by internalization (also known as translocation) of the protein substrate into the central protease cavity (step 3). Polypeptide unfolding and translocation are proposed to constitute the rate-limiting step of the reaction (Kenniston et al. 2003) and are often initiated at the recognition tag of the substrate. Both processes are hypothesized to be carried out by repetitive cycles of ATP hydrolysis that induce conformational changes within the ATP-dependent protease structure (Goldberg et al. 1994; Wang et al. 2001) via a threading mechanism (Licht and Lee 2008; Lee et al. 2001; Reid et al. 2001) (i.e., the polypeptide substrate is transported through the narrow central pore of the enzyme in a roughly linear conformation). In step 4, the translocated polypeptide substrates are sequestered within the central cavity of the protease machinery (the proteolytic chamber), where peptide bond cleavage occurs (Licht and Lee 2008; Ortega et al. 2000; Matouschek 2003; Singh et al.

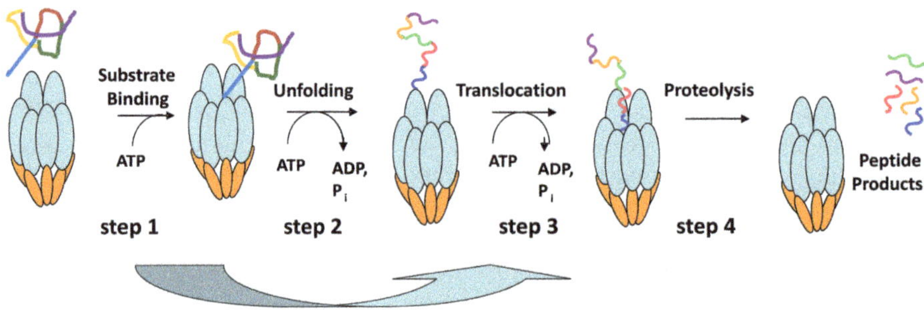

Fig. 8.2 General mechanism for ATP-dependent proteolysis

2000). For certain substrates of ATP-dependent proteases, proteolytic digestion generates peptides ranging from ~5 to 20 amino acids but not partially digested polypeptide intermediates, concluding that degradation proceeds processively (Maurizi 1987; Licht and Lee 2008; Nishii et al. 2005; Ondrovicova et al. 2005). The translocation and peptidase events can be isolated from substrate unfolding and selectively examined using unstructured substrates. Therefore, unstructured proteins such as λN, which is processively degraded by ELon in the presence of MgATP, are ideal substrates for investigating Lon's processive protein degradation mechanism (Mikita et al. 2013; Maurizi 1987). By incorporating fluorophores and corresponding fluorophore quenchers at specific sites along λN, the kinetic coordination of substrate translocation and different scissile site cleavage were determined by transient kinetic techniques to show that ELon initiates interaction with the carboxyl-terminal of λN. In the presence of MgATP, protein degradation occurs after the complete translocation of λN into the proteolytic chamber.

8.6 Lambda N Protein as a Substrate for ELon

The carboxyl terminal of λN interacts with ERNAP in the transcription complex to bypass the transcription termination site in the phage DNA template (Said et al. 2017; Krupp et al. 2019). In the ELon-mediated degradation, the carboxyl terminal of λN directs the initiation of substrate translocation into the proteolytic chamber of ELon, which is formed by six identical enzyme subunits to form a homooligomeric complex of approximately 540 kDa in size. Given ERNAP and ELon both interact with the carboxyl terminal of λN, it is plausible that the carboxyl terminal of λN serves as a signal to turn on and off the anti-termination transcription of RNA.

In vitro, ELon degrades purified λN in the presence of MgATP or MgAMPPNP (a non-hydrolyzable analog of ATP) (Maurizi 1987). Still, the rate of λN degradation is faster in the presence of ATP, and the same hydrolyzed peptide products were generated in both reactions (Maurizi 1987). The cleavage sites of λN by ELon is shown in Fig. 8.3. Since the unfolding of λN is not necessary, we will only need to consider the translocation and cleavage of the multiple peptide bonds within the substrate during its degradation. In vitro, full-length λN could be chemically modified or synthesized to become fluorogenic substrates for monitoring the kinetics of λN translocation and degradation at different scissile sites within the protein

[1]MDAQTRRRE [11]RRAQK AQWKA [21]ANPLLVGVSA [31]KPVNRP
ILSL [41]NRKPKSRVES [51]ALNPIDLTVLAEYHKQIESN [71]LQRI
ERKNQR [81]TWYSKPGERG [91]ITCSGRQKIK [101]GKSIPLI

Fig. 8.3 Amino acid sequence of λN. The arrows indicate the ELon cleavage sites

substrate. The K_d of ELon for full-length λN is 1.4 µM (Mikita et al. 2013). Lambda N lacking the carboxyl-terminal residues 99 to 107 is less efficiently degraded by ELon (Patterson-Ward et al. 2009). Since the K_d of ERNAP for λN is 5 µM whereas the K_d of ELon for λN is 1.4 µM, it is conceivable that Lon competes with ERNAP in binding to the carboxyl terminal of λN. The fate of λN is dictated by the condition that favors binding to one over the other. Demonstrating that ELon can effectively compete with ERNAP in the macromolecular transcription complex devoid of NUS factors and ancillary nucleic acid would lend proof of principle support for Lon's active role in regulating RNA transcription.

8.7 A Proof of Principle Study (Castro et al. 2025)

An intriguing aspect of the regulatory function of ATP-dependent proteases is the "timing" of substrate degradation; it is unclear how an ATP-dependent protease knows when to and not to degrade the substrate. To initiate protein degradation, an ATP-dependent protease needs to interact with at least one specific recognition element in a substrate, which may or may not be at the terminal of the substrate. Compared to the understanding of Lon functioning as a quality control protease, little is known about Lon functioning as a regulatory protease. To investigate the regulatory function of Lon, a protein substrate that functions as a timing protein must be used. To this end, we use the anti-termination transcription protein λN as the substrate. To evaluate the impact of ERNAP on ELon-mediated λN degradation, we compared the degradation profiles of full-length and a truncated λN lacking the carboxyl terminal (residues 99–107) by ELon in the presence versus the absence of the holo ERNAP, which contains the core RNAP complex and the sigma factor 70 (σ70). As noted in an earlier study, λN δ99–107 is degraded by ELon at a slower rate than full-length λN, with the K_d of the former for ELon at 5.2 µM and the latter at 1.4 µM (Mikita et al. 2013). As such, the time points for monitoring λN and λN δ99–107 are 0–4 min and 0–8 min, respectively. As shown in Fig. 8.4, holo ERNAP inhibits the degradation kinetics of N-terminal his-tagged full-length λN and N-terminal his-tagged λN δ99–107, but the inhibitory effect on the latter is more pronounced. This observation could be explained by ELon, but not ERNAP exhibiting weaker affinity for truncated λN and thus failing to effectively compete with ERNAP to degrade this protein.

Negative stain electron microscopy was used to visualize the altered interactions between ELon with full length versus the carboxyl terminal truncated λN. The EM images of ELon bound to full-length or truncated λN in the presence and absence of AMP-PNP were compared. The λN proteins were not visible in the EM images, presumably due to their small size. However, structural variations were discernable in the images, indicating that λN induced structural changes in the ELon particles. Figure 8.5 panels a–c show the EM data on ELon by itself, whereas Fig. 8.5 panels d and f show the MgAMPPNP bound ELon. For ELon by itself, 6844 particles were

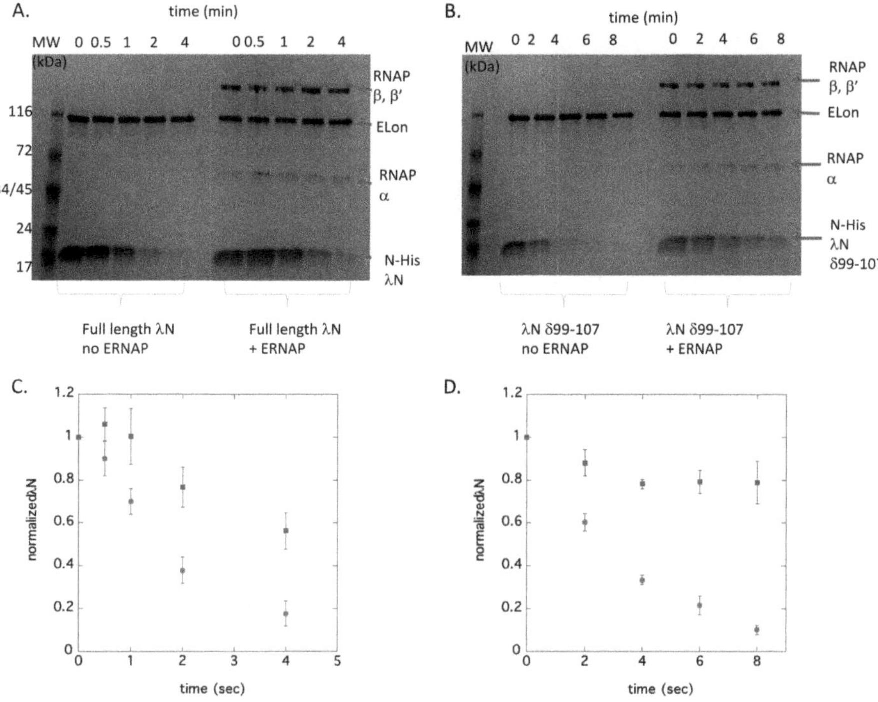

Fig. 8.4 (**a**) A representative SDS-PAGE result of ELon degradation of full-length λN at time = 0, 0.5, 1, 2, and 4 min. (**b**) A representative SDS-PAGE result of ELon degradation of λN δ99–107 at time = 0, 2, 4, 6, and 8 min. Same molecular weight standards as shown in (**a**). (**c**) The averaged time course of ELon degradation of full-length λN in the absence (red) versus the presence of ERNAP (blue). (**d**) The averaged time course of ELon degradation of λN δ99–107 in the absence (red) versus the presence of ERNAP (blue)

selected from 9 2D class averages to produce a low-resolution 3D volume (Fig. 8.5b, c). For the ELon MgAMPPNP complex, 5383 particles were selected from four 2D class averages (Fig. 8.5e). The 3D volume corresponding to ELon with MgAMPPNP seems to acquire a slightly tighter conformation (Fig. 8.5f). Comparing the images in Fig. 8.5c (ELon imaged in the absence of MgAMPPNP and λN) with Fig. 8.5f (ELon imaged in the presence of MgAMP-PNP but not λN) reveals the difference in the compactness of the particles, with the ELon:MgAMPPNP complex exhibiting a more compact structure. This finding is consistent with the limited tryptic footprinting study detecting a more compact structure in ELon when bound to MgATP or MgAMPPNP (Patterson-Ward et al. 2007).

Figure 8.6 panels a–f show the EM results of ELon incubated with full length versus delta 99–107 λN with MgAMPPNP. To generate the 3D volumes of ELon bound to full-length and truncated λN, 7347 and 9106 particles were selected from three 2D class averages each, respectively. Comparing Fig. 8.5c, f with Fig. 8.6c, f reveals that ELon adopts a more compact conformation when bound to the λN proteins even without a nucleotide. A survey of the micrographs shows that the

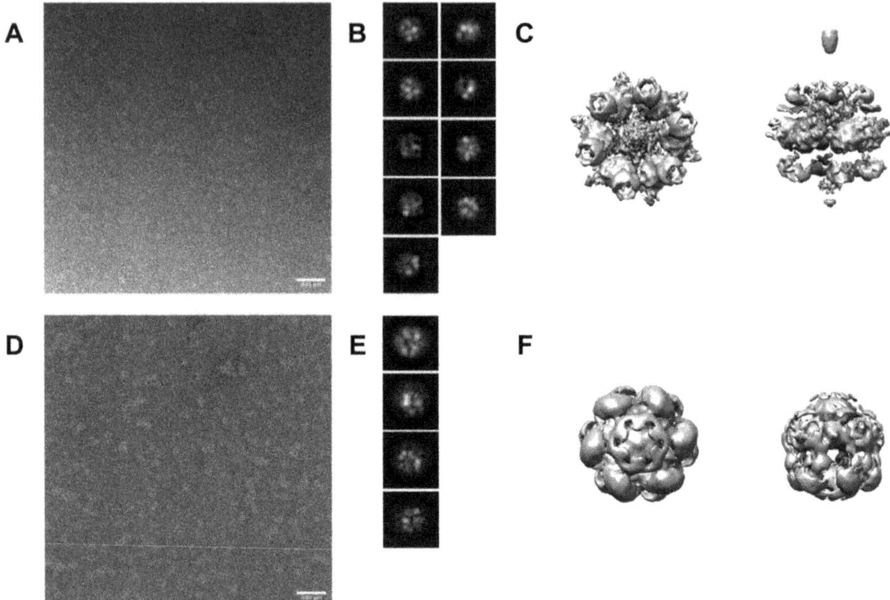

Fig. 8.5 (**a**) Negative stain micrograph of ELon without substrate. (**b**) 2D class averages of ELon without substrate. (**c**) 3D volume of ELon without substrate with C6 symmetry. (**d**) Negative stain micrograph of ELon bound to AMP-PNP. (**e**) 2D class averages of ELon bound to AMP-PNP. (**f**) 3D volume of ELon bound to AMP-PNP with C6 symmetry. The horizontal bar located at the bottom right of panels **a** and **d** represents 0.05 μm

particles containing ELon bound to full-length λN in the absence of MgAMPPNP have more identifiable, uniform particles (Fig. 8.6a) than the ELon: λN δ99–107 particles (Fig. 8.6d), presumably due to the weakened affinity of ELon for λN δ99–107 resulting in the formation of less complexes. This observation corroborates the activity findings shown in Fig. 8.4 that λN δ99–107 was more protected by ERNAP from ELon degradation than full-length:λN, thereby supporting the conclusion that the carboxyl terminal of:λN, flanked by residues 99–107, is crucial for ELon but not so much for ERNAP interaction.

Figure 8.7a–f show the EM data of the ELon:MgAMPPNP:λN complexes and the ELon:MgAMPPNP:λN δ99–107 complexes, respectively. The 3D volumes of ELon:MgAMPPNP bound to full-length versus λN δ99–107 were generated from 8138 and 10,020 particles selected from 6 and 7 2D class averages, respectively (Fig. 8.7b, c, e, f). In the presence of MgAMPPNP, the micrographs for full-length and truncated λN are similar, which may be due to more stability in the ELon:l:λN complexes gained from binding to MgAMPPNP, which promotes the translocation of the λN substrates into the proteolytic chamber (Fig. 8.7c, f). Since the full length and λN δ99–107 are degraded by ELon, albeit at a different rate (Fig. 8.4), the similarity in the EM images detected in Fig. 8.7 is likely attributed to the detection of ELon:λN:MgAMPPNP or ELon:λN δ99–107:MgAMPPNP translocation complex.

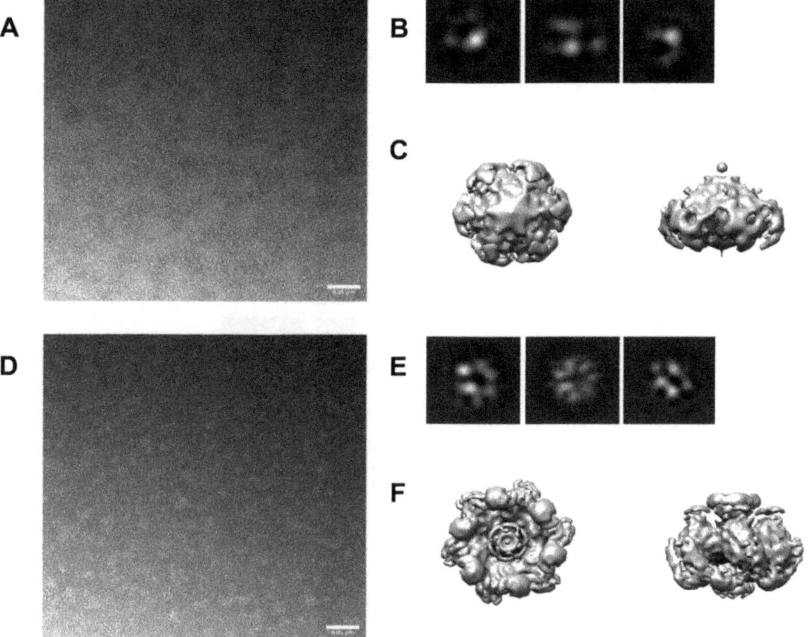

Fig. 8.6 (**a**) Negative stain micrograph of ELon bound to full-length lambda N. (**b**) 2D class averages of ELon bound to full-length lambda N. (**c**) 3D volume of ELon bound to full-length lambda N with C6 symmetry. (**d**) Negative stain micrograph of ELon bound to truncated lambda N. (**e**) 2D class averages of ELon bound to truncated lambda N. (**f**) 3D volume of ELon bound to truncated lambda N with C6 symmetry. The horizontal bar located at the bottom right of panels **a** and **d** represents 0.05 µm

For comparison, Fig. 8.7 shows the difference in the occupancy of the entrance pore of ELon bound to MgAMPPNP Fig. 8.7a, MgAMPPNP and λN Fig. 8.7b, and MgAMPPNP and:λN δ99–107. The tightest hexameric conformation, which also forms propeller-like particles, is found in ELon bound to AMP-PNP and full-length λN (Fig. 8.8b). Taken together, the EM images corroborate the proposal drawn from the λN degradation profiles summarized in Fig. 8.2 that λN lacking the carboxyl-terminal forms a weak complex with ELon and hence is more protected by ERNAP, whose affinity for the truncated:λN remains comparable with full-length λN.

8.8 Summary and Perspective

The holo ERNAP is 400 kDa, whereas ELon is a 534 kDa multimeric protein complex. Amazingly, the fate of the intrinsically unstructured protein λN depends on its carboxyl terminal, which constitutes about 30 amino acids, interacting with ERNAP or Elon. In this manuscript, we provide a review of the literature, including

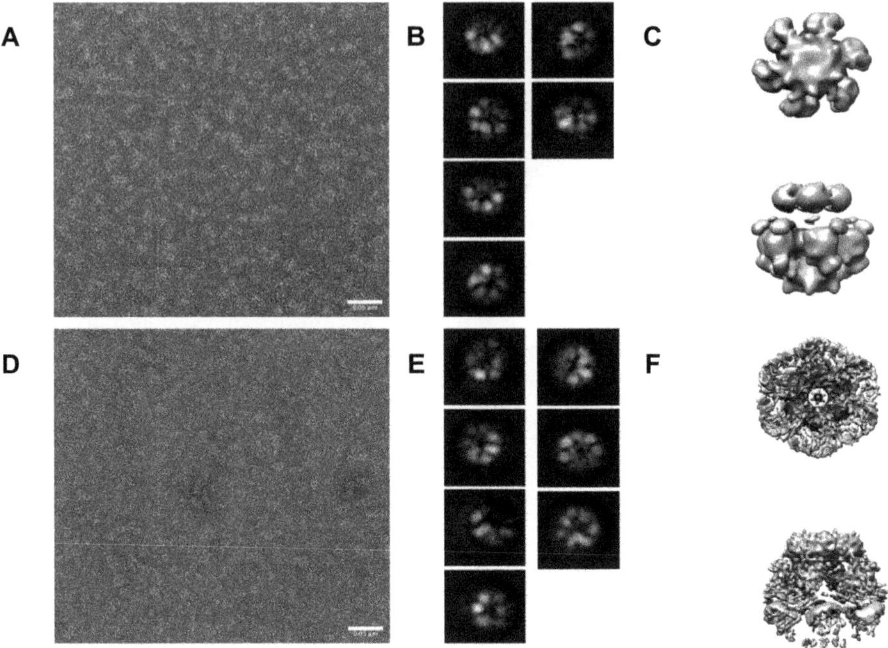

Fig. 8.7 (**a**) Negative stain micrograph of ELon bound to full-length lambda N and AMP-PNP. (**b**) 2D class averages of ELon bound to full-length lambda N and AMP-PNP. (**c**) 3D volume of ELon bound to full-length lambda N and AMP-PNP with C6 symmetry. (**d**) Negative stain micrograph of ELon bound to truncated lambda N and AMP-PNP. (**e**) 2D class averages of ELon bound to truncated lambda N and AMP-PNP. (**f**) 3D volume of ELon bound to truncated lambda N and AMP-PNP with C6 symmetry. The horizontal bar located at the bottom right of panels **a** and **d** represents 0.05 µm

newfound structural and kinetics data to support the idea that the λN-mediated anti-termination transcription process is an excellent system for investigating the regulatory function of Lon as a protease. The minimal component of the ERNA transcription complex containing ERNAP and the accessory protein, the σ factor, could inhibit λN degradation by ELon through SDS-PAGE, and the carboxyl-terminal of λN is important for Lon competing with RNAP interaction. Negative stain electron microscopy showed that λN lacking the carboxyl-terminal flanked by residues 99–107 interacted with ELon differently than full-length λN. The contribution of λN on anti-termination RNA transcription in *E. coli* has been quantitatively and structurally characterized for over 30 decades; any additional structural, binding, and activity studies to evaluate the impact of ELon-mediated degradation of λN will be readily incorporated into the existing mechanism framework for RNA transcription mediated by λN. Since the *E. coli* RNA transcription complex constitutes the N-utilizing substance A (NusA) and the NUT sequence that supplies the Box B RNA in addition to the holo RNAP complex, the study of ELon-mediated degradation of λN in the presence of the different RNA transcription components will

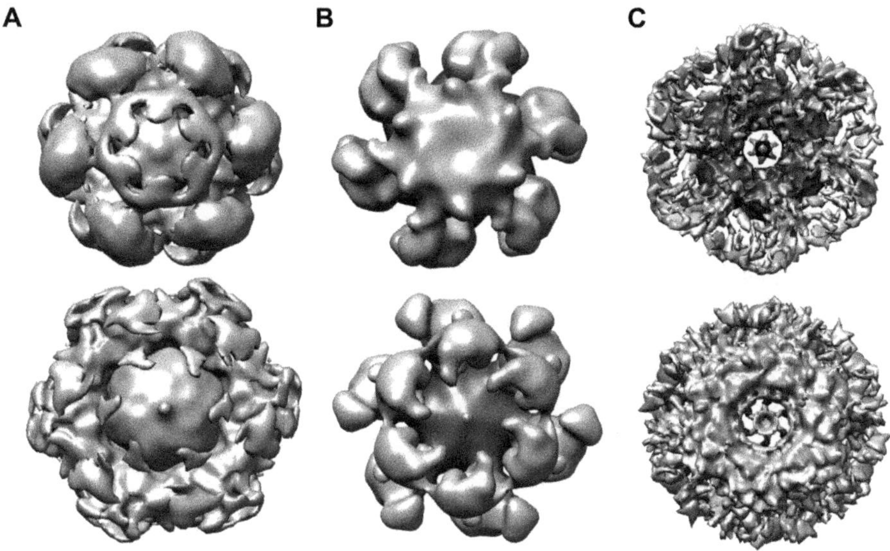

Fig. 8.8 Comparison of front and back view of the entrance pore in the ELon complexes. (**a**) ELon bound to AMP-PNP. (**b**) ELon bound to AMP-PNP and full-length lambda N. (**c**) ELon bound to AMP-PNP and truncated lambda N

provide mechanistic insights into how protein-protein interaction dictates the fate of λN, thereby affect the life cycle of lambda phage. As the structure and function relationship of λN functioning as an anti-termination transcription factor and as a Lon substrate have been quantitatively defined and reported in the literature, the dynamics connection of the two enzymatic processes should be readily deduced by the kinetic approaches the fluorogenic λN protein constructs designed as Lon protease substrates. Taken together, these investigations have provided a starting point for additional enzymology studies on the contribution of ELon toward RNA transcription by providing a means to allow quantitative assessment on whether ELon can degrade λN bound to the transcription complex, which will advance the existing understanding of the impact of protein-protein interaction between λN and the RNA transcription complex on regulatory protein turnover by an ATP-dependent protease. Knowing the role played by Lon will enhance our fundamental understanding of the physiological enzymology of this protease, which could be applied to study the regulatory functions of other ATP-dependent proteases, shedding new light into the regulatory rather than just quality control functions. Since intrinsically unstructured proteins as well as mitochondrial Lon play an important role in maintaining the functional integrity of mitochondria (Ito et al. 2012), a better understanding of how ELon degrades λN in the context of anti-termination transcription will aid the design of experiments to investigate comparable situations in mitochondria, which will benefit the investigation of mitochondria biology and diseases.

Acknowledgments The authors wish to thank Drs. Jason A. Mears, Kristy Rochon, and Kyle Whiddon for their assistance and guidance in executing the EM imaging, data acquisition, and analyses. The research activities of the authors were supported by an award from the National Science Foundation MCB 2210869.

Competing Interests The authors declare no competing financial interests.

References

Alberts B, Johnson A, Lewis J, et al. From DNA to RNA. In: Molecular biology of the cell. 4th ed. New York: Garland Science; 2002.

Amerik AY, et al. Site-directed mutagenensis of La protease: a catalytically active serine protease. FEBS Lett. 1991;287(1, 2):211–4.

Botos I, et al. The catalytic domain of Escherichia coli Lon protease has a unique fold and a Ser-Lys dyad in the active site. J Biol Chem. 2004a;279(9):8140–8.

Botos I, et al. Crystal structure of the AAA+ alpha domain of E. coli Lon protease at 1.9A resolution. J Struct Biol. 2004b;146(1–2):113–22.

Bulteau AL, Szweda LI, Friguet B. Mitochondrial protein oxidation and degradation in response to oxidative stress and aging. Exp Gerontol. 2006;41(7):653–7.

Castro M, Lee S, Lee I. Lambda N as a model substrate for studying the mechanism of Escherichia coli ATP-dependent protease Lon as a regulatory enzyme. bioRxiv. 2025. https://doi.org/10.1101/2025.01.24.634763

Cha SS, et al. Crystal structure of Lon protease: molecular architecture of gated entry to a sequestered degradation chamber. EMBO J. 2010;29(20):3520–30.

Choy JS, Aung LL, Karzai AW. Lon protease degrades transfer-messenger RNA-tagged proteins. J Bacteriol. 2007;189(18):6564–71.

Conant CR, et al. The anti-termination activity of bacteriophage lambda N protein is controlled by the kinetics of an RNA-looping-facilitated interaction with the transcription complex. J Mol Biol. 2008;384(1):87–108.

Das A, et al. Genetic and biochemical strategies to elucidate the architecture and targets of a processive transcription antiterminator from bacteriophage lambda. Methods Enzymol. 2003;371:438–59.

Davies KJ, Lin SW. Oxidatively denatured proteins are degraded by an ATP-independent proteolytic pathway in Escherichia coli. Free Radic Biol Med. 1988;5(4):225–36.

Duman RE, Lowe J. Crystal structures of Bacillus subtilis Lon protease. J Mol Biol. 2010;401(4):653–70.

Echols H. In: Hershey AD, editor. The bacteriophage lambda. Cold Spring Harbor: Cold Spring Harbor Laboratory; 1971. p. 247–70.

Fisher H, Glockshuber R. ATP hydrolysis is not stoichiometrically linked to proteolysis in the ATP-dependent protease La from Escherichia coli. J Biol Chem. 1993;268(30):22502–7.

Garcia-Nafria J, et al. Structure of the catalytic domain of the human mitochondrial Lon protease: proposed relation of oligomer formation and activity. Protein Sci. 2010;19(5):987–99.

Goff SA, Goldberg AL. An increased content of protease La, the lon gene product, increases protein degradation and blocks growth in Escherichia coli. J Biol Chem. 1987;262(10):4508–15.

Goldberg AL. The mechanism and functions of ATP-dependent proteases in bacterial and animal cells. Eur J Biochem. 1992;203(1–2):9–23.

Goldberg AL, et al. ATP-dependent protease La (Lon) from Escherichia coli. Methods Enzymol. 1994;244:350–75.

Gonzalez M, et al. Lon-mediated proteolysis of the Escherichia coli UmuD mutagenesis protein: in vitro degradation and identification of residues required for proteolysis. Genes Dev. 1998;12(24):3889–99.

Goodson JR, Winkler WC. Processive anti-termination. Microbiol Spectrum. 2018;6(5):10–1128.

Gottesman S. Proteases and their targets in Escherichia coli. Annu Rev Genet. 1996;30:465–506.

Gottesman S. Proteolysis in bacterial regulatory circuits. Annu Rev Cell Dev Biol. 2003;19:565–87.

Gottesman S, Maurizi MR. Regulation by proteolysis: energy-dependent proteases and their targets. Microbiol Rev. 1992;56(4):592–621.

Gottesman S, et al. Protein degradation in E. coli: the lon mutation and bacteriophage lambda N and cII protein stability. Cell. 1981a;24(1):225–33.

Gottesman S, Wickner S, Maurizi MR. Protein quality control: triage by chaperones and proteases. Genes Dev. 1997;11:815–23.

Gur E, Sauer RT. Recognition of misfolded proteins by Lon, a AAA(+) protease. Genes Dev. 2008;22(16):2267–77.

Hanson PI, Whiteheart SW. AAA+ proteins: have engine, will work. Nat Rev Mol Cell Biol. 2005;6(7):519–29.

Ishii Y, Amano F. Regulation of SulA cleavage by Lon protease by the C-terminal amino acid of SulA, histidine. Biochem J. 2001;358(Pt 2):473–80.

Ishii Y, et al. Regulatory role of C-terminal residues of SulA in its degradation by Lon protease in Escherichia coli. J Biochem. 2000;127(5):837–44.

Ito M, et al. Intrinsically disordered proteins in human mitochondria. Genes Cells. 2012;17(10):817–25.

Kaser M, Langer T. Protein degradation in mitochondria. Semin Cell Dev Biol. 2000;11(3):181–90.

Kenniston JA, et al. Linkage between ATP consumption and mechanical unfolding during the protein processing reactions of an AAA+ degradation machine. Cell. 2003;114(4):511–20.

Krupp F, et al. Structural basis for the action of an all-purpose transcription anti-termination factor. Mol Cell. 2019;74(1):143–157.e5.

Langer T, Neupert W. Regulated protein degradation in mitochondria. Experentia. 1996;52(12):1019–157.

Lee C, et al. ATP-dependent proteases degrade their substrates by processively unraveling them from the degradation signal. Mol Cell. 2001;7(3):627–37.

Legault P, et al. NMR structure of the bacteriophage lambda N peptide/boxB RNA complex: recognition of a GNRA fold by an arginine-rich motif. Cell. 1998;93(2):289–99.

Liao JH, et al. Binding and cleavage of E. coli HUbeta by the E. coli Lon protease. Biophys J. 2010;98(1):129–37.

Licht S, Lee I. Resolving individual steps in the operation of ATP-dependent proteolytic molecular machines: from conformational changes to substrate translocation and processivity. Biochemistry. 2008;47(12):3595–605.

Lu B, et al. Roles for the human ATP-dependent Lon protease in mitochondrial DNA maintenance. J Biol Chem. 2007;282(24):17363–74.

Matouschek A. Protein unfolding–an important process in vivo? Curr Opin Struct Biol. 2003;13(1):98–109.

Maupin-Furlow JA, et al. Archaeal proteasomes and other regulatory proteases. Curr Opin Microbiol. 2005;8(6):720–8.

Maurizi MR. Degradation in vitro of bacteriophage lambda N protein by Lon protease from Escherichia coli. J Biol Chem. 1987;262(6):2696–703.

Maurizi MR. Proteases and protein degradation in Escherichia coli. Experientia. 1992a;48(2):178–201.

Maurizi MR. Proteases and protein degradation in Escherichia coli. Experentia. 1992b;48:178–201.

Mikita N, et al. Processive degradation of unstructured protein by Escherichia coli Lon occurs via the slow, sequential delivery of multiple scissile sites followed by rapid and synchronized peptide bond cleavage events. Biochemistry. 2013;52(33):5629–44.

Mogridge J, et al. Independent ligand-induced folding of the RNA-binding domain and two functionally distinct anti-termination regions in the phage lambda N protein. Mol Cell. 1998;1(2):265–75.

Neuwald AF, et al. AAA+: a class of chaperone-like ATPases associated with the assembly, operation, and disassembly of protein complexes. Genome Res. 1999;9(1):27–43. http://www.genome.org/cgi/content/abstract/9/1/27.

Nishii W, et al. Cleavage mechanism of ATP-dependent Lon protease toward ribosomal S2 protein. FEBS Lett. 2005;579(30):6846–50.

Ondrovicova G, et al. Cleavage site selection within a folded substrate by the ATP-dependent Lon protease. J Biol Chem. 2005;280(26):25103–10.

Ortega J, et al. Visualization of substrate binding and translocation by the ATP-dependent protease, *ClpXP*. Mol Cell. 2000;6(6):1515–21.

Park SC, et al. Oligomeric structure of the ATP-dependent protease La (Lon) of Escherichia coli. Mol Cells. 2006;21(1):129–34.

Patterson J, et al. Correlation of an adenine-specific conformational change with the ATP-dependent peptidase activity of Escherichia coli Lon. Biochemistry. 2004;43(23):7432–42.

Patterson-Ward J, Huang J, Lee I. Detection and characterization of two ATP-dependent conformational changes in proteolytically inactive Escherichia coli Lon mutants by stopped flow kinetic techniques. Biochemistry. 2007;46(47):13593–605.

Patterson-Ward J, et al. Utilization of synthetic peptides to evaluate the importance of substrate interaction at the proteolytic site of Escherichia coli Lon protease. Biochim Biophys Acta. 2009;1794(9):1355–63.

Prasch S, et al. Interaction of the intrinsically unstructured phage lambda N protein with Escherichia coli NusA. Biochemistry. 2006;45(14):4542–9.

Rasulova FS, et al. The isolated proteolytic domain of Escherichia coli ATP-dependent protease Lon exhibits the peptidase activity. FEBS Lett. 1998;432(3):179–81.

Rees WA, et al. Bacteriophage lambda N protein alone can induce transcription anti-termination in vitro. Proc Natl Acad Sci USA. 1996;93(1):342–6.

Reid BG, et al. ClpA mediates directional translocation of substrate proteins into the ClpP protease. Proc Natl Acad Sci USA. 2001;98(7):3768–72.

Rep M, Grivell LA. The role of protein degradation in mitochondrial function and biogenesis. Curr Genet. 1996a;30:367–80.

Rep M, Grivell LA. The role of protein degradation in mitochondrial function and biogenesis. Curr Genet. 1996b;30(5):367–80.

Rotanova TV, et al. Classification of ATP-dependent proteases Lon and comparison of the active sites of their proteolytic domains. Eur J Biochem. 2004;271(23–24):4865–71.

Rudyak SG, Brenowitz M, Shrader TE. Mg2+-linked oligomerization modulates the catalytic activity of the Lon (La) protease from Mycobacterium smegmatis. Biochemistry. 2001;40(31):9317–23.

Said N, et al. Structural basis for λN-dependent processive transcription anti-termination. Nat Microbiol. 2017;2(7):17062.

Savel'ev AS, et al. ATP-dependent proteolysis in mitochondria. m-AAA protease and PIM1 protease exert overlapping substrate specificities and cooperate with the mtHsp70 system. J Biol Chem. 1998;273(32):20596–205602.

Shah IM, Wolf RE Jr. Sequence requirements for Lon-dependent degradation of the Escherichia coli transcription activator SoxS: identification of the SoxS residues critical to proteolysis and specific inhibition of in vitro degradation by a peptide comprised of the N-terminal 21 amino acid residues. J Mol Biol. 2006;357(3):718–31.

Shew JY, et al. C-terminal truncation of the retinoblastoma gene product leads to functional inactivation. Proc Natl Acad Sci USA. 1990;87:6–10.

Singh SK, et al. Unfolding and internalization of proteins by the ATP-dependent proteases ClpXP and ClpAP. Proc Natl Acad Sci USA. 2000;97(16):8898–903.

Starkova NN, et al. Mutations in the proteolytic domain of Escherichia coli protease Lon impair the ATPase activity of the enzyme. FEBS Lett. 1998;422(2):218–20.

Suzuki CK, et al. Requirement for the yeast gene LON in intramitochondrial proteolysis and maintenance of respiration. Science. 1994;264:273–6 and 891.

Suzuki CK, et al. ATP-dependent proteases that also chaperone protein biogenesis. Trends Biochem Sci. 1997a;22(4):118–23.

Suzuki CK, et al. ATP-dependent proteases that also chaperone protein biogenesis. Trends Biochem Sci. 1997b;22:118–23.

van Dijl JM, et al. The ATPase and protease domains of yeast mitochondrial Lon: roles in proteolysis and respiration-dependent growth. Proc Natl Acad Sci USA. 1998;95(18):10584–9.

van Dyck L, Langer T. ATP-dependent proteases controlling mitochondrial function in the yeast Saccharomyces cerevisiae. Cell Mol Life Sci. 1999;56(9–10):825–42.

van Dyck L, Pearce DA, Sherman F. PIM1 encodes a mitochondrial ATP-dependent protease that is required for mitochondrial function in the yeast Saccharomyces cerevisiae. J Biol Chem. 1994;269:238–42.

van Dyck L, Neupert W, Langer T. The ATP-dependent PIM1 protease is required for the expression of intron-containing genes in mitochondria. Genes Dev. 1998;12(10):1515–24.

Van Gilst MR, von Hippel PH. Assembly of the N-dependent anti-termination complex of phage lambda: NusA and RNA bind independently to different unfolded domains of the N protein. J Mol Biol. 1997;274(2):160–73.

Van Gilst MR, von Hippel PH. Quantitative dissection of transcriptional control system: N-dependent anti-termination complex of phage lambda as regulatory paradigm. Methods Enzymol. 2000;323:1–31.

Van Gilst MR, et al. Complexes of N anti-termination protein of phage lambda with specific and nonspecific RNA target sites on the nascent transcript. Biochemistry. 1997;36(6):1514–24.

Vasilyeva OV, et al. Domain structure and ATP-induced conformational changes in Escherichia coli protease Lon revealed by limited proteolysis and autolysis. FEBS Lett. 2002;526(1–3):66–70.

Wang J, et al. Crystal structures of the HslVU peptidase-ATPase complex reveal an ATP-dependent proteolysis mechanism. Structure. 2001;9(2):177–84.

Wlodawer A, et al. Structure and the mode of activity of Lon proteases from diverse organisms. J Mol Biol. 2022;434(7):167504.

Wright R, et al. Caulobacter Lon protease has a critical role in cell-cycle control of DNA methylation. Genes Dev. 1996;10(12):1532–42.

Chapter 9
The Pyruvate Dehydrogenase Complex: A 90-Year-Old Enigma Shaping the Future of Structural Enzymology

Toni K. Träger, Fotis L. Kyrilis, Greg Kafetzopoulos, Christian Tüting, and Panagiotis L. Kastritis

Abstract The ancient pyruvate dehydrogenase complex (PDHc) performs the "link reaction" of cellular respiration—a discovery from the 1930s that was central in the award of the 1953 Nobel Prize in Physiology and Medicine to Krebs and Lipmann. Fast forward to 2024, PDHc emerges with roles in Alzheimer's, cancer, and neuro-degeneration, as well as in obesity and aging processes. Due to these recent reports, structural analysis of PDHc, a 10-megadalton enzymatic complex, comes into focus—only now this analysis begins to unveil an enormous and challenging molecular complexity. Cutting-edge techniques and methods, such as cryo-electron microscopy (cryo-EM), cross-linking (XL) and mass spectrometry (MS), advanced molecular and biochemical analysis, and computational structural biology, powered by artificial intelligence (AI), converge to systematically probe the mechanistic details governing PDHc function. This chapter collects and updates the knowledge in PDHc structure and function and pinpoints unresolved questions, with the hope of not waiting another 90 years for their answer.

Keywords Keto acid dehydrogenase complex family · Citric acid cycle · Krebs cycle · TCA cycle · Glycolysis · Acetyl-CoA · Respirasome · Metabolic diseases · Structural biology · Enzyme regulation · Metabolon · Mitochondria · Nuclear

T. K. Träger · C. Tüting
Martin Luther University Halle-Wittenberg, Institute of Biochemistry and Biotechnology, Integrative Structural Biochemistry, Biozentrum, Halle, Germany

F. L. Kyrilis · G. Kafetzopoulos
National Hellenic Research Foundation (NHRF), Institute of Chemical Biology (ICB), Athens, Greece

P. L. Kastritis (✉)
Martin Luther University Halle-Wittenberg, Institute of Biochemistry and Biotechnology, Integrative Structural Biochemistry, Biozentrum, Halle, Germany

National Hellenic Research Foundation (NHRF), Institute of Chemical Biology (ICB), Athens, Greece
e-mail: panagiotis.kastritis@bct.uni-halle.de

A. M. Pedley (ed.), *Supramolecular Protein Assemblies In Cells*, Advances in Experimental Medicine and Biology 1514,
https://doi.org/10.1007/978-3-032-26629-3_9

function · Mitochondrial pyruvate carrier · Histone acetylation · Pyruvate oxidation regulation

9.1 Introduction

The pyruvate dehydrogenase complex (PDHc) stands among the most fundamental metabolic biomolecules bridging glycolysis, the second stage of catabolism, to the citric acid cycle, ensuring cells will adequately meet their energy requirements. Its underlying reaction has been investigated since the 1930s by Hans Adolf Krebs (Krebs and Johnson 1937), who was awarded the Nobel Prize for Physiology or Medicine with Fritz Lipmann (who discovered the coenzyme A) in 1953 for pathways directly connected to pyruvate oxidation. PDHc stands as a marvel of enzymatic complexity, offering ample room for significant advancements in our comprehension of its functions to this very day: unraveling the mechanistic intricacies of this enigmatic megadalton complex remains a challenging exploration. This is not only because of the progress in the fundamental understanding of how cells work and breathe; it is also due to the deciphering of the communication among hundreds of co-localized enzymes. Such advances are bound to unveil fundamental principles relevant across diverse metabolic pathways, which are regulated by intercommunicating, coordinated enzymes.

Of vital importance for the cell is the rate-limiting reaction catalyzed by PDHc, dubbed as the "link reaction," which underlies the oxidative decarboxylation of pyruvate leading to the formation of acetyl-CoA, CO_2, and NADH (H^+) (Reed 2001; Perham 1991; Patel and Roche 1990). PDHc, as a major supplier of acetyl-CoA, occupies a key position in the oxidation of glucose by linking the glycolytic pathway to the oxidative pathway of the tricarboxylic acid cycle and the cellular respiration.

Assigned with this role, PDHc has been identified in the mitochondrial matrix and close to the inner mitochondrial membrane (Smoly et al. 1970), where the substrate is imported from the cytosol and the end product acetyl-CoA is utilized in the Krebs cycle—fast forward to 2023, PDHc has been shown *in organello* to have preferential localization in proximity of the ATP synthase (Plokhikh et al. 2024) but also to localize in the nucleus with histone acetylation function (Nagaraj et al. 2017) (Fig. 9.1). Initial investigations regarding this metabolic gatekeeper employed early biochemical approaches and focused mainly on the enzymatic activity and regulatory principles of the complex. In addition, coarse morphological analysis was performed, mainly via electron microscopy with negatively stained endogenous and reconstituted PDH (sub)complexes (Ishikawa et al. 1966; Hayakawa et al. 1969). The results of those studies set the cornerstone of understanding the role, function, and regulation of such multienzyme complexes and set forth the hypothesis of the transient assembly of enzymes into unique functional entities—the metabolons (Robinson Jr. et al. 1987) (Fig. 9.2).

The notion of the formation of cellular metabolons was speculative, although very attractive, as enzyme proximity and organization into subcellular structures

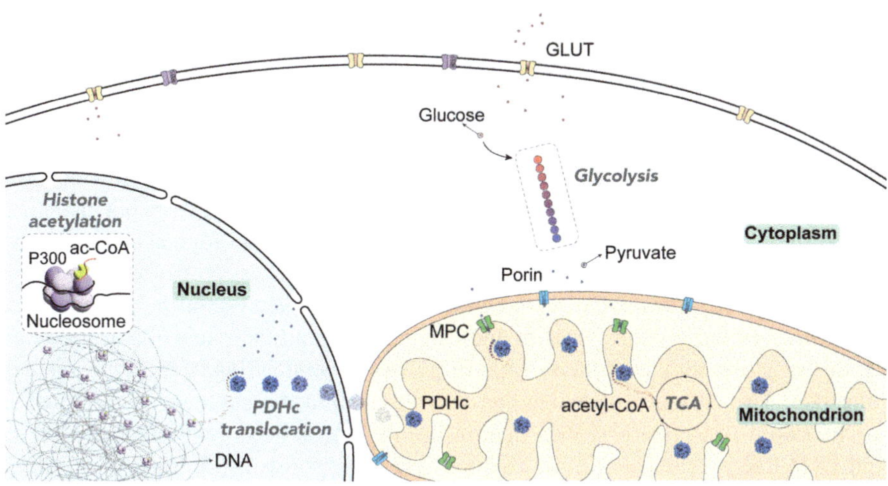

Fig. 9.1 Cellular localization of the pyruvate dehydrogenase complex in eukaryotes

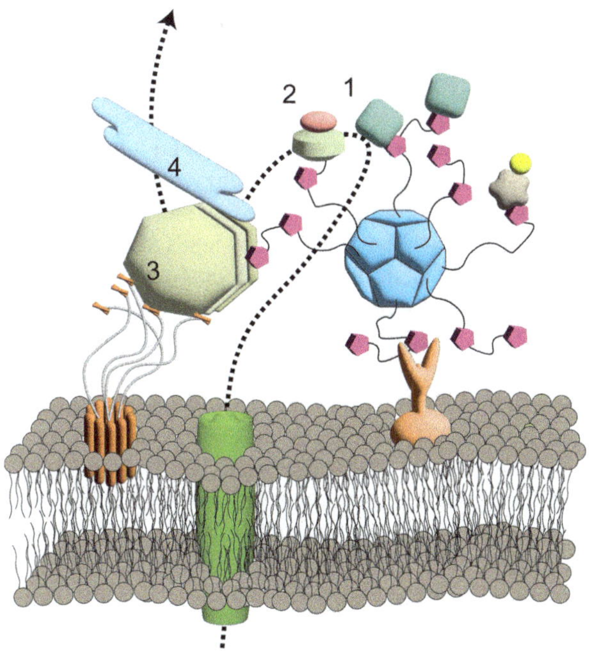

Fig. 9.2 Organization of metabolons as postulated by Kastritis et al. (2017)

provides explanations for how a cell could operate. Advances in computation and instrumentation, combined with groundbreaking progress in cryo-electron micros-copy (cryo-EM), enhanced the integration of modern structural studies, delineating the arrangement of individual proteins within the complex, their spatial orientations, and how they interact with each other (Tüting et al. 2021; Kyrilis et al. 2021). In addition to those, proteomic methodological advancements in mass spectrometry have determined PDHc composition (Kyrilis et al. 2021; Meinhold et al. 2024; Bosma et al. 1984). Downstream mass spectrometry techniques like protein-protein interaction mapping to elucidate the network of interactions within the complex, as well as cross-linking mass spectrometry (XL-MS), not only can potentially characterize the approximate stoichiometry of the embedded enzymes but also identify partners and coordinators of its activity at different physiological cellular states (Schweppe et al. 2017).

Glycolytic enzymes compose one of the first metabolic systems to be discovered with roles beyond glycolysis (Kim and Dang 2005), and their subcellular topological investigation remains fundamental in further understanding of their roles in cellular fate. The presence of metabolic enzymes within the nucleus was reported in the literature in the early 1960s demonstrating interactions with nuclear proteins and DNA (Kim and Dang 2005; Siebert and Humphrey 1965; Yang et al. 2012; Yogev et al. 2010). Interestingly, PDHc is not only localized in the mitochondria but was recently identified to also be localized in the nucleus (Sutendra et al. 2014). Briefly, a functional PDHc translocates to the nucleus, where it, in response to various signals, generates acetyl-CoA independently of mitochondria. Such moonlighting function of PDHc is crucial for histone acetylation, a modification deemed necessary in gene regulation and cell cycle progression.

Overall, PDHc occupies a central role in cellular metabolism, and due to this central role, pyruvate oxidation has also been dubbed "the link reaction," as mentioned above. Even slight deficiencies of the enzyme lead to severe encephalopathies and developmental disorders, i.e., causing the genetic disease called PDH deficiency (Patel et al. 2012a). PDHc deficiency leads to pyruvate accumulation and lactate production, causing lactic acidosis (Huckabee 1958). Biochemical alterations of PDHc or its regulatory proteins are also connected to various other pathological conditions (Patel et al. 2012b): this includes a role in neurodegeneration (Patel et al. 2012a; Robinson et al. 1987; Brown et al. 1988, 1989; Barnerias et al. 2010), obesity (Park et al. 2018; Koves et al. 2008), type 2 diabetes (Koves et al. 2008; Muoio and Neufer 2012; Kelley et al. 2002; Rocha et al. 2016), and other diseases (Soriano-Baguet et al. 2023; Zhu et al. 2020; Saunier et al. 2016). This makes PDHc and its associated interaction sub-network a critical target for therapeutic interventions in metabolic and inflammatory diseases, while treatments with phenylbutyrate (Ferriero et al. 2013) or dichloroacetate (Karissa et al. 2022) are still under evaluation (Karissa et al. 2022). The substrate flux through PDHc is tightly regulated in tissues under different metabolic conditions, and this feat is accomplished by covalent modification of the rate-limiting component of the PDHc, performed mainly by dedicated kinases and phosphatases that will be discussed in this chapter (Sect. 9.6.1). PDHc is also of substantial interest in regard to its role in cancer biology (Anwar et al. 2021). In many cancer types, a shift from oxidative metabolism to anaerobic glycolysis has been reported, a phenomenon which is dubbed as the "Warburg effect" (Schulze and Downward 2011; Kaplon et al. 2013; Sutendra and Michelakis 2013). PDHc involvement in disease has been investigated substantially (Anwar et al. 2021; Gray et al. 2014; Stacpoole and McCall 2023; Wang et al. 2021), elucidating its regulatory aspects in detail (Patel and Korotchkina 2001).

Therefore, to understand its central and multifaceted role in a plethora of cellular pathways, a detailed structural characterization is of paramount importance. This chapter collects recent knowledge in the biochemistry and structural analysis of PDHc and, through those, sheds light on yet-to-be-discovered aspects of the function of this enigmatic metabolon: a major player in cellular metabolism that remains elusive in various levels of organization, even after almost a century of research in pyruvate oxidation.

9.2   Biochemical Basis of the *Link Reaction*

Overall, the complex converts pyruvate to acetyl-CoA, CO_2, and NADH, utilizing three enzymes (E1, E2, E3) in the process; for the link reaction to occur, these enzymes use five cofactors, namely, CoA-SH, NAD+, FAD, ThDP, and the conjugate base of lipoic acid, the lipoate (Fig. 9.3).

9.2.1   Oxidative Decarboxylation of Pyruvate

The initial step, the oxidative decarboxylation of pyruvate, is carried out in a thiamine diphosphate (ThDP)-dependent catalytic cycle (Schellenberger 1998) (Fig. 9.4). Since the discovery of the C2-ThDP anion, our knowledge of the catalytic mechanism of ThDP-mediated enzymes has been massively advanced (Schellenberger 1998; Breslow 1957), revealing the importance of enzymatically guided reactions. This becomes evident when comparing the C^2-H deprotonation rates of free and enzyme-bound ThDP (Kern et al. 1997), as detailed ^{1}H NMR analysis revealed ylid formation of ThDP (the mechanism of ThDP enzymes which originates in the anionic (ylid) structure of the coenzyme) to be drastically enhanced in the substrate bound state, while favoring the imino-tautomer prior to C^2

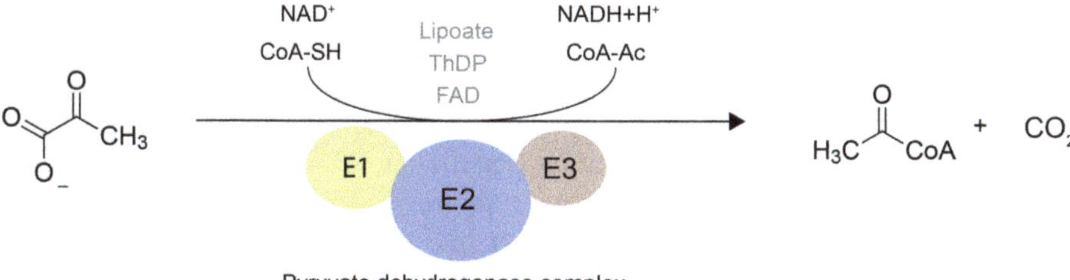

Fig. 9.3 The overall pyruvate dehydrogenase complex reaction

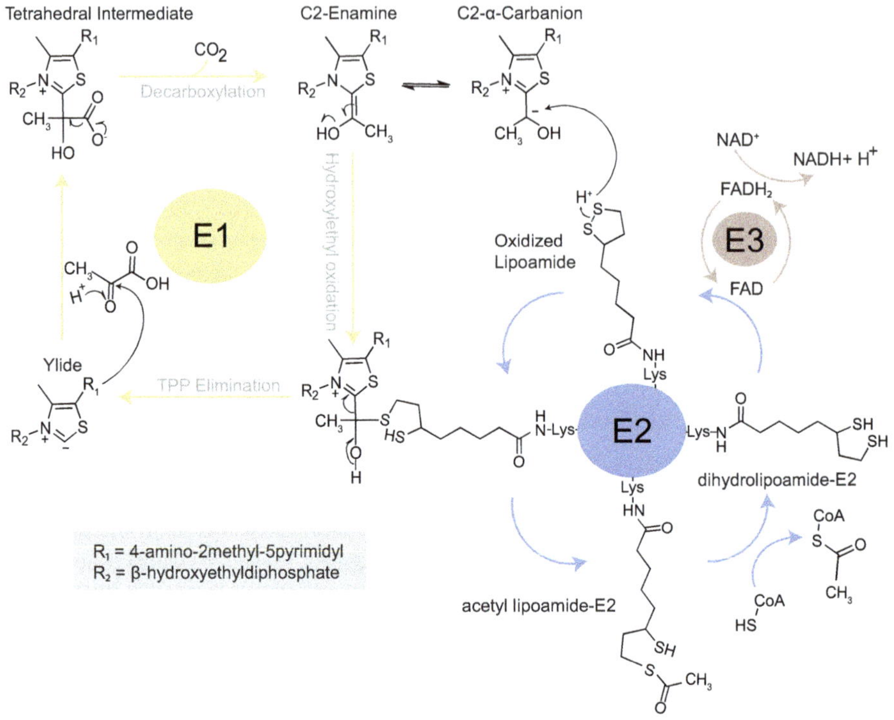

Fig. 9.4 Catalytic mechanism of the oxidative decarboxylation of pyruvate. The oxidative decarboxylation of pyruvate is carried out by three distinct reactions, each catalyzed by a specialized subdomain of the PDHc metabolon. Pyruvate is decarboxylated and transferred onto the LD domain of E2 in a ThDP-dependent reaction, utilizing an activated form of the cofactor for catalysis. The acetylated LD is then transported to the catalytic E2 core, where it undergoes transacetylation to generate acetyl-CoA. The reaction cycle is completed with the re-oxidation of the lipoyl group by the E3 domain, leveraging a coupled redox system of NAD and FAD in the process

deprotonation (Jordan et al. 2003; Nemeria et al. 2007). The aminopyrimidine ring of ThDP, in conjunction with a conserved glutamate residue, then acts as a proton acceptor (Lindqvist et al. 1992), deprotonating the C2 position and priming the cofactor for the nucleophilic attack on pyruvate. The result is the rate-limiting formation of a transient tetrahedral α-lactyl-ThDP intermediate that is readily converted to a C2 α-carbanion/enamine, by the elimination of carbon dioxide (Balakrishnan et al. 2012; Seifert et al. 2006). In the last step, 2-acetyl-ThDP transfers the acetyl group (Patel et al. 2014a; Gruys et al. 1989; Frey 1989) transfers the acetyl group onto the E2 LD-bound lipoyl moiety, coupling the E1 bound reaction to the next step via reductive acetylation (Patel et al. 2014b). Free lipoamide was shown to be an extremely poor substrate in its free form, displaying low affinities (K_m = 4 mM) and catalytic efficiencies (1.5 M^{-1} s^{-1}) for reductive acetylation in *Escherichia coli* (Graham et al. 1989). Compared to that, a profound increase in reactivity rates up to 60-fold was observed in *Azotobacter vinelandii* (Berg et al.

1998), for the complex-bound lipoyl group, a clear indication of diffusion limitation and efficiency of substrate channeling to the catalytic core.

9.2.2 Production of Acetyl-CoA

The catalytic mechanism of the acetyl-CoA production by the PDHc E2 subunit is believed to be analogous to the well-described chloramphenicol acetyl transferase (CAT) mechanism (Mattevi et al. 1992a; Russell and Guest 1991a; Shaw and Leslie 1991), which utilizes a catalytic histidine residue, acting as a general base, to subtract a proton from the thiol group of CoA (Mattevi et al. 1993). Following the activation of CoA, a tetrahedral intermediate is formed by the nucleophilic attack on the carbonyl carbon of the LD-bound acetyl moiety. Analysis of the *E. coli* E2 protein revealed that three catalytic residues are essential for transacetylation. Asp 606, Arg 440, and Ser 550 are crucial for (a) coordinating a hydrogen bond network and (b) the stabilization of the tetrahedral intermediate. These catalytic residues (D-H-R) are highly conserved in all E2 sequences across species (Russell and Guest 1991a) and have been validated by alanine scanning mutagenesis, which revealed drastic effects on the catalytic activity, reducing reaction rates to a fraction of the wild-type enzyme (Russell and Guest 1991b; Lewendon et al. 1990, 1988). The transition state is then terminated, resulting in the dissociation of free, acetyl-bound CoA from the active site of the E2 domain.

9.2.3 Regeneration of the Lipoyl

The regeneration of the LD-bound lipoyl group is catalyzed by the lipoyl dehydrogenase (E3), a flavin-containing enzyme which utilizes a redox-active disulfide in a catalytic mechanism comparable to the glutathione reductase (Ghisla and Massey 1989). Three distinct oxidation states for the reduction of diaphorase substrates, namely, (a) an oxidized state (Eox), (b) a two-electron reduced state (EH2), and (c) a four-electron reduced state (EH4), have been described to date (Argyrou et al. 2003). E3 contains both non-covalently bound flavin adenine dinucleotide (FAD) and the transiently bound nicotinamide adenine dinucleotide (NAD) as redox intermediates. The lipoyl dehydrogenase utilizes a histidine and a glutamate dyad to deprotonate the lipoyl group (Matthews et al. 1977; Westphal and de Kok 1988), enabling the nucleophilic attack on the active site disulfide and the subsequent re-oxidation of the LD (Benen et al. 1992; Kim and Patel 1992). The resulting thiol ion is then engaging in a charge transfer network with the FAD substrate, forming a reversible thiol C (Patel and Roche 1990) flavin adduct and catalyzing a nucleophilic thiolate interchange to complete the electron transfer (Benen et al. 1992). The electron transfer is then completed by the reduction of NAD^+ to NADH and H^+ via the reduced $FADH_2$, coupling the last step of the reaction to the cellular redox cycle.

Overall, the reactions within the component enzymes of PDHc, namely, the E1, E2, and E3, have been well described. However, vast differences in their catalytic efficiencies across organisms, samples, and physical-chemical conditions are described (Chang et al. 2021). It is of major interest how such measurements can be streamlined and approach closer-to-native conditions. High protein concentrations within the mitochondrial environment are sure to induce excluded volume effects, known to affect reaction rates (Ma and Nussinov 2013) and, of course, can induce localized concentration gradients (Zhou et al. 2008; Guigas and Weiss 2016), which may as well affect structural properties of each of the participating enzymes in an entirely distinct manner. Another bottleneck in understanding catalysis of the domains is the requirement of the lipoyl domain, which is not only present in the E2 in various organism-specific copy numbers (e.g., three LDs in human, but only one in yeast) but also present in the E3 binding protein (E3BP), a structural protein of eukaryotic PDHc. How the different, yet highly homologous, lipoyl domains orchestrate E1, E2, and E3 function remains under investigation.

9.3 Catalytic Domains of PDHc: Structure-Based Function

9.3.1 Structural Aspects of the E1 Subdomain

In all eukaryotes and gram-positive bacteria, E1 organizes as a heterotetramer ($\alpha_2\beta_2$) (Fig. 9.5a) (Frank et al. 2004), while gram-negative bacteria display a homodimer structure, combining both polypeptide heterologous chains in a single multi-domain protein (α_2) (Arjunan et al. 2002). Currently, we know that all variants of the native assembly contain two ThDP binding sites (Fig. 9.5b), coordinated between the dimeric interfaces and only exposing the C2 atoms to the solvent (Ciszak et al. 2003). The PDHc, in line with other ThDP utilizing enzymes, forces the cofactor upon binding in a highly energetic V-shaped conformation that presents a high-energy state crucial for enzymatic turnover (Guo et al. 1998; Pletcher et al. 1977; Jordan et al. 2005). This conformation is stabilized by a divalent cation in conjunction with multiple ionic interactions (Fig. 9.5b), stabilizing the pyrophosphate group (Jordan 2003), as well as hydrophobic side chains that reduce flexibility, creating an energetically favorable conformation, essential for enzyme function (Guo et al. 1998). Furthermore, a conserved network of hydrogen bonds stabilizes the nitrogen atoms of the aminopyrimidine ring of ThDP to increase the structural integrity and functional capacity of the enzyme (Fig. 9.5b) (Pei et al. 2008). ThDP analogs lacking these nitrogen atoms display significantly reduced activity, highlighting the critical nature of these interactions (Schellenberger et al. 1997). This motif consists of a highly conserved binding pattern (GDG..X_{26}..N(C)N) spanning ~30 residues that is observed in a plethora of ThDP associated enzymes such as the transketolase, pyruvate oxidase/decarboxylase, and acetolactate synthase (Muller et al. 1993; Hawkins et al. 1989). The substrate binding induces disorder-to-order transition of

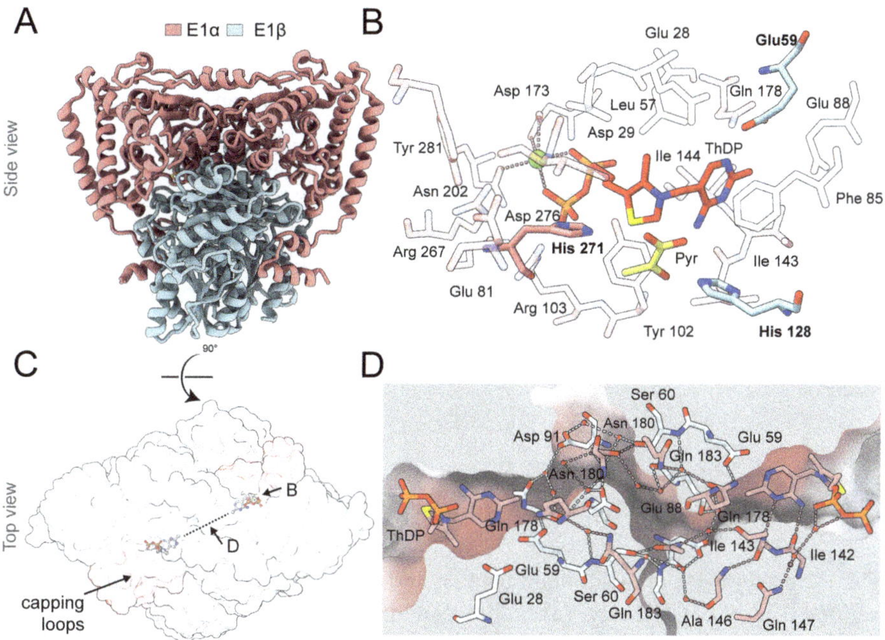

Fig. 9.5 Structure-based function of the E1 subdomain. (**a**) PDHc assembles into a heterotetramer ($\alpha_2\beta_2$) in presumably all organisms, except gram-negative bacteria (Frank et al. 2004). (**b**) ThDP is coordinated by an extensive hydrogen bonding network, including ionic stabilization provided by a divalent cation, that forces the cofactor into a high-energy V-conformation; catalytically relevant residues are highlighted. In *G. stearothermophilus* Glu 59 acts as the proton acceptor for ThDP C2 deprotonation, while His 271 is involved in proton transfer and stabilization of the transition state during decarboxylation of pyruvate. His 128 has been implicated to act as a general acid-base, protonating the dithiolate ring for the acetyl transfer, opening the disulfide bond for acetyl transfer (Fries et al. 2003). ThDP is furthermore stabilized by the binding of a divalent cation that is coordinated by multiple ionic interactions (Asp 173, Asn 202). (**c**) The E1 native assembly contains two distinct catalytic centers, each containing one reactive species of ThDP. Two flexible loop regions (red) guide substrate binding in the kinetically inequivalent reaction. (**d**) A network of hydrogen bonds inside an acidic water-filled cavity synchronizes the reaction cycles of each half site, allowing for active site communication via a hypothesized "proton wire" (Frank et al. 2004)

two loops, an inner and outer flexible loop (Fig. 9.5c, d). These loops actively contribute to the enzyme's catalytic process by enhancing substrate specificity, preventing nonspecific reactions, and facilitating communication within the active site (Kato et al. 2008; Kale et al. 2007). Although both active sites are chemically equivalent, there is a kinetic (Seifert et al. 2006) and thus functional half site inequivalence between the two ThDP-containing reaction centers, necessitating flexible domain interplay to facilitate the reaction cycle (Ciszak et al. 2003). Therefore, allosteric communication of E1 active sites must occur, possibly assisted by higher-order thermal fluctuation induced during asymmetric ligand binding (Cooper and Dryden 1984). Active site asymmetry additionally enables sequentially shifted reaction steps, previously biophysically studied (Seifert et al. 2006). This means that conformational changes of one active site induce conformational changes to the other. As a biochemical basis, a "proton wire" mechanism has been proposed (Fig. 9.5d), in which a network of hydrogen bonds along an acidic water tunnel communicates and coordinates active site states (Frank et al. 2004). Although major advancements regarding the catalytic mechanism, structural organization and enzymatic synchronization have been made in recent years, still unanswered questions regarding the E1 complex remain. Conformational changes during the catalytic landscape must be elucidated to validate previous mutational studies (Fries et al. 2003). Especially the dynamic binding of regulatory molecules and metabolites and their effect on the E1 structure-mediated function are bound to yield a potent research field for the targeted attenuation of E1 activity, which is crucial in cancer research.

9.3.2 Structural Aspects of the E2 Subdomain

The E2 protein is a multi-domain protein and contains three functionally different regions (Fig. 9.6a), namely, (a) a core domain that organizes the assembly of the inner catalytic core of the metabolon, (b) a peripheral subunit binding domain (PSBD) that mediates in eukaryotes the E1–E2 and in gram-negative bacteria, additionally, the E2–E3 interaction and (c) a lipoyl domain (LD) in which a lysine residue has been covalently modified with a lipoyl moiety for substrate transfer (Perham and Packman 1989). Therefore, two of the domains (core and LD) are purely catalytic, whereas the third domain is involved in protein-protein interactions. Domain interplay is guided by low complexity regions that allow for the free cycling of reaction intermediates along reaction centers. As expected, conservation of the E2 domains is much higher than that of low complexity regions. However, it is still a matter of ongoing research how function is connected to hypervariable amino acid residue regions (Ponting 2017). This complex organization of the E2 subdomains is essential for its function. Starting from the N-terminus, the number of lipoyl domains varies among orthologs. For example, the fungal E2 has one domain, the human E2 two, and *E. coli* utilizes three lipoyl domains (Russell and Guest 1991a). The question is, therefore, what effect do the domain repeats have on the structure and function of the assembled complex and how many can be accommodated across the tree of life. Again, such questions can now be approached utilizing modern structural biology tools that can leverage genomic output with structural data, ready to be mined (van Kempen et al. 2024).

Targeted mutational screening revealed that domain reduction had no detectable effect on catalytic activity, structural integrity, or complex formation (Guest et al. 1985). However, an increase in the number of domains enhanced their mobility, potentially expanding the coupling capabilities of the active site (Machado et al. 1993). This suggests that organisms might adapt by varying domain repeats to optimize metabolic fitness. Additional validations support this theory, leading to the

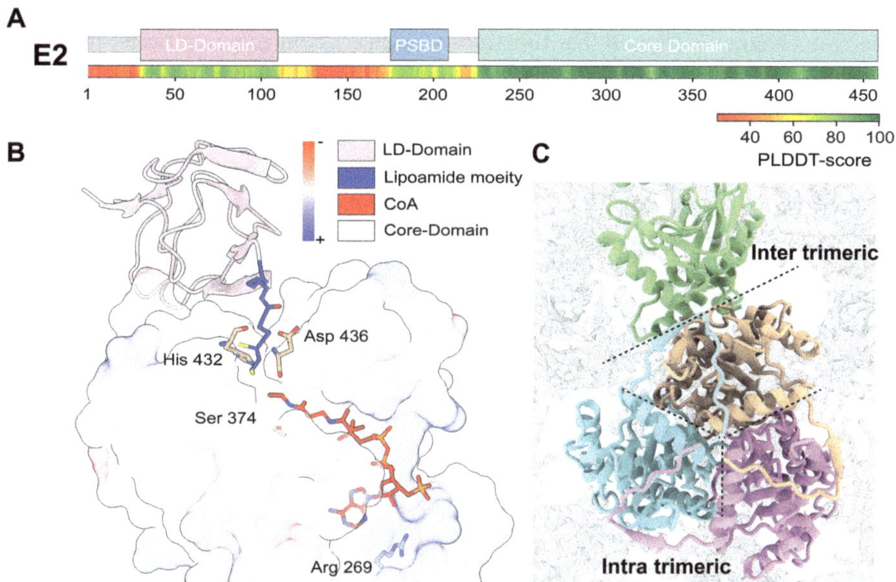

Fig. 9.6 Structural organization of the E2 subdomain. (**a**) The dihydrolipoyl transacetylase of E2 is a multi-domain protein that contains three domains connected by highly flexible linker regions (prediction confidence by AlphaFold is given as pLDDT score to highlight low complexity regions in red). (**b**) Each E2 monomer coordinates one equivalent of CoA (red) that is available for acetylation via the LD (pink) bound lipoyl moiety (blue). Electrostatic surface potential is displayed, showing an overall uncharged active site, while CoA phosphates are electrostatically stabilized by positively charged residues (Tüting et al. 2021). (**c**) The higher-order structure of the PDHc is guided by an interplay of inter- and intra-trimeric interface, assembling the catalytic core via conserved energetic forces (Tüting et al. 2021)

conclusion that E2 can fulfill specific requirements related to metabolic utilization (Dave et al. 1995; Guest et al. 1997). On the other hand, E3BP also includes LDs, and therefore, complexity increases (covered in more detail in Sect. 9.3.4). In E2, the second domain, connected to the LD by alanine and proline-rich regions, is the 5–7 kDa-sized PSBD, a small domain consisting of two α-helices. The PSBD functions as a selective recruiter of the peripheral domains, spatially connecting lipoyl transacetylation to the entire catalytic cycle (Patel et al. 2014b). Although initially characterized as a monomeric functioning complex in a wide range of structures (Frank et al. 2004; Pei et al. 2008; Arjunan et al. 2014), it has been recently shown that the PSBD is able to form a homodimer in *E. coli*, utilizing a network of hydrophobic interactions to guide PDHc stoichiometry (Meinhold et al. 2024). This specific work showed that the dimerization of PSBD is the major determinant of the overall stoichiometry of the bacterial PDHc, validated by structural and biochemical analysis (Meinhold et al. 2024).

The PSBD is then connected, via another flexible stretch of amino acids, to the catalytic core-forming domain (CD) that is responsible for the production of mitochondrial acetyl-CoA. Followed by the transient binding of the LD, and anchored by large-scale complementary electrostatic interactions (Skerlova et al. 2021), a reaction channel is formed that functions as a direct connection between the acetylated LD and the bound CoA. This channel facilitates the positioning of the substrate headgroups in proximity to the active site residues (see Sect. 9.2). The 30 Å long reaction channel is mostly uncharged, only anchoring CoA phosphates by a positively charged surface patch and positioning it close to the catalytic site (Fig. 9.6b) (Tüting et al. 2021). In the bovine branched-chain α-ketoacid dehydrogenase complex, a close relative of PDHc, active site availability is drastically influenced by the binding of CoA, drastically increasing LD affinity upon cofactor binding (Kato et al. 2006). Similar mechanisms may apply to the larger family of α-ketoacid dehydrogenase complexes across the tree of life, in which PDHc is the most well-studied representative.

Moving to the higher level of E2 organization, the basic building block of the catalytic core is the trimeric assembly of the acetyl transferase, or core-forming domain (CD), guided by two distinct intra- and intermolecular interfaces (Tüting et al. 2021) (Fig. 9.6c). Both the flexible domain organization and the multimeric assembly of the E2 protein are essential features for active site coupling across monomers. Both reducing electron and acetyl transfer mechanisms have been observed in the reconstituted E2 complex (Danson et al. 1978; Cate et al. 1980; Collins and Reed 1977; Song and Jordan 2012). It remains unclear, however, if this inter-chain transfer of substrates and redox equivalents is spatially limited, e.g., permitting only intra- but not inter-trimeric exchange. Additional analysis, especially of the crosstalk between domains and subunits in the PDHc, is therefore necessary to connect conformational changes observed within the trimeric assembly to catalytically relevant states.

9.3.3 Structural Aspects of the E3 Subdomain

The dihydrolipoamide dehydrogenase (E3), a disulfide oxidoreductase, forms a homodimer with four distinct domains: a FAD- and NAD$^+$-binding domain, positioned at the adjacent side of the monomer as well as a central and interface domain (Mattevi et al. 1992b) (Fig. 9.7). E3 has a multifunctional role, serving as a catalytic subunit for all α-ketoacid dehydrogenase complexes and is also crucial for the function of the glycine decarboxylation metabolon (Douce et al. 2001). It is of particular interest that the E3 is recruited by all α-ketoacid dehydrogenase complexes within mitochondria, and, therefore, questions arise about the availability of its unbound state, as well as its affinity to the different complexes in which it participates. The catalytically active form of the E3 is a homodimer that utilizes an intramolecular disulfide bridge for the transfer of redox equivalents from the LD to FAD and NAD$^+$. Higher oligomeric states have also been observed in relation to altered mitochondrial conditions (Klyachko et al. 2005). The previously described catalytic dyad, consisting of a histidine-glutamate pair, is located at the binding interface of both

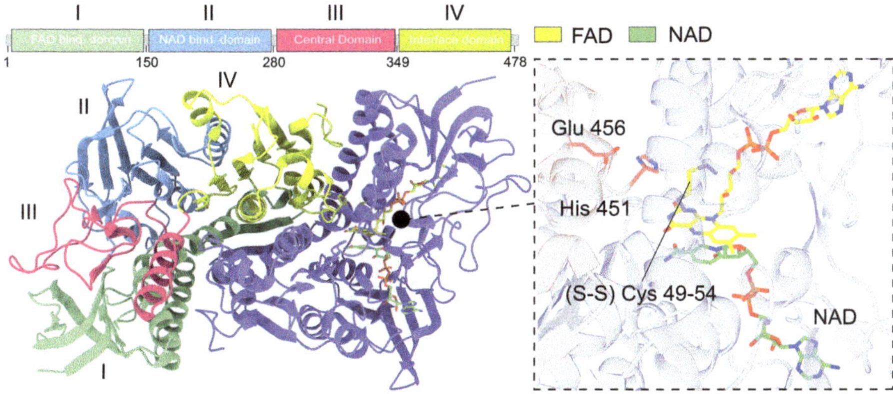

Fig. 9.7 Structure-based function of the E3 subdomain. The catalytically active form of E3 sub-domain of the PDHc is a homodimer (Mattevi et al. 1992b). The complex consists of four distinct domains for binding FAD and NAD$^+$ as well as forming the complex and its associated dimeric interface. In the native enzyme, each monomer binds NAD$^+$ and FAD for redox coupling of an intramolecular disulfide bond, while the catalytic dyad is provided by the adjacent monomer (Glasser et al. 2017)

monomers, requiring the interplay of both subunits for catalysis. The importance of dimerization is especially striking when analyzing destabilizing mutations at the dimeric interface: these mutations do not only affect overall PDHc activity but also induce new moonlighting catalytic activities, in which E3 can act as a protease or show increased diaphorase activity (Babady et al. 2007; Igamberdiev et al. 2004), highlighting the effect of PPIs for metabolic function. Recent works on other enzymes have shown new roles of enzymatic multimers and their filament forma-tion in cellular function (Lynch et al. 2020). It is therefore, exciting to probe the possibility that the high tendency of enzyme oligomerization (Traut 1994) is not just a tool for activity regulation but may also reveal cryptic, new enzymatic capabilities, E3 being a prominent example.

In the ortholog structure from *Pseudomonas aeruginosa* (Fig. 9.7), the active site cavity orients NADH on top of FAD, aligning both head groups for redox transfer. The head groups are additionally stabilized by the π-stacking of both heterocyclic groups (Glasser et al. 2017). Structural characterization of the oxidized NAD$^+$ revealed an alternative orientation of the nicotinamide headgroup, positioned ~12 Å away from the isoalloxazine moieties (Mattevi et al. 1992b; Brautigam et al. 2005). The structural basis for NAD$^+$ proton transfer remains unclear and requires addi-tional approaches to characterize the driving forces involved during LD regenera-tion. Especially the catalytic complex between LD and E3 remains elusive, most likely due to the transient nature of this protein-protein interaction. A similar sub-strate insertion model, observed for E1 and E2, has not yet been proposed and its elucidation remains of high importance for the larger field of α-ketoacid dehydroge-nase complexes.

9.3.4 Structural Organization of the E3BP Subdomain

Compared to the other domains of the PDHc, the E3BP, formerly described as protein X (De Marcucci et al. 1985), does not have an enzymatic role but is highly important for the coordination of the complex through the recruitment of E3. So far, it has only been characterized in mammalian and fungal enzymes (De Marcucci et al. 1985; Stoops et al. 1997) and shares the same domain architecture as the E2 (Fig. 9.8a). It is now accepted that bacteria do not possess an E3BP equivalent, and recruitment of E3 is mediated by E2 PSBD dimerization (Meinhold et al. 2024). E3BP also contains a LD, a PSBD, and a core binding domain (Forsberg 2023), and all domains are connected by flexible, low-complexity regions (Fig. 9.8b). The E3BP LD is homologous but not identical to the LD embedded in the E2, and therefore, it is of interest to know how this LD functions in comparison to the better-studied E2-embedded LD. Comparison of the fungal and human E3BP shows drastic differences in the core domain (Forsberg 2023) (Fig. 9.8c). In fungi (Tüting et al. 2021; Kyrilis et al. 2021; Forsberg 2023; Forsberg et al. 2020) E3BP forms a trimeric complex, whereas in mammals it is believed to be part of the central core (Prajapati et al. 2019). The latter, despite displaying a similar fold to the E2, shows no apparent transacetylase activity due to a nonfunctional CoA binding site (Harris et al. 1997). The PSBD of E3BP is responsible for binding the E3 subunit to the PDHc metabolon. The same questions posed for E2 regarding the flexible regions connecting its domains are also applicable for E3BP. However, truncations or

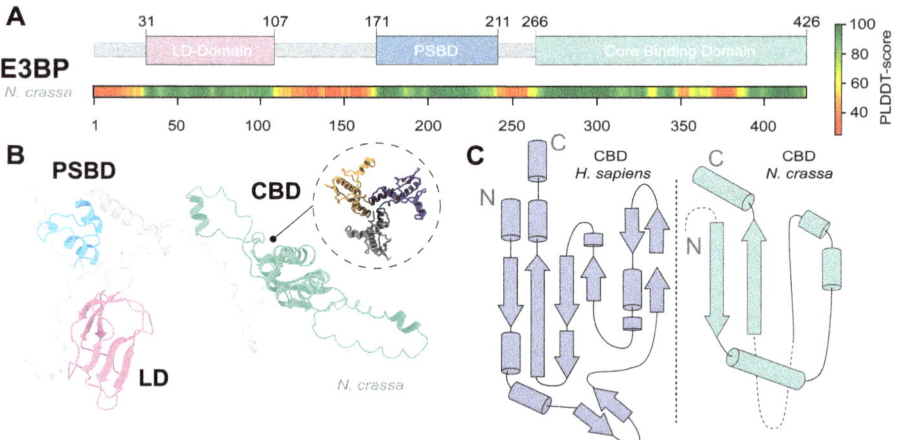

Fig. 9.8 Structural analysis of the E3BP subdomain. (**a, b**) E3BP contains a LD, a PSBD for the binding of E3, and a non-catalytic core binding domain (CBD) that are connected by highly flexible linker regions (prediction confidence by AlphaFold is given as pLDDT score to highlight low complexity regions in red) (Abramson et al. 2024). E3BP forms a homotrimer for the assembly within the E2 core (insert). (**c**) The fungal and human CBD of E3BP differ significantly in their overall architecture, displaying structural differences that lead to a migration of the non-catalytic domain within the E2 core

mutations of those flexible linkers or its ordered domains are limited in the literature, and therefore, no conclusion about its role, besides organizing the E3, is currently known.

9.4   Inter-subunit Interactions with PDHc

The abovementioned subcomplexes, E1, E2, E3, and E3BP, all present in polypeptide complexes, interact via various distinct interfaces to participate in pyruvate decarboxylation. In the reaction pathway, two distinct interaction patterns are present regarding E1 and E3: (a) a stable complex formation, involving the PSBD of E2 and E3BP, respectively, and (b) a transient complex formation during substrate shuttling, involving the highly mobile LD. Interaction surfaces are shown in Fig. 9.9.

9.4.1   Organization of E1 and E3 by Peripheral Subunit Binding Domains

E1 and E3 are stably, yet non-covalently, bound to the E2 core structure by interacting with the respective PSBDs (Chandrasekhar et al. 2013). In *E. coli* and other prokaryotes, the PSBD of E2 facilitates the binding of both subunits (Chandrasekhar et al. 2013), whereas in eukaryotes, the PSBD of E2 is only capable of binding E1, and the fourth protein, E3BP, is necessary to coordinate the E3 binding as mentioned above. In *E. coli*, the binding of the PSBD to the subunit is guided by ionic interactions, forming an "electrostatic zipper" (Mande et al. 1996; Frank et al. 2005). The binding is additionally complemented by hydrophobic interactions between the PSBD and the respective subunit. The major difference is in the underlying thermodynamics of binding and not the resulting binding constants. While the binding interface between the PSBD and E1 contains structural water molecules, the binding is driven by enthalpic forces. In contrast to this, the protein-protein interaction between the PSBD and E3 excludes solvent molecules, rendering this interaction entropic (Frank et al. 2005). The binding energies for the PSBD to E1 and E3 in *Bacillus stearothermophilus* are identical though, with −12.9 kcal/mol and −12.6 kcal/mol, respectively (Jung et al. 2002; Jung and Perham 2003). This would mean that entropic and enthalpic contributions are reversed, but of same magnitude, and therefore, E2 and E3BP recruit their respective interactors with the same affinities. Despite this striking observation, it is important to appreciate that new methods measuring thermodynamic properties and affinities in increasingly complex environments will open the route to a more accurate understanding of PDHc intra- and inter-subunit interactions, as well as how the physical-chemical environment contributes to the assembly of the complete metabolon.

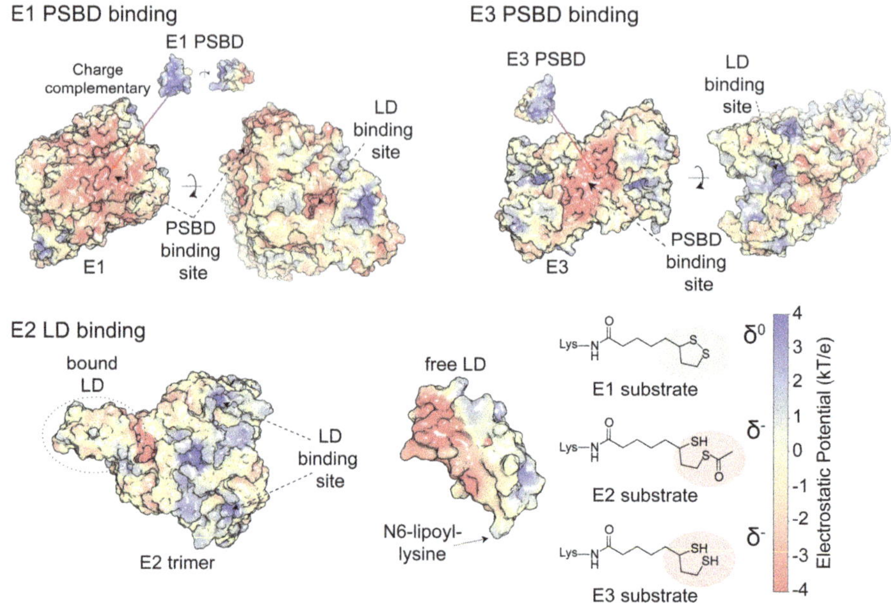

Fig. 9.9 Charge complementarity in the binding interfaces for the PSBD and LD in the PDHc metabolon. Electrostatic surface potential was calculated using APBS (Jurrus et al. 2018). The E1.PSBD binding site is from *Geobacillus stearothermophilus*, resolved by X-ray crystallography (PDB ID: 1 W85) (Frank et al. 2004). The E3.PSBD binding site is from *Homo sapiens*, resolved by X-ray crystallography (PDB ID: 2F5Z) (Brautigam et al. 2006). The E2.LD binding site is from *Escherichia coli*, resolved by cryo-EM (PDB ID: 7B9K) (Skerlova et al. 2021). The free LD is from human (PDB ID: 1FYC) (Howard et al. 1998), solved by solution NMR, including structural formulas with their partial charges for E1, E2, and E3 substrates (which are E3, E1, and E2 products). Charge complementarity is observed, with E1 binding a neutral oxidized lipoate, E2 binding an acetylated lipoate with a partially negative charge, and E3 binding a reduced lipoate with a partially negative charge, highlighting the role of electrostatic interactions in substrate stabilization and release. No experimental data are present for the E1 and E3 binding to the LD

The PSBD domain has a binary electrostatic surface potential (Fig. 9.9), with a negative and a positive surface, which plays an important role in surface charge complementary in binding. The binding of the PSBD to E1 occurs in the twofold axis of E1β (Pei et al. 2008), opposite of the E1α binding site. As only one binding site is available, the PSBD:E1 stoichiometry is 1:1 (translated to individual proteins, the complex E1α:E1β: PSBD is 2:2:1 (Meinhold et al. 2024)). Similarly, the PSBD binds E3 in the twofold symmetry axis, in which one PSBD is bound in a mutually exclusive manner (Ciszak et al. 2006). This binding could be independently proven by NMR spectroscopy (Kalia et al. 1993; Allen et al. 2005), as well as crystallography (Brautigam et al. 2006; Mande et al. 1996; Ciszak et al. 2006). Upon PSBD binding, no conformational change in E3 is visible, so the binding mechanism is thereby not characterized as induced fit (Allen et al. 2005). Both binding sites, in E1 and E3, respectively, are characterized by a highly negative electrostatic surface potential (Fig. 9.9), which is complemented by the positive electrostatic surface of the PSBD. PSBD binding to E1 and E3 has been well characterized for *E. coli* K12 and involves Arg 334, Arg 335, Arg 355, and Arg 358 in the PSBD (Chandrasekhar et al. 2013), which have also been probed by mutagenesis and activity read-outs. Additionally, Arg 350 appears to be the key residue to bind E1 because any mutation in this position (Arg350Ala, Arg350Lys, Arg350Glu) abolished E1 binding, whereas only Arg350Glu affects E3 binding and activity of the PDHc dropped dramatically to 8%, 5%, and 2.4%, respectively. In eukaryotes, binding of E1 and E3 is divided into the PSBDs of E2 and E3BP, respectively. In the human PSBDs of E2 and E3BP, key residues are distinct, rendering the binding to be exclusive. Aligned to the *E. coli* homologue, residue 335 which is an arginine in *E. coli*, is a lysine in the human E2 PSBD, and a glutamine in the E3BP PSBD. The Lys335Ala mutation in the E2 PSBD reduces the binding affinity for E1 by a factor of 86 (Korotchkina and Patel 2008). Analogously, the *E. coli* Arg 350, critical for E1 binding, is mutated to isoleucine in the E3BP PSBD and, however, remains conserved in E2 PSBD. Additionally, Arg 334, conserved in human E3BP, is also crucial for binding E3, as an alanine mutation reduced binding affinity for E3 by the factor of 230 (Brautigam et al. 2006). These results, together with the oligomerization interfaces and how these affect prokaryotic PDHc function (Meinhold et al. 2024) reveal an interplay of charge and hydrophobic interactions regulating preferential binding of E1 and E3 to their scaffold sequences.

9.4.2 The Transient Subcomplexes Involving the LD

Transient subcomplexes in PDHc involve specific LD binding during catalysis. The LD must traverse across the three active sites to complete a reaction cycle. Binding of the LD to E1 is weak, with a binding affinity >1 mM (Graham and Perham 1990). The ability of E2 to bind and process free acetylated lipoic acid (Reed et al. 1958) indicates that the binding of the acetylated LD is predominantly based on shape complementarity of the modified lipoate, and not based on domain complementarity. Restrained MD simulations with an E2-bound acetylated LD identified certain charge complementarity (Tüting et al. 2021) in the fungal PDHc. This electrostatic interaction interface was afterward experimentally verified in the prokaryotic counterpart (Skerlova et al. 2021). Conserved acidic residues across *E. coli* and other organisms suggest a common charge-complementarity mechanism.

Crystallization of a *B. stearothermophilus* LD-PSBD (excluding the core-forming domain of E2) di-domain, bound to *B. stearothermophilus* E3, failed to localize the LD in the electron density map and only revealed the PSBD bound (Mande et al. 1996). This again illustrates that LD binding is transient, and interaction might be driven by the complementarity of covalently bound lipoate. The LD binding sites of E1, E2, and E3 show different surface electrostatics, which correlate to the state of the LD (Fig. 9.9). The E1 LD binding site is negatively charged, with the oxidized lipoate as a neutral-charged substrate. Both E2 and E3 LD binding sites are positively charged, stabilizing the partially negative surface properties of the respective substrates (Fig. 9.9). The charged surface properties of the binding pockets also play an important role in modulating the dissociation constant (K_d), thereby ensuring an effective release of the product. This modulation of K_d by the electrostatic environment ensures that while the substrates bind effectively for catalysis, their affinity is sufficiently lowered to facilitate product release post-reaction.

LD complex formation for the related oxoglutarate dehydrogenase complex was rationalized by data-driven modeling (Skalidis et al. 2023). Particularly for those interactions, large surface areas with substantial electrostatic complementarity but also guiding surfaces were observed, where the LD samples to eventually localize to its respective site. Interestingly, the guiding surface of E1 was much more extended as compared to the one of E3. The correlation to the affinity is complex but must regulate LD on-rates.

Overall, knowledge of LD-bound complexes is limited, and this hampers our understanding of the embedded reactions. It is clearly a challenge to be undertaken, especially now that cryo-EM can resolve samples of increased complexity. Additionally, cross-linking MS has also made leaps in analyzing biomolecular complexes; therefore, models can be generated which are driven or/and agree with such experimental data, revealing operational principles of PDHc.

9.5 Architecture of the PDHc Metabolon

The endogenous PDHc is one of the biggest enzymatic complexes of the cell, with molecular weights up to 10 MDa. Despite knowledge of its individual components, the full assembly architecture is still unknown. This is mainly due to its sheer complexity, conformational and chemical heterogeneity, as well as its challenging purification, which not only requires isolating the native biomolecule and validating its native state but also retrieving enough quantity for biochemical and structural elucidations. In addition, across the tree of life, there are major differences in the overall architecture of this enormous protein complex which limits knowledge transfer from prokaryotic to eukaryotic counterparts.

The core-forming unit of the PDHc is the C-terminal domain of E2. In *E. coli* 24 copies of E2 form an octahedral core, highly similar to the eukaryotic oxoglutarate dehydrogenase complex (OGHDc) and branched-chain keto acid dehydrogenase (BCKDHc) (Reed and Hackert 1990). This core was also resolved from an *E. coli* lysate bound to 24 copies of the LD, in a "resting state" (Skerlova et al. 2021), while specific interactions need to be modulated to retrieve trimeric and functional E2s (Meinhold et al. 2024). In fungi and mammals, as well as other eukaryotes but also in the gram-positive bacterium *Bacillus stearothermophilus*, E2 forms an icosahedral core instead, composed of 60 E2 copies. In both cases, the building block of the E2 core is formed by a C3 symmetric trimeric assembly (Fig. 9.10). The determinant of the higher order—octahedral vs. icosahedral—is yet not fully understood, as the E2 protein of PDHc from different organisms, and all ketoacid dehydrogenase

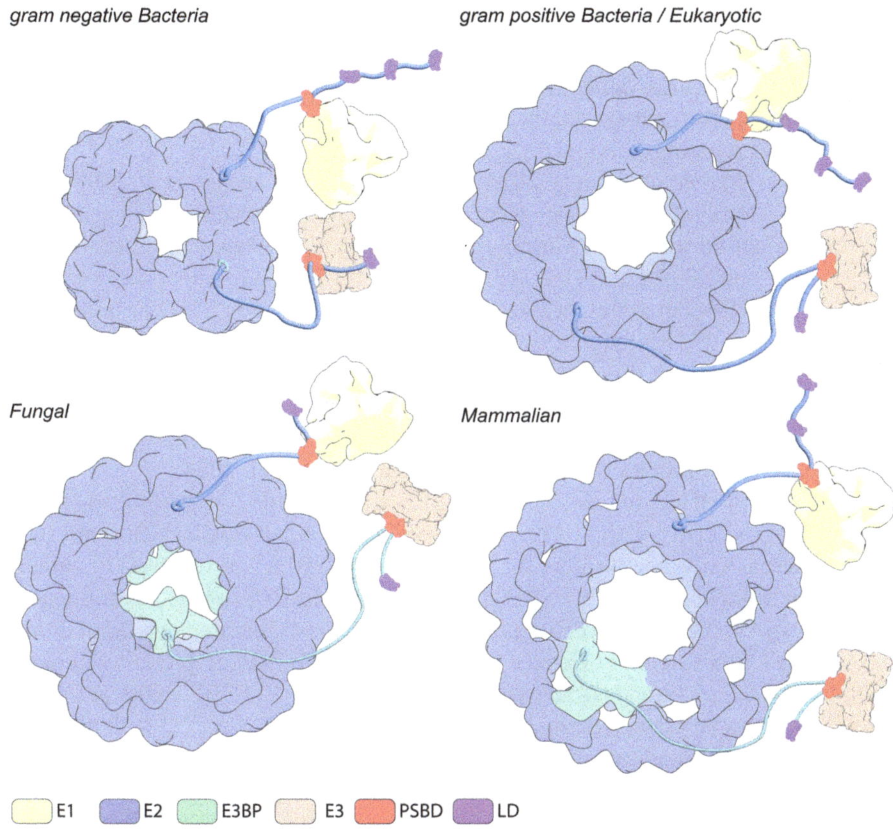

Fig. 9.10 Models and comparison of bacterial and eukaryotic E2 structures

complex family E2s share high sequence similarity. Additionally, eukaryotic PDHc contains the fourth protein E3BP. It has been recently reported by cryo-EM that, in the fungal PDHc, 12 E3BPs are bound with their C-terminal domain in the internal cavity of the icosahedral E2 core (Tüting et al. 2021; Kyrilis et al. 2021; Forsberg 2023; Forsberg et al. 2020).

Mammalian E3BP on the other hand is highly similar to E2 and substitutes E2 monomers in an either 48:12 (Hiromasa et al. 2004) or 40:20 (Brautigam et al. 2009) stoichiometry. Native MS analysis suggests a 2 E2:1 E3BP building block (Prajapati et al. 2019), but high-resolution atomic data remain elusive. Even the high-resolution structures of endogenous bovine PDHc recently reported (Liu et al. 2022) failed to localize the E3BP. As described in Sects. 9.3.2 and 9.3.4, the N-terminal PSBD of E2, and the one of E3BP both bind to the peripheral subcomplexes E1 and E3. *E. coli* E1 and E3 consist of homodimers (Arjunan et al. 2002), whereas in higher organisms and gram-positive bacteria, the E1 domains are split into two proteins, E1α and E1β, forming a heterotetramer (Frank et al. 2004) (Sect. 9.3.1 for details). Based on the subunit specificity of the PSBD, *E. coli*'s octahedral core can bind up to 24 peripheral subunits (E1 and E3); the fungal PDHc core can bind 60 E1 and 12 E3 subunits, and mammalian PDHc can bind either 40 E1 and 20 E3 or 48 E1 and 12 E3, depending on the assumed E2:E3BP ratio. The stoichiometry of the *E. coli* native complex has been proposed to have distinct ratios (E1:E2:E3), including 1:1:1 (24:24:24), 2:2:1 (24:24:12), 2:1:1 (48:24:24), and 4:3:2 (32:24:16) (Meinhold et al. 2024; Bosma et al. 1984; de Kok et al. 1998; Reed et al. 1975). Previous differences were explained by the variability of purification and quantification methods (Murphy and Jensen 2005). Considering the dimeric nature of E1 and E3, the first (1:1:1) and the last (4:3:2) stoichiometry fully saturate the available PSBD domains, whereas the two others exceed (2:1:1) or underscore (2:2:1) the available binding capability of the E2 core. Additionally, *E. coli* PSBD tends to dimerize, and dimeric PSBD binds two E1 dimers, while the free PSBD binds an E3 dimer, supporting the 4:3:2 subcomplex stoichiometry (Meinhold et al. 2024). Indeed, a recent publication clarified the stoichiometry utilizing an array of biochemical and structural biology methods (Meinhold et al. 2024).

In fungal PDHc, the E1:E2:E3:E3BP stoichiometry is estimated to be ~20:60:12:12 (Kyrilis et al. 2021); in this configuration, all E3BP PSBDs are saturated, whereas only one-third of the available E2 PSBDs are involved in subunit binding. It's unclear whether PSBD dimerization affects or not E1 recruitment in eukaryotes. Another controversial point is the complex stoichiometry in mammalian PDHc. Circumstantial evidence showed that if 20 E3BP are present within the complex, 20 E3 are bound (Brautigam et al. 2009), but if 12 E3BP are identified, 12 E3 are found to be bound (Hezaveh et al. 2018). In any case, the E1 stoichiometry appears to not fully saturate the available E2 PSBDs (Brautigam et al. 2009).

9.5.1 *Proposed Models for PDHc Metabolon Function*

Irrespectively of the complex composition, the spatial organization of the peripheral subunits is also unclear; PSBDs of E2 and E3BP are connected via a flexible (disordered) linker to the core domain. Recent reports have shown that the peripheral subunits E1 and E3 are tethered at a distance of 50–100 Å, relative to the core (Kyrilis et al. 2021). In this inter-subunit space, the LD domain, connected via a flexible linker N-terminal to the PSBD domain, must travel across the three active sites to perform the reaction cycle. Three different catalytic modes are hypothesized based on structural data (Fig. 9.11):

(i) The first model proposed for the catalytic mechanism of the cubic PDHc from *E. coli* was coined "multiple random coupling" (Hackert et al. 1983). This description is based on the abundance of available lipoyl domains, each equally capable of shuttling reaction intermediates. Consequently, higher-order interactions, such as the E3 distribution, are also affected by random processes. Thus, substrate channeling becomes a statistically guided process in which available LDs randomly oscillate between available domains, while progressing

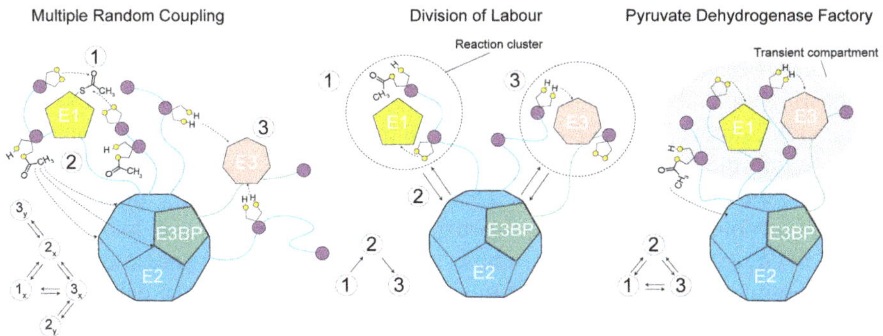

Fig. 9.11 Models explaining PDHc function considering the endogenous metabolon architecture. The proposed models aim to characterize the local and temporal organization of the pyruvate decarboxylation (Krebs and Johnson 1937), transacetylation (Reed 2001), and LD regeneration (Perham 1991)

on the next reaction coordinate. In light of the previously reported structural insights, the feasibility of a random distribution of E1 and E3 subdomains is questionable as E1 and E3 have shown specific localizations, yet are not highly accurate in terms of rotational freedom around the core (Kyrilis et al. 2021).

(ii) Another proposed model was the pseudo-atomic description of the "Division-of-Labor" (DOL) mechanism, suggesting a local clustering of E1 and E3 catalyzed reactions. Active site coupling via cross-acetylation and reduction would alleviate spatial constraints and connect these local clusters. This would prioritize the E2 LD for substrate channeling, limiting E3 LD based on their spatial restrictions and observed catalytic obsolescence (Prajapati et al. 2019; Rahmatullah et al. 1990).

(iii) Analysis of the endogenous fungal PDHc complex from *Thermochaetoides thermophila* suggested the "Pyruvate Dehydrogenase Factory." This mechanism is based on the observation of local clusters of E1 and E3, therefore further supporting a DOL-like mechanism. However, a diffuse formation of densities around the clusters was visualized, potentially accommodating other E1 and E3 proteins. Therefore, this higher-order structure must be composed of both catalytic and non-catalytic E1 and E3 subunits. Specifically, clusters of E1/E3 subunits can serve to form a transient nano-compartment, confining the catalytically active E1s and E3s (Kyrilis et al. 2021). This integrative description of the proposed metabolon-mediated catalytic mechanism expanded the previous quantitative description with spatial descriptions of active site distances, defining a 140 Å reaction exclusion zone to accommodate the LD domain. Instead of distinct catalytic clusters, as seen in the DOL mechanism, E1 and E3 spatially organize together in a shared cluster, carrying out the oxidative decarboxylation of pyruvate in a confined reaction chamber.

These proposed models, although reflecting currently available structural data, are hypothetical; this is because, to date, no PDHc structure of the entire metabolon has been determined. This is not only due to its immense complexity but also because integration of decades-long experimental data is not an easy feat. Given the structural redundancy of the PDHc, as well as the presence of disordered regions connecting the embedded structural domains, a structural description of the entire metabolon cannot capture PDHc function. In the metabolon, apart from global flexibility, conformational changes and allosteric events, as well as localized effects and changes, can orchestrate function. Therefore, it is highly unlikely that, given the inherent complexity of PDHc, the entire metabolon will be averaged. Instead, future endeavors must probe PDHc with single-molecule methods and integrate those data in a coherent model, which most likely will resemble the pyruvate dehydrogenase factory organization. However, reaching atomic details for the complex will be only a milestone in the long-awaited complete structure-function characterization of the PDHc metabolon; higher-order interactions and regulation of the complex are still in their infancy regarding their architectural analysis.

9.6 Genetic Basis and Regulation of the Pyruvate Dehydrogenase Complex (PDHc)

PDHc is a nuclear-encoded mitochondrial multienzyme complex; genetically, the E1 heterodimer is encoded by the PDHA1 and PDHB genes (Ducich et al. 2022), and the majority of PDHc deficiencies in genetically confirmed patients result from mutations in PDHA1 (Patel et al. 2012b; DeBrosse et al. 2012); more than 100 different mutations are described in the literature (Patel et al. 2012b; Barnerias et al. 2010; DeBrosse et al. 2012; Imbard et al. 2011; Lissens et al. 2000; Quintana et al. 2010). The 17 kb human PDHA1 is located on band p22.1. It contains ten introns, and despite its location on the X chromosome, similar numbers of affected males and females have been identified, with a wide spectrum of clinical pathologies (Dahl et al. 1990; Brown et al. 1994). However, the types of mutations are distributed unequally between genders (Patel et al. 2012b; DeBrosse et al. 2012; Lissens et al. 2000). For example, mutations in exons 10 and 11 (mostly deletion/insertion), which result in premature termination codons, are often observed in females, whereas missense/nonsense mutations in exons 3, 7, 8, and 11 are predominant in males (Lissens et al. 2000). Hemizygous males with PDHA1 mutations often express more severe clinical symptoms (including lethality in infancy), whereas females bearing the same mutation may show variable clinical symptoms and an overall greater survival due to skewed X-inactivation (Patel et al. 2012b; DeBrosse et al. 2012; Pithukpakorn 2005). Apart from PDHA1, an autosomal locus, PDHA2, showing significant cross-hybridization, has been detected on chromosome 4 and loci q22–q23 with a coding sequence similarity of 84% at the nucleotide level (Dahl et al. 1990). Postmeiotic spermatogenic cells express this intronless gene that was proven to be expressed also in the testis (Dahl et al. 1990). It has been shown that under prolonged hypoxia conditions, the E1β component acts in a regulatory way and is downregulated to inhibit PDHc activity (Yonashiro et al. 2018). This regulatory mechanism has been later related to the decreased PDHc presence in the nucleus upon hypoxic conditions (Eguchi and Nakayama 2019).

The E2 subunit of PDHc is mapped on the long arm of human chromosome 11 band q23.1, and several cytogenetic abnormalities including translocations have been associated to this region (Leung et al. 1993). Deregulation of the E2 gene's expression was quite recently related to E4 transcription factor 1 (E4F1) (Goguet-Rubio et al. 2016). In keratinocytes, this multifunctional protein regulates a metabolic program involved in pyruvate metabolism which is required to maintain skin homeostasis (Goguet-Rubio et al. 2016). Simultaneously, another work was published expanding E4F1's regulatory role to a set of four genes (PDHc E2, E3, mitochondrial pyruvate carrier 1 (Mpc1), and solute carrier family 25 member 19 (Slc25a19)) associated with pyruvate oxidation (Lacroix et al. 2016). In the same study, E4F1 was identified as a "master regulator" of PDHc because dysregulation of this factor could lead to 80% decrease of PDHc activity and thus alterations of pyruvate metabolism (Lacroix et al. 2016).

The E3BP paralog of E2 protein, employed to anchor E3 via the peripheral-subunit binding domain (PSBD), thus having a structural rather than a catalytic role (Harris et al. 1997), is encoded by PDX1 gene mapped on the short arm of chromosome 11 (Dey et al. 2002) and positions p12–13. It is of note that E2 and E3BP, despite their similarity, are encoded by genes located on different positions of chromosome 11.

The E3 subunit of the PDHc complex is encoded by a 20 kb long gene bearing 14 exons, ranging in size from 69 to 780 bp, and 13 introns, ranging in size from 93 bp to 7.0 kb (Feigenbaum and Robinson 1993). The gene has been localized on the long arm of chromosome 7 (Otulakowski et al. 1988) and specifically within bands q31 → q32 (Scherer et al. 1991) and its regulation is highly dependent on four protein binding domains (termed P1–4) located −322 to +47 bp of the promoter region (Yang et al. 2001). More specifically two elements, a CRE-like site −13 to −6 bp and the TACGAC direct repeat sequence located from −95 to −80 bp upstream the transcription start site were identified as required regions for the transcription of the gene (Yang et al. 2001).

9.6.1 *Regulation by the PDHc-Dedicated Kinases and Phosphatases*

As previously described, PDHc interconnects the glycolytic path to the subsequent TCA cycle. Thus, it occupies a central hub in eukaryotic metabolism; the regulation of PDHc activity substantially impacts energy conversion and utilization within cells. Therefore, during evolution, highly accurate PDHc regulation mechanisms have evolved in response to a wide range of stimuli, such as fasting and anoxia—and PDHc dysregulation is linked to a plethora of diseases which are described elsewhere (Anwar et al. 2021; Gray et al. 2014; Stacpoole and McCall 2023; Wang et al. 2021).

The main system that modulates PDHc's activity is the balance of phosphorylation-dephosphorylation of serine residues. Phosphorylation is an inhibitory signal, whereas dephosphorylation activates the complex. In humans, there are four kinases with the role of PDHc deactivation, named pyruvate dehydrogenase kinase 1–4 (PDK). Different tissues display a unique enrichment profile for PDK isoforms suggesting a specific, per-tissue, regulation profile. Regarding PDHc activation, there are two dedicated phosphatase enzymes, named PDP1/2 (pyruvate dehydrogenase phosphatase 1 or 2) (Patel et al. 2014b).

Structural and biochemical models can aid in understanding the roles of PDHc kinases and phosphatases (Fig. 9.12a–e). Human PDHc phosphorylation involves residues Ser 232, Ser 293, and Ser 300 (phosphorylation sites 3, 1, 2, respectively) of the E1 subunit (Rardin et al. 2009) (Fig. 9.12). The first position is favored both for kinase and phosphatase activity and thus contributes predominantly to PDHc regulation (Sale and Randle 1982) and can be phosphorylated and dephosphorylated by all mentioned isoforms. In contrast, the other two phosphorylation sites have lower specificity for the enzymes, while every site can be phosphorylated independently (Korotchkina and Patel 1995). Furthermore, only PDK1 can act on all three sites, whereas the other three isoforms can only modify the first two sites. Progressive phosphorylation of the three sites leads to inhibition enhancement and vice versa (Patel and Korotchkina 2006). Despite the common characteristics of the three Ser residues in E1, the mode of PDHc inhibition due to phosphorylation differs among them. The first position is located in the substrate channel of the enzyme, and its phosphorylation mainly inhibits the acetylation of CoA. According to the proposed mechanism, the phosphate group in position 1 deters the LD of E2 from interacting with the enzyme's active center (Gray et al. 2014; Ciszak et al. 2003). Presumably, the LD from the E3BP will also not be capable of binding. The other two sites are not proximal to the active center, and therefore, allosteric regulation is hypothesized here. For position 3 it has been specifically demonstrated that ThDP antagonizes the site's phosphorylation (Korotchkina and Patel 2001).

The mentioned PDKs and PDPs bind to PDHc E2 and E3BP lipoyl domains. In Mammals LD1 and LD2 are located in the N-terminal flexible regions of E2 and LD3 in the N-terminal region of E3BP. This binding regulates the catalytic activity of the modifying enzymes and restricts access to E1 phosphorylation sites. Similarly to the phosphorylation sites, PDK isoforms have different specificity for each LD leading to preferential binding. Furthermore, the degree of kinase activation due to LD binding also differs among isoforms, e.g., PDK3 is the mostly affected, whereas PDK1 is the least. Furthermore, the ratio of acetyl-CoA/CoA and NADH/NAD$^+$ regulates PDK activity in an LD-mediated mechanism. Specifically, the acetylation and reduction of LDs, by acetyl-CoA and NADH accordingly, increase the activity of kinases and binding to PDHc, thus deactivating it. This type of regulation is also isoform-specific with PDK2,4 being the most susceptible to it (Patel et al. 2014b; Roche et al. 2001).

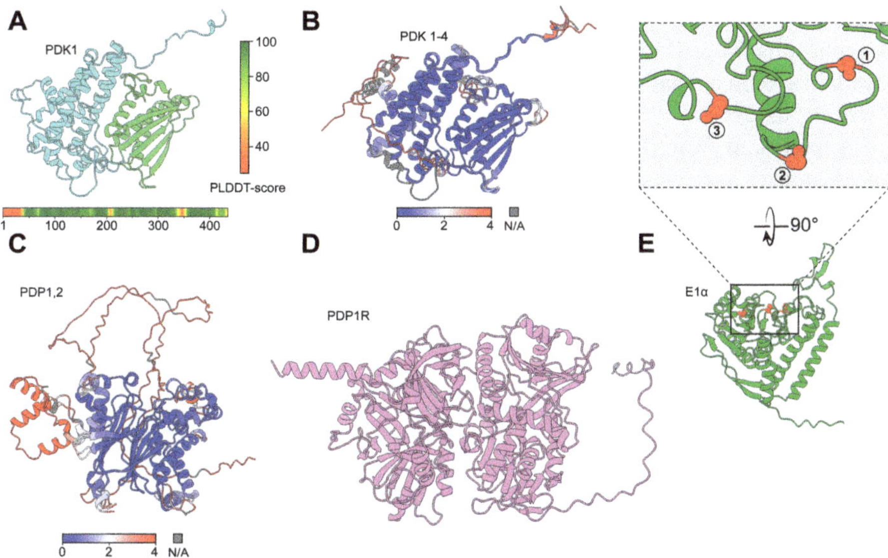

Fig. 9.12 Molecular regulation of PDHc via phosphorylation. AlphaFold 3 was utilized for the structural prediction of PDHc regulating kinases and phosphatases (Abramson et al. 2024). (**a**) The prediction for PDK1 is shown (UniProt ID Q15118): C-terminal catalytic domain colored in green, N-terminal domain colored in blue. (**b**) Predicted models for PDK1, 2, 3, and 4 (UniProt IDs Q15119, Q15120, Q16654) were aligned, and Cα RMSD was predicted. PDK isoforms entail a conserved structure which mostly differs in loops and unstructured regions. It is also visible that the C-terminal part is the most conserved. (**c**) Aligned structures of PDP1 and PDP2 catalytic subunits (UniProt IDs Q9P0J1, Q9P2J9). (**d**) Regulatory domain of PDP1 (UniProt ID Q8NCN5). (**e**) PDHc E1α model (UniProt ID P08559); the three phosphorylation sites appear in red and correspond to residues Ser 293 (position 1), Ser 300 (position 2), and Ser 232 (position 3)

PDKs share a conserved structure entailing an N-terminal and a C-terminal region connected by a loop (Fig. 9.12). In the active form, PDKs form homodimers in which the individual subunits are tethered through the C-terminal region. This domain also includes the active center of the enzyme whereas the N-terminal region binds to the LDs (Steussy et al. 2001). At any given time, only one or two PDK dimers are associated with PDHc. This raises the question how these enzymes reach and phosphorylate all of the E1 subunits included in PDHc. To address this, a hand-over-hand model has been proposed—according to the model, one PDK subunit is bound to a LD, whereas the N-terminal domain of the other subunit roams freely and is thus able to tether LDs of proximal E2 and E3BPs. This allows the PDK dimer to move across the PDHc periphery and thus phosphorylate multiple E1 subunits (Roche et al. 2003). In contrast to the kinases, PDP isoforms 1 and 2 differ significantly in their structure and activation (Fig. 9.12c, d). PDP1 is a heterodimer composed of a catalytic and a regulatory subunit, whereas for PDP2 only a catalytic subunit has been identified to date. PDP1, similarly to the PDKs, binds to the LDs and also requires calcium for its catalytic activity. Interestingly, PDP2 requires none of the above and can activate PDHc without being bound to its LD moieties (Karpova et al. 2003; Denton 2009). It is of interest that despite decades of knowledge regarding phosphatase and kinase interactions with LDs, few experimental structures are available, such as that of PDK3 with LD2 (Kato et al. 2005; Devedjiev et al. 2007; Green et al. 2008).

It is evident from the above that the high level of PDHc regulation by its dedicated kinases and phosphatases urgently needs structural and computational data to derive underlying mechanistic models; considering the high protein density within PDHc, its activation/deactivation equilibria must be regulated by fine-tuned mechanisms that can only be deduced utilizing integrative structural biology methods, combined with single-molecule approaches.

The phosphorylation/dephosphorylation equilibrium of PDHc is the main mediator for the regulation of its responses to environmental stimuli. There are various stimuli that modulate the activity of PDKs and PDPs, and those include hormones and signaling molecules (Behal et al. 1993). Given its role in energy flux and conversion, PDHc activation needs to be regulated by the availability of nutrients and energy balance of the cell. This is manifested through substrate and end-product regulation via PDKs and PDPs. Specifically, pyruvate is an induction signal for the enzyme and has an inhibitory effect on all PDK isoforms, thus enhancing PDH activity (Roche et al. 2001). This enables PDHc to respond accordingly to an abundance of carbohydrates, and thus pyruvate, and process them further toward the synthesis of ATP. In a similar fashion, high substrate concentration ratios of CoA and NAD^+ to acetyl-CoA and reduced NADH induce dehydrogenase activity by inhibiting PDKs. This occurs through the acetylation/reduction state of LDs as previously discussed (Bao et al. 2004). Furthermore, the ATP/ADP concentration ratio affects PDHc activity in a similar fashion, as ADP inhibits PDK1, 2, and 4. Therefore, when the cell requires energy in the form of ATP, cellular respiration is enhanced via increase in pyruvate oxidation. Finally, ions, including Mg^{2+} and Ca^{2+}, induce PDP1, thus activating PDHc (Roche et al. 2001). Ultimately, how and to what extent PDHc structurally alters in response to stimuli that may activate or deactivate it has not been systematically probed or addressed, despite decades of research in PDHc regulation. Such a research field is highly promising, if we aim to truly understand cell metabolism in the years to come.

9.7 Mitochondrial Pyruvate Transport and the Dynamic Localization of PDHc

9.7.1 Mitochondrial Pyruvate Transport

PDHc is known to primarily localize inside the mitochondrial matrix. This means that the substrate (pyruvate) must be transported through the double membrane of the mitochondrion for the reaction to take place. Regarding the permeability of the outer membrane, the transport of pyruvate through pore-forming voltage-dependent ion channels may offer an explanation. However, for the inner membrane, a specific mitochondrial transporter or a channel for the translocation is required, with the majority belonging to the SLC25 (solute carrier 25) mitochondrial carrier family (Kunji et al. 2020). In the 1970s, researchers demonstrated the existence of a mitochondrial protein facilitating this transportation; this protein displayed favorable characteristics such as saturation kinetics and sensitivity to sulfhydryl reagents (Papa et al. 1971) and could be blocked by small molecules (Halestrap and Denton 1974). Despite these early findings, it was only in 2012 that MPC proteins were officially assigned to this role (Herzig et al. 2012; Bricker et al. 2012). Many isoforms of these proteins exist across species, and they play a similar, yet distinct, role in regulating and transporting pyruvate into mitochondria. Initially, these proteins were observed to form hetero-complexes of approximately 150 kDa on blue native electrophoresis (Bricker et al. 2012), but recently it was revealed that the functional MPC is actually a hetero-dimeric complex (Tavoulari et al. 2019, 2022). Despite previously identified as a protein member of the SLC25, MPC belongs to a new protein family called SLC54 (Gyimesi and Hediger 2020).

Concerning the transport mechanism, it was initially proposed that pyruvate might diffuse across the membrane (Bakker and van Dam 1974; Klingenberg 1970). However, precise mechanistic details remain unclear, primarily due to the absence of the identification of a specific inhibitor for pyruvate transport. Papa and colleagues demonstrated that pyruvate translocation is directly proton-coupled (Papa et al. 1971). As a result of this proton symport pattern, the increase of pH was observed to be related to increased pyruvate import. Intramitochondrial pyruvate exchanges with extramitochondrial through a process of exchange diffusion, moving in and out of the mitochondrial matrix (Papa et al. 1971). Additionally, a small molecule (a-cyano-4-hydroxycinnamate), capable of inhibiting pyruvate oxidation at low concentrations (<0.2 mM), was identified. It is speculated that it functions by blocking mitochondrial pyruvate transport in intact mitochondria (Halestrap and Denton 1974). In contrast, when the same compound was incubated with mitochondrial lysates at concentrations of up to 2 mM, it had no inhibitory effects on the enzymatic activity of related enzymes, providing further evidence for the identification of a pyruvate carrier (Halestrap and Denton 1974). An alternative pathway for pyruvate to enter mitochondria was provided through experiments where upon MPC inhibition, despite the significant decrease in glucose metabolism in the TCA cycle, glucose metabolism was not completely halted, which could be due to passive diffusion or the presence of an alternate or non-facilitated transport mechanism (Vacanti et al. 2014). A significant aspect of pyruvate flux in mitochondria is its dependence on the ΔpH gradience with the maximum rate of transport also being affected by the presence of exchangeable substrates (Halestrap 1978). The overall concept for pyruvate transport, as described above, was initially proposed in the 1970s using isolated rat mitochondria (Papa et al. 1971; Papa and Paradies 1974; Halestrap 1975), and recent studies have confirmed it conclusively in proteoliposomes containing purified yeast and human MPC heterodimers (Tavoulari et al. 2019, 2022).

Regardless, 10 years after the molecular identification of MPC proteins, the exact mechanism of transport remains poorly understood due to the lack of structural insights. A functional assembly relies on both the MPC1 and MPC2 for the formation of the approximately 150 kDa-sized oligomer (Bricker et al. 2012). Recently, the recombinant human carrier was studied but only involving a functional MPC2 homodimer and not the heterodimer (Nagampalli et al. 2018), probably due to the lack (in the model system) of a critical and yet unidentified protein component for stable complex formation of the human MPC1:MPC2 complex (Hanlon et al. 1990). This investigation, despite confirming prior knowledge that pyruvate transport by the MPC is a rapid and specific process that depends on co-proton import and redox balance (Halestrap 1975; Gray et al. 2016), revealed seven times lower specificity for pyruvate in the MPC2-MPC2 compared to Km values reported in the literature for the MPC1-MPC2 heterodimer (Nagampalli et al. 2018), suggesting a more efficient transport in the latter case. The predicted (AlphaFold 3) model of a heterodimer is composed of one MPC1 and one MPC2 with three transmembrane α-helices each and an extra α-helix protruding out of the lipid bilayer (Fig. 9.13a, b). Contrary to prior predictions (McCommis and Finck 2015), the two monomers do not exhibit opposing topological orientations.

9.7.2 Mitochondrial Import and Export of PDHc

Several studies reveal PDHc pathways toward the mitochondria or the nucleus (Nagaraj et al. 2017; Sutendra et al. 2014; Chen et al. 2018) with thc latter being less elucidated. The transfer of proteins/polypeptide chains inside the mitochondrial matrix and their final destinations takes place via the translocase of the outer/inner membrane (TOM/TIM) and dedicated translocases in a manner that is well understood, at least at the cellular level. Initially, the Oxa-1 insertase located in the inner mitochondrial membrane (IMM) acts as an import point (Hildenbeutel et al. 2012), and assists the insertion of mitochondrial genome encoded respiratory chain complexes that are synthesized within the matrix into the IMM (Stuart 2002). A mitochondrial-localization sequence (MLS) drives the four subunits of PDHc to enter the mitochondria, where they are properly folded and assembled in the inner membrane by mitochondrial chaperones (Martin 1997), and the MLS is removed by the mitochondrial matrix peptidase (Chacinska et al. 2009).

The exit pathways of PDHc must be more complex; the intricate architecture and the MDa size qualify PDHc as one of the largest known enzymatic complexes of the cell. The presence of a double membrane in organelles such as the mitochondrion poses an additional challenge in comprehending the entry and exit pathways for molecules like PDHc. First indications for PDHc mobility from the inner to the outer mitochondrial membrane, came around 2011 when intact and functional PDHc, was shown to translocate across mitochondrial membranes from the matrix to the outer mitochondrial membrane (Hitosugi et al. 2011). In contrast to the better-studied import mechanism, the mechanism of the export pathway and the following

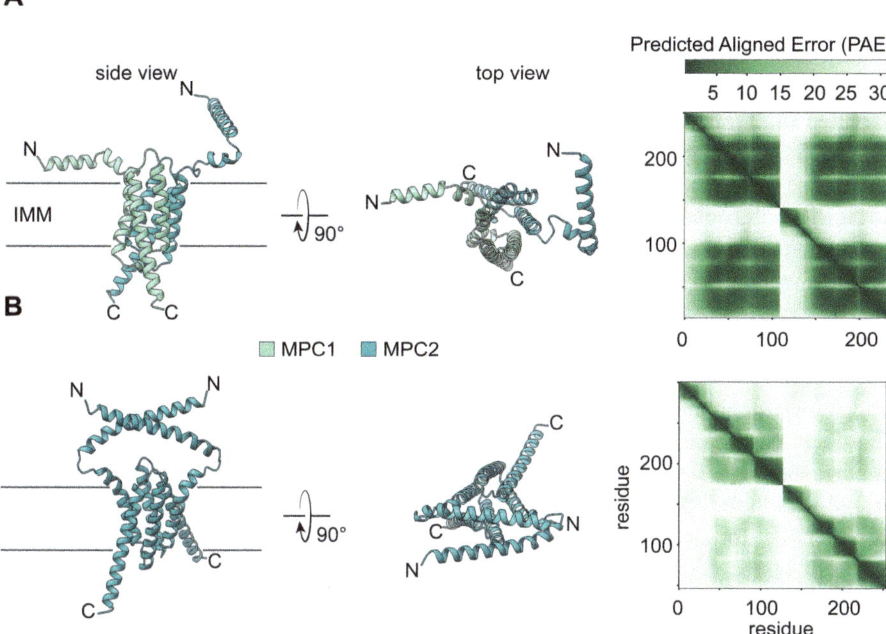

Fig. 9.13 Modes of MPC organization. (**a**) Heterodimeric- and (**b**) homodimeric states of the MPC have been predicted using AlphaFold 3 (Abramson et al. 2024) (model quality reported via predicted aligned error is reported)

nuclear translocation remain elusive (Sutendra et al. 2014). Indications pointed to the existence of an export mechanism through which the full complex (or its components) could be exported from mitochondria, using a stress-induced mitochondria-derived vesicle (MDV)-mediated mechanism (McLelland et al. 2014; Soubannier et al. 2012). The latter has been proven as a mechanism to carry damaged mitochondrial proteins into the lysosomes for degradation (Soubannier et al. 2012). However, those vesicles despite their 70–100 nm diameters are not favored to encapsulate an intact 10 MDa, ~45 nm-sized PDHc (Sumegi et al. 1987); each must accommodate 1–2 PDHc molecules, a transport pathway which may not be energy efficient.

9.8   Nuclear Localization and Nuclear Import Pathways of PDHc

Several metabolic enzymes have been observed to perform moonlighting functions in the nucleus, under conditions of cellular stress (Kim and Dang 2005; Siebert and Humphrey 1965; Yang et al. 2012; Yogev et al. 2010). *Sutendra* et al. were the first to identify the role of PDHc for acetyl-CoA-mediated acetylation of core histones within the nucleus, a process triggered upon exposure to proliferative stimuli like serum and epidermal growth factor (EGF) or hypoxia (Sutendra et al. 2014). They identified an active assembly of PDHc and observed decreased de novo acetyl-CoA synthesis and acetylation of core histones upon knockdown of its components in isolated functional nuclei (Sutendra et al. 2014). The fact that nuclear PDHc lacks an MLS, while simultaneously lacking a similar nuclear-localization sequence, gives rise to the explanation of a mitochondrial origin, where PDHc enzymes lost their MLS and subsequently moved toward the nucleus as an active enzymatic complex (Sutendra et al. 2014). Such a hypothesis is exciting, as a clear mitochondrial export mechanism is currently unknown for the PDHc (see Sect. 9.7.2); in addition, due to the sheer size of the metabolon, extensive membrane re-organization and assembly of even larger biomolecular complexes must be involved. In addition to the absence of MLS, there are no kinases detected within the nucleus, proposing a regulatory mechanism different to the one present in mitochondria (Sutendra et al. 2014). In particular, upon cell cycle-dependent increase in the nuclear PDHc levels, a corresponding decrease in mitochondrial PDHc abundance occurs, favoring the translocation theory of the complex from the mitochondria to the nucleus (Sutendra et al. 2014). In parallel, nuclear PDHc inhibition leads to decreased acetylation of specific lysine residues on histones important for G1-S phase progression and expression of S phase markers (Sutendra et al. 2014). It is clear that this dynamic translocation of mitochondrial PDHc to the nucleus reveals a mechanism of nuclear acetyl-CoA synthesis involved in later steps of histone acetylation and, thereby, epigenetic regulation (Sutendra et al. 2014).

In addition to that, a recent sequence data comparison revealed putative canonical nuclear localization signals (NLS) for all TCA cycle enzymes (including PDHc), with the exception of citrate synthase (CS) (Kafkia et al. 2022). It was demonstrated that the TCA cycle and thus PDHc are also partially operational in the nucleus, and proximity labeling mass spectrometry results revealed proximity of the involved enzymes with core nuclear proteins (Kafkia et al. 2022).

Despite its known flexibility and dynamic nature, the nuclear pore complex (NPC) shares low chances to act as a pathway for nuclear import of PDHc (Paci et al. 2020; Pante and Kann 2002), although it may transfer even HIV particles with dedicated assembly/disassembly mechanisms, as very recently described (Kreysing et al. 2025). PDHc escaping NPC-mediated transport is rather unusual as most nucleo-cytoplasmic communication is regulated by the NPC (Beck and Hurt 2017). Recently, a non-canonical pathway for nuclear PDHc translocation circumventing the nuclear pore complex (NPC) was described (Zervopoulos et al. 2022) (Fig. 9.14). The proposed mechanism supports that PDHc is released from nucleus-tethered mitochondria and deposited on the lamin layer before entering the nucleoplasm in a process involving mitofusin-2 (MFN2) (Zervopoulos et al. 2022). MFN2 is not only present in the nuclear envelope and the nucleoplasm but upon proliferative stimuli can change the balance toward the nuclear envelope abundance and enhance tethering of mitochondria to the nuclear envelope (Zervopoulos et al. 2022). An analogous MFN2-mediated mechanism has been described for mitochondria tethering to

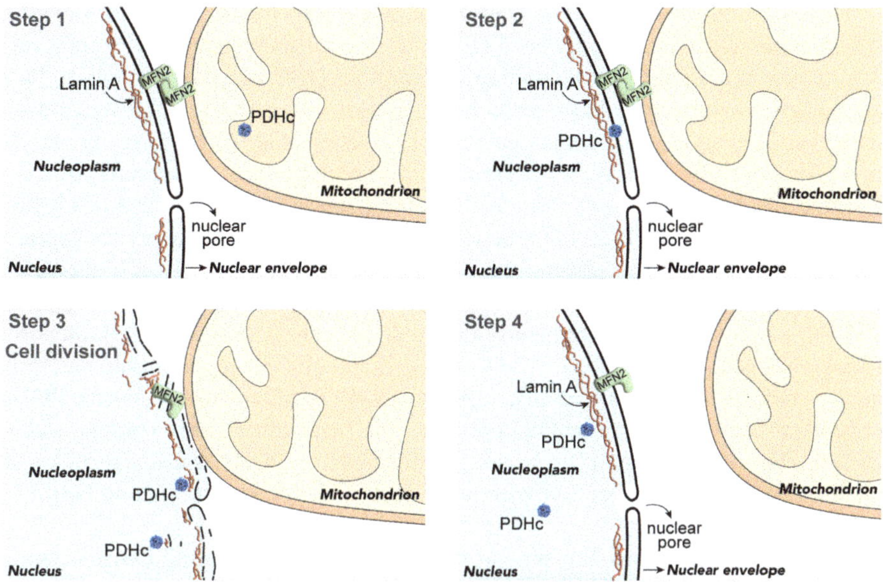

Fig. 9.14 Graphical representation of the non-canonical pathway of PDHc translocation from mitochondria to the nucleus

the endoplasmic reticulum (ER) (de Brito and Scorrano 2008; Murley and Nunnari 2016).

9.8.1 Nuclear PDHc Function

Beyond its traditional biosynthetic and bioenergetic roles in cellular functions, many TCA cycle enzymes are also crucial for the spatiotemporal regulation of gene expression and genome repair (Carey et al. 2015; TeSlaa et al. 2016; Gonzalez-Angulo et al. 2012; Liu et al. 2017; Raffel et al. 2017; Morris et al. 2019; Jiang et al. 2015; Sivanand et al. 2017; Sulkowski et al. 2020; Dai et al. 2020). In that sense, nuclear translocation of the PDHc plays a critical role in multiple cases of transcription regulation pathways (Rodriguez et al. 2001; Zheng et al. 2003; Ghosh et al. 1999; Subramanian and Miller 2000; Feo et al. 2000). During early embryogenesis and zygotic genome activation (ZGA), some TCA cycle enzymes transiently relocate to the nucleus, a process essential for epigenetic remodeling (Nagaraj et al. 2017). This process relies on PDHc presence, and any failure of the enzyme to enter the nucleus leads to the loss of specific histone modifications, blocking ZGA. Notably, in humans, pyruvate dehydrogenase temporarily localizes to the nucleus at the 4/8-cell stage, aligning with embryonic genome activation and suggesting a conserved mechanism in early development.

Histone modifications, such as acetylation using acetyl-CoA, produced by PDHc, significantly underscore the role of PDHc in transcription regulation. It is important to note here that PDHc is not a unique source of acetyl-CoA in the nucleus and that also other enzymes, like ATP-citrate lyase, are known to link cellular metabolism with histone acetylation (Sivanand et al. 2017; Wellen et al. 2009). Histone acetylation is mediated by the reversible transfer of the acetyl moiety of acetyl-CoA to the ε-amino group of a lysine, resulting in the neutralization of a positive charge (Allfrey et al. 1964). In the nucleus, the reaction is catalyzed by histone acetyltransferases (HATs) and reversed by histone deacetylases (HDACs) (McCullough and Grant 2010). There are highly conserved sites at the amino-terminal tails of histones H3 and H4, where lysines are frequently targeted for acetylation-related modifications (Shogren-Knaak et al. 2006). Additionally, metabolic reprogramming, interconnected with histone modifications, supports $CD4^+$ T-cell activation, where PDHc-dependent acetyl-CoA generation stands as the rate-limiting step (Mocholi et al. 2023). This process, apart from the involvement of PDHc translocation to the nucleus, requires its interaction with p300 acetyltransferase and histone H3K27ac (Mocholi et al. 2023).

Recently, a similar mechanism was revealed in plants, triggering PDHc accumulation inside the nucleus, connected to the production of acetyl-CoA which acts as a substrate for the ethylene-mediated histone acetylation (Shao et al. 2024). In detail, EIN2 (ethylene-insensitive protein 2) known to be involved in transcription regulation (Zhang et al. 2017) was detected to co-precipitate with PDHc in the nuclear fraction and clear evidence of colocalization for all subunits of the PDHc complex with EIN2 in response to ethylene was provided (Shao et al. 2024). In addition, the signal of nuclear presence for E1, E2, and E3 was clearly elevated upon ethylene treatment, whereas the cytosolic signal was significantly decreased (Shao et al. 2024).

9.9 Other Functional Roles and Interactions of PDHc

The pyruvate dehydrogenase complex presents several additional functions apart from its central role in the decarboxylation of pyruvate to acetyl-CoA (Kato et al. 2008). Compartmentalized activities of PDHc generate building blocks for lipid structures and regulate lipogenic gene expression to foster prostate tumorigenesis (Chen et al. 2018) by regulating the expression of SREBP-targeted genes through histone acetylation (Chen et al. 2018). The E2 subunit of PDHc also participates in the induction of the gene expression, encoding protoxin in *Bacillus thuringiensis (Bt)*. The proteinaceous crystals produced by *Bt* mainly consist of Cry proteins which are in a majority toxic to specific insects (Bravo et al. 2011). As of that, *Bt* has been widely used as a biopesticide for more than 60 years (Bravo et al. 2011; Sanahuja et al. 2011). It has been shown that in the presence of soluble E2, there is a connection between sugar metabolism, spore formation, and toxin production. Its recruitment for regulating the plasmid-encoded *cry* genes provides a mechanism for enhancing selectively their transcription (Walter and Aronson 1999). During infection with *Mycoplasma pneumoniae*, the E2 protein of PDHc can act as a structural component via binding fibronectins of the extracellular matrix, leading to the adhesion of bacterial cells to host cells (Dallo et al. 2002).

Within the Plantae kingdom, PDHc, mentioned in the literature as PDC, can be encountered in mitochondria as well as in plastids (Thelen et al. 1998; Taylor et al. 1992). Genetic information of *Arabidopsis thaliana* PDHc components is encoded by two E1α genes (IAR4 and IAR4L) (Quint et al. 2009; Luethy et al. 1995), one E1β gene (MAB1) (Luethy et al. 1994), three E2 genes (mtE2-1, mtE2-2, and mtE2-3) (Guan et al. 1995; Thelen et al. 1999; Taylor et al. 2004), and two E3 genes (At1g48030 and At3g17240) (Lutziger and Oliver 2001). Plant mitochondrial PDHc possesses a pretty similar role as in other kingdoms, driving carbon entry to the citric acid cycle, whereas in plastids it provides acetyl-CoA and NADH for fatty acid and isoprenoid biosynthesis (Camp and Randall 1985). Mitochondrial and chloroplast PDHc also have different substrate specificities, Mg^{2+} and p*H* requirements (Miernyk and Randall 1987). Chloroplast PDHc in particular has been shown to have different regulatory mechanisms related to dark-light transitions (Perham and Packman 1989; Miernyk and Randall 1987) and not the covalent modification motif of specific residue phosphorylation met in the mammalian and mitochondrial complex (Randall et al. 1981). The overall architecture of plant PDHc is less studied than the one found in mammals or bacteria, and limited structural information, mainly from AlphaFold predictions, is currently available for the three enzymatic components of the complex. Despite that, experimental data suggest that subunits of PDHc can be biochemically and structurally discriminated (Taylor et al. 1992; Camp and Randall 1985). On another note, in the green alga Chlamydomonas, the dihydrolipoamide acetyltransferase subunit (DLA2) of the chloroplast PDHc participates in chloroplast gene expression (Neusius et al. 2022). Recently, experimental evidence revealed a connection between the mitochondrial PDHc E1β component and polar auxin transport during tissue development, suggesting a functional link between mitochondrial metabolism and organ formation (Ohbayashi et al. 2019).

9.9.1 Examples of Validated Protein Interactors of PDHc

The crowded environment of a mitochondrion imposes proximity for many enzymes involved in the citric acid cycle as well as for their products. For example, citrate synthase (CS) and the pyruvate dehydrogenase complex-citrate synthase interactions (PDHc-CS) revealed that kinetic parameters of the coupled enzyme system were decreased when compared to values observed in the individual enzyme reactions, pointing to an association between the two enzymes (Sumegi et al. 1980). Experiments with active enzyme gel chromatography validated that CS interacts with PDHc in its functioning state in a motif where CS binds to the transacetylase core of PDHc. However, upon binding, all components of the complex are participating in the interaction (Sumegi and Alkonyi 1983). This observation prompts consideration of the dynamic compartmentalization of acetyl-CoA within mitochondria, leading to the preferential flow of acetyl-CoA from PDHc toward CS (Sumegi and Alkonyi 1983).

The cytoplasmic STAT3 upon Ser 727-phosphorylation has been shown to translocate inside mitochondria where it associates with PDHc E1, accelerating the conversion of pyruvate to acetyl-CoA, elevating the mitochondrial membrane potential, and promoting ATP synthesis (Xu et al. 2016). In a similar pattern, PDHc E2 component has previously been reported to interact with STAT family members in a cytokine stimulation-dependent manner (Chueh et al. 2011). This process has been reported also for nuclear translocation of STAT3 which is reversed by dephosphorylation (Hazan-Halevy et al. 2010; Dasgupta et al. 2014; Lin et al. 2002; Zhang et al. 1995). In the case of nucleus translocation of STAT3, the phenomenon is highly tyrosine and serine phosphorylation dependent (Liu et al. 2005; Aziz et al. 2010), whereas in the case of mitochondrial translocation, it depends on acetylation (Xu et al. 2016). Finally, the mitochondrial metabolic activation is reversed by SIRT5 which deacetylates STAT3 (Xu et al. 2016).

To unveil more direct structural interactions of PDHc, cross-linking mass spectrometry is an ideal tool, as it can be applied from isolated systems to tissues and organs (Piersimoni et al. 2022; Graziadei and Rappsilber 2022). Current cross-linking data point to interactions of PDHc with various other biomolecules within mitochondria, and a recent in vivo determination of a PDHc binder was validated with cross-linking, AlphaFold modeling, and cellular assays (O'Reilly et al. 2023). In short, the E1 subunit of the PDHc was identified to be inhibited by PDHl, a specialized protein in the gram-positive bacterium *Bacillus subtilis*—studies toward this direction for eukaryotes will undoubtedly shed light on the, yet elusive, PDHc interaction networks, their organism-specific adaptations, and their changes upon perturbation.

9.10   Concluding Remarks

This book chapter covered the basics and recent findings regarding the molecular and structural biology of the 10-megadalton PDHc metabolon. Although substantial strides have been made to understand its intricate architecture, and at the same time, it is true that only now do we begin to understand its immense complexity. A major question that lingers, and waits for an answer in the years to come, is how the hundreds of subunits; that in concert catalyze the link reaction, inter-communicate, and are regulated either in an individual or in a concerted fashion. It is definitely expected that current advances in proteomics, structural biology, computational biology augmented by artificial intelligence, and the availability of genomic and transcriptomic data must be integrated and go hand-in-hand with functional assays, driven by systematic investigations of the pyruvate dehydrogenase complexes, to ultimately characterize the link reaction at unprecedented detail. Such structure-function studies will bring us closer not only to understand how the cell operates but also to the molecular basis of various diseases in which PDHc stars as a central player.

Acknowledgments This work was supported by the Federal Ministry of Education and Research (BMBF, ZIK program) (grant nos. 03Z22HN23, 03Z22HI2, and 03COV04), the European Regional Development Funds (EFRE) for Saxony-Anhalt (grant nos. ZS/2016/04/78115 and ZS/2024/05/187255), the Deutsche Forschungsgemeinschaft (project numbers 391498659, RTG 2467, and 514901783, CRC 1664), the European Union through funding of the Horizon Europe ERA Chair "hot4cryo" project number 101086665, and the Martin Luther University Halle-Wittenberg.

Conflict of Interest   The authors have no conflicts of interest to declare that are relevant to the content of this chapter.

References

Abramson J, et al. Accurate structure prediction of biomolecular interactions with AlphaFold 3. Nature. 2024;630:493–500.

Allen MD, Broadhurst RW, Solomon RG, Perham RN. Interaction of the E2 and E3 components of the pyruvate dehydrogenase multienzyme complex of Bacillus stearothermophilus. Use of a truncated protein domain in NMR spectroscopy. FEBS J. 2005;272:259–68.

Allfrey VG, Faulkner R, Mirsky AE. Acetylation and methylation of histones and their possible role in the regulation of Rna synthesis. Proc Natl Acad Sci USA. 1964;51:786–94.

Anwar S, Shamsi A, Mohammad T, Islam A, Hassan MI. Targeting pyruvate dehydrogenase kinase signaling in the development of effective cancer therapy. Biochim Biophys Acta Rev Cancer. 2021;1876:188568.

Argyrou A, Sun G, Palfey BA, Blanchard JS. Catalysis of diaphorase reactions by Mycobacterium tuberculosis lipoamide dehydrogenase occurs at the EH4 level. Biochemistry. 2003;42:2218–28.

Arjunan P, et al. Structure of the pyruvate dehydrogenase multienzyme complex E1 component from Escherichia coli at 1.85 A resolution. Biochemistry. 2002;41:5213–21.

Arjunan P, et al. Novel binding motif and new flexibility revealed by structural analyses of a pyruvate dehydrogenase-dihydrolipoyl acetyltransferase subcomplex from the Escherichia coli pyruvate dehydrogenase multienzyme complex. J Biol Chem. 2014;289:30161–76.

Aziz MH, et al. Protein kinase Cvarepsilon mediates Stat3Ser727 phosphorylation, Stat3-regulated gene expression, and cell invasion in various human cancer cell lines through integration with MAPK cascade (RAF-1, MEK1/2, and ERK1/2). Oncogene. 2010;29:3100–9.

Babady NE, Pang YP, Elpeleg O, Isaya G. Cryptic proteolytic activity of dihydrolipoamide dehydrogenase. Proc Natl Acad Sci USA. 2007;104:6158–63.

Bakker EP, van Dam K. The movement of monocarboxylic acids across phospholipid membranes: evidence for an exchange diffusion between pyruvate and other monocarboxylate ions. Biochim Biophys Acta. 1974;339:285–9.

Balakrishnan A, Nemeria NS, Chakraborty S, Kakalis L, Jordan F. Determination of pre-steady-state rate constants on the Escherichia coli pyruvate dehydrogenase complex reveals that loop movement controls the rate-limiting step. J Am Chem Soc. 2012;134:18644–55.

Bao H, Kasten SA, Yan X, Roche TE. Pyruvate dehydrogenase kinase isoform 2 activity limited and further inhibited by slowing down the rate of dissociation of ADP. Biochemistry. 2004;43:13432–41.

Barnerias C, et al. Pyruvate dehydrogenase complex deficiency: four neurological phenotypes with differing pathogenesis. Dev Med Child Neurol. 2010;52:e1–9.

Beck M, Hurt E. The nuclear pore complex: understanding its function through structural insight. Nat Rev Mol Cell Biol. 2017;18:73–89.

Behal RH, Buxton DB, Robertson JG, Olson MS. Regulation of the pyruvate dehydrogenase multienzyme complex. Annu Rev Nutr. 1993;13:497–520.

Benen J, et al. Lipoamide dehydrogenase from Azotobacter vinelandii: site-directed mutagenesis of the His450-Glu455 diad. Kinetics of wild-type and mutated enzymes. Eur J Biochem. 1992;207:487–97.

Berg A, Westphal AH, Bosma HJ, de Kok A. Kinetics and specificity of reductive acylation of wild-type and mutated lipoyl domains of 2-oxo-acid dehydrogenase complexes from Azotobacter vinelandii. Eur J Biochem. 1998;252:45–50.

Bosma HJ, de Kok A, Westphal AH, Veeger C. The composition of the pyruvate dehydrogenase complex from Azotobacter vinelandii. Does a unifying model exist for the complexes from gram-negative bacteria? Eur J Biochem. 1984;142:541–9.

Brautigam CA, Chuang JL, Tomchick DR, Machius M, Chuang DT. Crystal structure of human dihydrolipoamide dehydrogenase: NAD+/NADH binding and the structural basis of disease-causing mutations. J Mol Biol. 2005;350:543–52.

Brautigam CA, et al. Structural insight into interactions between dihydrolipoamide dehydrogenase (E3) and E3 binding protein of human pyruvate dehydrogenase complex. Structure. 2006;14:611–21.

Brautigam CA, Wynn RM, Chuang JL, Chuang DT. Subunit and catalytic component stoichiometries of an in vitro reconstituted human pyruvate dehydrogenase complex. J Biol Chem. 2009;284:13086–98.

Bravo A, Likitvivatanavong S, Gill SS, Soberon M. Bacillus thuringiensis: a story of a successful bioinsecticide. Insect Biochem Mol Biol. 2011;41:423–31.

Breslow R. Rapid deuterium exchange in Thiazolium Salts1. J Am Chem Soc. 1957;79:1762–3.

Bricker DK, et al. A mitochondrial pyruvate carrier required for pyruvate uptake in yeast, Drosophila, and humans. Science. 2012;337:96–100.

Brown GK, et al. "Cerebral" lactic acidosis: defects in pyruvate metabolism with profound brain damage and minimal systemic acidosis. Eur J Pediatr. 1988;147:10–4.

Brown GK, Brown RM, Scholem RD, Kirby DM, Dahl HH. The clinical and biochemical spectrum of human pyruvate dehydrogenase complex deficiency. Ann N Y Acad Sci. 1989;573:360–8.

Brown GK, Otero LJ, LeGris M, Brown RM. Pyruvate dehydrogenase deficiency. J Med Genet. 1994;31:875–9.

Camp PJ, Randall DD. Purification and characterization of the pea chloroplast pyruvate dehydrogenase complex: a source of acetyl-CoA and NADH for fatty acid biosynthesis. Plant Physiol. 1985;77:571–7.

Carey BW, Finley LW, Cross JR, Allis CD, Thompson CB. Intracellular alpha-ketoglutarate maintains the pluripotency of embryonic stem cells. Nature. 2015;518:413–6.

Cate RL, Roche TE, Davis LC. Rapid intersite transfer of acetyl groups and movement of pyruvate dehydrogenase component in the kidney pyruvate dehydrogenase complex. J Biol Chem. 1980;255:7556–62.

Chacinska A, Koehler CM, Milenkovic D, Lithgow T, Pfanner N. Importing mitochondrial proteins: machineries and mechanisms. Cell. 2009;138:628–44.

Chandrasekhar K, et al. Insight to the interaction of the dihydrolipoamide acetyltransferase (E2) core with the peripheral components in the Escherichia coli pyruvate dehydrogenase complex via multifaceted structural approaches. J Biol Chem. 2013;288:15402–17.

Chang A, et al. BRENDA, the ELIXIR core data resource in 2021: new developments and updates. Nucleic Acids Res. 2021;49:D498–508.

Chen J, et al. Compartmentalized activities of the pyruvate dehydrogenase complex sustain lipogenesis in prostate cancer. Nat Genet. 2018;50:219–28.

Chueh FY, et al. Nuclear localization of pyruvate dehydrogenase complex-E2 (PDC-E2), a mitochondrial enzyme, and its role in signal transducer and activator of transcription 5 (STAT5)-dependent gene transcription. Cell Signal. 2011;23:1170–8.

Ciszak EM, Korotchkina LG, Dominiak PM, Sidhu S, Patel MS. Structural basis for flip-flop action of thiamin pyrophosphate-dependent enzymes revealed by human pyruvate dehydrogenase. J Biol Chem. 2003;278:21240–6.

Ciszak EM, et al. How dihydrolipoamide dehydrogenase-binding protein binds dihydrolipoamide dehydrogenase in the human pyruvate dehydrogenase complex. J Biol Chem. 2006;281:648–55.

Collins JH, Reed LJ. Acyl group and electron pair relay system: a network of interacting lipoyl moieties in the pyruvate and alpha-ketoglutarate dehydrogenase complexes from Escherichia coli. Proc Natl Acad Sci USA. 1977;74:4223–7.

Cooper A, Dryden DT. Allostery without conformational change. A plausible model. Eur Biophys J. 1984;11:103–9.

Dahl HH, Brown RM, Hutchison WM, Maragos C, Brown GK. A testis-specific form of the human pyruvate dehydrogenase E1 alpha subunit is coded for by an intronless gene on chromosome 4. Genomics. 1990;8:225–32.

Dai Z, Ramesh V, Locasale JW. The evolving metabolic landscape of chromatin biology and epigenetics. Nat Rev Genet. 2020;21:737–53.

Dallo SF, Kannan TR, Blaylock MW, Baseman JB. Elongation factor Tu and E1 beta subunit of pyruvate dehydrogenase complex act as fibronectin binding proteins in mycoplasma pneumoniae. Mol Microbiol. 2002;46:1041–51.

Danson MJ, Fersht AR, Perham RN. Rapid intramolecular coupling of active sites in the pyruvate dehydrogenase complex of Escherichia coli: mechanism for rate enhancement in a multimeric structure. Proc Natl Acad Sci USA. 1978;75:5386–90.

Dasgupta M, et al. Critical role for lysine 685 in gene expression mediated by transcription factor unphosphorylated STAT3. J Biol Chem. 2014;289:30763–71.

Dave E, Guest JR, Attwood MM. Metabolic engineering in Escherichia coli: lowering the lipoyl domain content of the pyruvate dehydrogenase complex adversely affects the growth rate and yield. Microbiology. 1995;141(Pt 8):1839–49.

de Brito OM, Scorrano L. Mitofusin 2 tethers endoplasmic reticulum to mitochondria. Nature. 2008;456:605–10.

de Kok A, Hengeveld AF, Martin A, Westphal AH. The pyruvate dehydrogenase multi-enzyme complex from gram-negative bacteria. Biochim Biophys Acta. 1998;1385:353–66.

De Marcucci O, Lindsay JG, Component X. An immunologically distinct polypeptide associated with mammalian pyruvate dehydrogenase multi-enzyme complex. Eur J Biochem. 1985;149:641–8.

DeBrosse SD, et al. Spectrum of neurological and survival outcomes in pyruvate dehydrogenase complex (PDC) deficiency: lack of correlation with genotype. Mol Genet Metab. 2012;107:394–402.

Denton RM. Regulation of mitochondrial dehydrogenases by calcium ions. Biochim Biophys Acta. 2009;1787:1309–16.

Devedjiev Y, Steussy CN, Vassylyev DG. Crystal structure of an asymmetric complex of pyruvate dehydrogenase kinase 3 with lipoyl domain 2 and its biological implications. J Mol Biol. 2007;370:407–16.

Dey R, Aral B, Abitbol M, Marsac C. Pyruvate dehydrogenase deficiency as a result of splice-site mutations in the PDX1 gene. Mol Genet Metab. 2002;76:344–7.

Douce R, Bourguignon J, Neuburger M, Rebeille F. The glycine decarboxylase system: a fascinating complex. Trends Plant Sci. 2001;6:167–76.

Ducich NH, Mears JA, Bedoyan JK. Solvent accessibility of E1alpha and E1beta residues with known missense mutations causing pyruvate dehydrogenase complex (PDC) deficiency: impact on PDC-E1 structure and function. J Inherit Metab Dis. 2022;45:557–70.

Eguchi K, Nakayama K. Prolonged hypoxia decreases nuclear pyruvate dehydrogenase complex and regulates the gene expression. Biochem Biophys Res Commun. 2019;520:128–35.

Feigenbaum AS, Robinson BH. The structure of the human dihydrolipoamide dehydrogenase gene (DLD) and its upstream elements. Genomics. 1993;17:376–81.

Feo S, Arcuri D, Piddini E, Passantino R, Giallongo A. ENO1 gene product binds to the c-myc promoter and acts as a transcriptional repressor: relationship with Myc promoter-binding protein 1 (MBP-1). FEBS Lett. 2000;473:47–52.

Ferriero R, et al. Phenylbutyrate therapy for pyruvate dehydrogenase complex deficiency and lactic acidosis. Sci Transl Med. 2013;5:175ra131.

Forsberg BO. The structure and evolutionary diversity of the fungal E3-binding protein. Commun Biol. 2023;6:480.

Forsberg BO, Aibara S, Howard RJ, Mortezaei N, Lindahl E. Arrangement and symmetry of the fungal E3BP-containing core of the pyruvate dehydrogenase complex. Nat Commun. 2020;11:4667.

Frank RA, Titman CM, Pratap JV, Luisi BF, Perham RN. A molecular switch and proton wire synchronize the active sites in thiamine enzymes. Science. 2004;306:872–6.

Frank RA, Pratap JV, Pei XY, Perham RN, Luisi BF. The molecular origins of specificity in the assembly of a multienzyme complex. Structure. 2005;13:1119–30.

Frey PA. 2-Acetylthiamin pyrophosphate: an enzyme-bound intermediate in thiamin pyrophosphate-dependent reactions. Biofactors. 1989;2:1–9.

Fries M, Jung HI, Perham RN. Reaction mechanism of the heterotetrameric (alpha2beta2) E1 component of 2-oxo acid dehydrogenase multienzyme complexes. Biochemistry. 2003;42:6996–7002.

Ghisla S, Massey V. Mechanisms of flavoprotein-catalyzed reactions. Eur J Biochem. 1989;181:1–17.

Ghosh AK, Steele R, Ray RB. Functional domains of c-myc promoter binding protein 1 involved in transcriptional repression and cell growth regulation. Mol Cell Biol. 1999;19:2880–6.

Glasser NR, Wang BX, Hoy JA, Newman DK. The pyruvate and alpha-ketoglutarate dehydrogenase complexes of pseudomonas aeruginosa catalyze pyocyanin and phenazine-1-carboxylic acid reduction via the subunit dihydrolipoamide dehydrogenase. J Biol Chem. 2017;292:5593–607.

Goguet-Rubio P, et al. E4F1-mediated control of pyruvate dehydrogenase activity is essential for skin homeostasis. Proc Natl Acad Sci USA. 2016;113:11004–9.

Gonzalez-Angulo AM, et al. Gene expression, molecular class changes, and pathway analysis after neoadjuvant systemic therapy for breast cancer. Clin Cancer Res. 2012;18:1109–19.

Graham LD, Perham RN. Interactions of lipoyl domains with the E1p subunits of the pyruvate dehydrogenase multienzyme complex from Escherichia coli. FEBS Lett. 1990;262:241–4.

Graham LD, Packman LC, Perham RN. Kinetics and specificity of reductive acylation of lipoyl domains from 2-oxo acid dehydrogenase multienzyme complexes. Biochemistry. 1989;28:1574–81.

Gray LR, Tompkins SC, Taylor EB. Regulation of pyruvate metabolism and human disease. Cell Mol Life Sci. 2014;71:2577–604.

Gray LR, Rauckhorst AJ, Taylor EB. A method for multiplexed measurement of mitochondrial pyruvate carrier activity. J Biol Chem. 2016;291:7409–17.

Graziadei A, Rappsilber J. Leveraging crosslinking mass spectrometry in structural and cell biology. Structure. 2022;30:37–54.

Green T, et al. Structural and functional insights into the molecular mechanisms responsible for the regulation of pyruvate dehydrogenase kinase 2. J Biol Chem. 2008;283:15789–98.

Gruys KJ, Datta A, Frey PA. 2-Acetylthiamin pyrophosphate (acetyl-TPP) pH-rate profile for hydrolysis of acetyl-TPP and isolation of acetyl-TPP as a transient species in pyruvate dehydrogenase catalyzed reactions. Biochemistry. 1989;28:9071–80.

Guan Y, Rawsthorne S, Scofield G, Shaw P, Doonan J. Cloning and characterization of a dihydrolipoamide acetyltransferase (E2) subunit of the pyruvate dehydrogenase complex from Arabidopsis thaliana. J Biol Chem. 1995;270:5412–7.

Guest JR, Lewis HM, Graham LD, Packman LC, Perham RN. Genetic reconstruction and functional analysis of the repeating lipoyl domains in the pyruvate dehydrogenase multienzyme complex of Escherichia coli. J Mol Biol. 1985;185:743–54.

Guest JR, et al. Enzymological and physiological consequences of restructuring the lipoyl domain content of the pyruvate dehydrogenase complex of Escherichia coli. Microbiology. 1997;143(Pt 2):457–66.

Guigas G, Weiss M. Effects of protein crowding on membrane systems. Biochim Biophys Acta. 2016;1858:2441–50.

Guo F, Zhang D, Kahyaoglu A, Farid RS, Jordan F. Is a hydrophobic amino acid required to maintain the reactive V conformation of thiamin at the active center of thiamin diphosphate-requiring enzymes? Experimental and computational studies of isoleucine 415 of yeast pyruvate decarboxylase. Biochemistry. 1998;37:13379–91.

Gyimesi G, Hediger MA. Sequence features of mitochondrial transporter protein families. Biomolecules. 2020;10:1611.

Hackert ML, Oliver RM, Reed LJ. Evidence for a multiple random coupling mechanism in the alpha-ketoglutarate dehydrogenase multienzyme complex of Escherichia coli: a computer model analysis. Proc Natl Acad Sci USA. 1983;80:2226–30.

Halestrap AP. The mitochondrial pyruvate carrier. Kinetics and specificity for substrates and inhibitors. Biochem J. 1975;148:85–96.

Halestrap AP. Pyruvate and ketone-body transport across the mitochondrial membrane. Exchange properties, pH-dependence and mechanism of the carrier. Biochem J. 1978;172:377–87.

Halestrap AP, Denton RM. Specific inhibition of pyruvate transport in rat liver mitochondria and human erythrocytes by alpha-cyano-4-hydroxycinnamate. Biochem J. 1974;138:313–6.

Hanlon TE, Nurco DN, Kinlock TW, Duszynski KR. Trends in criminal activity and drug use over an addiction career. Am J Drug Alcohol Abuse. 1990;16:223–38.

Harris RA, Bowker-Kinley MM, Wu P, Jeng J, Popov KM. Dihydrolipoamide dehydrogenase-binding protein of the human pyruvate dehydrogenase complex. DNA-derived amino acid sequence, expression, and reconstitution of the pyruvate dehydrogenase complex. J Biol Chem. 1997;272:19746–51.

Hawkins CF, Borges A, Perham RN. A common structural motif in thiamin pyrophosphate-binding enzymes. FEBS Lett. 1989;255:77–82.

Hayakawa T, et al. Mammalian alpha-keto acid dehydrogenase complexes. V. Resolution and reconstitution studies of the pig heart pyruvate dehydrogenase complex. J Biol Chem. 1969;244:3660–70.

Hazan-Halevy I, et al. STAT3 is constitutively phosphorylated on serine 727 residues, binds DNA, and activates transcription in CLL cells. Blood. 2010;115:2852–63.

Herzig S, et al. Identification and functional expression of the mitochondrial pyruvate carrier. Science. 2012;337:93–6.

Hezaveh S, Zeng AP, Jandt U. Full enzyme complex simulation: interactions in human pyruvate dehydrogenase complex. J Chem Inf Model. 2018;58:362–9.

Hildenbeutel M, et al. The membrane insertase Oxa1 is required for efficient import of carrier proteins into mitochondria. J Mol Biol. 2012;423:590–9.

Hiromasa Y, Fujisawa T, Aso Y, Roche TE. Organization of the cores of the mammalian pyruvate dehydrogenase complex formed by E2 and E2 plus the E3-binding protein and their capacities to bind the E1 and E3 components. J Biol Chem. 2004;279:6921–33.

Hitosugi T, et al. Tyrosine phosphorylation of mitochondrial pyruvate dehydrogenase kinase 1 is important for cancer metabolism. Mol Cell. 2011;44:864–77.

Howard MJ, et al. Three-dimensional structure of the major autoantigen in primary biliary cirrhosis. Gastroenterology. 1998;115:139–46.

Huckabee WE. Relationships of pyruvate and lactate during anaerobic metabolism. I. Effects of infusion of pyruvate or glucose and of hyperventilation. J Clin Invest. 1958;37:244–54.

Igamberdiev AU, Bykova NV, Ens W, Hill RD. Dihydrolipoamide dehydrogenase from porcine heart catalyzes NADH-dependent scavenging of nitric oxide. FEBS Lett. 2004;568:146–50.

Imbard A, et al. Molecular characterization of 82 patients with pyruvate dehydrogenase complex deficiency. Structural implications of novel amino acid substitutions in E1 protein. Mol Genet Metab. 2011;104:507–16.

Ishikawa E, Oliver RM, Reed LJ. Alpha-Keto acid dehydrogenase complexes, V. Macromolecular organization of pyruvate and alpha-ketoglutarate dehydrogenase complexes isolated from beef kidney mitochondria. Proc Natl Acad Sci USA. 1966;56:534–41.

Jiang Y, et al. Local generation of fumarate promotes DNA repair through inhibition of histone H3 demethylation. Nat Cell Biol. 2015;17:1158–68.

Jordan F. Current mechanistic understanding of thiamin diphosphate-dependent enzymatic reactions. Nat Prod Rep. 2003;20:184–201.

Jordan F, et al. Dual catalytic apparatus of the thiamin diphosphate coenzyme: acid-base via the 1′,4′-iminopyrimidine tautomer along with its electrophilic role. J Am Chem Soc. 2003;125:12732–8.

Jordan F, Nemeria NS, Sergienko E. Multiple modes of active center communication in thiamin diphosphate-dependent enzymes. Acc Chem Res. 2005;38:755–63.

Jung HI, Perham RN. Prediction of the binding site on E1 in the assembly of the pyruvate dehydrogenase multienzyme complex of Bacillus stearothermophilus. FEBS Lett. 2003;555:405–10.

Jung HI, Bowden SJ, Cooper A, Perham RN. Thermodynamic analysis of the binding of component enzymes in the assembly of the pyruvate dehydrogenase multienzyme complex of Bacillus stearothermophilus. Protein Sci. 2002;11:1091–100.

Jurrus E, et al. Improvements to the APBS biomolecular solvation software suite. Protein Sci. 2018;27:112–28.

Kafkia E, et al. Operation of a TCA cycle subnetwork in the mammalian nucleus. Sci Adv. 2022;8:eabq5206.

Kale S, Arjunan P, Furey W, Jordan F. A dynamic loop at the active center of the Escherichia coli pyruvate dehydrogenase complex E1 component modulates substrate utilization and chemical communication with the E2 component. J Biol Chem. 2007;282:28106–16.

Kalia YN, et al. The high-resolution structure of the peripheral subunit-binding domain of dihydrolipoamide acetyltransferase from the pyruvate dehydrogenase multienzyme complex of Bacillus stearothermophilus. J Mol Biol. 1993;230:323–41.

Kaplon J, et al. A key role for mitochondrial gatekeeper pyruvate dehydrogenase in oncogene-induced senescence. Nature. 2013;498:109–12.

Karissa P, et al. Comparison between Dichloroacetate and Phenylbutyrate treatment for pyruvate dehydrogenase deficiency. Br J Biomed Sci. 2022;79:10382.

Karpova T, Danchuk S, Kolobova E, Popov KM. Characterization of the isozymes of pyruvate dehydrogenase phosphatase: implications for the regulation of pyruvate dehydrogenase activity. Biochim Biophys Acta. 2003;1652:126–35.

Kastritis PL, et al. Capturing protein communities by structural proteomics in a thermophilic eukaryote. Mol Syst Biol. 2017;13:936.

Kato M, Chuang JL, Tso SC, Wynn RM, Chuang DT. Crystal structure of pyruvate dehydrogenase kinase 3 bound to lipoyl domain 2 of human pyruvate dehydrogenase complex. EMBO J. 2005;24:1763–74.

Kato M, et al. A synchronized substrate-gating mechanism revealed by cubic-core structure of the bovine branched-chain alpha-ketoacid dehydrogenase complex. EMBO J. 2006;25:5983–94.

Kato M, et al. Structural basis for inactivation of the human pyruvate dehydrogenase complex by phosphorylation: role of disordered phosphorylation loops. Structure. 2008;16:1849–59.

Kelley DE, He J, Menshikova EV, Ritov VB. Dysfunction of mitochondria in human skeletal muscle in type 2 diabetes. Diabetes. 2002;51:2944–50.

Kern D, et al. How thiamine diphosphate is activated in enzymes. Science. 1997;275:67–70.

Kim JW, Dang CV. Multifaceted roles of glycolytic enzymes. Trends Biochem Sci. 2005;30:142–50.

Kim H, Patel MS. Characterization of two site-specifically mutated human dihydrolipoamide dehydrogenases (His-452—Gln and Glu-457—Gln). J Biol Chem. 1992;267:5128–32.

Klingenberg M. Mitochondria metabolite transport. FEBS Lett. 1970;6:145–54.

Klyachko NL, et al. pH-dependent substrate preference of pig heart lipoamide dehydrogenase varies with oligomeric state: response to mitochondrial matrix acidification. J Biol Chem. 2005;280:16106–14.

Korotchkina LG, Patel MS. Mutagenesis studies of the phosphorylation sites of recombinant human pyruvate dehydrogenase. Site-specific regulation. J Biol Chem. 1995;270:14297–304.

Korotchkina LG, Patel MS. Probing the mechanism of inactivation of human pyruvate dehydrogenase by phosphorylation of three sites. J Biol Chem. 2001;276:5731–8.

Korotchkina LG, Patel MS. Binding of pyruvate dehydrogenase to the core of the human pyruvate dehydrogenase complex. FEBS Lett. 2008;582:468–72.

Koves TR, et al. Mitochondrial overload and incomplete fatty acid oxidation contribute to skeletal muscle insulin resistance. Cell Metab. 2008;7:45–56.

Krebs HA, Johnson WA. Metabolism of ketonic acids in animal tissues. Biochem J. 1937;31:645–60.

Kreysing JP, et al. Passage of the HIV capsid cracks the nuclear pore. Cell. 2025;188:930–943.e21.

Kunji ERS, King MS, Ruprecht JJ, Thangaratnarajah C. The SLC25 carrier family: important transport proteins in mitochondrial physiology and pathology. Physiology (Bethesda). 2020;35:302–27.

Kyrilis FL, et al. Integrative structure of a 10-megadalton eukaryotic pyruvate dehydrogenase complex from native cell extracts. Cell Rep. 2021;34:108727.

Lacroix M, et al. E4F1 controls a transcriptional program essential for pyruvate dehydrogenase activity. Proc Natl Acad Sci USA. 2016;113:10998–1003.

Leung PS, et al. Chromosome localization and RFLP analysis of PDC-E2: the major autoantigen of primary biliary cirrhosis. Autoimmunity. 1993;14:335–40.

Lewendon A, Murray IA, Kleanthous C, Cullis PM, Shaw WV. Substitutions in the active site of chloramphenicol acetyltransferase: role of a conserved aspartate. Biochemistry. 1988;27:7385–90.

Lewendon A, Murray IA, Shaw WV, Gibbs MR, Leslie AG. Evidence for transition-state stabilization by serine-148 in the catalytic mechanism of chloramphenicol acetyltransferase. Biochemistry. 1990;29:2075–80.

Lin J, Tang H, Jin X, Jia G, Hsieh JT. p53 regulates Stat3 phosphorylation and DNA binding activity in human prostate cancer cells expressing constitutively active Stat3. Oncogene. 2002;21:3082–8.

Lindqvist Y, Schneider G, Ermler U, Sundstrom M. Three-dimensional structure of transketolase, a thiamine diphosphate dependent enzyme, at 2.5 A resolution. EMBO J. 1992;11:2373–9.

Lissens W, et al. Mutations in the X-linked pyruvate dehydrogenase (E1) alpha subunit gene (PDHA1) in patients with a pyruvate dehydrogenase complex deficiency. Hum Mutat. 2000;15:209–19.

Liu L, McBride KM, Reich NC. STAT3 nuclear import is independent of tyrosine phosphorylation and mediated by importin-alpha3. Proc Natl Acad Sci USA. 2005;102:8150–5.

Liu PS, et al. Alpha-ketoglutarate orchestrates macrophage activation through metabolic and epigenetic reprogramming. Nat Immunol. 2017;18:985–94.

Liu S, Xia X, Zhen J, Li Z, Zhou ZH. Structures and comparison of endogenous 2-oxoglutarate and pyruvate dehydrogenase complexes from bovine kidney. Cell Discov. 2022;8:126.

Luethy MH, Miernyk JA, Randall DD. The nucleotide and deduced amino acid sequences of a cDNA encoding the E1 beta-subunit of the Arabidopsis thaliana mitochondrial pyruvate dehydrogenase complex. Biochim Biophys Acta. 1994;1187:95–8.

Luethy MH, Miernyk JA, Randall DD. The mitochondrial pyruvate dehydrogenase complex: nucleotide and deduced amino-acid sequences of a cDNA encoding the Arabidopsis thaliana E1 alpha-subunit. Gene. 1995;164:251–4.

Lutziger I, Oliver DJ. Characterization of two cDNAs encoding mitochondrial lipoamide dehydrogenase from Arabidopsis. Plant Physiol. 2001;127:615–23.

Lynch EM, Kollman JM, Webb BA. Filament formation by metabolic enzymes-a new twist on regulation. Curr Opin Cell Biol. 2020;66:28–33.

Ma B, Nussinov R. Structured crowding and its effects on enzyme catalysis. Top Curr Chem. 2013;337:123–37.

Machado RS, Guest JR, Williamson MP. Mobility in pyruvate dehydrogenase complexes with multiple lipoyl domains. FEBS Lett. 1993;323:243–6.

Mande SS, Sarfaty S, Allen MD, Perham RN, Hol WG. Protein-protein interactions in the pyruvate dehydrogenase multienzyme complex: dihydrolipoamide dehydrogenase complexed with the binding domain of dihydrolipoamide acetyltransferase. Structure. 1996;4:277–86.

Martin J. Molecular chaperones and mitochondrial protein folding. J Bioenerg Biomembr. 1997;29:35–43.

Mattevi A, et al. Atomic structure of the cubic core of the pyruvate dehydrogenase multienzyme complex. Science. 1992a;255:1544–50.

Mattevi A, Obmolova G, Sokatch JR, Betzel C, Hol WG. The refined crystal structure of pseudomonas putida lipoamide dehydrogenase complexed with NAD+ at 2.45 A resolution. Proteins. 1992b;13:336–51.

Mattevi A, Obmolova G, Kalk KH, Teplyakov A, Hol WG. Crystallographic analysis of substrate binding and catalysis in dihydrolipoyl transacetylase (E2p). Biochemistry. 1993;32:3887–901.

Matthews RG, Ballou DP, Thorpe C, Williams CH Jr. Ion pair formation in pig heart lipoamide dehydrogenase: rationalization of pH profiles for reactivity of oxidized enzyme with dihydrolipoamide and 2-electron-reduced enzyme with lipoamide and iodoacetamide. J Biol Chem. 1977;252:3199–207.

McCommis KS, Finck BN. Mitochondrial pyruvate transport: a historical perspective and future research directions. Biochem J. 2015;466:443–54.

McCullough SD, Grant PA. Histone acetylation, acetyltransferases, and ataxia--alteration of histone acetylation and chromatin dynamics is implicated in the pathogenesis of polyglutamine-expansion disorders. Adv Protein Chem Struct Biol. 2010;79:165–203.

McLelland GL, Soubannier V, Chen CX, McBride HM, Fon EA. Parkin and PINK1 function in a vesicular trafficking pathway regulating mitochondrial quality control. EMBO J. 2014;33:282–95.

Meinhold S, Zdanowicz R, Giese C, Glockshuber R. Dimerization of a 5-kDa domain defines the architecture of the 5-MDa gammaproteobacterial pyruvate dehydrogenase complex. Sci Adv. 2024;10:eadj6358.

Miernyk JA, Randall DD. Some properties of pea mitochondrial phospho-pyruvate dehydrogenase-phosphatase. Plant Physiol. 1987;83:311–5.

Mocholi E, et al. Pyruvate metabolism controls chromatin remodeling during CD4(+) T cell activation. Cell Rep. 2023;42:112583.

Morris JP t, et al. alpha-Ketoglutarate links p53 to cell fate during tumour suppression. Nature. 2019;573:595–9.

Muller YA, et al. A thiamin diphosphate binding fold revealed by comparison of the crystal structures of transketolase, pyruvate oxidase and pyruvate decarboxylase. Structure. 1993;1:95–103.

Muoio DM, Neufer PD. Lipid-induced mitochondrial stress and insulin action in muscle. Cell Metab. 2012;15:595–605.

Murley A, Nunnari J. The emerging network of mitochondria-organelle contacts. Mol Cell. 2016;61:648–53.

Murphy GE, Jensen GJ. Electron cryotomography of the E. coli pyruvate and 2-oxoglutarate dehydrogenase complexes. Structure. 2005;13:1765–73.

Nagampalli RSK, et al. Human mitochondrial pyruvate carrier 2 as an autonomous membrane transporter. Sci Rep. 2018;8:3510.

Nagaraj R, et al. Nuclear localization of mitochondrial TCA cycle enzymes as a critical step in mammalian zygotic genome activation. Cell. 2017;168:210–23. e211.

Nemeria N, et al. The 1′,4′-iminopyrimidine tautomer of thiamin diphosphate is poised for catalysis in asymmetric active centers on enzymes. Proc Natl Acad Sci USA. 2007;104:78–82.

Neusius D, et al. Lysine acetylation regulates moonlighting activity of the E2 subunit of the chloroplast pyruvate dehydrogenase complex in Chlamydomonas. Plant J. 2022;111:1780–800.

O'Reilly FJ, et al. Protein complexes in cells by AI-assisted structural proteomics. Mol Syst Biol. 2023;19:e11544.

Ohbayashi I, et al. Mitochondrial pyruvate dehydrogenase contributes to auxin-regulated organ development. Plant Physiol. 2019;180:896–909.

Otulakowski G, Robinson BH, Willard HF. Gene for lipoamide dehydrogenase maps to human chromosome 7. Somat Cell Mol Genet. 1988;14:411–4.

Paci G, Zheng T, Caria J, Zilman A, Lemke EA. Molecular determinants of large cargo transport into the nucleus. elife. 2020;9:e55963.

Pante N, Kann M. Nuclear pore complex is able to transport macromolecules with diameters of about 39 nm. Mol Biol Cell. 2002;13:425–34.

Papa S, Paradies G. On the mechanism of translocation of pyruvate and other monocarboxylic acids in rat-liver mitochondria. Eur J Biochem. 1974;49:265–74.

Papa S, Francavilla A, Paradies G, Meduri B. The transport of pyruvate in rat liver mitochondria. FEBS Lett. 1971;12:285–8.

Park S, et al. Role of the pyruvate dehydrogenase complex in metabolic remodeling: differential pyruvate dehydrogenase complex functions in metabolism. Diabetes Metab J. 2018;42:270–81.

Patel MS, Korotchkina LG. Regulation of mammalian pyruvate dehydrogenase complex by phosphorylation: complexity of multiple phosphorylation sites and kinases. Exp Mol Med. 2001;33:191–7.

Patel MS, Korotchkina LG. Regulation of the pyruvate dehydrogenase complex. Biochem Soc Trans. 2006;34:217–22.

Patel MS, Roche TE. Molecular biology and biochemistry of pyruvate dehydrogenase complexes. FASEB J. 1990;4:3224–33.

Patel KP, O'Brien TW, Subramony SH, Shuster J, Stacpoole PW. The spectrum of pyruvate dehydrogenase complex deficiency: clinical, biochemical and genetic features in 371 patients. Mol Genet Metab. 2012a;106:385–94.

Patel KP, O'Brien TW, Subramony SH, Shuster J, Stacpoole PW. The spectrum of pyruvate dehydrogenase complex deficiency: clinical, biochemical and genetic features in 371 patients. Mol Genet Metab. 2012b;105:34–43.

Patel H, Nemeria NS, Andrews FH, McLeish MJ, Jordan F. Identification of charge transfer transitions related to thiamin-bound intermediates on enzymes provides a plethora of signatures useful in mechanistic studies. Biochemistry. 2014a;53:2145–52.

Patel MS, Nemeria NS, Furey W, Jordan F. The pyruvate dehydrogenase complexes: structure-based function and regulation. J Biol Chem. 2014b;289:16615–23.

Pei XY, Titman CM, Frank RA, Leeper FJ, Luisi BF. Snapshots of catalysis in the E1 subunit of the pyruvate dehydrogenase multienzyme complex. Structure. 2008;16:1860–72.

Perham RN. Domains, motifs, and linkers in 2-oxo acid dehydrogenase multienzyme complexes: a paradigm in the design of a multifunctional protein. Biochemistry. 1991;30:8501–12.

Perham RN, Packman LC. 2-Oxo acid dehydrogenase multienzyme complexes: domains, dynamics, and design. Ann N Y Acad Sci. 1989;573:1–20.

Piersimoni L, Kastritis PL, Arlt C, Sinz A. Cross-linking mass spectrometry for investigating protein conformations and protein-protein interactions horizontal line a method for all seasons. Chem Rev. 2022;122:7500–31.

Pithukpakorn M. Disorders of pyruvate metabolism and the tricarboxylic acid cycle. Mol Genet Metab. 2005;85:243–6.

Pletcher J, Sax M, Blank G, Wood M. Stereochemistry of intermediates in thiamine catalysis. 2. Crystal structure of DL-2-(alpha-hydroxybenzyl)thiamine chloride hydrochloride trihydrate. J Am Chem Soc. 1977;99:1396–403.

Plokhikh KS, et al. Association of 2-oxoacid dehydrogenase complexes with respirasomes in mitochondria. FEBS J. 2024;291:132–41.

Ponting CP. Biological function in the twilight zone of sequence conservation. BMC Biol. 2017;15:71.

Prajapati S, et al. Structural and functional analyses of the human PDH complex suggest a "division-of-labor" mechanism by local E1 and E3 clusters. Structure. 2019;27:1124–36. e1124.

Quint M, Barkawi LS, Fan KT, Cohen JD, Gray WM. Arabidopsis IAR4 modulates auxin response by regulating auxin homeostasis. Plant Physiol. 2009;150:748–58.

Quintana E, et al. Mutational study in the PDHA1 gene of 40 patients suspected of pyruvate dehydrogenase complex deficiency. Clin Genet. 2010;77:474–82.

Raffel S, et al. BCAT1 restricts alphaKG levels in AML stem cells leading to IDHmut-like DNA hypermethylation. Nature. 2017;551:384–8.

Rahmatullah M, Radke GA, Andrews PC, Roche TE. Changes in the core of the mammalian-pyruvate dehydrogenase complex upon selective removal of the lipoyl domain from the trans-acetylase component but not from the protein X component. J Biol Chem. 1990;265:14512–7.

Randall DD, Williams M, Rapp BJ. Phosphorylation-dephosphorylation of pyruvate dehydrogenase complex from pea leaf mitochondria. Arch Biochem Biophys. 1981;207:437–44.

Rardin MJ, Wiley SE, Naviaux RK, Murphy AN, Dixon JE. Monitoring phosphorylation of the pyruvate dehydrogenase complex. Anal Biochem. 2009;389:157–64.

Reed LJ. A trail of research from lipoic acid to alpha-keto acid dehydrogenase complexes. J Biol Chem. 2001;276:38329–36.

Reed LJ, Hackert ML. Structure-function relationships in dihydrolipoamide acyltransferases. J Biol Chem. 1990;265:8971–4.

Reed LJ, Leach FR, Koike M. Studies on a lipoic acid-activating system. J Biol Chem. 1958;232:123–42.

Reed LJ, et al. Reconstitution of the Escherichia coli pyruvate dehydrogenase complex. Proc Natl Acad Sci USA. 1975;72:3068–72.

Robinson JB Jr, Inman L, Sumegi B, Srere PA. Further characterization of the Krebs tricarboxylic acid cycle metabolon. J Biol Chem. 1987;262:1786–90.

Robinson BH, MacMillan H, Petrova-Benedict R, Sherwood WG. Variable clinical presentation in patients with defective E1 component of pyruvate dehydrogenase complex. J Pediatr. 1987;111:525–33.

Rocha M, et al. Mitochondrial dysfunction and endoplasmic reticulum stress in diabetes. Curr Pharm Des. 2016;22:2640–9.

Roche TE, et al. Distinct regulatory properties of pyruvate dehydrogenase kinase and phosphatase isoforms. Prog Nucleic Acid Res Mol Biol. 2001;70:33–75.

Roche TE, et al. Essential roles of lipoyl domains in the activated function and control of pyruvate dehydrogenase kinases and phosphatase isoform 1. Eur J Biochem. 2003;270:1050–6.

Rodriguez A, De La Cera T, Herrero P, Moreno F. The hexokinase 2 protein regulates the expression of the GLK1, HXK1 and HXK2 genes of Saccharomyces cerevisiae. Biochem J. 2001;355:625–31.

Russell GC, Guest JR. Sequence similarities within the family of dihydrolipoamide acyltransferases and discovery of a previously unidentified fungal enzyme. Biochim Biophys Acta. 1991a;1076:225–32.

Russell GC, Guest JR. Site-directed mutagenesis of the lipoate acetyltransferase of Escherichia coli. Proc Biol Sci. 1991b;243:155–60.

Sale GJ, Randle PJ. Occupancy of phosphorylation sites in pyruvate dehydrogenase phosphate complex in rat heart in vivo. Relation to proportion of inactive complex and rate of re-activation by phosphatase. Biochem J. 1982;206:221–9.

Sanahuja G, Banakar R, Twyman RM, Capell T, Christou P. Bacillus thuringiensis: a century of research, development and commercial applications. Plant Biotechnol J. 2011;9:283–300.

Saunier E, Benelli C, Bortoli S. The pyruvate dehydrogenase complex in cancer: an old metabolic gatekeeper regulated by new pathways and pharmacological agents. Int J Cancer. 2016;138:809–17.

Schellenberger A. Sixty years of thiamin diphosphate biochemistry. Biochim Biophys Acta. 1998;1385:177–86.

Schellenberger A, Hubner G, Neef H. Cofactor designing in functional analysis of thiamin diphosphate enzymes. Methods Enzymol. 1997;279:131–46.

Scherer SW, Otulakowski G, Robinson BH, Tsui LC. Localization of the human dihydrolipoamide dehydrogenase gene (DLD) to 7q31→q32. Cytogenet Cell Genet. 1991;56:176–7.

Schulze A, Downward J. Flicking the Warburg switch-tyrosine phosphorylation of pyruvate dehydrogenase kinase regulates mitochondrial activity in cancer cells. Mol Cell. 2011;44:846–8.

Schweppe DK, et al. Mitochondrial protein interactome elucidated by chemical cross-linking mass spectrometry. Proc Natl Acad Sci USA. 2017;114:1732–7.

Seifert F, et al. Direct kinetic evidence for half-of-the-sites reactivity in the E1 component of the human pyruvate dehydrogenase multienzyme complex through alternating sites cofactor activation. Biochemistry. 2006;45:12775–85.

Shao Z, et al. Nuclear pyruvate dehydrogenase complex regulates histone acetylation and transcriptional regulation in the ethylene response. Sci Adv. 2024;10:eado2825.

Shaw WV, Leslie AG. Chloramphenicol acetyltransferase. Annu Rev Biophys Biophys Chem. 1991;20:363–86.

Shogren-Knaak M, et al. Histone H4-K16 acetylation controls chromatin structure and protein interactions. Science. 2006;311:844–7.

Siebert G, Humphrey GB. Enzymology of the nucleus. Adv Enzymol Relat Areas Mol Biol. 1965;27:239–88.

Sivanand S, et al. Nuclear acetyl-CoA production by ACLY promotes homologous recombination. Mol Cell. 2017;67:252–65. e256.

Skalidis I, et al. Structural analysis of an endogenous 4-megadalton succinyl-CoA-generating metabolon. Commun Biol. 2023;6:552.

Skerlova J, Berndtsson J, Nolte H, Ott M, Stenmark P. Structure of the native pyruvate dehydrogenase complex reveals the mechanism of substrate insertion. Nat Commun. 2021;12:5277.

Smoly JM, Kuylenstierna B, Ernster L. Topological and functional organization of the mitochondrion. Proc Natl Acad Sci USA. 1970;66:125–31.

Song J, Jordan F. Interchain acetyl transfer in the E2 component of bacterial pyruvate dehydrogenase suggests a model with different roles for each chain in a trimer of the homooligomeric component. Biochemistry. 2012;51:2795–803.

Soriano-Baguet L, et al. Pyruvate dehydrogenase fuels a critical citrate pool that is essential for Th17 cell effector functions. Cell Rep. 2023;42:112153.

Soubannier V, et al. A vesicular transport pathway shuttles cargo from mitochondria to lysosomes. Curr Biol. 2012;22:135–41.

Stacpoole PW, McCall CE. The pyruvate dehydrogenase complex: life's essential, vulnerable and druggable energy homeostat. Mitochondrion. 2023;70:59–102.

Steussy CN, et al. Structure of pyruvate dehydrogenase kinase. Novel folding pattern for a serine protein kinase. J Biol Chem. 2001;276:37443–50.

Stoops JK, et al. On the unique structural organization of the Saccharomyces cerevisiae pyruvate dehydrogenase complex. J Biol Chem. 1997;272:5757–64.

Stuart R. Insertion of proteins into the inner membrane of mitochondria: the role of the Oxa1 complex. Biochim Biophys Acta. 2002;1592:79–87.

Subramanian A, Miller DM. Structural analysis of alpha-enolase. Mapping the functional domains involved in down-regulation of the c-myc protooncogene. J Biol Chem. 2000;275:5958–65.

Sulkowski PL, et al. Oncometabolites suppress DNA repair by disrupting local chromatin signalling. Nature. 2020;582:586–91.

Sumegi B, Alkonyi I. A study on the physical interaction between the pyruvate dehydrogenase complex and citrate synthase. Biochim Biophys Acta. 1983;749:163–71.

Sumegi B, Gyocsi L, Alkonyi I. Interaction between the pyruvate dehydrogenase complex and citrate synthase. Biochim Biophys Acta. 1980;616:158–66.

Sumegi B, Liposits Z, Inman L, Paull WK, Srere PA. Electron microscopic study on the size of pyruvate dehydrogenase complex in situ. Eur J Biochem. 1987;169:223–30.

Sutendra G, Michelakis ED. Pyruvate dehydrogenase kinase as a novel therapeutic target in oncology. Front Oncol. 2013;3:38.

Sutendra G, et al. A nuclear pyruvate dehydrogenase complex is important for the generation of acetyl-CoA and histone acetylation. Cell. 2014;158:84–97.

Tavoulari S, et al. The yeast mitochondrial pyruvate carrier is a hetero-dimer in its functional state. EMBO J. 2019;38:EMBJ2018100785.

Tavoulari S, et al. Key features of inhibitor binding to the human mitochondrial pyruvate carrier hetero-dimer. Mol Metab. 2022;60:101469.

Taylor AE, Cogdell RJ, Lindsay JG. Immunological comparison of the pyruvate dehydrogenase complexes from pea mitochondria and chloroplasts. Planta. 1992;188:225–31.

Taylor NL, Heazlewood JL, Day DA, Millar AH. Lipoic acid-dependent oxidative catabolism of alpha-keto acids in mitochondria provides evidence for branched-chain amino acid catabolism in Arabidopsis. Plant Physiol. 2004;134:838–48.

TeSlaa T, et al. Alpha-Ketoglutarate accelerates the initial differentiation of primed human pluripotent stem cells. Cell Metab. 2016;24:485–93.

Thelen JJ, Miernyk JA, Randall DD. Partial purification and characterization of the maize mitochondrial pyruvate dehydrogenase complex. Plant Physiol. 1998;116:1443–50.

Thelen JJ, et al. The dihydrolipoamide S-acetyltransferase subunit of the mitochondrial pyruvate dehydrogenase complex from maize contains a single lipoyl domain. J Biol Chem. 1999;274:21769–75.

Traut TW. Dissociation of enzyme oligomers: a mechanism for allosteric regulation. Crit Rev Biochem Mol Biol. 1994;29:125–63.

Tüting C, et al. Cryo-EM snapshots of a native lysate provide structural insights into a metabolon-embedded transacetylase reaction. Nat Commun. 2021;12:6933.

Vacanti NM, et al. Regulation of substrate utilization by the mitochondrial pyruvate carrier. Mol Cell. 2014;56:425–35.

van Kempen M, et al. Fast and accurate protein structure search with Foldseek. Nat Biotechnol. 2024;42:243–6.

Walter T, Aronson A. Specific binding of the E2 subunit of pyruvate dehydrogenase to the upstream region of bacillus thuringiensis protoxin genes. J Biol Chem. 1999;274:7901–6.

Wang X, Shen X, Yan Y, Li H. Pyruvate dehydrogenase kinases (PDKs): an overview toward clinical applications. Biosci Rep. 2021;41:BSR20204402.

Wellen KE, et al. ATP-citrate lyase links cellular metabolism to histone acetylation. Science. 2009;324:1076–80.

Westphal AH, de Kok A. Lipoamide dehydrogenase from Azotobacter vinelandii. Molecular cloning, organization and sequence analysis of the gene. Eur J Biochem. 1988;172:299–305.

Xu YS, et al. STAT3 undergoes acetylation-dependent mitochondrial translocation to regulate pyruvate metabolism. Sci Rep. 2016;6:39517.

Yang HS, et al. Human dihydrolipoamide dehydrogenase gene transcription is mediated by cAMP-response element-like site and TACGAC direct repeat. Int J Biochem Cell Biol. 2001;33:902–13.

Yang W, et al. PKM2 phosphorylates histone H3 and promotes gene transcription and tumorigenesis. Cell. 2012;150:685–96.

Yogev O, et al. Fumarase: a mitochondrial metabolic enzyme and a cytosolic/nuclear component of the DNA damage response. PLoS Biol. 2010;8:e1000328.

Yonashiro R, Eguchi K, Wake M, Takeda N, Nakayama K. Pyruvate dehydrogenase PDH-E1beta controls tumor progression by altering the metabolic status of cancer cells. Cancer Res. 2018;78:1592–603.

Zervopoulos SD, et al. MFN2-driven mitochondria-to-nucleus tethering allows a non-canonical nuclear entry pathway of the mitochondrial pyruvate dehydrogenase complex. Mol Cell. 2022;82:1066–77. e1067.

Zhang X, Blenis J, Li HC, Schindler C, Chen-Kiang S. Requirement of serine phosphorylation for formation of STAT-promoter complexes. Science. 1995;267:1990–4.

Zhang F, et al. EIN2 mediates direct regulation of histone acetylation in the ethylene response. Proc Natl Acad Sci USA. 2017;114:10274–9.

Zheng L, Roeder RG, Luo Y. S phase activation of the histone H2B promoter by OCA-S, a coactivator complex that contains GAPDH as a key component. Cell. 2003;114:255–66.

Zhou HX, Rivas G, Minton AP. Macromolecular crowding and confinement: biochemical, biophysical, and potential physiological consequences. Annu Rev Biophys. 2008;37:375–97.

Zhu X, et al. Stimulating pyruvate dehydrogenase complex reduces itaconate levels and enhances TCA cycle anabolic bioenergetics in acutely inflamed monocytes. J Leukoc Biol. 2020;107:467–84.

Chapter 10
Enzyme Assemblies in Nucleotide Metabolism: Structure, Regulation, and Disease Implications

Jack P. Boylan, Timothy D. Iles, Alexis Nguyen, and Anthony M. Pedley ⓘ

Abstract Nucleotide biosynthesis is essential for cell growth and relies on the coordination of both salvage and de novo pathways to satisfy intracellular needs. While traditionally, the regulation of these pathways has been attributed to factors like substrate availability, genetic rewiring, and metabolite-driven feedback inhibition, recent findings have unveiled a new, more complex layer of regulation. Enzymes within these pathways assemble into dynamic supramolecular protein assemblies, such as metabolons and filaments, which findings suggest might influence enzymatic activity and regulate nucleotide flux. Advances in fluorescence microscopy and cryo-electron microscopy have enhanced our ability to characterize the spatial and temporal dynamics, as well as the biophysical properties, of these assemblies. In this chapter, we explore the diverse higher-order purine and pyrimidine metabolic enzyme assemblies, highlight how state-of-the-art microscopy has transformed our understanding of their structures and regulatory roles in nucleotide biosynthesis, and discuss their implications in human disease.

Jack P. Boylan and Timothy D. Iles contributed equally with all other contributors.

J. P. Boylan
Molecular, Cellular, and Integrative Biosciences Graduate Program, The Huck Institutes of the Life Sciences, The Pennsylvania State University, University Park, PA, USA

T. D. Iles
Department of Biochemistry and Molecular Biology, Roy J. and Lucille A. Carver College of Medicine, University of Iowa, Iowa City, IA, USA

A. Nguyen
Department of Biochemistry and Molecular Biology, The Pennsylvania State University, University Park, PA, USA

A. M. Pedley (✉)
Department of Biochemistry and Molecular Biology, Roy J. and Lucille A. Carver College of Medicine, Holden Comprehensive Cancer Center, Fraternal Order of Eagles Diabetes Research Center, University of Iowa, Iowa City, Iowa, USA
e-mail: anthony-pedley@uiowa.edu

A. M. Pedley (ed.), *Supramolecular Protein Assemblies In Cells*, Advances in Experimental Medicine and Biology 1514,
https://doi.org/10.1007/978-3-032-26629-3_10

Keywords Metabolism · Nucleotide · Purine · Pyrimidine · Regulation · Protein complexes · Biomolecular condensates · Metabolon · Filaments

10.1   Introduction

Cellular metabolism is the collection of chemical reactions cells use to generate the necessary biomolecules required for its growth and proliferation. Among these biomolecules are nucleic acids, classified by either their purine (adenine, guanine) or pyrimidine (cytosine, thymidine, uracil) base. Commonly, these bases are enzymatically attached to sugar-phosphate backbones to form nucleotides that are essential for the synthesis and repair of genetic material.

Beyond their role as DNA and RNA building blocks, nucleotides serve as energy carriers, signaling molecules, and cofactors essential for various enzymatic functions. The nucleotide adenosine 5′-triphosphate (ATP) is particularly indispensable for powering otherwise slow and thermodynamically unfavorable reactions; thus, it is often termed the "energy currency" of the cell. Each cell can use up to ten million ATP molecules per second to support key functions such as signal transduction, active transport, and cell motility (Flamholz et al. 2014). On an organismal and tissue level, nucleotide monophosphates can adopt cyclic forms, such as cyclic adenosine 5′-monophosphate (cAMP) and cyclic guanosine 5′-monophosphate (cGMP) to function as secondary messengers that regulate hormone and neurotransmitter signaling (Zaccolo et al. 2021; Sutherland and Rall 1958; Hardman et al. 1971) as well as smooth muscle relaxation and vasodilation (Hardman et al. 1971; Hardman 1984; Murray 1990). Last, nucleotides can serve as cofactors to enhance enzymatic activities necessary for the transformation of biomolecules and to balance the cell's overall redox state. It is estimated that 13% of the human proteome uses purine cofactors such as nicotinamide adenine dinucleotide (NAD) and acetyl coenzyme A (Haystead 2006; Murray and Bussiere 2009), while pyrimidine cofactors such as thiamine pyrophosphate are heavily used for carbohydrate, glycogen, and lipid metabolism, respectively.

For decades, researchers have questioned how individual enzymes in multistep metabolic processes coordinate their activities in a complex and dynamic cellular environment for efficient biomass and energy production. Emerging evidence within nucleotide metabolism has started to shed light on this mystery by revealing that purine and pyrimidine biosynthetic enzymes compartmentalize into higher-order assemblies (Chitrakar et al. 2017; Schmitt and An 2017; Sweetlove and Fernie 2018). These entities resemble biomolecular condensates in the form of metabolons and filamentous structures, each with distinct properties and regulatory roles. This chapter explores the various structural forms that nucleotide metabolic enzymes can adopt, their role in regulating nucleotide metabolism, and their connection to human disease.

10.2 Nucleotide Metabolic Pathways

Nucleotides are produced by two complementary pathways, de novo and salvage biosynthesis, where the extent that each pathway is utilized is reflected by the overall cellular demand. Under normal physiological conditions, nucleotide salvage pathways are dominant (Lane and Fan 2015; Murray 1971; Nyhan 2005). However, when demand is increased, often the salvage pathways become insufficient, and cells activate and rely on the energy-intensive de novo pathways to maintain nucleotide pools (Lane and Fan 2015; Pareek et al. 2021a; Villa et al. 2019).

The synthesis of nucleotides is directly influenced by other metabolic pathways, such as glycolysis, the citric acid cycle, and the folate cycle. These other pathways provide necessary precursors, substrates, and cofactors for nucleotide metabolism. Essential for the formation of nucleotides is the production of phosphoribosyl pyrophosphate (PRPP), the sugar phosphate backbone that the nitrogenous base is assembled onto (Fox and Kelley 1971; Hove-Jensen et al. 2017). For PRPP production, cells direct glucose-6-phosphate from glycolysis through the pentose phosphate pathway to generate ribose 5′-monophosphate (R5P). R5P is further converted into PRPP through an ATP-driven reaction catalyzed by phosphoribosyl pyrophosphate synthetase (PRPS, EC 2.7.6.1) (Tatibana et al. 1995). Successful nucleotide synthesis also relies on the adequate supply of substrates such as amino acids (glutamine, glycine, and aspartate), bicarbonate, and ATP (Ahn and Metallo 2015; Spinelli and Haigis 2018; Tedeschi et al. 2013). Although redundant cytoplasmic processes can also produce these vital elements, most are either produced and exported from mitochondria or acquired from the cellular microenvironment (Tibbetts and Appling 2010). Additionally, mitochondria produce the necessary cofactors 10-formyl-tetrahydrofolate (10f-THF) and NAD+ for de novo purine synthesis and coenzyme Q (ubiquinone) for de novo pyrimidine synthesis. The 10f--THF cofactor is generated after mitochondria-derived formate is released into the cytoplasm and integrated into tetrahydrofolate by MTHFD1 (EC 6.3.4.3) (Ducker and Rabinowitz 2017; Petrova et al. 2023; Yang and Vousden 2016), whereas NAD+ is most commonly generated from tryptophan in the liver or from dietary nicotinic acid (Katsyuba and Auwerx 2017). For pyrimidines, the essential cofactor, coenzyme Q, is localized to the inner mitochondrial membrane and is an essential component in the electron transport chain (Alcazar-Fabra et al. 2016). Therefore, it is not surprising that nucleotide metabolism is highly sensitive to the downregulation of mitochondria-localized metabolic processes such as those encountered during hypoxic growth conditions (Chen et al. 2005; Doigneaux et al. 2020; Wheaton et al. 2011).

In the following section, we will cover the purine and pyrimidine nucleotide salvage and de novo biosynthetic pathways. While many of these pathways are conserved across all living species, we will primarily focus our attention on the biochemical transformations and enzymes observed in humans. For detailed discussions on the differences among organisms, we highly recommend several other great review articles on this topic (Ayoub et al. 2024; Chua and Fraser 2020; Moffatt and Ashihara 2002).

10.2.1 Nucleotide Salvage Synthesis

Nucleotide salvage refers to the process where free nucleic acids and/or nucleosides are recycled back into their corresponding nucleotide monophosphate. For free bases, this process is often assisted by base-specific phosphoribosyl transferases, where the free base is incorporated onto PRPP (Murray 1971; Nyhan 2005). In purine salvage, hypoxanthine-guanine phosphoribosyl transferase (HPRT1, EC 2.4.2.8) converts hypoxanthine and guanine to inosine 5′-monophosphate (IMP) and guanosine 5′-monophosphate (GMP), respectively, while adenine phosphoribosyl transferase (APRT, EC 2.4.2.7) generates AMP. Similarly, the pyrimidine base uracil can be readily converted into uridine 5′-monophosphate (UMP) by uracil phosphoribosyl transferase (UPRT, EC 2.4.2.9).

Additionally, free uracil and thymine can be transformed into their nucleotide monophosphate in a two-step mechanism. First, a pyrimidine nucleoside phosphorylase exchanges the phosphate group on ribose 1-phosphate (or the 2′-deoxy form for thymine) with a free base to generate the corresponding nucleoside. The nucleoside then is phosphorylated on the 5′-hydroxyl by either uridine-cytidine kinase (uridine, cytidine, EC 2.7.1.48) or thymidine kinase (thymidine, EC 2.7.1.21) to afford the pyrimidine monophosphate. Additionally, cytidine can be irreversibly deaminated into uridine. It should be noted that a similar processing of purine nucleotide monophosphates from their corresponding nucleoside is not as well established in mammalian systems aside from the phosphorylation of adenosine into AMP by adenosine kinase (EC 2.7.1.20) (Vannoni et al. 2004).

10.2.2 De Novo Purine Biosynthesis

The de novo biosynthesis of purines in humans is an energy-intensive ten-step process, requiring five molecules of ATP to produce one molecule of IMP from PRPP (Fig. 10.1). First characterized by Buchannan and Greenberg (Buchanan and Hartman 1959; Goldthwait et al. 1956; Greenberg 1951; Hartman and Buchanan 1959), these transformations are catalyzed by six enzymes and have been the subject of many detailed review articles (Lane and Fan 2015; Pareek et al. 2021a; Pedley and Benkovic 2017). Briefly, the first committed step in the pathway is the amidation of PRPP into 5-phosphoribosyl amine (5-PRA) by PPAT (EC 2.4.2.14). 5-PRA is then converted into GAR by the first of three nonsequential reactions catalyzed by the GART enzyme (phosphoribosyl glycinamide synthetase or GARS domain, EC 6.3.4.13). In the third step, the transformylase (GAR Tfase, EC 2.1.2.2) domain of GART catalyzes the transfer of a formyl group from 10f-THF, derived

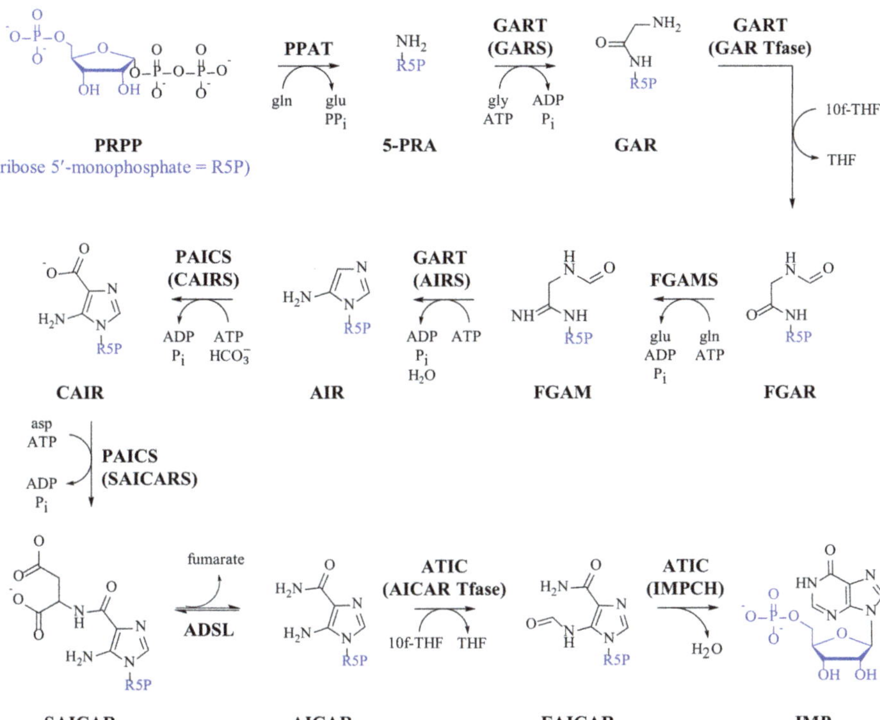

Fig. 10.1 The de novo purine biosynthetic pathway. In humans, the de novo purine biosynthetic pathway converts PRPP into IMP in a ten-step reaction cascade catalyzed by six enzymes. Enzyme names are in presented in bold font above the arrows of the reaction(s) they catalyze. In the case of multifunctional enzymes, the specific domains responsible for the transformation are shown in parentheses. The common names of pathway intermediates are presented in bold font under their respective chemical structure. Atoms in the structures of pathway intermediates colored blue represent the ribose 5′-monophosphate (R5P) backbone

from mitochondria-exported formate, to GAR, resulting in the production of formyl-GAR (FGAR). Then FGAR is handed off to FGAMS (also referred to as PFAS, EC 6.3.5.3), a glutamine amidotransferase that shuttles an ammonia molecule from its glutaminase domain to the synthetase domain via a proposed intramolecular tunnel (Massiere and Badet-Denisot 1998; Raushel et al. 1999) (Fig. 10.3), where FGAR is amidated to produce FGAM. FGAM is then transferred back to GART to catalyze its third and final transformation. In this transformation, the phosphoribosyl amino-imidazole synthetase (AIRS, EC 6.3.3.1) domain of GART produces AIR. The next two steps in the pathway are performed by the AIR carboxylase (EC 4.1.1.21) and SAICAR synthetase (EC 6.3.2.6) domains of PAICS to generate SAICAR, which then is reversibly converted into AICAR by ADSL (EC 4.3.2.2). The final two steps in de novo purine biosynthesis are catalyzed by ATIC. First, its transformylase (AICAR Tfase, EC 2.1.2.3) domain uses the 10f-THF cofactor to formylate AICAR generating formyl-AICAR (FAICAR) followed by the IMP cyclo-hydrolase

(IMPCH, EC 3.5.4.10) domain rapidly cyclizing FAICAR to generate IMP, the precursor to purine nucleotides AMP and GMP.

IMP is further transformed into either AMP or GMP through the adenylate or guanylate biosynthetic branch, respectively (Hershfield and Seegmiller 1976; Hinzpeter et al. 2019). In the adenylate branch, adenylosuccinate is generated from IMP by ADSS (EC 6.3.4.4) followed by the incorporation of fumarate in a second reaction catalyzed by ADSL to generate AMP. The guanylate branch is also a two-step process. Here, IMP dehydrogenase (IMPDH, EC 1.1.1.205) uses NAD+ as an electron donor to generate xanthosine 5′-monophosphate (XMP), which is further amidated into GMP by the glutamine amidotransferase GMP synthetase (GMPS, EC 6.3.5.2) in a similar mechanism as FGAMS (Oliver et al. 2014). While the kinetics of the individual enzymes downstream of IMP would suggest that the guanylate branch would be preferred (Zhao et al. 2015), most tissues showed the opposite (Pizzichini et al. 1989). This discrepancy has been postulated to be a result of a proposed feedback inhibitory mechanism (Hershfield and Seegmiller 1976).

10.2.3 *De Novo Pyrimidine Biosynthesis*

The de novo pyrimidine biosynthetic pathway consists of six steps carried out by three enzymes in humans (Fig. 10.2). The first three steps in the pathway are catalyzed by the cytosolic trifunctional CAD protein, a hexamer possessing carbamoyl phosphate synthetase (CPSase, EC 6.3.5.5), aspartate transcarbamylase (ATCase, EC 2.1.3.2), and dihydroorotase (DHOase, EC 3.5.2.3) activities (Li et al. 2021). For efficient production of carbamoyl phosphate, reaction intermediates in CPSase are passed through intramolecular tunnels connecting active sites (Thoden et al. 1997). Carbamoyl phosphate is then converted into carbamoyl aspartate by ATCase and further acted upon by the zinc metalloenzyme DHOase, which catalyzes the reversible condensation reaction between carbamoyl aspartate and dihydroorotate (Kelly et al. 1986).

The product of CAD, dihydroorotate, is then transported into the mitochondrial inner membrane space where dihydroorotic dehydrogenase (DHODH, EC 1.3.5.2) oxidizes dihydroorotate into orotate through the reduction of coenzyme Q (Chen and Jones 1976; Jones 1980; Boukalova et al. 2020). Orotate is then released back into the cytoplasm and attached to available PRPP forming orotate 5′-monophosphate (OMP) via UMP synthase (UMPS or OPRT, EC 4.1.1.23, 2.4.2.10). The final step in the pathway uses UMPS again to decarboxylate OMP into UMP, the final product of de novo pyrimidine biosynthesis.

From UMP, other pyrimidine nucleotides such as cytosine 5′-triphosphate (CTP) and deoxythymidine 5′-monophosphate (dTMP) are produced. To maintain CTP levels, UMP is sequentially phosphorylated into uridine triphosphate (UTP) by UMP and UDP kinases (EC 2.7.4.22 and 2.7.4.6) and further processed by the glutamine amidotransferase CTP synthetase (CTPS, EC 6.3.4.2) to synthesize CTP. The generation of dTMP for DNA synthesis is a multistep process. First, the 2′-hydroxyl

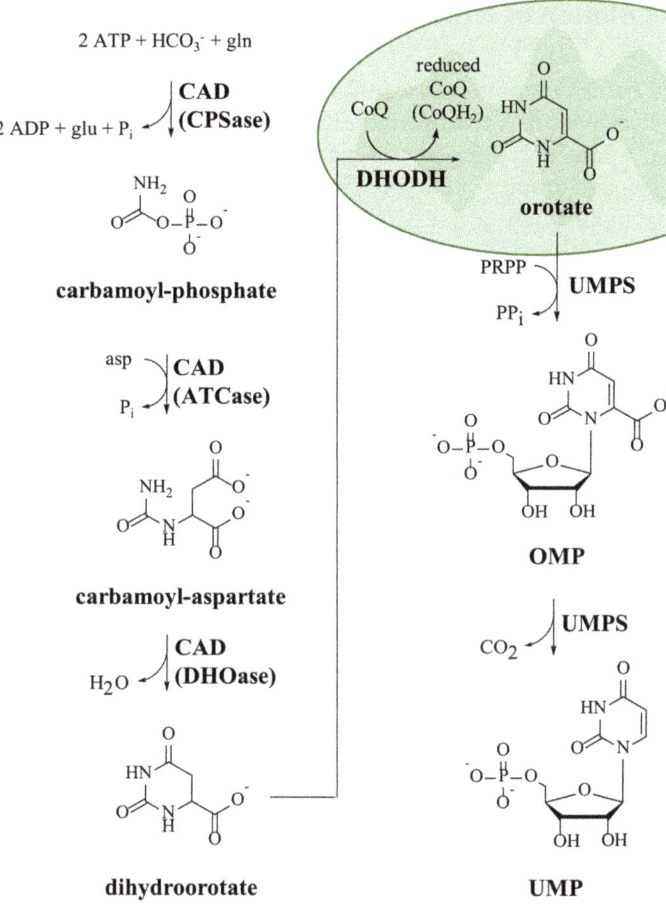

Fig. 10.2 The de novo pyrimidine biosynthetic pathway. In humans, the generation of UMP from ATP, bicarbonate, and glutamine is a six-step process catalyzed by three enzymes and relies on the efficient transport of dihydroorotate in and orotate out of mitochondria (green). Enzyme names are in presented in bold font above the arrows of the reaction(s) they catalyze. In the case of multifunctional enzymes, the specific domains responsible for the transformation are shown in parentheses. The common names of pathway intermediates are presented in bold font under their respective chemical structure

group on the ribose sugar of UDP is removed by ribonucleotide reductase (EC 1.17.4.1) resulting in dUDP. dUDP is phosphorylated again to generate dUTP and then readily dephosphorylated into dUMP by a dUTPase. dUMP is also made through the deamination of dCMP by dCMP deaminase (EC 3.5.4.12). dUMP is then converted into dTMP in a 5,10-methylene-THF-dependent methylation reaction catalyzed by thymidylate synthase. Additional phosphorylations of dTMP can occur to synthesize the DNA building block 2′-deoxythymidine 5′-triphosphate (dTTP).

10.3   Regulation of Nucleotide Metabolism

To maintain the appropriate balance of purines and pyrimidines for vital cell processes, similar conditions that increase purine demand also increase the need for pyrimidines. Anabolic cell growth during the G1 and S phases exhibit the highest nucleotide demand to support DNA replication (Fridman et al. 2013; Sigoillot et al. 2003) as exemplified by a fivefold increase in de novo purine biosynthesis in HCT116 human colorectal carcinoma cells (Fridman et al. 2013). Historically, the upregulation in nucleotide metabolism in these phases has been linked to heightened intracellular levels of rate-limiting substrates, such as phosphoribosyl pyrophosphate (PRPP), the precursor for both purine and pyrimidine biosynthesis (Jing et al. 2019). However, recent studies suggest that additional regulatory mechanisms, including posttranslational modifications and allosteric activation, also play significant roles in controlling nucleotide metabolism in a cell cycle-dependent manner. For instance, the upregulation of cyclin D1 during the G1 phase activates pyrimidine biosynthesis in hepatocytes via phosphorylation of the trifunctional CAD protein (Wu et al. 2023). This phosphorylated state of CAD has been shown to influence pyrimidine production resulting in increased proliferation. For example, MAPK-mediated phosphorylation of CAD in late G1 results in a 1.9-fold increase in BHK 165-23 baby hamster kidney cells (Sigoillot et al. 2003).

Additionally, the upregulation of certain oncogenes can increase nucleotide production. As a master regulator of cell metabolism, *MYC* has been shown to contribute to over 40% of human cancers (Dang et al. 2009). To enhance dNTP pools needed to fuel these cancers, c-Myc directly increases the availability of the PRPP precursor (Cunningham et al. 2014; Mannava et al. 2008) and various substrates (Liu et al. 2019a; Morrish et al. 2009; Nikiforov et al. 2002) as well as upregulate genes encoding enzymes within purine (*PPAT, PAICS, IMPDH1/2*) (Mannava et al. 2008; Agarwal et al. 2020; Barfeld et al. 2015; Wang et al. 2017) and pyrimidine (*CAD, DHODH, TYMS*) production (Mannava et al. 2008; Dong et al. 2020). Additionally, *KRAS* and *PTEN* mutations associated with highly fatal cancers are highly dependent on de novo pyrimidine biosynthesis, making these cancer cells sensitive to pathway inhibition (Santana-Codina et al. 2018; Wang et al. 2021).

Environmental stimuli, such as hypoxia and nutrient availability, can also trigger significant swings in nucleotide metabolism. Hypoxia inhibits many mitochondrial processes, thereby reducing the production of key substrates required for purine and pyrimidine biosynthesis (Chen et al. 2005; Doigneaux et al. 2020; Wheaton et al. 2011). Similarly, when faced with nutrient deficiencies such as glutamine deprivation, cells lack the nitrogen source for nucleotide metabolism (Yoo et al. 2020). In de novo purine biosynthesis, the lack of glutamine impairing the activity of the rate-limiting enzyme, PPAT and has shown to decrease small cell lung cancer growth (Kodama et al. 2020). Conversely, in high glucose environments, a common cancer adaptation, PRPP synthesis is heightened and thereby, increasing nucleotide levels (Ali and Ben-Sahra 2023). Together, these findings highlight our growing understanding of the complexities of nucleotide regulation in disease that align nucleotide production with the changes in cellular demand.

10.4 Supramolecular Protein Assemblies in Nucleotide Metabolism

Over the past two decades, there has been growing support for a new level of metabolic regulation in cells. In response to changes in metabolite demand, metabolic enzymes compartmentalize into higher-ordered structures. The discovery of this new level of regulation has rewritten how we view metabolic regulation in cells and reinvigorated the field of metabolism once again. Although best exemplified by enzymes in nucleotide metabolism, other higher-ordered structures comprised of human glycolytic (Kohnhorst et al. 2017; Puchulu-Campanella et al. 2013; Webb et al. 2017), tricarboxylic acid cycle (Bulutoglu et al. 2016; Velot et al. 1997; Wu and Minteer 2015; Wu et al. 2015), and serine biosynthetic enzymes (Rabattoni et al. 2023) have been observed.

Most of these higher-ordered metabolic enzyme assemblies are best defined as biomolecular condensates. These membrane-less organelles have been a topic of wide discussion in cell biology as their characterization and prevalence have become better appreciated (Banani et al. 2017; Bracha et al. 2019; Lyon et al. 2021). While the cellular advantage of forming metabolic condensates is not fully understood, we can envision scenarios where this compartmentalization can promote unique and ideal chemical environments that can either accelerate metabolic flux through a metabolic pathway or serve as a reservoir of essential proteins and/or metabolites that can be easily accessed to provide the cell with a quick source of biomolecules when needed. The argument that compartmentalization accelerates flux is founded on the principle that the enhanced proximity of pathway enzymes limits intermediate diffusion by promoting substrate channeling. The nucleolus is one such example with three levels of compartmentalization that vary in protein and ribosomal RNA (rRNA) composition as well as density and viscoelasticity (Brangwynne et al. 2011; Feric et al. 2016; Lafontaine et al. 2021). It has been proposed that the different surface tensions at each layer help pass along only properly assembled intermediates while the increased local concentration enhances the rate of reaction for generating the rRNA subunits (Feric et al. 2016). Many supramolecular assemblies comprised of enzymes outside of metabolism have also been best defined as reservoirs that can either sequester biomolecules for later use and/or protect the cell from accumulating toxic or reactive intermediates in the bulk cytosol. Stress granules are an example, where energy-intensive protein synthesis machinery sequestered and stalled during periods of stress and releasing them again when the stress has abated (Protter and Parker 2016).

Most condensates within metabolism adopt one of two types of structures: metabolons and filaments. In the following sections, we will discuss the discovery, properties, and regulation of several prominent examples of higher-ordered enzyme assemblies formed by purine and pyrimidine metabolic enzymes.

10.4.1 Metabolons

The earliest suggestion of metabolons is embedded in the Oparin-Haldane theory, where chemical reactions in the primordial soup are physically separated into coacervates (Haldane 1929; Oparin 1967). Fifty years later, investigations into mitochondrial protein complexes led biochemists to consider the possibility of multienzyme complexes within cellular metabolism whereby organizations of pathway enzymes could influence the rate of metabolic reactions (Reed 1974; Reed and Cox 1966). Wilson further demonstrated the existence of distinct pools of glycolytic and mitochondrial enzymes that have different kinetic properties between soluble and membrane-bound forms (Wilson 1978). Continued investigations into the partitioning of mitochondrial enzymes led Srere to reveal that enzymes within the tricarboxylic acid cycle assemble into metabolons or "supramolecular complex[es] of sequential metabolic enzymes and cellular structural elements" (Srere 1985). Metabolons have been observed in many metabolic processes; however, recent studies have suggested that metabolon formation is not conserved across all species (Narayanaswamy et al. 2009; Noree et al. 2019; O'Connell et al. 2014). Some of the best characterized metabolons are the purinosome in de novo purine biosynthesis (see Sect. 10.4.1.1), the glucosome (or G-bodies) in glycolysis (Kohnhorst et al. 2017; Puchulu-Campanella et al. 2013), and the pioneering metabolon assembly comprised of tricarboxylic acid cycle enzymes (Velot et al. 1997; Barnes and Weitzman 1986), with many new and exciting metabolons emerging such as the serinosome in serine biosynthesis (Rabattoni et al. 2023) and the pyrimidinosome in de novo pyrimidine biosynthesis (see Sect. 10.4.1.2).

Metabolons are highly dynamic condensates where their ability to form and dissolve has been correlated with changes in cellular cues and metabolic demands (Sweetlove and Fernie 2018). Therefore, the advantage of forming metabolons has been hypothesized to assist in the channeling of substrates between sequential enzymes in a metabolic pathway to enhance metabolite flux (Pareek et al. 2021b). Despite being first proposed over 40 years ago, metabolons have been proven to be difficult to reconstitute and characterize on a molecular level outside their cellular environment. Part of this can be attributed to their weak and transient associations, but it is also possible that there are unrecognized critical dependencies on their assembly. However, new innovative technologies and more detailed in vitro reconstitution protocols are starting to overcome these limitations to unlock a deeper understanding of metabolons and other condensates (Alberti et al. 2019; Deng and Wan 2024; He et al. 2022). In the following sections, the characterization of two metabolons in nucleotide metabolism, the purinosome (de novo purine biosynthesis) and the pyrimidinosome (de novo pyrimidine biosynthesis) will be discussed.

10.4.1.1   The Purinosome

First speculations that enzymes within the de novo purine biosynthetic pathway assemble into structures was proposed by Buchanan (Hartman and Buchanan 1959). Rowe and McCairns further substantiated this idea based on the co-purification of pathway enzymes from pigeon liver and human lymphocytes (McCairns et al. 1983; Rowe et al. 1978). The case for the formation of these complexes was significantly strengthened by a series of notable kinetic studies. First, a short half-life of the first intermediate in de novo purine biosynthesis, 5-PRA, was measured, and heavily suggested complexation must occur to facilitate substrate channeling and efficient IMP production (Schendel et al. 1988; Antle et al. 1996). To support this hypothesis, data derived from the enzyme kinetics of bacterial PPAT and GARS disagreed with a free diffusion model for 5-PRA suggesting an interaction between sequential enzymes in the pathway (Rudolph and Stubbe 1995). Additional speculation grew after more careful biochemical and structural investigations into the relationship between GART and FGAMS were performed (Zhang et al. 2012). GART catalyzes nonsequential steps (second, third, and fifth reactions) in de novo purine biosynthesis and must coordinate activities with FGAMS (fourth reaction) to transform 5-PRA into AIR. Combined with the hypothesized channeling of 5-PRA, a model was proposed that connected the first three enzymes within de novo purine biosynthesis (Fig. 10.3).

Although experimental findings hinted at direct interactions, no evidence of stable associations was found. Fluorescence microscopy later provided the first clues as to how these enzymes are organized within human cells. In these studies, pathway enzymes were expressed as fluorescent protein chimeras in nutrient depleted HeLa cells, an experimental condition known to activate de novo purine biosynthesis (An et al. 2008). All purine pathway enzymes co-clustered with FGAMS, and this clustering was reversible in response to purine supplementation. Since the original discovery, the composition of the purinosome has expanded to include enzymes downstream of IMP formation as well as a host of key regulators that have been extensively reviewed (Schmitt and An 2017; Sweetlove and Fernie 2018; Pareek et al. 2021a; Ayoub et al. 2024; Pedley and Benkovic 2017; Pedley et al. 2022a). Importantly, the clustering was deemed not an artifact of protein overexpression based on its inability to co-cluster with noted stress granule and aggresome markers (Chan et al. 2015; French et al. 2013) or a result of translational repression (Pedley et al. 2022b).

Purinosomes have been detected across many cell types (Doigneaux et al. 2020; An et al. 2008; Fu et al. 2015; Baresova et al. 2012; Mangold et al. 2018; Williamson et al. 2017; Yamada et al. 2020) and analyzed using a wide variety of techniques including immunofluorescence (Baresova et al. 2012, 2016, 2018; Pedley and Benkovic 2018), protein ligation assays (PLAs) (Doigneaux et al. 2020; Sha and Benkovic 2024), and mass spectral imaging (Pareek et al. 2020) (Fig. 10.4). These newly applied methods deviate from the traditional overexpression of fluorescent protein chimeras to allow for purinosome detection using endogenous proteins. Researchers at Charles University first paved the way forward by widely adopting

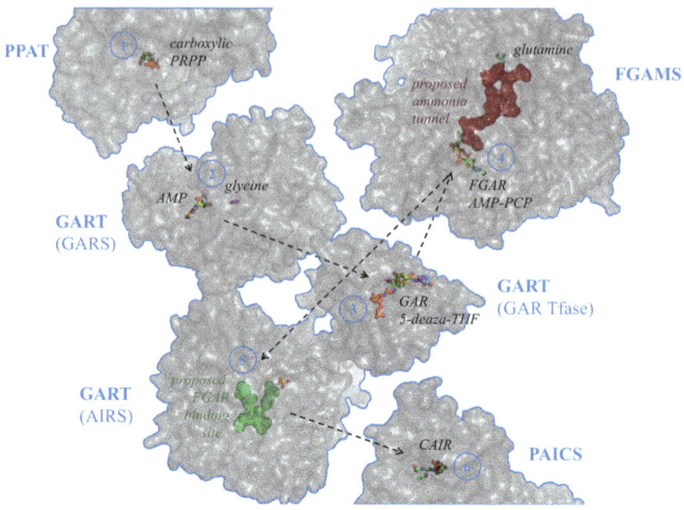

Fig. 10.3 Hypothesized model for substrate channeling between de novo purine biosynthetic enzymes. Kinetic arguments suggest that efficient purine production through the de novo purine biosynthetic pathway requires the coordination of sequential pathway enzymes (gray surface renderings: PPAT, GART, FGAMS, PAICS) to facilitate substrate (sticks rendering) channeling. The dashed lines indicate the transit of substrates between enzyme active sites including the use of an ammonia tunnel (red) in FGAMS. Human structural predictions for PPAT, GART, and FGAMS were obtained from AlphaFold (Jumper et al. 2021; Varadi et al. 2024). Pathway intermediates and cofactors were extracted from deposited PDB IDs after aligning the deposited protein X-ray structure to the human structural predictions or available human X-ray structure (PAICS) [PDB IDs: 1ECC (PPAT: carboxylic PRPP), 2YS6 (GARS: AMP and glycine), 1CDE (GAR Tfase: 5-deaza-tetrahydrofolate), 1CLI (AIRS: Asp65, Asp94, His190 to denote proposed active site), 2HS4 (FGAMS: FGAR, AMP-PCP), 1T3T (FGAMS: cysteine-glutamine crosslink), and 6YB8 (PAICS: CAIR)]. The placement of these ligands is to illustrate their relative binding pockets and might not reflect the true active site. All structures were rendered in PyMOL (Schrodinger 2024)

immunofluorescence to correlate purinosome formation with clinically relevant disease-causing mutations in PAICS, ADSL, and ATIC (Baresova et al. 2012, 2016, 2018; Weng et al. 2024). Other antibody-based detection strategies such as PLAs have also demonstrated enhanced localization (<40 nm) between pathway enzymes such as ADSL and FGAMS under cellular growth conditions favoring purinosome formation (Doigneaux et al. 2020; Sha and Benkovic 2024). Lastly, mass spectral imaging has been leveraged to identify localized hot spots of metabolic activity through the accumulation of the pathway intermediate AICAR in frozen-hydrated purine-depleted HeLa cells rather than looking for pathway enzyme colocalization (Pareek et al. 2020). To supplement many of the image-based detection approaches, biochemical evidence through split-complementation assays and co-immunoprecipitation, where associations are stabilized through the supplementation of excess pathway substrates, supports the notion of at least pairwise associations between pathway enzymes (He et al. 2022).

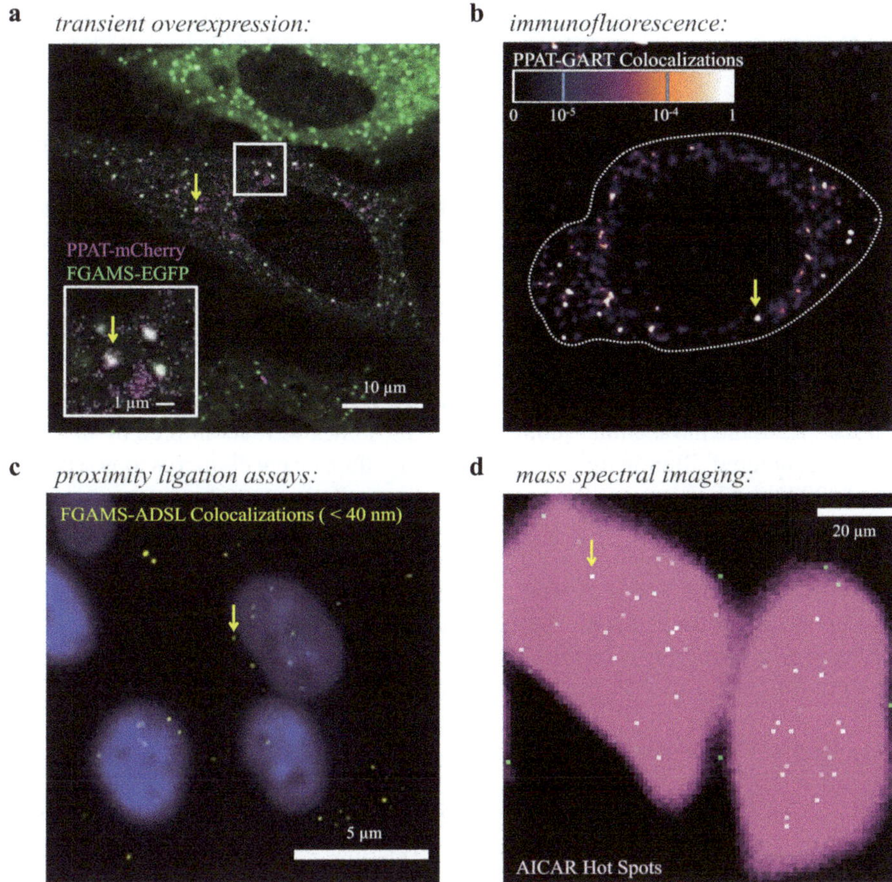

Fig. 10.4 Methods for purinosome detection. Since its discovery, purinosome detection has started to transition from (**a**) the traditional transient overexpressed detection method to now detect endogenous purinosome assemblies by (**b**) immunofluorescence, (**c**) protein proximity ligation assays, and (**d**) mass spectral imaging. The yellow arrows indicate a representative purinosome. (Figure Permissions: **a** was modified from Pedley and Benkovic (2018) and reproduced with permission from Springer Nature; immunofluorescence data used to generate **b** was originally published in Baresova et al. (2018) and reproduced with permission from V. Baresova and M. Zikanova; **c** was originally published in Doigneaux et al. (2020); and **d** was modified from Pareek et al. (2020) and reproduced with permission from the American Association for the Advancement of Science (AAAS/*Science*))

Purinosomes as Condensates

Purinosomes are largely ascribed to create unique microenvironments that allow for clustered channeling (Pareek et al. 2021b) between pathway enzymes. The compartmentalization allows the cell to overcome many kinetic and thermodynamic barriers as well as limit the diffusion of metabolite regulators such as SAICAR

(Keller et al. 2014; Yan et al. 2016) and AICAR (Corton et al. 1995; Kim et al. 2016) that can modulate other metabolic processes. Recently, the purinosome has been accepted as part of a growing number of membrane-less condensates observed within living cells (Bracha et al. 2019). In the following paragraph and section, we will review the efforts used to classify purinosomes as condensates and the proposed mechanism for how their liquid-like behavior can be regulated.

Three criteria were used to benchmark the purinosome against other reported biomolecular condensates (e.g., liquid droplets). First, the fluidity of the purinosome was revealed by fluorescence recovery after photobleaching (FRAP) where rapid exchange of and internal mixing of de novo purine biosynthetic enzymes within the purinosome was observed (An et al. 2008; Kyoung et al. 2015). Second, lattice light sheet fluorescence microscopy showed that the purinosome, as denoted by FGAMS-mCherry clusters, is spherical like other liquid droplets (Pedley et al. 2022b). Last, the intracellular motions of purinosomes were monitored by instantaneous structured illumination high-resolution fluorescence microscopy (Pedley et al. 2022b; Chan et al. 2018). In numerous instances, purinosomes displayed directed motion along microtubules and merged with static purinosomes forming a larger condensate (Pedley et al. 2022b). These characteristics have supported the claim that the purinosome as well as analogous metabolic assemblies (Sweetlove and Fernie 2018; Bracha et al. 2019; An et al. 2019) are part of the growing catalog of liquid-like condensates in cells.

Regulation of Purinosome Formation

Purinosome formation is a highly dynamic process that responds to changes in both extracellular purine availability and in overall cellular energy and biomass demands. Since its discovery, numerous cellular conditions have been shown to favor purinosome formation or its disassembly. A current listing of those factors that influence purinosome formation is summarized in Table 10.1.

The earliest investigations into the molecular signals that regulate purinosome formation used a shRNA loss-of-function kinase library in a dynamic mass redistribution (DMR) assay to evaluate those signaling pathways and kinases within them that alter purinosome formation (French et al. 2016; Verrier et al. 2011). In independent investigations, the DMR assay revealed a dependency on Gαi (via the α2A receptor) (Verrier et al. 2011), MAPK, and mTOR signaling with purinosome formation (French et al. 2016). While G-protein coupled receptor (GPCR) and mitogen-activated protein kinase (MAPK) signaling has largely been unexplored, mTOR activity has been correlated with purinosome-mitochondria association, whereby inhibition of mTOR by rapamycin resulted in a dose-dependent decrease in the percentage of purinosomes (measured as FGAMS assemblies) localized near mitochondria (French et al. 2016).

The observations made between cell signaling and purinosome formation suggest that enzyme compartmentalization might be driven by a phosphorylation event. In line with that hypothesis, a targeted posttranslational modification (PTM) study

Table 10.1 Cellular conditions that influence purinosome formation

Favoring purinosome formation	Ref(s)	Disrupting purinosome formation and/or stability	Ref(s)
Nutrient and oxygen availability		*Nutrient availability*	
Purine depletion (nutrient starvation)[a, b]	An et al. (2008)	Purine supplementation[a, b]	An et al. (2008)
Hypoxia[c]	Doigneaux et al. (2020)		
Genetic alterations		*Genetic alterations*	
HPRT deficiencies (e.g., Lesch-Nyhan disease)[a, b]	Fu et al. (2015)	In born error mutations in *PAICS, ADSL, ATIC*[a, b]	Baresova et al. (2012), Weng et al. (2024), and Pelet et al. (2019)
Activation of cell processes		*Inhibition of cell processes*	
Epinephrine (G$_{\alpha i}$ pathway)[a] (MAPK pathway)[a]	Verrier et al. (2011) French et al. (2016)	Nocodazole (microtubules)[a]	Chan et al. (2018) and An et al. (2010a)
PAICS polyubiquitination (recruitment of UBAP2)	Chou et al. (2023)	Rapamycin (mTOR)[a] STA9090, NVP-AUY922, 17-AAG (Hsp90)[a]	French et al. (2016) French et al. (2013) and Pedley et al. (2022b)
Inhibition of cellular processes			
GSK2334470, Gö6893 (Akt-independent PDK1 signaling)[a, d]	Schmitt et al. (2018)		
Casein kinase II	An et al. (2010b)		

The cellular conditions that either promote or disrupt purinosome formation are largely tied to nutrient and oxygen availability, genetic alterations resulting in enzyme deficiencies, and the modulation of cellular processes such as cell signaling, cytoskeletal formation, and proteostasis
[a]As observed by FGAMS overexpression as a purinosome marker
[b]As observed with endogenous protein detection methods
[c]Purinosome formation was observed but shown to be in an inactive state
[d]Purinosome core formation only

cataloged 174 different modifications across the 6 pathway enzymes under conditions that favor purinosome formation as well as those that result in its disassembly (Liu et al. 2019b). Over two-thirds of the modifications detected were only observed in one growth conditions, largely encompassing the reported serine or threonine phosphorylation modifications. One example is residue Thr397 on PPAT, where the AKT-mediated phosphorylation likely occurs under conditions that disfavor purinosome formation. The involvement of the PI3K/Akt signaling pathway on regulating de novo purine biosynthesis has been long established (Saha et al. 2014; Wang et al. 2009); however, evidence that pathway enzyme compartmentalization is dependent on this pathway has been inconclusive and additional studies need to be performed to better understand this relationship.

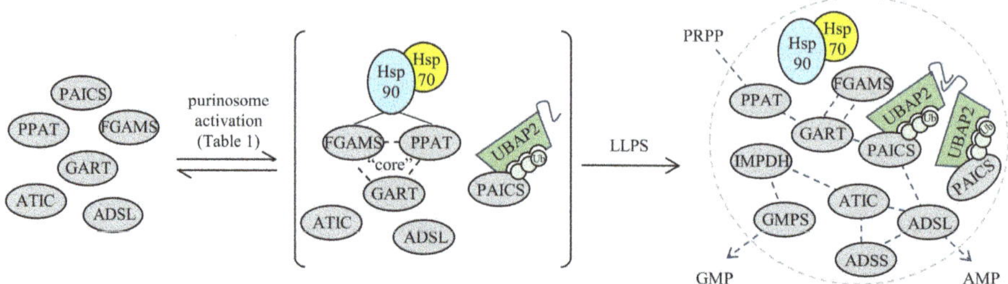

Fig. 10.5 The proposed mechanism(s) of purinosome assembly. The formation of purinosomes has often been correlated by activation of the de novo purine biosynthetic pathway through a variety of triggers outlined in Table 10.1. The initial trigger is hypothesized to form a "core" scaffold comprised of the first three pathway enzymes and might be regulated by molecular chaperones. Recently, a mechanism where the polyubiquitination of PAICS induced binding of UBAP2, an adapter protein with a C-terminal intrinsically disordered region (Chou et al. 2023). It is through this region that the liquid-liquid phase separation (LLPS) occurs. It is believed that the complete purinosome assembly encompasses all pathway enzymes, and their possible regulators, to efficiently convert PRPP into GMP and AMP

More recently, two additional mechanisms have been proposed to promote purinosome formation by inducing the liquid-liquid phase transition (Fig. 10.5). The first mechanism prescribes a new role for the molecular chaperones Hsp70 and Hsp90 in the direct regulation of purine metabolism. Two of the pathway enzymes, PPAT and FGAMS, were identified as clients of Hsp70 and Hsp90 (French et al. 2013; Pedley et al. 2018) and shown to localize to the purinosome in HeLa cells (French et al. 2013). Inhibition of Hsp90 chaperoning activity with ganetespib (STA9090) decreased client-Hsp90 interactions and enhanced client protein degradation through the proteasome (Pedley et al. 2018). Live cell imaging of the purinosome indicated that inhibition of Hsp90 activity also resulted in the time-dependent decrease in purinosome sphericity where the resulting entity was void of other purinosome enzymes (Pedley et al. 2022b). Combined, these results suggest that Hsp90 is serving a role in preserving the liquid-like properties and that inhibiting its function can transition the purinosome into a more aggregated insoluble state. The second reported mechanism is through the ubiquitination of PAICS. Recently, PAICS was shown to be polyubiquitinated by the cullin-5/ankyrin repeat and SOCS box containing 11(Cul5/ASB11) ubiquitin ligase (Chou et al. 2023). This ubiquitinated form of PAICS recruits ubiquitin-associated protein 2 (UBAP2) to promote the liquid-liquid phase transition, triggered by intrinsically disordered regions of UBAP2 as discussed in the next section (Sect. 10.4.1.1.3).

Proposed Stepwise Assembly of Purinosome Formation

Several cell-based assays have inferred that the assembly of the purinosome is a stepwise process and have hinted at possible scaffolds (Fig. 10.5). Early on, a modified Tango GPCR assay was adapted to monitor the pairwise associations between pathway enzymes under conditions that favor and disfavor purinosome formation (Deng et al. 2012). Here, significant associations were observed among the first three enzymes (PPAT, GART, FGAMS) within the pathway under purinosome-forming conditions, whereas the other remaining enzymes showed weaker associations among themselves as well as the first three enzymes. From this study, the first three enzymes to strongly associate were coined the purinosome "core," a sub-complex scaffold that facilitates its association with the downstream pathway enzymes. Several other reports have also further supported this notion of a purinosome "core." First, a detailed account of the activity of phosphoinositide-dependent protein kinase 1 (PDK1, EC 2.7.11.1) on purinosome formation showed that inhibition of PDK1 can induce the assembly of the purinosome "core" void of the other enzymes (Schmitt et al. 2018). Additionally, the presence of a core was inferred from calculated diffusion coefficients of pathway enzymes between the purinosome and the bulk cytoplasm. Similarities among enzyme diffusion coefficients suggested the likeliness that sub-complexes form and could migrate together to the purinosome (Kyoung et al. 2015). These sub-complexes included the "core" (PPAT-GART-FGAMS) as well as PAICS-ADSL. Together, these three investigations suggested that the associations among the "core" purinosome enzymes might be more stable than originally thought and further support the notion that regulation of purinosome formation might rest on the structure and/or function of those enzymes.

In addition to these studies, PAICS, an enzyme downstream of the first three pathway enzymes, has been recently proposed as a suitable scaffold. Outside of its ability to directly bind the intrinsically disordered protein UBAP2 (Chou et al. 2023), its multivalency could also arise from its homo-octameric structure (Li et al. 2007). Through co-immunoprecipitation studies, PAICS was shown to interact with other purinosome enzymes, regardless of cellular conditions, indicating that specific sub-complexes with PAICS might exist even prior to complete purinosome formation and, thereby, could be responsible for organizing the purinosome (He et al. 2022).

The early characterization of the purinosome composition has expanded to include purine biosynthetic enzymes downstream of IMP as well as a variety of ancillary proteins. Metabolic flux studies were the first to hint at the possibility that enzymes downstream of IMP were also part of the purinosome (Zhao et al. 2015), and later stable isotope incorporation assays designed to look at substrate channeling underscored their importance in enhancing the efficiency of making AMP and GMP (Pareek et al. 2020). IMPDH and ADSS were further shown to co-cluster with FGAMS to validate their inclusion in the purinosome (Zhao et al. 2015). In addition to metabolic enzymes, affinity purification mass spectrometry experiments revealed other possible ancillary proteins inside the metabolon, such as tubulin and mitochondrial membrane-bound proteins (French et al. 2013, 2016). It is still to be determined how these proteins are integrated into the process of formation and the inclusion of these proteins generated hypotheses about their subcellular localization.

Subcellular Localization

The finding that FGAMS under purinosome-forming growth conditions associates with mitochondrial proteins is not that surprising (French et al. 2013, 2016). FGAMS and other enzymes within the de novo purine biosynthetic pathway require a handful of substrates and cofactors, such as glycine, aspartate, glutamine, and formate (to generate the essential 10f-THF cofactor). Therefore, it has been widely hypothesized that purinosomes localize to the source of these substrates for efficient uptake of these materials to achieve higher metabolic flux.

The relationship between purinosomes and mitochondria was first probed by three-dimensional stochastic optical reconstruction microscopy (3D STORM) (French et al. 2016). Colocalization analysis showed a high degree of association in both purine-depleted HeLa cells (~ 65%) (French et al. 2016) and dermal fibroblasts derived from patients deficient in purine salvage (~81%) (Chan et al. 2018) (Fig. 10.6a). Using PLA, a tenfold increase in the number of HeLa cells showing associations between FGAMS and the outer mitochondrial pre-protein import protein TOMM20 upon purine-depletion was observed (Doigneaux et al. 2020). In addition to FGAMS, ADSL also co-purified with isolated mitochondria extracted from purine-depleted HeLa cells (French et al. 2016) further supporting the hypothesis that the complete purinosome is mitochondria-localized.

Recent efforts in identifying the direct interaction between mitochondria and purinosomes that stabilizes their subcellular localization showed purinosomes to be localized near the mitochondrial transporters SLC25A13 and SLC25A38, which distribute pathway substrates aspartate/glutamate and glycine, respectively from the inner mitochondria membrane space (Sha and Benkovic 2024). In purine-deficient conditions, roughly 72% of cells were positive for a PFAS-ADSL PLA signal, compared to 43% in purine-rich conditions. Following SLC25A38 knockdown, only 28% of cells retained a positive association under purine-deficient conditions. This

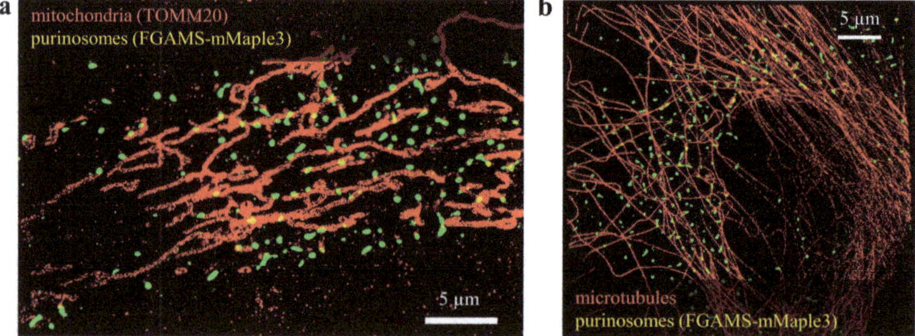

Fig. 10.6 Purinosomes (FGAMS-mMaple3) colocalize with (**a**) mitochondria (TOMM20) and (**b**) microtubules (SiR-tubulin) in HPRT-deficient human fibroblasts. Purinosomes are shown in green, whereas mitochondria or microtubules are red. (Images were originally published in Chan et al. (2018) and reproduced with permission from The National Academy of Science (Copyright 2018))

suggests a mechanism where the direct transfer of mitochondria-derived substrates into the purinosome can stabilize purinosome assembly and increase its processing of purine nucleotides.

Purinosome-mitochondria associations have also been shown to be stabilized through their association with microtubules. Almost all purinosomes colocalize with microtubules (Chan et al. 2018; An et al. 2010a) (Fig. 10.6b) and three-color imaging of purinosomes, mitochondria, and microtubules revealed that purinosomes nestle at the microtubule-mitochondria interface (Chan et al. 2018). Purinosomes not at the interface were shown by time lapse imaging to be trafficked along microtubules toward mitochondria. Further, time-dependent destabilization of the purinosome-mitochondria association was observed when microtubules were depolymerized via nocodazole treatment. This disruption was previously reported to result in a 36% decrease in purine production and a loss of the purinosome foci phenotype (An et al. 2010a).

10.4.1.2 The Pyrimidinosome

Since the discovery of the purinosome, there has been speculation as to whether an analogous metabolon comprised of enzymes within de novo pyrimidine biosynthesis is formed. One of the hurdles in this logic is that the second enzyme in the pathway, DHODH, is localized in the inner mitochondrial membrane, requiring its substrate and product to shuttle across mitochondrial membranes (Fig. 10.2). Recently, a series of microscopy, mass spectrometry, co-immunoprecipitation, and stable isotope tracing analyses has suggested the presence of such an assembly to coordinate the biochemical transformations to efficiently produce UMP (Yang et al. 2023). In the following sections, we will briefly present our current understanding of this metabolon, coined the pyrimidinosome, in human cell lines.

Pyrimidinosome Composition and Proposed Metabolite Flux

The spatial arrangement and composition of the pyrimidinosome were determined through co-fractionation of HEK293T cell lysate by size exclusion chromatography and sucrose density gradient centrifugation assays (Yang et al. 2023). These data showed the pyrimidinosome was comprised of at least five proteins: the three pathway enzymes that catalyze de novo pyrimidine biosynthesis (CAD, DHODH, UMPS), glutamine oxaloacetate transaminase 1 (GOT1, EC 2.6.1.1), and voltage-dependent anion-selective channel protein 3 (VDAC3) (Fig. 10.7). GOT1 catalyzes the reversible reaction converting glutamate and oxaloacetate into aspartate and α-ketoglutarate. The cytosolic GOT1 produced aspartate is then transferred to the ATCase domain of CAD through a direct association, where it is integrated into the carbamoyl aspartate substrate needed for DHO production. To facilitate transport of DHO into the inner mitochondrial membrane space, CAD associates with VDAC3, where DHODH transforms it into orotate. Orotate is then shuttled back through

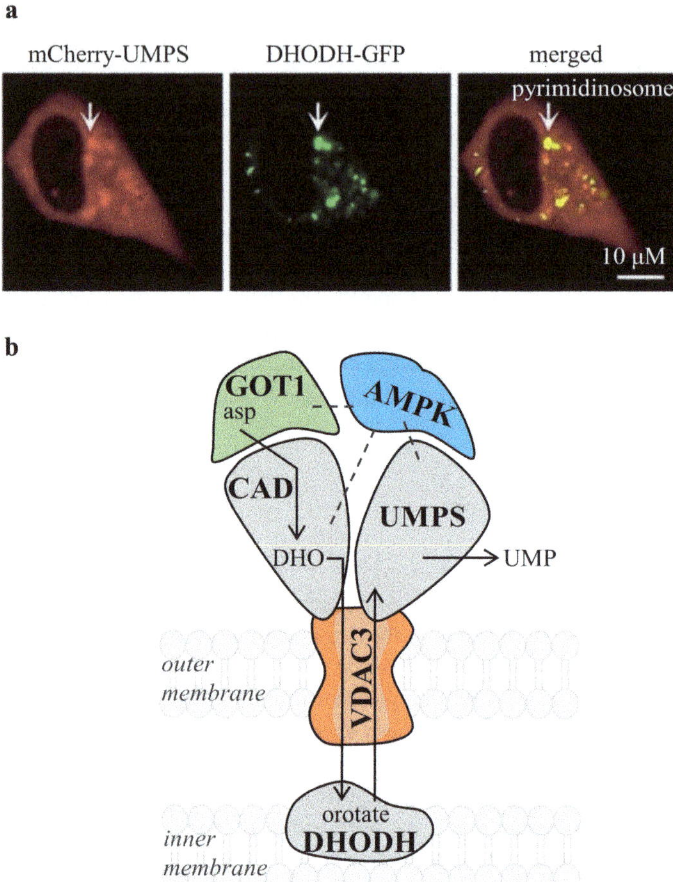

Fig. 10.7 The proposed model of the pyrimidinosome. (**a**) Fluorescently labeled proteins in the de novo pyrimidine biosynthetic pathway (mCherry-UMPS and DHODH-GFP) co-cluster into discrete granular bodies called pyrimidinosomes. (Images were originally published in Yang et al. (2023) and reproduced with permission from Springer Nature (Copyright 2023)). (**b**) Biochemical studies showed that the pyrimidinosome consists of all three de novo pyrimidine biosynthetic enzymes (CAD, DHODH, and UMPS), GOT1, and the outer mitochondrial transporter VDAC3. The assembly and regulation of UMP production are influenced by AMPK, which interacts with GOT1, the dihydroorotase domain of CAD, and UMPS

VDAC3 and received by cytosolic UMPS to ultimately make the product of the pathway, UMP. Depletion of extracellular pyrimidines increased GOT1-CAD and UMPS-DHODH associations and induced clustering of these associations with exogenous labeled VDAC3 in both biochemical and microscopy analyses. Therefore, these results suggest that VDAC3 has both a functional and structural role in the pyrimidinosome.

Pyrimidinosome Regulation and Insights Into Its Function

The catalytic subunit of AMPK was shown to directly interact with CAD through its DHO domain (Yang et al. 2023). Additionally, biochemical co-immunoprecipitation assays revealed that AMP-activated protein kinase (AMPK, EC 2.7.11.1) subunits also bind the N-terminal region of GOT1 and the OPRT domain of UMPS. During energy stress, AMPK is activated by phosphorylation, leading to its dissociation from the pyrimidinosome. This release of AMPK increases the de novo biosynthesis of dihydroorotate and orotate; however, this this does not directly correlate with an increase in UMP levels. The buildup of intracellular dihydroorotate and orotate is a direct consequence of UMPS inactivation upon AMPK dissociation and has shown to enhance the DOHD-mediated ferroptosis defense mechanism (Yang et al. 2023).

10.4.2 Cytoophidia (Filamentous Structures)

In addition to phase-separated metabolons, metabolic enzymes have also been observed to assemble into dynamic filamentous structures (Park and Horton 2019). Among the earliest identified examples are IMPDH and CTPS, enzymes which catalyze critical steps in the de novo purine and pyrimidine biosynthesis pathways, respectively. Early observations described these structures as unknown "rods and rings" measuring approximately 2–10 µm in the cytoplasm and less than 3 µm in the nucleus of cells from hepatitis C patients (Carcamo et al. 2011; Covini et al. 2012; Juda et al. 2014). Five years later, in 2010, similar IMPDH and CTPS structures were observed in *Escherichia coli*, yeast, and Drosophila (Ingerson-Mahar et al. 2010; Liu 2010; Noree et al. 2010). Due to their snake-like appearance, the larger structures were named "cytoophidia" derived from Greek for cellular snake (Liu 2010).

Since their discovery, cytoophidia have been observed in a variety of other organisms, including zebrafish (Chang et al. 2021), mouse embryonic stem cells (Carcamo et al. 2011; Chang et al. 2015; Peng et al. 2024), and both cancerous (Chang et al. 2017) and noncancerous human cells (Calise et al. 2018; Duong-Ly et al. 2018). As originally discovered, these structures typically arise in response to changing nucleotide demands, such as nutrient starvation or cellular proliferation (Chang et al. 2015; Gou et al. 2014; Petrovska et al. 2014; Wu and Liu 2019; Zhang and Liu 2024). This suggests that cytoophidia play a meaningful role in regulating nucleotide biosynthesis (Hvorecny and Kollman 2023; Lynch et al. 2020; Yin et al. 2024). Cytoophidia of CTPS and IMPDH can form independently of each other, and the assembly of one is not required for the other. However, the two are often found aligned side-by-side, and the overexpression of CTPS increases the number of IMPDH cytoophidia and might be the result of both structures responding to fluctuating nucleotide demand (Chang et al. 2018). Super-resolution imaging of these cytoophidia has suggested that these structures are reticular and might be interwoven CTPS and IMPDH filaments as well as provide a scaffold for other proteins (Chang et al. 2018; Fang et al. 2022).

In the following sections, we will focus on these filament-forming metabolic enzymes, examining how their supramolecular assemblies form and how they are regulated to confer nucleotide synthesis.

10.4.2.1 IMPDH Filaments

Inosine monophosphate dehydrogenase (IMPDH) catalyzes the conversion of IMP into XMP, the first committed and rate-limiting step in the guanylate branch of purine biosynthesis (Fig. 10.8a). In humans, there are two ubiquitously expressed isoforms of IMPDH, IMPDH1 and IMPDH2 (Natsumeda et al. 1990). While IMPDH1 is typically expressed at low levels in most cells, IMPDH2 is often the dominant isoform (Senda and Natsumeda 1994) and is expressed at higher levels in many forms of cancer where there is heightened IMPDH activity (Jackson et al. 1975; Nagai et al. 1991, 1992; Xu et al. 2020).

In response to changes in GTP demand, IMPDH readily forms filaments. They can be induced by glutamine or folate starvation (Calise et al. 2016) and dissolve during embryonic tissue differentiation or when the demand for GTP subsides (Peng et al. 2024; Calise et al. 2018). While IMPDH is conserved across species, these structures have only been observed in vertebrates, most notably in mouse embryonic stem cells and human T cells during an immune response (Carcamo

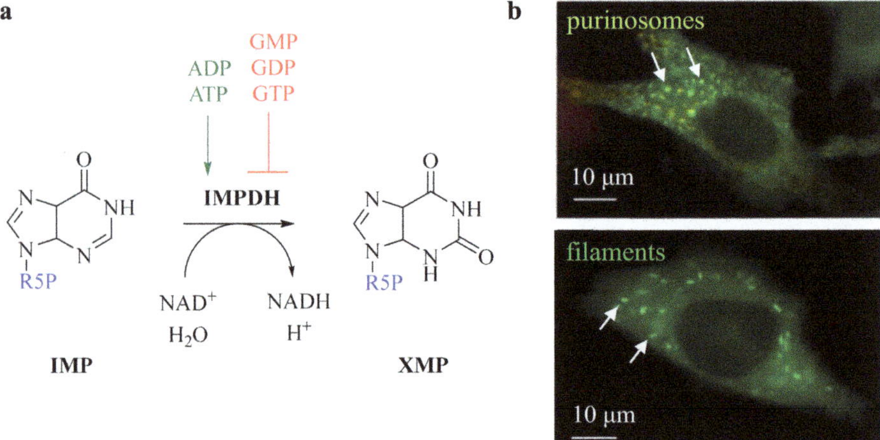

Fig. 10.8 IMPDH assemblies in HeLa cells. (**a**) IMPDH catalyzes the conversion of IMP to XMP. Filament formation can be promoted through the binding of adenine nucleotides (ADP, ATP) and feedback inhibited with guanine nucleotides (GMP, GDP, GTP). (**b**) Transient overexpression of IMPDH2 in purine-depleted HeLa cells can be present in both purinosomes as denoted by their co-clustering with FGAMS and (**b**) filamentous structures. Arrows point to representative structures. (Images were modified from Zhao et al. (2015))

et al. 2011; Calise et al. 2018). The intracellular localization of IMPDH filaments suggest that these assemblies are not enclosed in membranes, associated with organelles such the Golgi apparatus, mitochondria, or the endoplasmic reticulum, and are not comprised of cytoskeletal proteins known to form filaments (Carcamo et al. 2011; Juda et al. 2014; Ji et al. 2006). Although IMPDH has been reported to be a component of other condensates such as the purinosome (Zhao et al. 2015), fluorescence microscopy studies confirmed that these filaments are distinct assemblies and can exist in both the cytoplasm and nucleus (Ahangari et al. 2021; Woulfe et al. 2024) (Fig. 10.8b).

Unlike metabolons, our understanding of the filamentous structures of IMPDH has greatly benefited from the ability to reconstitute these assemblies outside of cells. As a result, a vast array of structural data has been generated to gain a better appreciation for the structural features that might regulate guanidine nucleotide production. In the following sections, we will present the key features of these higher-ordered IMPDH structures to propose how these structures might directly affect nucleotide metabolism.

Regulation of Higher-Ordered IMPDH States

IMPDH consists of 514 amino acids folded into 2 domains: a catalytic domain and a regulatory Bateman domain, connected by a flexible linker (Hedstrom 2009). IMPDH monomers readily assemble into tetramers through interactions in their catalytic domains, and these tetramers can further associate into octamers (a dimer of tetramers) through associations between their Bateman domains (Buey et al. 2017; Labesse et al. 2013). These octameric units can then be incorporated into filaments and larger bundle structures (Chang et al. 2022).

The Bateman domain in humans has three canonical nucleotide binding sites that assist in regulating the catalytic activity of IMPDH (Buey et al. 2022; Fernandez-Justel et al. 2019) (Fig. 10.9a). Adenine and guanine nucleotides compete for binding to Site 1 and Site 2. When ATP occupies both sites, the octamer undergoes two structural shifts. First, the octamer is extended to create a hollow interior exposing the catalytic domains to the bulk solvent. Second, each tetramer is flattened relative to the fourfold symmetry axis (Johnson and Kollman 2020). These changes align key catalytic residues to create the fully active IMPDH octamer. GTP feedback inhibits IMPDH activity exclusively by binding Site 3. A high degree of cooperation exists between Site 2 and Site 3, so when GTP rather than ATP binds site 2, an allosteric change promotes the binding of GTP to Site 3. The presence of GTP at both Site 2 and Site 3 causes the octamer to compress, and the tetrameric unit becomes bent. The bowing of the tetramers and the compression of the octamer contribute to the full inhibition of IMPDH by GTP (Buey et al. 2022; Johnson and Kollman 2020). This feedback mechanism helps maintain the proper nucleotide levels within the cell, whereby disruption of this regulatory domain leads to cell cycle arrest until nucleotide levels are restored (Pimkin et al. 2009). Additionally, the allosteric regulation of IMPDH octameric structure allows a mechanism by which the formation

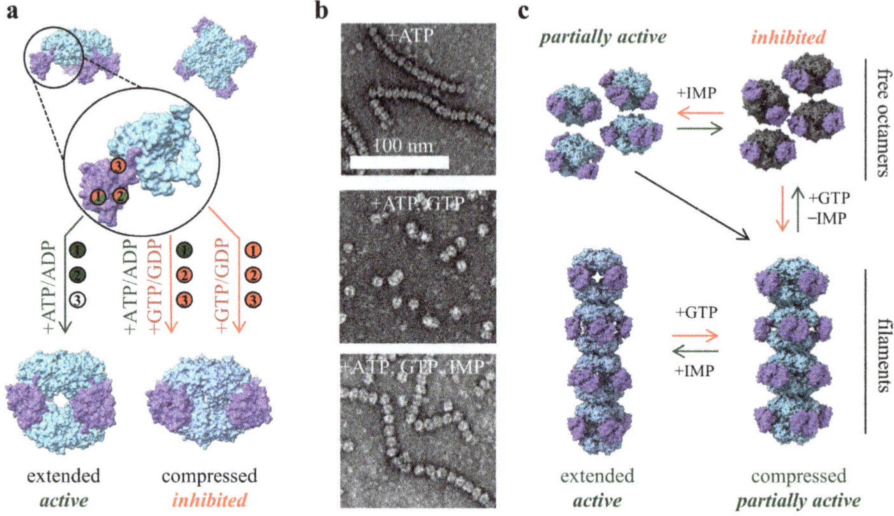

Fig. 10.9 Structure and conformation of IMPDH2 filaments. IMPDH is regulated by allosteric binding and filament formation. (**a**) IMPDH2 tetramer has three allosteric binding sites in the Bateman domain (magenta). The binding of ATP/ADP in Site 1 and Site 2 creates an extended active conformation (PDB ID: 6U8N), whereas binding of GTP/GDP at Site 2 and Site 3 can induce a compressed inhibited conformation (PDB ID: 6UC2). (**b**) Negative stain electron micrographs showing IMPDH2 filament formation in buffers with ATP, GTP, and/or IMP. Filaments form in the presence of ATP but dissolve as GTP levels rise. Addition of IMP can restore filament formation. (Images originally published by Johnson and Kollman (2020) and used with permission from Kollman). (**c**) Schematic of allosteric binding effects on activity and filamentation of IMPDH2 (PDB ID: 6UAJ, 6U9O, 6U8N, and 6UC2). Activating effects are denoted by green whereas inhibiting effects are denoted by red

of filaments can be dictated based on the conformation of its basic protomeric units (Fernandez-Justel et al. 2019).

In vitro reconstitutions imaged by negative-stain electron microscopy showed the behavior of IMPDH filamentation with various purine nucleotides (Johnson and Kollman 2020) (Fig. 10.9b). In the apo state, purified IMPDH assembles into short filaments. However, upon the addition of ATP and its transition to the extended octameric state, filaments measuring hundreds of nanometers long form for both IMPDH1 and IMPDH2. However, the effects that IMPDH1 and IMPDH2 filaments have upon exposure to GTP differ. For IMPDH2, increasing GTP concentrations triggers a dose-dependent dissolution of the filaments into their individual compressed octamers, even at high ATP concentrations. Interestingly, GTP-induced IMPDH2 filament dissolution can be rescued with small concentrations of its substrate IMP. On the other hand, IMPDH1 filaments remain intact when exposed to GTP despite a shift in the protomers to the compressed conformation. This observation was especially intriguing since both IMPDH isoforms share 84% sequence homology (Natsumeda et al. 1990) and yet can behave very differently when exposed to GTP.

High-resolution models from cryo-EM imaging helped mechanistically explain these differences. Both filaments of human IMPDH1 and IMPDH2 soaked in its substrate exhibit a right-handed helical symmetry with a rise of ~110 Å and a twist of ~30° (Johnson and Kollman 2020; Burrell et al. 2022). Despite incorporating some compressed octamers/tetramers, even in substrate only conditions, the polymerization interface between octamers was preserved through opposing octamer's catalytic domains through a tyrosine-arginine interaction (Y12-R356'). Despite nearly identical polymerization interfaces, IMPDH1 filaments were more homogenous in the extended conformation while IMPDH2 filaments incorporated a mixture of extended and compressed octamers (Burrell et al. 2022).

However, when exposed to GTP, the structures of IMPDH1 and IMPDH2 filaments are strikingly different. For IMPDH2, the octamers compress and begin to dissociate from the filament. Those units that remained intact kept the same polymerization interface as in the substrate-bound state; however, despite compressing, the polymerized octamers maintained a "flat" tetramer conformation compared to the "bowed" tetramer conformation of the octamers when they were released (Johnson and Kollman 2020) (Fig. 10.9c). When IMP was added, the otherwise disordered active sites loop became ordered. This rigidity resists the transition from "flat" to "bowed" tetramers and maintains the octamers in a filament-competent conformation. This observation helps to explain how the IMPDH2 filaments are rescued and reformed by IMP by promoting the "flat" tetramer confirmation (Johnson and Kollman 2020). IMPDH1, on the other hand, does not keep the same polymerization contacts when exposed to GTP. Although its octamers are still compressed, it adopts new interactions dominated by Y12/E487' and E15/K489' that shift the helical symmetry (Burrell et al. 2022). These new interactions create a conformation that permits tetramers to be 'bowed' while polymerized, thereby explaining how IMPDH1 filaments remain intact upon addition of GTP.

Insights Into IMPDH Filament Function

Under physiological conditions, base recycling via the salvage pathway is sufficient to meet the cellular GTP demand and IMP production via the de novo pathway is downregulated (Murray 1971). Under these conditions, IMPDH exists predominantly in the free, compressed state but shifts in response to changes in the ATP/GTP ratio. As ATP levels rise, ATP outcompetes GTP for binding at Site 2, activating IMPDH and transiently shifting it into the filament-competent extended conformation. Small filaments may form here but are dissolved raising GTP levels and promoting the inactive, bowed tetramer compressed state (Johnson and Kollman 2020). In this way, the ATP/GTP ratio regulates IMPDH activity and IMP fate in the same way as other non-filament-forming enzymes. IMPDH filaments' impact on nucleotide metabolism regulation is minimal here. During periods of high nucleotide demand such as proliferation or nutrient starvation, the salvage pathway alone is insufficient for meeting demand, and de novo synthesis pathways are activated, resulting in increased IMP. It has been shown that increased IMP levels promote

IMPDH filament assembly (Keppeke et al. 2018). Therefore, IMPDH filament formation appears to be a result of increased IMP generation through the de novo purine biosynthetic pathway.

Mutations within catalytic domain of IMPDH have provided insights into the function of IMPDH filaments. Cells expressing the Y12C point mutation in mouse embryonic stem cells disfavored filament formation. These cells experienced decreased proliferation by not satisfying the cell's overall guanine demand (Peng et al. 2024). When this mutation was combined with the filament promoting S275L mutation, the activity of IMPDH remained the same when IMPDH was in the filament versus free in solution (Keppeke et al. 2018; Anthony et al. 2017). Although polymerization does not increase IMPDH activity, polymerization does make IMPDH less sensitive to GTP inhibition, especially when IMP is present. The binding affinity of GTP to polymerized IMPDH2 is approximately twofold higher than that of free IMPDH2 (Johnson and Kollman 2020). As the cryo-EM models showed, when IMP is bound, IMPDH2 filaments resist tetramer bowing. Although the octamer is compressed, without tetramer bowing, IMPDH2's activity is not fully inhibited (Johnson and Kollman 2020).

IMPDH1 does not experience decreased sensitivity as seen with IMPDH2 (Burrell et al. 2022). The ability of the IMPDH1 filament to accommodate the fully inhibited octamer means the octamers in IMPDH1 filaments behave identical to those free in solution. In the retina, IMPDH1 is the dominant isoform expressed as two splice variants with additional N-terminal and/or C-terminal residues (Hedstrom 2009; Spellicy et al. 2007). Similarly, both variants assemble into filaments that do exhibit reduced GTP inhibition sensitivity (Burrell et al. 2022). Structural rearrangement due to the additional N-terminal residues prevents tetramer bowing like observed with IMPDH2, while the additional C-terminal residues destabilize the compressed conformation. These effects are only observed when IMPDH1 is in the filament and not when it is a free octamer. Therefore, these variants allow IMPDH1 to remain active even in high GTP environments.

10.4.2.2 CTPS Filaments

CTP synthetase (CTPS) is a glutamine amidotransferase that catalyzes the ATP-dependent conversion of UTP to CTP, which is widely regarded as the rate-limiting step in the biosynthesis of cytidine nucleotides (Fig. 10.10). CTPS has two major domains: an N-terminal glutaminase (GAT) domain and a C-terminal amidoligase (AL) domain, connected by a helical linker. When active, the GAT domain hydrolyzes glutamine and shuttles the resulting ammonia through an intramolecular tunnel to the active site in the AL domain where it reacts with the phosphorylated UTP to produce CTP (Zhou et al. 2021).

Achieving an active conformation of CTPS requires nucleotide-mediated oligomerization and allosteric conformational changes. In the absence of nucleotides, CTPS exists as an inactive homodimer bound through the AL domain (Noree et al. 2014). Upon binding ATP and UTP, it transitions into an X-shaped homotetramer

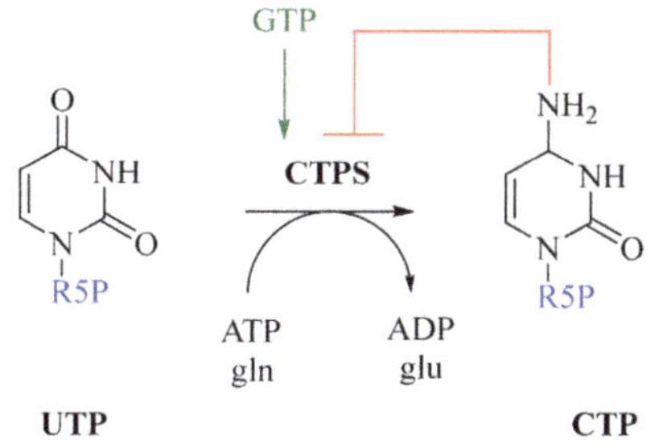

Fig. 10.10 The glutamine amidotransferase CTPS catalyzes the conversion of UTP to CTP in an ATP-dependent manner. Filamentation of CTP polymeric units can be activated upon GTP binding and feedback inhibited upon CTP binding

(dimer of dimers) and presents an allosteric GTP binding site on the GAT domains (Zhou et al. 2021). When GTP binds, the resulting conformation promotes the hydrolysis of glutamine in the GAT domain as well as stabilizes the intramolecular ammonia tunnel permitting catalysis (Bearne et al. 2022). The product CTP can also compete with ATP and UTP for binding that either triggers formation of an inactive tetramer or inhibits an active tetramer (Zhou et al. 2021). Therefore, it is possible that the conformational state and activity of CTPS could be governed based on the relative abundance of each of the major nucleotides.

A wide variety of species have shown that CTPS tetramers can spontaneously assemble into filamentous structures (Carcamo et al. 2011; Ingerson-Mahar et al. 2010; Liu 2010; Noree et al. 2010; Chang et al. 2021; Daumann et al. 2018; Zhou et al. 2020). While the molecular triggers governing filamentation are still largely unknown, recent studies have linked their formation to mTOR signaling (Sun and Liu 2019a; Andreadis et al. 2019), impaired glutamine metabolism (Ingerson-Mahar et al. 2010; Barry et al. 2014), and enhanced cell proliferation (Chang et al. 2017). When extracellular glutamine levels dropped below 0.5 mM, the number of HeLa and mouse hepatocellular carcinoma cells expressing CTP filaments rose threefold (Gou et al. 2014). In addition, an array of 81 cancer tissues representing 11 cancer types showed that 44% of the samples displayed high degrees of cytoophidia formation, whereas neighboring noncancerous tissue showed none (Chang et al. 2017). Combined, these findings suggest that in humans these structures might only be present in highly proliferating cells, such as those in tumors, to assist with meeting the increase in nucleotide demand.

Depending on the species they are found in and the shifts in metabolic demands, CTPS cytoophidia can either inactivate CTPS and protect it from degradation or become hyperactivated to boost CTP production. This wide range in catalytic activities upon filamentation is noted among higher-ordered *E. coli* and human CTPS (ecCTPS and hCTPS) structures. In the following sections, the various forms of CTPS filaments will be contrasted to better understand the structural features that lead to drastically different outcomes in metabolism.

Structure and Function of *E. coli* CTPS (ecCTPS) Filaments

The filamentation of ecCTPS tetramers occurs during periods of high intracellular CTP levels and consist exclusively of tetramers in their inactive product-bound state (Barry et al. 2014; Lynch et al. 2017) (Fig. 10.11). In this state, the tetramers interact with each other via alpha helices located in both the linker (residues 274–284) and GAT (residues 330–336) domains. Mutations at these interfaces such as E277R in the linker domain abolished polymerization and showed a fivefold reduction in activity regardless if exposed to its substrate UTP (Barry et al. 2014). Interestingly, cryo-EM revealed that while the substrate binding sites on ecCTPS filaments are not sterically blocked and remain solvent accessible, no density for UTP was found in the models, indicating that filamentation may decrease its substrate binding affinity (Barry et al. 2014; Lynch et al. 2017). As CTP levels dissipate, the filaments dissolve into free, active tetramers, and CTP production rises. This reversibility demonstrates a key role of compartmentalization: sequestration and storage. When CTP levels are sufficient, *E. coli* stores inactive CTPS in filaments protecting it from ubiquitination and degradation (Sun and Liu 2019b). Tetramers can be rapidly released in seconds and activated to respond to CTP demands without needing to transcribe and translate new protein (Barry et al. 2014; Bar-Peled and Kory 2022). Therefore, filamentation itself enhances the inhibition and adds another layer regulation beyond allosteric inhibition. The prevailing explanation is that filamentation stabilizes conformations with inherently less UTP affinity regardless of CTP binding (Lynch et al. 2017).

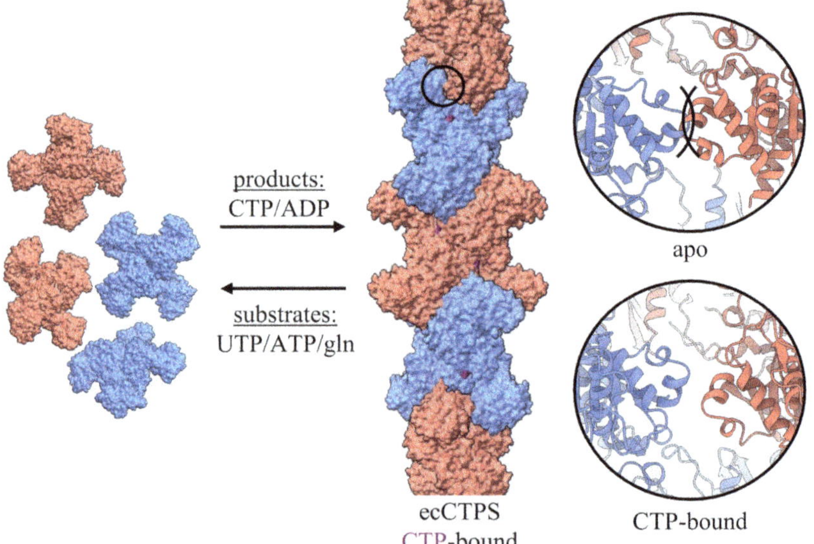

Fig. 10.11 Structure and conformation of *E. coli* CTPS (ecCTPS) filaments. ecCTPS filaments form from free tetramers upon product (CTP/ADP) production, whereas substrates (UTP/ATP/glutamine (gln)) dissociate filaments. The filament polymerization interface shows steric clash between tetrameric units in the apo state (PDB ID: 5U3C), which is absent in the CTP-bound conformation (PDB ID: 5U05)

Structure and Function of Human CTPS (hCTPS) Filaments

Humans have two CTPS isoforms, hCTPS1 and hCTPS2, which share approximately 75% sequence homology. CTPS is generally expressed at low levels but is often upregulated in proliferating cell populations such as T cells (Martin et al. 2014) and most cancer cell lines (Huang et al. 2024; Kizaki et al. 1980; Lin et al. 2022; Williams et al. 1978). Both human isoforms readily form active tetramers that can polymerize into filaments and other higher-ordered bundled structures. Given the numerous correlations of higher-ordered CTPS structures with human disease, significant efforts have been made by researchers to better understand how the structure of higher-ordered CTPS assemblies can provide clues into how they might regulate cytidine nucleotide production.

hCTPS1 filaments are constructed from active, substrate-bound tetramers and assembled during periods of high nucleotide demand (Fig. 10.12a, b). Unlike

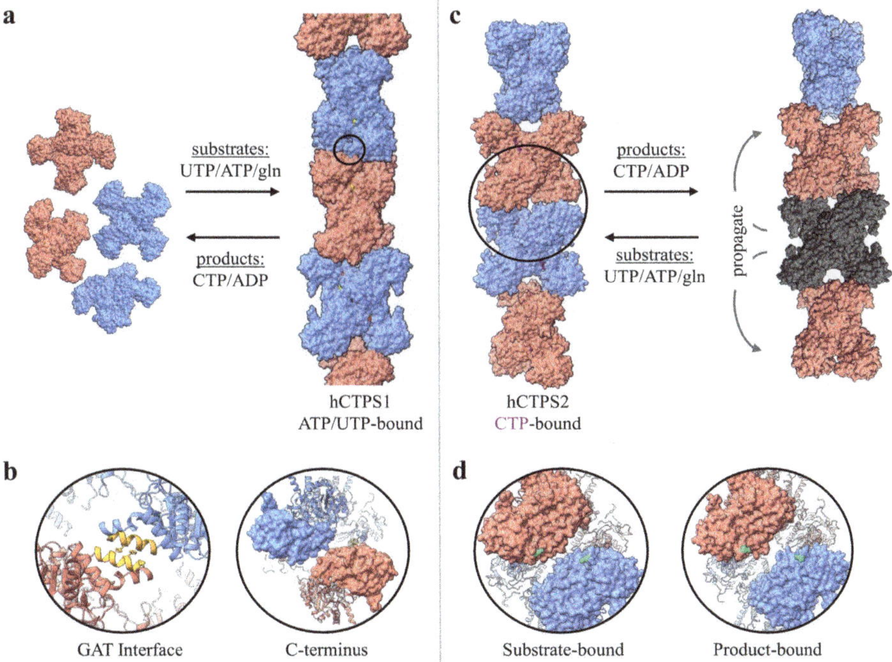

Fig. 10.12 Structure and conformation of (**a, b**) hCTPS1 and (**c, d**) hCTPS2 filaments. Schematic showing the behavior of (**a**) hCTPS1 and (**c**) hCTPS2 filaments. (**a**) hCTPS1 filaments are formed upon substrate (ATP/UTP) binding. (**b**) The polymerization interface occurs through a helix (orange) in the GAT domain. The primary contact H355 is shown. Stabilized C-terminal residues contribute to strengthening the filament assembly (PDB ID: 5U03). (**c, d**) hCTPS2 forms filaments from any conformational (substrate- or product-bound) state and can rapidly propagate a switch in conformation across all tetramers in the filament. C-terminal residues remain disordered in both substrate and product-bound conformations. These findings reveal a mechanism for the tolerance of conformation change in filament tetramers (PDB ID: 6PK7)

ecCTPS compartmentalization, which acts as a reservoir of metabolic potential, hCTPS1 filaments are hyperactive showing sixfold increased activity over free tetramers (Lynch et al. 2021). Cryo-EM studies revealed that hCTPS1 filaments form upon addition of UTP and ATP by locking each tetramer in the active conformation as noted by a right-handed helix with a rise of ~104 Å and a twist of ~61° (Lynch et al. 2017). When CTP levels increase, as illustrated by soaking hCTPS1 filaments with glutamine, the tetramers adopt an inactive conformation leading to filament disassembly.

Although there is a high sequence homology between prokaryotic and eukaryotic CTPS, the polymerization interface is not identical (Lynch et al. 2017). Eukaryotic CTPS has a unique helix in the GAT domain that is absent in prokaryotic CTPS (Guo et al. 2024). This helix mediates tetramer stacking, and mutations within the helical interface such as H355A disfavor hCTPS1 polymerization (Lynch et al. 2017) (Fig. 10.12b). Additionally, contacts between disordered C-terminal tails of hCTPS1 help stabilize the polymerization (Lynch and Kollman 2020). In the product-bound state, these tails are likely sterically incompatible with filament assembly. As the C-terminal tails have numerous sites for phosphorylation, filamentation has also been proposed to be phosphoregulated (Choi and Carman 2007; Kassel et al. 2010).

Filaments of hCTPS2 can form from either the product- or substrate-bound conformations of the tetramer and transition the tetramers while still polymerized (Lynch and Kollman 2020) (Fig. 10.12c, d). High UTP substrate levels generate substrate-bound active filaments, while high CTP product levels will form product inhibited filaments. Like hCTPS1, the novel eukaryotic GAT helix insertion serves as the site of polymerization. However, slight sequence differences between the isoforms, especially in the C-terminal tails that share only 41% homology, eliminates the C-terminal contacts while strengthening the primary H355 interaction site (Lynch and Kollman 2020). In the substrate-bound filament, the C-terminal residues are 36 Å apart, and this distance increases upon the transition to the CTP-bound state (Fig. 10.12d). The transition between tetramers is highly cooperative, and the resulting tight conformational switch allows for the filaments to go between active and inactive states without dissolving and reforming (Hvorecny and Kollman 2023; Lynch and Kollman 2020).

10.5   Disease Relevance and Possible Therapeutic Targets

To maintain accuracy in cellular processes such as DNA replication, cells maintain a highly regulated balance of purine and pyrimidine nucleotides. In disease, either the need for nucleotides is increased to support the pathological state or the balance is impaired that substantially impacts diseased cell viability and proliferation. For most cancers, this dependency generally makes cells more sensitive to perturbation in their biosynthetic pathways. Therefore, targeting these processes has led to the development and FDA approval of several purine and pyrimidine antimetabolites currently used in the treatment of solid tumors (Parker 2009). Targeting nucleotide metabolism, whether direct or indirect, can reduce the levels of any given nucleotide inside these diseased cells to inhibit their proliferation. This imbalance can also result in the accumulation of genetic mutations that might further pause cell growth. Additionally, targeting nucleotide metabolism increases the levels of secreted nucleotides. This is best illustrated with adenosine, where increased nucleoside levels outcompete natural ligands for binding to immune cell receptors and activates the immune system (Han et al. 2024; Hasko et al. 2005). While targeting nucleotide metabolism can be quite beneficial, it can also impede lymphocyte growth and elicit a significant undesired secondary effect on immune cells.

In rare cases, genetic mutations encoded in nucleotide biosynthetic enzymes give rise to disease. These mutations present themselves across a wide spectrum of clinical phenotypes. Most notable are inborn errors in purine biosynthesis that result in neuromuscular and neurodevelopment defects and often result in neonatal death (Jurecka and Tylki-Szymanska 2022). In these cases, anabolic and catabolic processes are impacted resulting in accumulation of likely toxic intermediates and decrease in AMP and GMP production.

The impact that these disease phenotypes have on nucleotide metabolism has been widely correlated with the presence or absence of supramolecular protein assemblies. These observations have provided a more in-depth understanding of their cellular function(s) and brought to light the possibility of targeting them with small molecules to treat many disease states where nucleotide production is enhanced. In this section, we will examine the evidence that links these assemblies to specific diseases and discuss strategies that have been used to target them to assist in the generation of a new class of therapeutics.

10.5.1 Purinosome

Since its discovery in HeLa cells (An et al. 2008), purinosome formation has been observed in a wide range of cell lines including breast cancer (Doigneaux et al. 2020; An et al. 2008; Schmitt et al. 2018), liver cancer (French et al. 2013; Baresova et al. 2012), and melanoma cell lines (Chou et al. 2023). Hypoxia, a common pathological state, showed that purinosomes readily form and localize near mitochondria like those in nutrient-depleted normoxic growth conditions (Doigneaux et al. 2020). However, these growth conditions do not permit purines from being produced through the de novo pathway, suggesting that pathway activation is not necessary to induce the LLPS to form purinosomes. A similar observation was observed within the pyrimidinosome, where hypoxia induced metabolon formation, through the dissociation of AMPK, yet UMPS activity and subsequent UMP production were inhibited (Yang et al. 2023).

Enhanced purinosome formation was also observed in cells deficient in HPRT and purine salvage. There are over 600 reported mutations in the HPRT1 gene that present a broad range of clinical phenotypes ranging from hyperuricemia and gout to Lesch-Nyhan disease (LND) (Fu et al. 2014). Disease severity mutants in *HPRT1* correlated with deficiency in HPRT enzyme activity and purinosome formation, whereby LND fibroblasts displayed the highest levels of purinosomes (Fu et al. 2015). The intracellular purine levels remained relatively stable in these diseased fibroblasts, later to be verified by an increase in de novo purine biosynthesis. Additionally, there was a notable increase in extracellular purine metabolites suggesting that free bases and nucleosides unable to be recycled are secreted. Other cell models where *HPRT1* was knocked out have been generated and illustrated similar effects (Baresova et al. 2018).

There are a host of other diseases caused by inborn errors and other genetic defects that deregulate de novo purine biosynthesis. Loss of function mutations within *ADSL* and *PAICS* have been linked to a variety of complications in children (Jurecka and Tylki-Szymanska 2022; Balasubramaniam et al. 2014; Nassogne et al. 2024). ADSL deficiency presents a spectrum of symptoms ranging from developmental delays, autism, and reduced muscle tone to severe prenatal complications resulting in fatal brain damage (Spiegel et al. 2006; Maaswinkel-Mooij et al. 1997; Van den Berghe et al. 1997). Less common are *PAICS* missense mutations. In a recent studied case, siblings were presented with severe malformation and eventual neonatal death (Weng et al. 2024; Pelet et al. 2019). Both *ADSL* and *PAICS* inborn errors cause structural changes in the enzyme's catalytic site, diminishing its activity and causing metabolic imbalances. Cell models of these loss-of-function mutations showed decreases in purinosome formation (Baresova et al. 2012; Weng et al. 2024; Pelet et al. 2019). In addition to the loss of the purinosome as being a result of these diseases, these studies indicated that enzymes downstream of the purinosome core also play an imperative role in the overall stability of the assembly.

Despite a growing number of inhibitors of de novo purine biosynthesis, only a few have been evaluated for their ability to alter purinosome dynamics. Compound 14 was developed as an inhibitor of ATIC dimerization and showed to impact purinosome formation in both nutrient-depleted and hypoxia-induced purinosome formation (Doigneaux et al. 2020). In addition, the use of mycophenolic acid to inhibit IMPDH activity resulted in a loss in the substrate channeled phenotype implying that the purinosome architecture was disrupted (Pareek et al. 2020). Other pathway inhibitors such as lometrexol as well as folate cycle disruptors have had significant impact on purine metabolism (Zarou et al. 2021; Chen et al. 2023; Robinson et al. 2020) but have not been tested for their ability to break apart the purinosome.

10.5.2 Filamentous Structures

Mutations within IMPDH and CTPS have also been tied to various diseases. Dysregulation of IMPDH1, attributed to diminished GTP feedback inhibition, can result in rare retinal diseases such as retinitis pigmentosa and Leber congenital

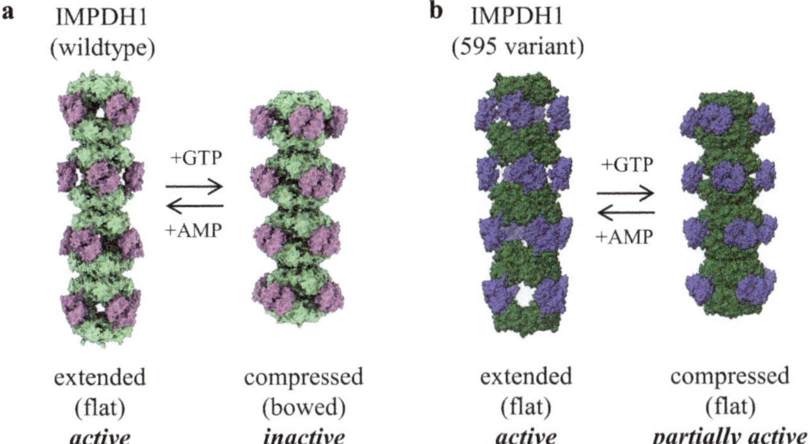

Fig. 10.13 The retinal disease IMPDH1(595) variant is less sensitive to GTP feedback inhibition. Unlike (**a**, left) wild-type IMPDH1 (PDB ID: 7RES), disordered terminal extensions on the (**b**, left) retina disease variant IMPDH(595) (PDB ID: 7RFH) promote a flat conformation of the octameric units (**b**, right; PDB ID: 7RGD) in the presence of GTP, rather than a bowed conformation shown in (**a**, right) wild-type IMPDH1 filaments (PDB ID: 7RFG). By maintaining the flat conformation, IMPDH1(595) variant filaments are less sensitive to GTP feedback inhibition

amaurosis (Bowne et al. 2002, 2006; Kennan et al. 2002; Wada et al. 2005). Structural insights into this altered sensitivity to GTP revealed that disease-promoting splice variants such as IMPDH1(595) introduce disordered stretches of amino acids to the termini. Upon filament formation, a compressed filament is still adopted upon GTP binding (Fig. 10.13). However, biochemical characterizations revealed that the GTP-bound IMPDH1(595) variant is six times less sensitive to GTP inhibition than wild type and might be attributed to differences in the structural conformation of the octamer units. One noted change was the inability of the octameric units to adopt a bowed conformation that leads to total inhibition. Further, the most studied retinopathy mutations within these splice variants are located near the allosteric or interdomain hinge region and do not promote the compressed filament conformation (Burrell et al. 2022; Buey et al. 2015). Therefore, the inability of the IMPDH1 octameric unit to adopt a bowed conformation when in the compressed filament has been proposed to be a structural mechanism surrounding GTP dysregulation (O'Neill et al. 2023).

IMPDH2 overexpression has been correlated with many brain cancers and leukemias (Senda and Natsumeda 1994; Nagai et al. 1991; Collart et al. 1992; Konno et al. 1991) and has led to the development of many IMPDH inhibitors (Naffouje et al. 2019). Additionally, mutations within IMPDH2 such as L254P and K238R have been correlated with dystonia and other neurodevelopmental disorders (Zech et al. 2020). Structural investigations by cryo-EM into how the L254 mutant can promote diseases indicated that the equilibrium between the active extended and the inactive compressed states is disrupted to favor a slightly compressed state that more closely resembles the active extended conformation (O'Neill et al. 2023). In this state, IMPDH2 is rendered insensitive to GTP feedback inhibition even in conditions of high GTP. Although most of the identified IMPDH2 mutations tied to dystonia are in or near the Bateman domain (O'Neill et al. 2023; Zech et al. 2020), they do not show perturb the enzyme and its associated filament in the same manner (O'Neill et al. 2023). This observation suggests that there might be specific mutations that are more severe than others; however, the correlation between mutation state and severity of dystonia has not been determined.

Lastly, CTPS cytoophidia is widely implicated in cancer metabolism, particularly in hepatocellular carcinomas (Chang et al. 2017). The generation of CTP in cells is critical for a variety of cellular functions but becomes especially vital in cancer cells, where there is a shift in metabolic pathways to support proliferation. Increased CTPS filaments within cancer cells enhance its catalytic activity and stabilize the assembly, allowing for maintained CTP production. This underscores the significant role of nucleotide metabolism in disease progression, particularly in cancer and genetic disorders related to purine metabolism.

10.6 Concluding Remarks

The discovery of supramolecular protein complexes has completely transformed our understanding of how biochemical processes are regulated. While this chapter focuses on nucleotide metabolism, many other metabolic processes adopt similar higher-ordered structures that bear analogous features. These observations have led many researchers, including us, to propose a more generalized hypothesis that metabolic processes are compartmentalized into discrete structures to more effectively regulate metabolite production. Although initially hypothesized by kinetic arguments, recent biochemical and structural evidence have started to provide a more concrete molecular understanding of how these assemblies respond to changes in metabolic demand.

Many of our breakthroughs in this area are a result of sophisticated microscopy-based technologies. Super- and high-resolution fluorescence microscopy has allowed us to visualize the finer details of assembly organization inside of cells including their subcellular localizations and dynamics. Furthermore, biosensors and other live cell reporters can now monitor the biochemical and physicochemical properties of these assemblies to provide insights into their microenvironment and how changes in these features perturb complexes. As illustrated in this chapter with IMPDH and CTPS filaments, cryo-EM has exponentially increased our understanding of the formation and dynamics of these structures. We are now aware that the activity of these filaments is regulated through allosteric changes in the conformation of polymeric units. Other emerging technologies such as mass spectral imaging and cryogenic electron tomography hold tremendous promise to complement and further these efforts.

Despite all the advances, there are many questions that are still outstanding. Metabolons remain elusive assemblies in part due to their difficulty in reconstituting outside a cellular environment. Fluorescence imaging has allowed us to identify specific stimuli that trigger metabolon formation, but the structural features that induce protein-protein interactions and substrate channeling are unknown. Developing new tools to stabilize and characterize transient interactions in conjunction with structural determinations would provide the necessary insights into tackling this conundrum. We also do not have a great appreciation of how these metabolons are organized or localized in cells. Why is purinosome formation dependent on microtubules and how do enzymes within this metabolon associate with mitochondria? Lessons from the analogous pyrimidinosome suggest that mitochondrial localization might be through transporter proteins that assist in shuttling intermediates across the membranes. Insights gathered from looking at purinosome associations with amino acid transporters have invigored this possibility, yet further biochemical and functional studies need to be performed to verify this current hypothesis.

The presence of these assemblies has been observed in many diseases and pathological states. However, there has not been a clear evaluation of whether changes in assembly properties brought on by disease-promoting mutations or splice variants contribute to disease progression. In the case of IMPDH2, the presence of specific mutations in the Bateman domain can alter its sensitivity to GTP feedback inhibition. However, not all mutations attributed to disease show the same degree of GTP sensitivity. These findings raise the question of whether any of these mutations can serve as prognostic indicators of disease onset or severity. Furthermore, there is still a question of whether targeting these assemblies with small molecules can present a viable avenue for the development of novel therapeutics. By uncovering distinct molecular features within these assemblies that correlate with disease, we can start to apply medicinal chemistry tools to develop new therapeutics that selectively disrupt enzyme compartmentalization.

In conclusion, the past two decades have ignited new interest in nucleotide metabolism through the discovery of metabolons and filaments. Although we still do not have a complete picture of how these higher-ordered complexes are assembled and their biological function and significance, we are excited about what will come next, both on a scientific and technological front, to help us better understand the regulation of nucleotide metabolism.

Acknowledgments We thank Antoine Gedeon for his careful reading of the chapter and for his helpful comments. We acknowledge the following individuals for permission to reproduce and use data in this chapter: V. Baresova and M. Zikanova for Fig. 10.4b, V. Pareek and S.J. Benkovic for Fig. 10.4d, and J. Kollman for Fig. 10.9b. Financial and/or research support was provided by The National Institutes of Health (R35GM150481 to AMP, T32GM149417 to JPB), The Department of Biochemistry and Molecular Biology at the University of Iowa (to AMP), and the National Cancer Institute of the National Institutes of Health (Award Number P30CA086862 to the University of Iowa Carver College of Medicine and the Holden Comprehensive Cancer Center).

Competing Interests The authors declare no competing financial interests.

References

Agarwal S, et al. PAICS, a purine nucleotide metabolic enzyme, is involved in tumor growth and the metastasis of colorectal cancer. Cancers (Basel). 2020;12(4):772.

Ahangari N, et al. Nuclear IMPDH filaments in human gliomas. J Neuropathol Exp Neurol. 2021;80(10):944–54.

Ahn CS, Metallo CM. Mitochondria as biosynthetic factories for cancer proliferation. Cancer Metab. 2015;3(1):1.

Alberti S, Gladfelter A, Mittag T. Considerations and challenges in studying liquid-liquid phase separation and biomolecular condensates. Cell. 2019;176(3):419–34.

Alcazar-Fabra M, Navas P, Brea-Calvo G. Coenzyme Q biosynthesis and its role in the respiratory chain structure. Biochim Biophys Acta. 2016;1857(8):1073–8.

Ali ES, Ben-Sahra I. Regulation of nucleotide metabolism in cancers and immune disorders. Trends Cell Biol. 2023;33(11):950–66.

An S, et al. Reversible compartmentalization of de novo purine biosynthetic complexes in living cells. Science. 2008;320(5872):103–6.

An S, et al. Microtubule-assisted mechanism for functional metabolic macromolecular complex formation. Proc Natl Acad Sci USA. 2010a;107(29):12872–6.

An S, et al. Dynamic regulation of a metabolic multi-enzyme complex by protein kinase CK2. J Biol Chem. 2010b;285(15):11093–9.

An S, et al. Phase-separated condensates of metabolic complexes in living cells: purinosome and glucosome. Methods Enzymol. 2019;628:1–17.

Andreadis C, et al. The TOR pathway modulates cytoophidium formation in Schizosaccharomyces pombe. J Biol Chem. 2019;294(40):14686–703.

Anthony SA, et al. Reconstituted IMPDH polymers accommodate both catalytically active and inactive conformations. Mol Biol Cell. 2017;28(20):2600–8.

Antle VD, et al. Substrate specificity of glycinamide ribonucleotide synthetase from chicken liver. J Biol Chem. 1996;271(14):8192–5.

Ayoub N, Gedeon A, Munier-Lehmann H. A journey into the regulatory secrets of the de novo purine nucleotide biosynthesis. Front Pharmacol. 2024;15:1329011.

Balasubramaniam S, Duley JA, Christodoulou J. Inborn errors of purine metabolism: clinical update and therapies. J Inherit Metab Dis. 2014;37(5):669–86.

Banani SF, et al. Biomolecular condensates: organizers of cellular biochemistry. Nat Rev Mol Cell Biol. 2017;18(5):285–98.

Baresova V, et al. Mutations of ATIC and ADSL affect purinosome assembly in cultured skin fibroblasts from patients with AICA-ribosiduria and ADSL deficiency. Hum Mol Genet. 2012;21(7):1534–43.

Baresova V, et al. CRISPR-Cas9 induced mutations along de novo purine synthesis in HeLa cells result in accumulation of individual enzyme substrates and affect purinosome formation. Mol Genet Metab. 2016;119(3):270–7.

Baresova V, et al. Study of purinosome assembly in cell-based model systems with de novo purine synthesis and salvage pathway deficiencies. PLoS One. 2018;13(7):e0201432.

Barfeld SJ, et al. Myc-dependent purine biosynthesis affects nucleolar stress and therapy response in prostate cancer. Oncotarget. 2015;6(14):12587–602.

Barnes SJ, Weitzman PD. Organization of citric acid cycle enzymes into a multienzyme cluster. FEBS Lett. 1986;201(2):267–70.

Bar-Peled L, Kory N. Principles and functions of metabolic compartmentalization. Nat Metab. 2022;4(10):1232–44.

Barry RM, et al. Large-scale filament formation inhibits the activity of CTP synthetase. elife. 2014;3:e03638.

Bearne SL, Guo CJ, Liu JL. GTP-dependent regulation of CTP synthase: evolving insights into allosteric activation and NH(3) translocation. Biomolecules. 2022;12(5):647.

Boukalova S, et al. Dihydroorotate dehydrogenase in oxidative phosphorylation and cancer. Biochim Biophys Acta Mol Basis Dis. 2020;1866(6):165759.

Bowne SJ, et al. Mutations in the inosine monophosphate dehydrogenase 1 gene (IMPDH1) cause the RP10 form of autosomal dominant retinitis pigmentosa. Hum Mol Genet. 2002;11(5):559–68.

Bowne SJ, et al. Spectrum and frequency of mutations in IMPDH1 associated with autosomal dominant retinitis pigmentosa and leber congenital amaurosis. Invest Ophthalmol Vis Sci. 2006;47(1):34–42.

Bracha D, Walls MT, Brangwynne CP. Probing and engineering liquid-phase organelles. Nat Biotechnol. 2019;37(12):1435–45.

Brangwynne CP, Mitchison TJ, Hyman AA. Active liquid-like behavior of nucleoli determines their size and shape in Xenopus laevis oocytes. Proc Natl Acad Sci USA. 2011;108(11):4334–9.

Buchanan JM, Hartman SC. Enzymic reactions in the synthesis of the purines. In: Advances in enzymology and related areas of molecular biology; Wiley (Hoboken, NJ), 1959. p. 199–261.

Buey RM, et al. Guanine nucleotide binding to the Bateman domain mediates the allosteric inhibition of eukaryotic IMP dehydrogenases. Nat Commun. 2015;6:8923.

Buey RM, et al. A nucleotide-controlled conformational switch modulates the activity of eukaryotic IMP dehydrogenases. Sci Rep. 2017;7(1):2648.

Buey RM, et al. The gateway to guanine nucleotides: allosteric regulation of IMP dehydrogenases. Protein Sci. 2022;31(9):e4399.

Bulutoglu B, et al. Direct evidence for metabolon formation and substrate channeling in recombinant TCA cycle enzymes. ACS Chem Biol. 2016;11(10):2847–53.

Burrell AL, et al. IMPDH1 retinal variants control filament architecture to tune allosteric regulation. Nat Struct Mol Biol. 2022;29(1):47–58.

Calise SJ, et al. 'Rod and ring' formation from IMP dehydrogenase is regulated through the one-carbon metabolic pathway. J Cell Sci. 2016;129(15):3042–52.

Calise SJ, et al. Immune response-dependent assembly of IMP dehydrogenase filaments. Front Immunol. 2018;9:2789.

Carcamo WC, et al. Induction of cytoplasmic rods and rings structures by inhibition of the CTP and GTP synthetic pathway in mammalian cells. PLoS One. 2011;6(12):e29690.

Chan CY, et al. Purinosome formation as a function of the cell cycle. Proc Natl Acad Sci USA. 2015;112(5):1368–73.

Chan CY, et al. Microtubule-directed transport of purine metabolons drives their cytosolic transit to mitochondria. Proc Natl Acad Sci USA. 2018;115(51):13009–14.

Chang CC, et al. Cytoophidium assembly reflects upregulation of IMPDH activity. J Cell Sci. 2015;128(19):3550–5.

Chang CC, et al. CTP synthase forms the cytoophidium in human hepatocellular carcinoma. Exp Cell Res. 2017;361(2):292–9.

Chang CC, et al. Interfilament interaction between IMPDH and CTPS cytoophidia. FEBS J. 2018;285(20):3753–68.

Chang CC, et al. CTPS forms the cytoophidium in zebrafish. Exp Cell Res. 2021;405(2):112684.

Chang CC, et al. Molecular crowding facilitates bundling of IMPDH polymers and cytoophidium formation. Cell Mol Life Sci. 2022;79(8):420.

Chen JJ, Jones ME. The cellular location of dihydroorotate dehydrogenase: relation to de novo biosynthesis of pyrimidines. Arch Biochem Biophys. 1976;176(1):82–90.

Chen KF, et al. Transcriptional repression of human cad gene by hypoxia inducible factor-1alpha. Nucleic Acids Res. 2005;33(16):5190–8.

Chen J, et al. De novo nucleotide biosynthetic pathway and cancer. Genes Dis. 2023;10(6):2331–8.

Chitrakar I, et al. Higher order structures in purine and pyrimidine metabolism. J Struct Biol. 2017;197(3):354–64.

Choi MG, Carman GM. Phosphorylation of human CTP synthetase 1 by protein kinase A: identification of Thr455 as a major site of phosphorylation. J Biol Chem. 2007;282(8):5367–77.

Chou MC, et al. PAICS ubiquitination recruits UBAP2 to trigger phase separation for purinosome assembly. Mol Cell. 2023;83(22):4123–4140 e12.

Chua SM, Fraser JA. Surveying purine biosynthesis across the domains of life unveils promising drug targets in pathogens. Immunol Cell Biol. 2020;98(10):819–31.

Collart FR, et al. Increased inosine-5′-phosphate dehydrogenase gene expression in solid tumor tissues and tumor cell lines. Cancer Res. 1992;52(20):5826–8.

Corton JM, et al. 5-aminoimidazole-4-carboxamide ribonucleoside. A specific method for activating AMP-activated protein kinase in intact cells? Eur J Biochem. 1995;229(2):558–65.

Covini G, et al. Cytoplasmic rods and rings autoantibodies developed during pegylated interferon and ribavirin therapy in patients with chronic hepatitis C. Antivir Ther. 2012;17(5):805–11.

Cunningham JT, et al. Protein and nucleotide biosynthesis are coupled by a single rate-limiting enzyme, PRPS2, to drive cancer. Cell. 2014;157(5):1088–103.

Dang CV, Le A, Gao P. MYC-induced cancer cell energy metabolism and therapeutic opportunities. Clin Cancer Res. 2009;15(21):6479–83.

Daumann M, et al. Characterization of filament-forming CTP synthases from Arabidopsis thaliana. Plant J. 2018;96(2):316–28.

Deng B, Wan G. Technologies for studying phase-separated biomolecular condensates. Adv Biotechnol. 2024;2(1):10.

Deng Y, et al. Mapping protein-protein proximity in the purinosome. J Biol Chem. 2012;287(43):36201–7.

Doigneaux C, et al. Hypoxia drives the assembly of the multienzyme purinosome complex. J Biol Chem. 2020;295(28):9551–66.

Dong Y, et al. Regulation of cancer cell metabolism: oncogenic MYC in the driver's seat. Signal Transduct Target Ther. 2020;5(1):124.

Ducker GS, Rabinowitz JD. One-carbon metabolism in health and disease. Cell Metab. 2017;25(1):27–42.

Duong-Ly KC, et al. T cell activation triggers reversible inosine-5′-monophosphate dehydrogenase assembly. J Cell Sci. 2018;131(17):jcs223289.

Fang YF, et al. Super-resolution imaging reveals dynamic reticular cytoophidia. Int J Mol Sci. 2022;23(19):11698.

Feric M, et al. Coexisting liquid phases underlie nucleolar subcompartments. Cell. 2016;165(7):1686–97.

Fernandez-Justel D, et al. A nucleotide-dependent conformational switch controls the polymerization of human IMP dehydrogenases to modulate their catalytic activity. J Mol Biol. 2019;431(5):956–69.

Flamholz A, Phillips R, Milo R. The quantified cell. Mol Biol Cell. 2014;25(22):3497–500.

Fox IH, Kelley WN. Phosphoribosylpyrophosphate in man: biochemical and clinical significance. Ann Intern Med. 1971;74(3):424–33.

French JB, et al. Hsp70/Hsp90 chaperone machinery is involved in the assembly of the purinosome. Proc Natl Acad Sci USA. 2013;110(7):2528–33.

French JB, et al. Spatial colocalization and functional link of purinosomes with mitochondria. Science. 2016;351(6274):733–7.

Fridman A, et al. Cell cycle regulation of purine synthesis by phosphoribosyl pyrophosphate and inorganic phosphate. Biochem J. 2013;454(1):91–9.

Fu R, et al. Genotype-phenotype correlations in neurogenetics: Lesch-Nyhan disease as a model disorder. Brain. 2014;137(Pt 5):1282–303.

Fu R, et al. Clinical severity in Lesch-Nyhan disease: the role of residual enzyme and compensatory pathways. Mol Genet Metab. 2015;114(1):55–61.

Goldthwait DA, Peabody RA, Greenberg GR. On the mechanism of synthesis of glycinamide ribotide and its formyl derivative. J Biol Chem. 1956;221(2):569–77.

Gou KM, et al. CTP synthase forms cytoophidia in the cytoplasm and nucleus. Exp Cell Res. 2014;323(1):242–53.

Greenberg GR. De novo synthesis of hypoxanthine via inosine-5-phosphate and inosine. J Biol Chem. 1951;190(2):611–31.

Guo C, Wang Z, Liu JL. Filamentation and inhibition of prokaryotic CTP synthase with ligands. mLife. 2024;3(2):240–50.

Haldane JBS. The origin of life. Rationalist Annual. 1929;148:3–10.

Han Y, et al. Unlocking the adenosine receptor mechanism of the tumour immune microenvironment. Front Immunol. 2024;15:1434118.

Hardman JG. Cyclic nucleotides and regulation of vascular smooth muscle. J Cardiovasc Pharmacol. 1984;6(Suppl 4):S639–45.

Hardman JG, Robison GA, Sutherland EW. Cyclic nucleotides. Annu Rev Physiol. 1971;33:311–36.

Hartman SC, Buchanan JM. Nucleic acids, purines, pyrimidines (nucleotide synthesis). Annu Rev Biochem. 1959;28:365–410.

Hasko G, et al. Adenosine receptor signaling in the brain immune system. Trends Pharmacol Sci. 2005;26(10):511–6.

Haystead TA. The purinome, a complex mix of drug and toxicity targets. Curr Top Med Chem. 2006;6(11):1117–27.

He J, et al. Multienzyme interactions of the de novo purine biosynthetic protein PAICS facilitate purinosome formation and metabolic channeling. J Biol Chem. 2022;298(5):101853.

Hedstrom L. IMP dehydrogenase: structure, mechanism, and inhibition. Chem Rev. 2009;109(7):2903–28.

Hershfield MS, Seegmiller JE. Regulation of de novo purine biosynthesis in human lymphoblasts. Coordinate control of proximal (rate-determining) steps and the inosinic acid branch point. J Biol Chem. 1976;251(23):7348–54.

Hinzpeter F, Tostevin F, Gerland U. Regulation of reaction fluxes via enzyme sequestration and co-clustering. J R Soc Interface. 2019;16(156):20190444.

Hove-Jensen B, et al. Phosphoribosyl diphosphate (PRPP): biosynthesis, enzymology, utilization, and metabolic significance. Microbiol Mol Biol Rev. 2017;81(1):e00040–16.

Huang H, et al. Cytidine triphosphate synthase 1-mediated metabolic reprogramming promotes proliferation and drug resistance in multiple myeloma. Heliyon. 2024;10(13):e33001.

Hvorecny KL, Kollman JM. Greater than the sum of parts: mechanisms of metabolic regulation by enzyme filaments. Curr Opin Struct Biol. 2023;79:102530.

Ingerson-Mahar M, et al. The metabolic enzyme CTP synthase forms cytoskeletal filaments. Nat Cell Biol. 2010;12(8):739–46.

Jackson RC, Weber G, Morris HP. IMP dehydrogenase, an enzyme linked with proliferation and malignancy. Nature. 1975;256(5515):331–3.

Ji Y, et al. Regulation of the interaction of inosine monophosphate dehydrogenase with mycophenolic acid by GTP. J Biol Chem. 2006;281(1):206–12.

Jing X, et al. Cell-cycle-dependent phosphorylation of PRPS1 fuels nucleotide synthesis and promotes tumorigenesis. Cancer Res. 2019;79(18):4650–64.

Johnson MC, Kollman JM. Cryo-EM structures demonstrate human IMPDH2 filament assembly tunes allosteric regulation. elife. 2020;9:e53243.

Jones ME. Pyrimidine nucleotide biosynthesis in animals: genes, enzymes, and regulation of UMP biosynthesis. Annu Rev Biochem. 1980;49:253–79.

Juda P, et al. Ultrastructure of cytoplasmic and nuclear inosine-5′-monophosphate dehydrogenase 2 "rods and rings" inclusions. J Histochem Cytochem. 2014;62(10):739–50.

Jumper J, et al. Highly accurate protein structure prediction with AlphaFold. Nature. 2021;596(7873):583–9.

Jurecka A, Tylki-Szymanska A. Inborn errors of purine and pyrimidine metabolism: a guide to diagnosis. Mol Genet Metab. 2022;136(3):164–76.

Kassel KM, et al. Regulation of human cytidine triphosphate synthetase 2 by phosphorylation. J Biol Chem. 2010;285(44):33727–36.

Katsyuba E, Auwerx J. Modulating NAD(+) metabolism, from bench to bedside. EMBO J. 2017;36(18):2670–83.

Keller KE, et al. SAICAR induces protein kinase activity of PKM2 that is necessary for sustained proliferative signaling of cancer cells. Mol Cell. 2014;53(5):700–9.

Kelly RE, Mally MI, Evans DR. The dihydroorotase domain of the multifunctional protein CAD. Subunit structure, zinc content, and kinetics. J Biol Chem. 1986;261(13):6073–83.

Kennan A, et al. Identification of an IMPDH1 mutation in autosomal dominant retinitis pigmentosa (RP10) revealed following comparative microarray analysis of transcripts derived from retinas of wild-type and Rho(−/−) mice. Hum Mol Genet. 2002;11(5):547–57.

Keppeke GD, et al. IMP/GTP balance modulates cytoophidium assembly and IMPDH activity. Cell Div. 2018;13:5.

Kim J, et al. AMPK activators: mechanisms of action and physiological activities. Exp Mol Med. 2016;48(4):e224.

Kizaki H, et al. Increased cytidine 5′-triphosphate synthetase activity in rat and human tumors. Cancer Res. 1980;40(11):3921–7.

Kodama M, et al. A shift in glutamine nitrogen metabolism contributes to the malignant progression of cancer. Nat Commun. 2020;11(1):1320.

Kohnhorst CL, et al. Identification of a multienzyme complex for glucose metabolism in living cells. J Biol Chem. 2017;292(22):9191–203.

Konno Y, et al. Expression of human IMP dehydrogenase types I and II in Escherichia coli and distribution in human normal lymphocytes and leukemic cell lines. J Biol Chem. 1991;266(1):506–9.

Kyoung M, et al. Dynamic architecture of the purinosome involved in human de novo purine biosynthesis. Biochemistry. 2015;54(3):870–80.

Labesse G, et al. MgATP regulates allostery and fiber formation in IMPDHs. Structure. 2013;21(6):975–85.

Lafontaine DLJ, et al. The nucleolus as a multiphase liquid condensate. Nat Rev Mol Cell Biol. 2021;22(3):165–82.

Lane AN, Fan TW. Regulation of mammalian nucleotide metabolism and biosynthesis. Nucleic Acids Res. 2015;43(4):2466–85.

Li SX, et al. Octameric structure of the human bifunctional enzyme PAICS in purine biosynthesis. J Mol Biol. 2007;366(5):1603–14.

Li G, et al. Pyrimidine biosynthetic enzyme CAD: its function, regulation, and diagnostic potential. Int J Mol Sci. 2021;22(19):10253.

Lin Y, et al. CTPS1 promotes malignant progression of triple-negative breast cancer with transcriptional activation by YBX1. J Transl Med. 2022;20(1):17.

Liu JL. Intracellular compartmentation of CTP synthase in Drosophila. J Genet Genomics. 2010;37(5):281–96.

Liu T, et al. MYC predetermines the sensitivity of gastrointestinal cancer to antifolate drugs through regulating TYMS transcription. EBioMedicine. 2019a;48:289–300.

Liu C, et al. Mapping post-translational modifications of de novo purine biosynthetic enzymes: implications for pathway regulation. J Proteome Res. 2019b;18(5):2078–87.

Lynch EM, Kollman JM. Coupled structural transitions enable highly cooperative regulation of human CTPS2 filaments. Nat Struct Mol Biol. 2020;27(1):42–8.

Lynch EM, et al. Human CTP synthase filament structure reveals the active enzyme conformation. Nat Struct Mol Biol. 2017;24(6):507–14.

Lynch EM, Kollman JM, Webb BA. Filament formation by metabolic enzymes-a new twist on regulation. Curr Opin Cell Biol. 2020;66:28–33.

Lynch EM, et al. Structural basis for isoform-specific inhibition of human CTPS1. Proc Natl Acad Sci USA. 2021;118(40):e2107968118.

Lyon AS, Peeples WB, Rosen MK. A framework for understanding the functions of biomolecular condensates across scales. Nat Rev Mol Cell Biol. 2021;22(3):215–35.

Maaswinkel-Mooij PD, et al. Adenylosuccinase deficiency presenting with epilepsy in early infancy. J Inherit Metab Dis. 1997;20(4):606–7.

Mangold CA, et al. Expression of the purine biosynthetic enzyme phosphoribosyl formylglycinamidine synthase in neurons. J Neurochem. 2018;144(6):723–35.

Mannava S, et al. Direct role of nucleotide metabolism in C-MYC-dependent proliferation of melanoma cells. Cell Cycle. 2008;7(15):2392–400.

Martin E, et al. CTP synthase 1 deficiency in humans reveals its central role in lymphocyte proliferation. Nature. 2014;510(7504):288–92.

Massiere F, Badet-Denisot MA. The mechanism of glutamine-dependent amidotransferases. Cell Mol Life Sci. 1998;54(3):205–22.

McCairns E, et al. De novo purine synthesis in human lymphocytes. Partial co-purification of the enzymes and some properties of the pathway. J Biol Chem. 1983;258(3):1851–6.

Moffatt BA, Ashihara H. Purine and pyrimidine nucleotide synthesis and metabolism. In: Arabidopsis Book, vol. 1; American Society of Plant Biologists (Rockville, MD), 2002. p. e0018.

Morrish F, et al. c-Myc activates multiple metabolic networks to generate substrates for cell-cycle entry. Oncogene. 2009;28(27):2485–91.

Murray AW. The biological significance of purine salvage. Annu Rev Biochem. 1971;40:811–26.

Murray KJ. Cyclic AMP and mechanisms of vasodilation. Pharmacol Ther. 1990;47(3):329–45.

Murray JM, Bussiere DE. Targeting the purinome. Methods Mol Biol. 2009;575:47–92.

Naffouje R, et al. Anti-tumor potential of IMP dehydrogenase inhibitors: a century-long story. Cancers (Basel). 2019;11(9):1346.

Nagai M, et al. Selective up-regulation of type II inosine 5′-monophosphate dehydrogenase messenger RNA expression in human leukemias. Cancer Res. 1991;51(15):3886–90.

Nagai M, Natsumeda Y, Weber G. Proliferation-linked regulation of type II IMP dehydrogenase gene in human normal lymphocytes and HL-60 leukemic cells. Cancer Res. 1992;52(2):258–61.

Narayanaswamy R, et al. Widespread reorganization of metabolic enzymes into reversible assemblies upon nutrient starvation. Proc Natl Acad Sci USA. 2009;106(25):10147–52.

Nassogne MC, Marie S, Dewulf JP. Neurological presentations of inborn errors of purine and pyrimidine metabolism. Eur J Paediatr Neurol. 2024;48:69–77.

Natsumeda Y, et al. Two distinct cDNAs for human IMP dehydrogenase. J Biol Chem. 1990;265(9):5292–5.

Nikiforov MA, et al. A functional screen for Myc-responsive genes reveals serine hydroxymethyltransferase, a major source of the one-carbon unit for cell metabolism. Mol Cell Biol. 2002;22(16):5793–800.

Noree C, et al. Identification of novel filament-forming proteins in Saccharomyces cerevisiae and Drosophila melanogaster. J Cell Biol. 2010;190(4):541–51.

Noree C, et al. Common regulatory control of CTP synthase enzyme activity and filament formation. Mol Biol Cell. 2014;25(15):2282–90.

Noree C, et al. A quantitative screen for metabolic enzyme structures reveals patterns of assembly across the yeast metabolic network. Mol Biol Cell. 2019;30(21):2721–36.

Nyhan WL. Nucleotide synthesis via salvage pathway. In: eLS; 2005. https://doi.org/10.1038/npg.els.0003909.

O'Connell JD, et al. A proteomic survey of widespread protein aggregation in yeast. Mol Biosyst. 2014;10(4):851–61.

Oliver JC, et al. Conformational changes involving ammonia tunnel formation and allosteric control in GMP synthetase. Arch Biochem Biophys. 2014;545:22–32.

O'Neill AG, et al. Neurodevelopmental disorder mutations in the purine biosynthetic enzyme IMPDH2 disrupt its allosteric regulation. J Biol Chem. 2023;299(8):105012.

Oparin AI. Proiskhozhdenie zhizny. In: Bernal JD, editor. The origin of life. London: Weidenfeld and Nicholson; 1967.

Pareek V, et al. Metabolomics and mass spectrometry imaging reveal channeled de novo purine synthesis in cells. Science. 2020;368(6488):283–90.

Pareek V, Pedley AM, Benkovic SJ. Human de novo purine biosynthesis. Crit Rev Biochem Mol Biol. 2021a;56(1):1–16.

Pareek V, et al. Metabolic channeling: predictions, deductions, and evidence. Mol Cell. 2021b;81(18):3775–85.

Park CK, Horton NC. Structures, functions, and mechanisms of filament forming enzymes: a renaissance of enzyme filamentation. Biophys Rev. 2019;11(6):927–94.

Parker WB. Enzymology of purine and pyrimidine antimetabolites used in the treatment of cancer. Chem Rev. 2009;109(7):2880–93.

Pedley AM, Benkovic SJ. A new view into the regulation of purine metabolism: the purinosome. Trends Biochem Sci. 2017;42(2):141–54.

Pedley AM, Benkovic SJ. Detecting purinosome metabolon formation with fluorescence microscopy. Methods Mol Biol. 2018;1764:279–89.

Pedley AM, et al. Role of HSP90 in the regulation of de novo purine biosynthesis. Biochemistry. 2018;57(23):3217–21.

Pedley AM, Pareek V, Benkovic SJ. The purinosome: a case study for a mammalian metabolon. Annu Rev Biochem. 2022a;91:89–106.

Pedley AM, et al. Purine biosynthetic enzymes assemble into liquid-like condensates dependent on the activity of chaperone protein HSP90. J Biol Chem. 2022b;298(5):101845.

Pelet A, et al. PAICS deficiency, a new defect of de novo purine synthesis resulting in multiple congenital anomalies and fatal outcome. Hum Mol Genet. 2019;28(22):3805–14.

Peng M, et al. The IMPDH cytoophidium couples metabolism and fetal development in mice. Cell Mol Life Sci. 2024;81(1):210.

Petrova B, et al. Regulatory mechanisms of one-carbon metabolism enzymes. J Biol Chem. 2023;299(12):105457.

Petrovska I, et al. Filament formation by metabolic enzymes is a specific adaptation to an advanced state of cellular starvation. elife. 2014;3:e02409.

Pimkin M, Pimkina J, Markham GD. A regulatory role of the Bateman domain of IMP dehydrogenase in adenylate nucleotide biosynthesis. J Biol Chem. 2009;284(12):7960–9.

Pizzichini M, et al. Purine de novo synthesis and inosinic branch point in vivo in different tissues: a biomathematical model. In: Adv Exp Med Biol, vol. 253B; 1989. p. 43–5.

Protter DSW, Parker R. Principles and properties of stress granules. Trends Cell Biol. 2016;26(9):668–79.

Puchulu-Campanella E, et al. Identification of the components of a glycolytic enzyme metabolon on the human red blood cell membrane. J Biol Chem. 2013;288(2):848–58.

Rabattoni V, et al. The human phosphorylated pathway: a multienzyme metabolic assembly for l-serine biosynthesis. FEBS J. 2023;290(15):3877–95.

Raushel FM, Thoden JB, Holden HM. The amidotransferase family of enzymes: molecular machines for the production and delivery of ammonia. Biochemistry. 1999;38(25):7891–9.

Reed LJ. Multienzyme complexes. Acc Chem Res. 1974;7(2):40–6.

Reed LJ, Cox DJ. Macromolecular organization of enzyme systems. Annu Rev Biochem. 1966;35(1):57–84.

Robinson AD, Eich ML, Varambally S. Dysregulation of de novo nucleotide biosynthetic pathway enzymes in cancer and targeting opportunities. Cancer Lett. 2020;470:134–40.

Rowe PB, et al. De novo purine synthesis in avian liver. Co-purification of the enzymes and properties of the pathway. J Biol Chem. 1978;253(21):7711–21.

Rudolph J, Stubbe J. Investigation of the mechanism of phosphoribosylamine transfer from glutamine phosphoribosylpyrophosphate amidotransferase to glycinamide ribonucleotide synthetase. Biochemistry. 1995;34(7):2241–50.

Saha A, et al. Akt phosphorylation and regulation of transketolase is a nodal point for amino acid control of purine synthesis. Mol Cell. 2014;55(2):264–76.

Santana-Codina N, et al. Oncogenic KRAS supports pancreatic cancer through regulation of nucleotide synthesis. Nat Commun. 2018;9(1):4945.

Schendel FJ, et al. Characterization and chemical properties of phosphoribosylamine, an unstable intermediate in the de novo purine biosynthetic pathway. Biochemistry. 1988;27(7):2614–23.

Schmitt DL, An S. Spatial organization of metabolic enzyme complexes in cells. Biochemistry. 2017;56(25):3184–96.

Schmitt DL, et al. Spatial alterations of De Novo purine biosynthetic enzymes by Akt-independent PDK1 signaling pathways. PLoS One. 2018;13(4):e0195989.

Schrodinger LLC. The PyMOL molecular graphics system, Version 3.1. 2024.

Senda M, Natsumeda Y. Tissue-differential expression of two distinct genes for human IMP dehydrogenase (E.C.1.1.1.205). Life Sci. 1994;54(24):1917–26.

Sha Z, Benkovic SJ. Purinosomes spatially co-localize with mitochondrial transporters. J Biol Chem. 2024;300(9):107620.

Sigoillot FD, et al. Cell cycle-dependent regulation of pyrimidine biosynthesis. J Biol Chem. 2003;278(5):3403–9.

Spellicy CJ, et al. Characterization of retinal inosine monophosphate dehydrogenase 1 in several mammalian species. Mol Vis. 2007;13:1866–72.

Spiegel EK, Colman RF, Patterson D. Adenylosuccinate lyase deficiency. Mol Genet Metab. 2006;89(1–2):19–31.

Spinelli JB, Haigis MC. The multifaceted contributions of mitochondria to cellular metabolism. Nat Cell Biol. 2018;20(7):745–54.

Srere PA. The metabolon. Trends Biochem Sci. 1985;10(3):109–10.

Sun Z, Liu JL. mTOR-S6K1 pathway mediates cytoophidium assembly. J Genet Genomics. 2019a;46(2):65–74.

Sun Z, Liu JL. Forming cytoophidia prolongs the half-life of CTP synthase. Cell Discov. 2019b;5:32.

Sutherland EW, Rall TW. Fractionation and characterization of a cyclic adenine ribonucleotide formed by tissue particles. J Biol Chem. 1958;232(2):1077–91.

Sweetlove LJ, Fernie AR. The role of dynamic enzyme assemblies and substrate channelling in metabolic regulation. Nat Commun. 2018;9(1):2136.

Tatibana M, et al. Mammalian phosphoribosyl-pyrophosphate synthetase. Adv Enzym Regul. 1995;35:229–49.

Tedeschi PM, et al. Contribution of serine, folate and glycine metabolism to the ATP, NADPH and purine requirements of cancer cells. Cell Death Dis. 2013;4(10):e877.

Thoden JB, et al. Structure of carbamoyl phosphate synthetase: a journey of 96 a from substrate to product. Biochemistry. 1997;36(21):6305–16.

Tibbetts AS, Appling DR. Compartmentalization of Mammalian folate-mediated one-carbon metabolism. Annu Rev Nutr. 2010;30:57–81.

Van den Berghe G, Vincent MF, Jaeken J. Inborn errors of the purine nucleotide cycle: adenylosuccinase deficiency. J Inherit Metab Dis. 1997;20(2):193–202.

Vannoni D, et al. Enzyme activities controlling adenosine levels in normal and neoplastic tissues. Med Oncol. 2004;21(2):187–95.

Varadi M, et al. AlphaFold Protein Structure Database in 2024: providing structure coverage for over 214 million protein sequences. Nucleic Acids Res. 2024;52(D1):D368–75.

Velot C, et al. Model of a quinary structure between Krebs TCA cycle enzymes: a model for the metabolon. Biochemistry. 1997;36(47):14271–6.

Verrier F, et al. GPCRs regulate the assembly of a multienzyme complex for purine biosynthesis. Nat Chem Biol. 2011;7(12):909–15.

Villa E, et al. Cancer cells tune the signaling pathways to empower de novo synthesis of nucleotides. Cancers (Basel). 2019;11(5):688.

Wada Y, et al. Screen of the IMPDH1 gene among patients with dominant retinitis pigmentosa and clinical features associated with the most common mutation, Asp226Asn. Invest Ophthalmol Vis Sci. 2005;46(5):1735–41.

Wang W, et al. The phosphatidylinositol 3-kinase/akt cassette regulates purine nucleotide synthesis. J Biol Chem. 2009;284(6):3521–8.

Wang X, et al. Purine synthesis promotes maintenance of brain tumor initiating cells in glioma. Nat Neurosci. 2017;20(5):661–73.

Wang W, et al. Targeting pyrimidine metabolism in the era of precision cancer medicine. Front Oncol. 2021;11:684961.

Webb BA, et al. The glycolytic enzyme phosphofructokinase-1 assembles into filaments. J Cell Biol. 2017;216(8):2305–13.

Weng WC, et al. Expanding clinical spectrum of PAICS deficiency: comprehensive analysis of two sibling cases. Eur J Hum Genet. 2024;33:870.

Wheaton WW, Chandel NS. Hypoxisa. 2. Hypoxia regulates cellular metabolism. Am J Physiol Cell Physiol. 2011;300(3):C385–93.

Williams JC, et al. Increased CTP synthetase activity in cancer cells. Nature. 1978;271(5640):71–3.

Williamson J, et al. Purine biosynthesis enzymes in hippocampal neurons. NeuroMolecular Med. 2017;19(4):518–24.

Wilson JE. Ambiquitous enzymes: variation in intracellular distribution as a regulatory mechanism. Trends Biochem Sci. 1978;3(2):124–5.

Woulfe J, et al. Inosine monophosphate dehydrogenase intranuclear inclusions are markers of aging and neuronal stress in the human substantia nigra. Neurobiol Aging. 2024;134:43–56.

Wu Z, Liu JL. Cytoophidia respond to nutrient stress in Drosophila. Exp Cell Res. 2019;376(2):159–67.

Wu F, Minteer S. Krebs cycle metabolon: structural evidence of substrate channeling revealed by cross-linking and mass spectrometry. Angew Chem Int Ed Engl. 2015;54(6):1851–4.

Wu F, Pelster LN, Minteer SD. Krebs cycle metabolon formation: metabolite concentration gradient enhanced compartmentation of sequential enzymes. Chem Commun (Camb). 2015;51(7):1244–7.

Wu H, et al. Cyclin D1 extensively reprograms metabolism to support biosynthetic pathways in hepatocytes. J Biol Chem. 2023;299(12):105407.

Xu H, et al. IMPDH2 promotes cell proliferation and epithelial-mesenchymal transition of non-small cell lung cancer by activating the Wnt/beta-catenin signaling pathway. Oncol Lett. 2020;20(5):219.

Yamada S, Sato A, Sakakibara SI. Nwd1 regulates neuronal differentiation and migration through purinosome formation in the developing cerebral cortex. iScience. 2020;23(5):101058.

Yan M, et al. Succinyl-5-aminoimidazole-4-carboxamide-1-ribose 5′-phosphate (SAICAR) activates pyruvate kinase isoform M2 (PKM2) in its dimeric form. Biochemistry. 2016;55(33):4731–6.

Yang M, Vousden KH. Serine and one-carbon metabolism in cancer. Nat Rev Cancer. 2016;16(10):650–62.

Yang C, et al. De novo pyrimidine biosynthetic complexes support cancer cell proliferation and ferroptosis defence. Nat Cell Biol. 2023;25(6):836–47.

Yin Y, et al. Cytoophidia: a conserved yet promising mode of enzyme regulation in nucleotide metabolism. Mol Biol Rep. 2024;51(1):245.

Yoo HC, et al. Glutamine reliance in cell metabolism. Exp Mol Med. 2020;52(9):1496–516.

Zaccolo M, Zerio A, Lobo MJ. Subcellular organization of the cAMP signaling pathway. Pharmacol Rev. 2021;73(1):278–309.

Zarou MM, Vazquez A, Vignir Helgason G. Folate metabolism: a re-emerging therapeutic target in haematological cancers. Leukemia. 2021;35(6):1539–51.

Zech M, et al. Monogenic variants in dystonia: an exome-wide sequencing study. Lancet Neurol. 2020;19(11):908–18.

Zhang Y, Liu JL. The impact of developmental and metabolic cues on cytoophidium formation. Int J Mol Sci. 2024;25(18):10058.

Zhang QC, et al. Structure-based prediction of protein-protein interactions on a genome-wide scale. Nature. 2012;490(7421):556–60.

Zhao H, et al. Quantitative analysis of purine nucleotides indicates that purinosomes increase de novo purine biosynthesis. J Biol Chem. 2015;290(11):6705–13.

Zhou S, Xiang H, Liu JL. CTP synthase forms cytoophidia in archaea. J Genet Genomics. 2020;47(4):213–23.

Zhou X, et al. Structural basis for ligand binding modes of CTP synthase. Proc Natl Acad Sci USA. 2021;118(30):e2026621118.

Chapter 11
Mammalian Respiratory Chain Complex Assemblies and Their Links to Mitochondria Stress-Induced Human Diseases

Runyu Guo and Maojun Yang

Abstract Mitochondria are considered the central organelle in cellular energy metabolism and an integral platform for signal transduction. Respiratory chain complexes are the most abundant and critical protein machines in mitochondria. Thanks to advancing technologies such as cryo-EM, molecular dynamics simulation, and FRET-based live imaging, though still under hot debate, we have now gained a much deeper insight into the organization, regulation, and functional mechanism of the respiratory chain. Accordingly, developing novel compounds targeting mitochondria is particularly appealing, for mitochondria dysfunction might be the underlying cause of many annoying human diseases, including metabolic syndromes, neurodegenerative diseases, cardiovascular diseases, and tumors.

Keywords Respiratory chain complexes · Supercomplexes · Electron transport chain · Complex I · Mitochondria · Structural biology · Mitochondrial disorders · Signaling · Drug discovery

R. Guo
Ministry of Education Key Laboratory of Protein Science, Beijing Advanced Innovation Center for Structural Biology, Beijing Frontier Research Center for Biological Structure, Tsinghua-Peking Center for Life Sciences, School of Life Sciences, Tsinghua University, Beijing, China

Phytovent Biopharma, Changfa International Precision Medicine Accelerator Center, Life Science Park, Changping District, Beijing, China

M. Yang (✉)
Ministry of Education Key Laboratory of Protein Science, Beijing Advanced Innovation Center for Structural Biology, Beijing Frontier Research Center for Biological Structure, Tsinghua-Peking Center for Life Sciences, School of Life Sciences, Tsinghua University, Beijing, China
e-mail: rohickey@coh.org

11.1 Introduction

After more than 100 years of research, mitochondria are now considered crucial organelle in modern cell biology and medical research. The earliest records of mitochondria-like membrane structures can be traced back to the 1840s, and it was not until 1890 that R. Altmann identified this ubiquitous membrane structure in all kinds of cells as a separate class of organelles, called "protoplasts," and proposed for the first time that this organelle may originate from bacteria (Altmann 1890). Unfortunately, this hypothesis was not recognized at the time, and it was not until 1970 that L. Margulis presented more detailed evidence, citing the many structural and genetic similarities between mitochondria and bacteria, that the "endosymbiosis hypothesis" gradually became popular (Margulis 1970). In 1898, C. Benda renamed the organelle as "mitochondria," because he observed that the structure was sometimes thread-like (Greek mitos for "thread"), and sometimes granular (Greek khondrion for "little granule") during spermatogenesis, and the name has remained (Ernster and Schatz 1981).

In 1955, A. L. Lehninger et al. identified three phosphorylation coupling sites in the respiratory chain (Nielsen and Lehninger 1955), the first site consisting of NADH, FMN, and UQ, the second site consisting of cytochrome b and cytochrome c_1, and the third site consisting of cytochrome c, cytochrome a, and cytochrome a_3. In the following study, the phosphorylated-coupling sites in the respiratory chain were found to be in the fixed protein structure, and FAD with iron-sulfur clusters was also found to be involved in electron transport. Since 1960, Y. Hatefi and other researchers in the laboratory of D. E. Green have isolated the respiratory chain protein complex I (NADH dehydrogenase, the first coupling site, CI), complex II (succinate dehydrogenase, CII), complex III (cytochrome c reductase, the second coupling site, CIII), and complex IV (cytochrome c oxidase, the third coupling site, CIV), which laid the foundation for the subsequent study of mitochondrial respiratory chain complexes (RCCs) (Green and Hatefi 1961).

To explain how respiratory chain electron transport is coupled to oxidative phosphorylation, in 1961, P. Mitchell proposed the famous chemo-osmotic hypothesis (Mitchell 1961) and described his theory in detail in 1966 (Mitchell 1966). The concept of "proton driving force" was first proposed. The chemo-osmotic hypothesis contains four aspects: (1) ATP synthase is required to transport protons to synthesize ATP, (2) The electron transport respiratory chain requires simultaneous proton transport for coupled phosphorylation, (3) the simultaneous localization of ATP synthase and respiratory chain on the non-permeable membrane structure is fundamental for the maintenance and utilization of the proton driving force, and (4) there are some transporters on the membrane structure, which can transport ions while consuming the proton driving force. A large number of subsequent studies have continuously confirmed the validity of the chemo-osmotic hypothesis, and the research on the mitochondrial respiratory chain has gradually focused on the molecular mechanism of electron transport and proton transport.

40 years after Y. Hatefi and colleagues isolated CI, CII, CIII, and CIV of the respiratory chain from mammalian mitochondria, in 2000, another discovery profoundly changed the understanding of the composition of the mitochondrial respiratory chain. In this year, H. Shagger et al. developed a new technique for protein gel electrophoresis, Blue Native PAGE (Schägger and Pfeiffer 2000). Using this technique, proteins with molecular weights ranging from 100 kDa to 3000 kDa can be effectively separated and remain active in the gel, which is very suitable for the analysis of macromolecular protein machinery in mitochondria. Their results show that there are supercomplexes (SC) in the form of $SCIII_2IV_1$ and $SCIII_2IV_2$ in yeast mitochondria, and two major supercomplexes ($SCI_1III_2IV_1$ and SCI_1III_2) and two minor supercomplexes ($SCI_1III_2IV_2$ and $SCI_1III_2IV_3$) in mitochondria from bovine heart.

Up to now, the supramolecular assembly of mitochondria RCCs is widely recognized, and the number of articles reporting the relevance between metabolic disruption (or gene mutation), supramolecular assembly of the respiratory chain, and complicated situation of diseases is surging (Braun 2020; Cuillerier et al. 2021; Li et al. 2021; Signorile et al. 2022; Ukolova et al. 2020). Our lab was the first to report the complete high-resolution 3D structure of mammalian respiratory chain $SCI_1III_2IV_1$ (Gu et al. 2016; Wu et al. 2016), megacomplex $MCI_2III_2IV_2$ Guo et al. 2017), and IF1-inhibited ATP synthases tetramer (Gu et al. 2019). These structures updated our knowledge of the interactions among respiratory chain complexes, the compartmentation and channeling of substrates, the bending of the membrane by protein, and the mechanism of coupling between proton pumping and electron transfer (Guo et al. 2016, 2018). Recently, our lab reported many new details of the structure of RCCS, including CI combined with six different substrates and thus in different conformations (Gu et al. 2022), and $MCI_2III_2IV_2$ with fic different states in which the conformations of CI, CIII, and CIV change coordinately (Zhang et al. 2024).

Disruption of mitochondrial biological function can lead to systemic cellular disorders, including abnormal production of reactive oxygen species (ROS), dysregulation of calcium homeostasis, defective mitochondrial synthesis, disruption of mitochondrial homeostasis and quality control, necrotizing cell death induced by mitochondrial permeability pore (MPTP), inappropriate activation or inhibition of apoptosis, decreased cellular ATP/ADP ratio, decreased NAD^+ levels, activating inflammation, and alterations in mitochondrial signaling pathways (AMPK, mTOR, Hif1α, etc.) (Padhy et al. 2024; Gandhi et al. 2024; Qiang 2020; Choi et al. 2024; Hong et al. 2024), which means mitochondria could be the "Achilles' Heel" of many kinds of diseases. In support, a soaring number of researches have demonstrated that mitochondria morphology, mitochondria quality control, mitochondria ROS, and mitochondria metabolites are deeply involved in the development of various kinds of diseases (Ulfig and Jakob 2024; Santos 2021; Al Amir Dache and Thierry 2023; Johnson et al. 2024), such as cardiovascular disease, neurodegenerative disease, metabolic stress, chronic inflammation, and aging. So, the purpose of this chapter is to depict the state-of-the-art structural details of the respiratory chain and to explain why manipulating the respiratory chain supramolecular assemblies can help to conquer those notorious diseases. Based on these structural details, our lab has been working on developing new drugs targeting key parts of RCCs.

11.2 Basic Structural Concepts of Respiratory Chain Complexes

11.2.1 Basic Structural Concepts of CI

There are 45 subunits in mammalian CI, among which the 14 core subunits are directly involved in electron transport and proton transport, and the other 31 accessory subunits are proven to be necessary for proper CI assembly and function. Seven core subunits are located in the hydrophilic arm, responsible for electron transfer, and the other seven core subunits in the transmembrane arm for proton pumping. The overall structure of CI is L-shaped.

The electron transporters in CI are all bound in the hydrophilic arm, including the immobilized FMN and eight iron-sulfur centers, seven of which are involved in electron transport, as well as NADH and Q which are continuously bound and released. The FMN and NADH binding sites are within the NDUFV1 subunit, which is also the binding site for the N3 ([4Fe4S]) iron-sulfur center. The other seven iron-sulfur centers were N1a ([2Fe2S]), N1b ([2Fe2S]), N2 ([4Fe4S]), N4 ([4Fe4S]), N5 ([4Fe4S]), N6a ([4Fe4S]), and N6b ([4Fe4S]), binding to NDUFS1, NDUFV2, NDUFS7, and NDUFS8 subunits, respectively, and N1a ([2Fe2S]) is not involved in electron transport. The binding site of Q is located between NDUFS2, NDUFS7, and ND1. The head of Q penetrates the hydrophilic arm of complex I, where it directly interacts with Y87 and H38 in the NDUFS2 subunit and is located approximately 12 Å away from the N2 ([4Fe4S]) iron-sulfur center in NDUFS7 and approximately 15 Å away from the inner mitochondrial membrane surface. The tail chain of Q is long, starting from the head and passing through a semi-closed tube enclosed by NDUFS2, NDUFS7, and ND1, directly into the inner mitochondrial membrane. The surface of the half-closed tube is mostly charged or polar amino acids, facilitating the passage of the Q polar head. This half-closed tube is referred to as the "Q chamber" (Fig. 11.1a).

The channels that transport protons across the membrane in CI are all located within the transmembrane arm. At present, the mainstream view is that the transmembrane arm contains four proton channels located in ND5, ND4, ND2, and (ND1 + ND3 + ND4L + ND6). The distribution of transmembrane helices of ND5, ND4, and ND2 is nearly identical, and their structures are in turn very similar to those of the known antiporter proteins MrpA and MrpD, so these three subunits are considered antiporters in CI. Each subunit contains 13 transmembrane helices, and ND5 also has a long lateral helix (HL) (Fig. 11.1b).

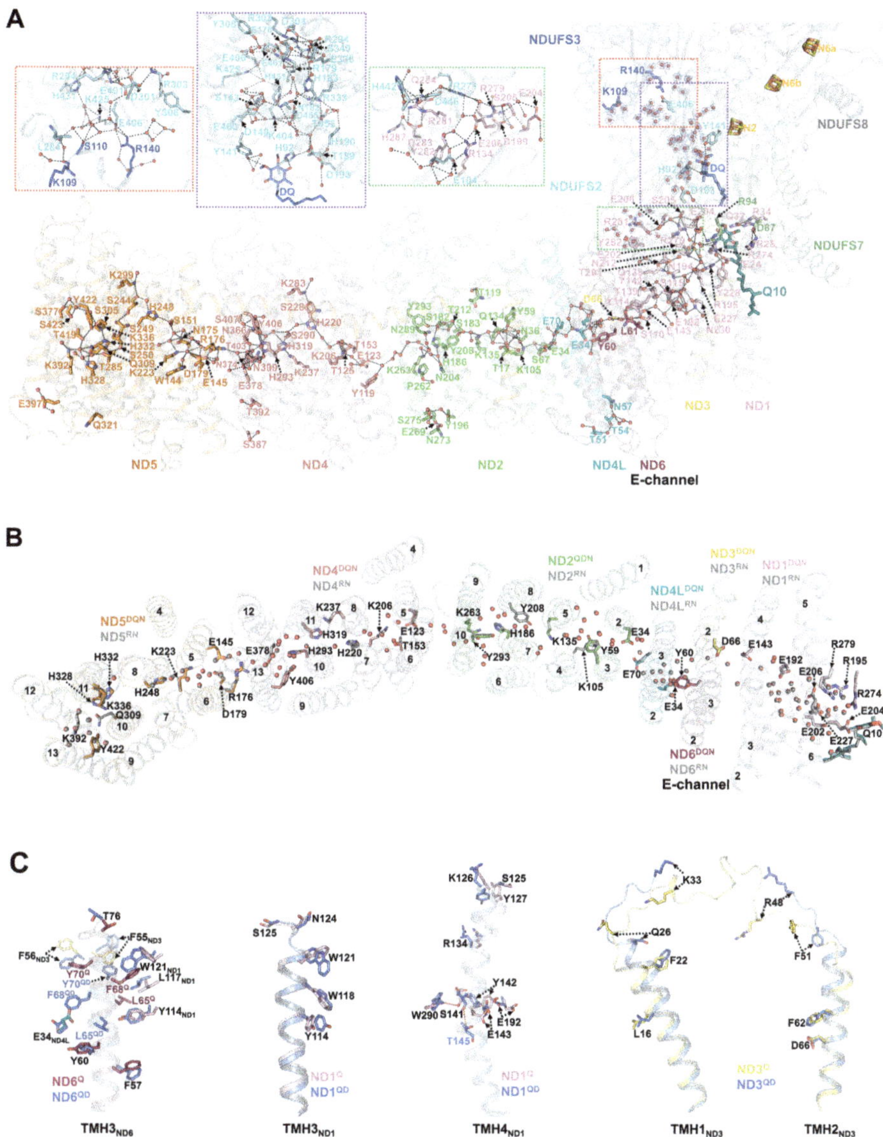

Fig. 11.1 Key features and active/deactive transition of CI. (**a**) Key polar residues in the E-channel, three antiporter-like subunits and NDUFS2 shown as sticks. Associated water molecules are shown as spheres. H-bonds are shown as black dashed lines. The left and middle insets zoom in on an H-bond network formed between ordered water molecules and polar residues within NDUFS2, pointing to the mitochondrial matrix. The right inset zooms in on an H-bond network formed between ordered water molecules and polar residues at the interface of ND1 and NDUFS2, connecting the mitochondrial matrix and the E-channel. PDB accession code: 5GUP. (**b**) Conserved polar residues and water molecules on the central hydrophilic axis of the E-channel, ND2, ND4, and ND5. PDB accession code: 5GUP. (**c**) Typical conformational changes between active CI and deactive CI. PDB accession code: 5GUP

Of these 13 transmembrane helices, TM4-8 forms one half-channel connecting the mitochondrial matrix and is denoted as the N channel, while TM9-13 forms the other half-channel connecting the mitochondrial membrane space and is denoted as the P channel. TM7 and TM12 have a break in the middle of the helix, and both have a conserved Lys at the break (Glu at the ND4_TM12 break), denoted TM7Lys and TM12Lys. TM8 has a bend in the middle of the helix and a conserved Lys at the bend (His at the bend of ND5_TM8), denotcd TM8Lys. TM5, although not broken, also has a conserved Glu in its mid-helix, denoted TM5Glu. The N channel and P channel are connected by TM8Lys to form a complete proton transport channel (Fig. 11.1b).

In addition to ND5, ND4, and ND2, a fourth proton channel is thought to be composed of ND1, ND3, ND4L, and ND6 together. In the middle of their trans-membrane helices, E213, E163, and E130 (not conserved) in ND1, D72, and E74 in ND3, and E32 and E67 in ND4L form a "glutamate channel," with one end connected to TM5Glu in ND2 and the other to Glu in the Q chamber. This glutamate channel is not only considered part of the fourth proton channel of complex I but also plays a role in connecting the Q reaction site with the proton transport subunit (Fig. 11.1a).

11.2.2 Basic Structural Concepts of CIII

Mammalian CIII is a homodimer composed of a total of 22 subunits. In each monomer, three conserved subunits, cytochrome b, cytochrome c_1, and Rieske protein (ISP), form the core subunits for electron transport and proton transport. Cytochrome b contains two hemes, heme b_H (near the mitochondrial matrix, with a maximum absorption peak at 562 nm) and heme b_L (near the mitochondrial membrane space, with a maximum absorption peak at 566 nm), cytochrome c_1 contains heme c_1, and Rieske protein contains a [2Fe2S] center. These are the electron transporters in CIII. The prevailing view is that the cytochrome b subunit contains two Q/QH_2 binding sites, a Q_i site near heme b_H and a Q_o site near heme b_L. Cytochrome c binds to cytochrome c_1, is oxidized by the latter, and leaves CIII into the IMM (inner mitochondria membrane) (Fig. 11.2a).

The conformation of the two core subunits, cytochrome b and cytochrome c_1, is fixed, while the structures of CIII with different inhibitors revealed that the hydrophilic end of the third core subunit, Rieske protein, can undergo drastic conformational changes, possibly related to the release and binding of QH2 at the Q_o site. The two Rieske proteins are in a crossed state in the CIII dimer, with its N-terminal transmembrane region anchored to one monomer of CIII, whereas its C-terminal hydrophilic region is bound between the cytochrome b and cytochrome c_1 subunits of another CIII monomer, with the [2Fe2S] center in the Rieske protein located in the hydrophilic region. This phenomenon indicates that CIII must function as a dimer (Fig. 11.2a).

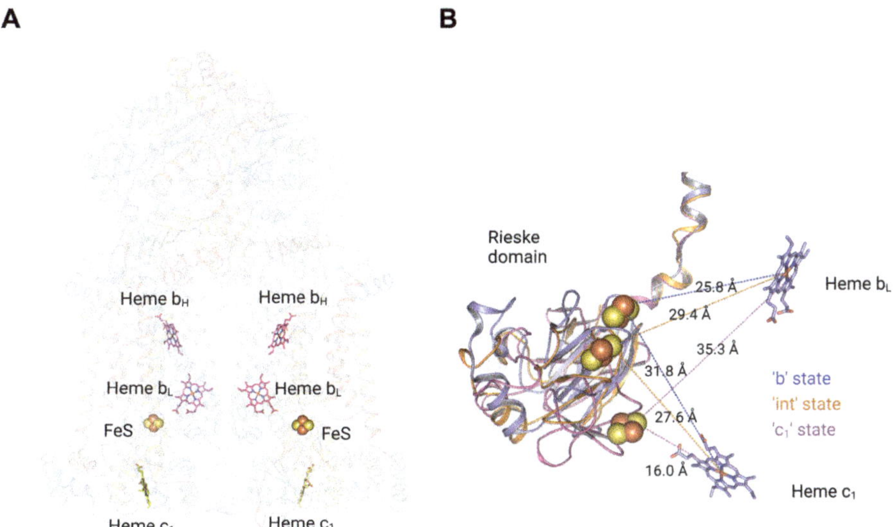

Fig. 11.2 Key features of CIII. (**a**) Electron carriers in the CIII dimer, including two 6 hemes and 2 FeS clusters. PDB accession code: 2A06. (**b**) the movement of Rieske domain in the three states of CIII. The distance between the FeS cluster in the Rieske domain and the two hemes are measured and labeled. PDB accession codes: b state (2A06), int state (1BGY), c1 state (1BE3)

CIII can bind a wide variety of inhibitors, most of which specifically distinguish between the Q_i and Q_o sites. For example, antimycin A binds specifically to the Q_i site, while stigmatellin and myxothiazol bind specifically to the Q_o site. This indicates that the Q_i and Q_o sites must be different in their electrochemical properties. Stigmatellin and myxothiazol, when bound to the Q_o site, can cause Rieske protein to be in different conformations, changing the distance between the center of Rieske protein [2Fe2S] and heme b_L and heme c_1. Upon stigmatellin binding, Rieske [2Fe2S] comes close to heme b_L, whereas upon myxothiazol binding, Rieske [2Fe2S] is released from heme b_L. According to the change in the central position of Rieske [2Fe2S], three different states of CIII, namely, "b_L," "c_1," and "int," can be identified. This property of the Rieske protein suggests that it may transfer electrons between heme b_L and heme c_1 (Fig. 11.2b).

11.2.3 Basic Structural Concepts of CIV

The crystal structure of mammalian CIV is in the form of a homodimer. Each monomer contains 13 subunits, of which 3 subunits (SU1, SU2, SU3) are encoded by mitochondrial genes and are the core subunits for electron and proton transport. SU1 contains two hemes, heme a and heme a_3, and also contains a Cu center (Cu_B) consisting of two copper atoms. SU2 contains another Cu center (Cu_A), again composed of two Cu atoms. SU3 contains seven transmembrane helices and three stable phospholipid molecules, which directly bind to SU1 to deliver oxygen molecules to SU1 and protect the redox reaction site. The other ten subunits, which are encoded by nuclear genes, are auxiliary subunits. The monomer of complex IV has a molecular weight of 210 kDa and a total of 14 transmembrane helices. There are at least two classes of proton channels in complex IV, one that delivers protons to the redox reaction site and the other that transports protons directly from the matrix into the membrane space.

CIV includes two copper centers (Cu_A and Cu_B), and two heme (heme a and heme a_3), wherein the copper atom in CuB and the iron atom in heme a_3 form a binuclear reaction center (iron-copper center), and the molecular oxygen is bound between the two atoms. Cytochrome c binds to the IMM side of CIV and transfers an electron to the Cu_A/heme a site, which then transfers the electron to the oxygen-oxidized heme a_3/Cu_B [FeCu] center.

In the process of CIV transferring four electrons to oxygen molecules to generate water, four protons need to be absorbed from the substrate side to be transferred to oxygen. At the same time, protons are transported directly from the matrix into the membrane space through allosteric modulation of heme a site coupling by redox reaction (conformational change of Asp51). Proton transport in complex IV is mediated by a network of hydrogen bonds between a large number of acid (base) amino acid residues and immobilized water molecules.

These hydrogen bond networks can be summarized into three proton transport pathways, the H pathway, the K pathway, and the D pathway. The K pathway is responsible for transporting two of the four protons required for the oxygen reduction reaction, delivering the protons within the matrix via the conserved Lys319 to the conserved Tyr244 that covalently binds heme a_3. The D pathway is responsible for transporting the other two protons required for the oxygen reduction reaction, transferring protons within the matrix via the conserved Asp91 to the nearby conserved Glu242 and ultimately to OH^-, to form water. The H-pathway transfers protons from the matrix to the conserved Arg38 via His413 and heme a and then to the IMM through the hydrogen bond network of Arg38, heme a, and Asp51.

11.2.4 Basic Structural Concepts of SC

Since its first report in 2000, BN-PAGE has been the most authoritative and widely used method to analyze the assembly form of SCs. On the BN-PAGE, bands above the CI band are considered as some kinds of SCs, and the number and position of the SC bands could vary according to different species. The most classic form is $SCI_1III_2IV_1$, while the number of CIV and the connecting subunit of CIV (COX7A2L or COX7A2) could vary, leading to different bands. Besides, more and more evidence suggests the existence of SC forms with two CIs.

To describe the assembly state of RCCs, three models were proposed. First, the fluid model, where all redox components are independent diffusible particles with the small electron carriers shuttling between the huge respiratory complexes I–IV;

hence electron transport is considered a multi-collisional, obstructed, and long-range diffusional process. Second, the solid model, where CI, CIII, and CIV must assemble into different forms of SCs to be functional, including $SCI_1III_2IV_1$, $SCI_1III_2IV_2$, $SCI_1III_2IV_4$, $SCIII_2IV_1$, and $SCIII_2IV_2$. Those detected free forms of RCCs were interpreted as assembly intermediates. Third, the plastic model, where the organization of OXPHOS complexes is very flexible. Both the assembled SCs and free individual complexes can perform their function, with SCs being more efficient in energy generation and less active in ROS production. The ratio of free complexes integrating into SCs is very likely under elaborate regulation, to accommodate to different demands of the cell environment. Up to now, most structural and kinetic studies support the plastic model.

11.3 The Hotly Debated Coupling Mechanism of CI

The matrix arm of CI is responsible for electron transfer from NADH to Q, which is a process with common consensus; the membrane arm of CI is responsible for proton pumping, whose detailed proton channel is not certain; as for the coupling mechanism between electron transfer and proton pumping, things are far more complicated, and many mechanisms are proposed (Hirst 2013; Kampjut and Sazanov 2020; Djurabekova et al. 2022; Kampjut and Sazanov 2022).

In the matrix arm, there are eight or nine FeS clusters, depending on different species, and normally seven of them form the mainstream of electron transfer from NADH to Q. In the steady state with superfluous NADH available, it was estimated that every other FeS cluster in the mainstream was reduced with one electron, and most of the still reducible clusters are nearly equipotential with the $NAD^+/NADH$ pair at around -250 mV (Bridges et al. 2012; Euro et al. 2008). The only cluster with a higher potential is the final N2 cluster at around -150 mV, and the largest drop in redox potential occurs between N2 and the quinone/quinol pair ($+100$ mV), indicating that the crucial energy-releasing step is the quinone reduction, protonation, or perhaps even its release out of the binding cavity. However, no massive conformational change enough to trigger the initiation of proton pumping was observed from any of the resolved structures of CI at this Q binding site.

In the membrane arm, three subunits, ND2, ND4, and ND5, evolutionarily related to the Mrp Na^+/H^+ antiporter were naturally considered as proton channels (Li et al. 2020; Steiner and Sazanov 2020), and a fourth proton channel was proposed to be encircled by ND1, ND3, ND6, and ND4L. The Lys or Glu residues at the middle of transmembrane helices of these subunits were observed to form a hydrophilic line within the middle of the hydrophobic membrane and were considered critical for proton translocation (Efremov and Sazanov 2011). A Glu-rich tunnel was observed to link this hydrophilic line with the Q reduction site and proposed to initiate the proton pumping (Fig. 11.1). However, recently, some groups claim that based on water molecules and hydrogen network observed in the latest structure of the CI membrane arm, ND5 could probably be the only proton channel that is pumping protons, while the other channels are all less hydrated and not likely to be pumping protons (Kampjut and Sazanov 2020; Mühlbauer et al. 2020; Parey et al. 2021).

Vaguely two kinds of models of coupling mechanism were proposed by recent studies, an electrostatic wave model mostly based on molecular dynamic simulation (Warnau et al. 2018; Hoias Teixeira and Menegon Arantes 2019; Friedman et al. 2021), and an open-close turnover model mostly derived from structural studies showing significant conformational change (Kampjut and Sazanov 2020). In the electrostatic wave model, based on single-particle Heisenberg equations of motion, two-particle Coulomb interaction for the exchange energy, and phenomenological Langevin equations for the moving parts (shuttles and conformation changes), the authors proposed that the piston movement of electrons in the matrix arm could ignite an electrostatic wave across the membrane domain, and energy released from electron transfer is thus converted to $\Delta\Psi m$ by coupled proton pumping (Friedman et al. 2021). However, no solid structural evidence could support this assumption. In the open-close turnover model, the authors determined two conformations of CI experimentally, "open state" and "close state." The binding and releasing of Q in the Q chamber induced conformational changes in the whole CI structure, and these changes led to the pumping of protons (Kampjut and Sazanov 2020). However, many argued that the "open/close state" of CI is hard to distinguish from the "deactive/active state" of CI (Fig. 11.1c), which is not related to coupling between electron transfer and proton pumping but to ischemia-reperfusion (Blaza et al. 2018; Vial et al. 2019; Roca et al. 2022). Our structural data support the "deactive/active" theory of CI conformational change, and we proposed a "Double Q model" based on our CI structures in different substrate binding states, which will be discussed below (Gu et al. 2022).

11.4 High-Resolution Structure of CI in Different Substrate Binding States

Since most CI are assembled into supercomplexes (SCs) in mammalian mitochondria, we thought that the CI structure purified from $SCI_1III_2IV_1$ might be closer to the in vivo conformation. Therefore, we prepared porcine heart $SCI_1III_2IV_1$ cryo-EM samples under six treatment conditions, including Q10, Q10 + NADH, rotenone, rotenone+NADH, Q1 + NADH, and decyl-ubiquinone (DQ) + NADH, to capture CI in different Q binding states (CI^Q, CI^{QN}, CI^R, CI^{RN}, CI^{Q1N}, CI^{DQN}, respectively) (Gu et al. 2022). In the Q chamber of the six active CI structures, we successfully identified three different Q10 binding sites, namely, Site 1, Site 1F, and Site 2. The second Q binding site in the Q chamber has been detected by many groups. In addition, we detected a Q10 binding at Site 3 in the Q chamber of the deactive state CI. Notably, we identified a second Q-binding cavity formed by subunits NDUFA9 and ND3, in which Q10 binding was also detected at Site 4.

In CI^Q, one native Q10 was detectable in the previously described Q chamber. We defined the position of the Q10 head group in CI^Q as Site 1, where the methoxy group of Q10 forms a hydrogen bond with His92$_{S2}$ (Fig. 11.3a–c). Site 1 can not only bind UQ10 but also rotenone, DQ, and Q1. Most of the residues that interact with rotenone are also involved in UQ10 binding, suggesting that rotenone can mimic the binding mode of UQ10 and thus be a competitive inhibitor of CI. The binding mode of Q1 at Site 1 is almost identical to that of Q10, but a second flipped Q1 was unexpectedly observed in Site 1^F. The flipped Q1 in Site 1^F indicates that the short-chain Q1 has greater conformational freedom than DQ and Q10, which is consistent with MD simulations and associated kinetic data (Fedor et al. 2017; Haapanen et al. 2019).

In addition to Site 1 and Site 1^F, we identified Site 2 in the Q chamber. In CI^{Q1N}, CI^R, CI^{RN}, CI^{Q1N}, and CI^{DQN}, we all detected a rod-like density extending from the charged middle region to the exit of the Q chamber, suggesting that the native Q10 does not leave the Q chamber after the addition of rotenone, NADH, Q1, or DQ. The position of the Q10 head group in CI^{Q1N} is defined as Site 2, in the charged middle region of the Q chamber (Fig. 11.3d–f).

Oxidized Q is reduced to its reduced form QH$_2$ at Site 1. To investigate the details of this redox reaction, we prepared CIQN by adding NADH to native CIQ. Despite the background noise and discontinuous density observed in Site 1 of CIQN, the density of Q10 in Site 2 was surprisingly clear and uninterrupted, suggesting that QH2 easily moved from Site 1 to Site 2. More specifically, our CIQN structure suggests that native Q10 will not leave the Q cavity after being reduced at Site 1 (Fig. 11.3d–f).

Site 2 is surrounded by a large number of polar residues from ND1 and NDUFS7, and there is a complex network of hydrogen bonds connecting to Site 2 from two directions. In one direction, polar residues from ND1 and NDUFS2 link Site 2 to the mitochondrial matrix. In the other direction, polar residues from ND1 link Site 2 to the glutamate channel, which can be further linked to key residues in the transmembrane proton pumps. The carefully organized network of hydrogen bonds connected to Site 2 suggests that this site may be important in coupling electron transport to proton pumping (Gu et al. 2022) (Fig. 11.3d–f).

CI can switch between active and deactive states (Blaza et al. 2018; Vial et al. 2019; Roca et al. 2022) (Fig. 11.1c). In contrast to the active CI, in the Q chamber of the deactive CI, there was no reasonable substrate or inhibitor density at Site 1, 1F, or 2, whereas there was a clear Q10 head density near the exit of the Q chamber. We define the position of this Q10 head in the deactivated CI structure as Site 3, and Q10 at this site is not involved in electron transfer.

In our CI^Q structure, we unexpectedly found another density of UQ10 in the gap of the C-terminal domain of NDUFA9 and defined the position of the Q10 head group within it as Site 4 (Gu et al. 2022). NDUFA9 plays an important role in CI assembly and is essential for stabilizing the connection between the matrix and the membrane arm. In deactivated CI^{QD}, Arg212$_{A9}$ wobbles about 5 Å to bind the NADPH molecule, and the density near NDUFA9$_{helix9-10}$ is disrupted, so that the second Q-binding pocket cannot be detected. Therefore, we suggest that Q10 bound

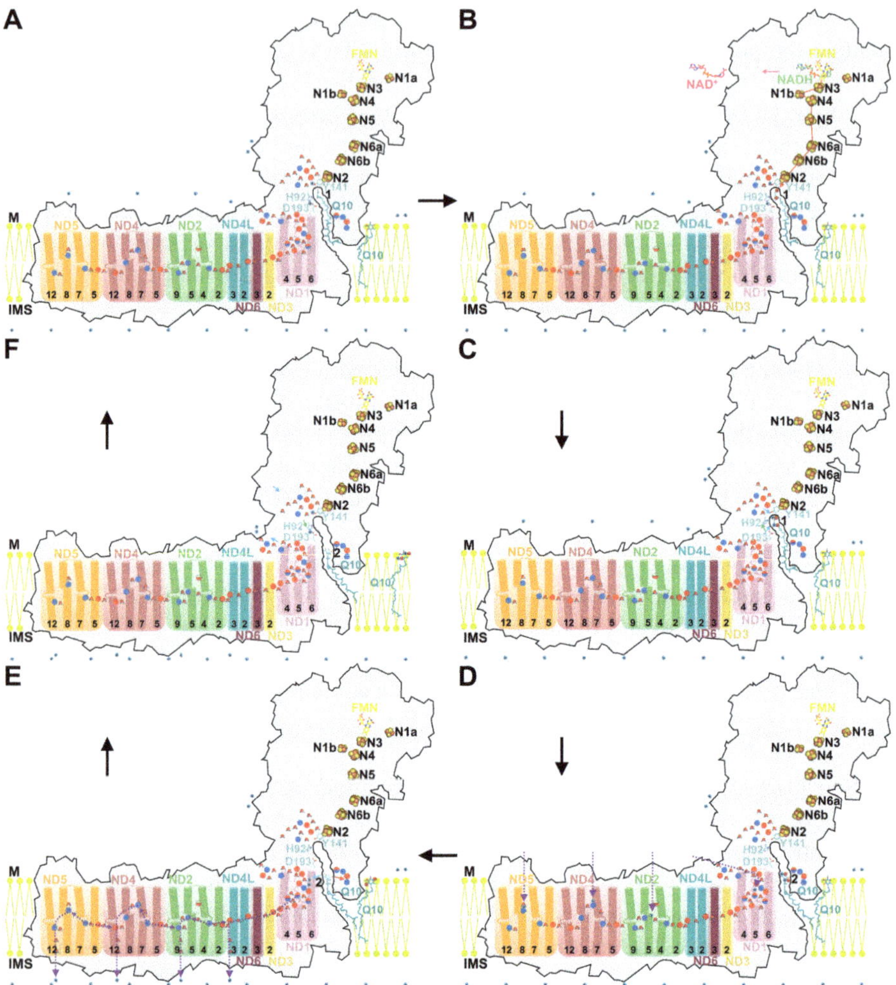

Fig. 11.3 The two-Q model of CI. Transmembrane helices of membrane core subunits involved in proton translocation are labeled by numbers. Proton translocation pathways are indicated by purple dashed arrows. Q-binding sites 1 and 2 are labeled 1 and 2. The blue circle is a positively charged residue; the red circle is a negatively charged residue; the blue circle with "+" is a proton; the red circle with "−" is an electron; and the bent red and gray circles are water. The red arrow shows electron transfer, the blue arrow proton transfer, and the green arrow conformational change. FMN, flavin mononucleotide; IMS, intermembrane space; M, matrix. Active-state CI invariably binds one Q10 molecule in its Q chamber. (**a**) At first, oxidized Q10 binds at Site 1. (**b**) Q10 is reduced to Q102− at Site 1 by NADH. (**c**) This Q102− is then protonated by both His92S2 and Tyr141S2 nearby, generating QH2-10. Deprotonated His92S2 undergoes a conformational change away from Asp193S2. (**d**) QH2-10 moves from Site 1 to Site 2. Four protons enter the proton half-channels from the matrix side. (**e**), At Site 2, two electrons are transferred by QH2-10 to a secondary Q10, reducing it to Q102−. The protons released from QH2-10 trigger proton pumping from the E-channel, ND2, ND4, and ND5. (**f**) His92S2 and Tyr141S2 are subsequently re-protonated by the protons from the mitochondrial matrix. The secondary Q102− is protonated by protons from mitochondrial matrix and moves away from CI. Oxidized Q10 at Site 2 moves to Site 1 for the next catalytic cycle

to Site 4 may play a role in assisting CI assembly and maintaining CI in the active state.

In summary, we report the cryo-EM structures of porcine CI under six treatment conditions with resolutions ranging from 2.37 to 3.51 Å, where each condition can be classified as active and deactive (Gu et al. 2022). These structures shed light on how hydrophobic Q10 with a long isopentenoid tail is bound and reduced in a narrow Q-binding chamber. Conformational changes during the active/deactive transition occur primarily around the Q chamber, and the deactive state may represent the offline state (Fig. 11.1c). The membrane-anchored C-terminal domain of NDUFA9 binds to another Q10 molecule, which can act as a sensor to detect Q10 molecules dissolved in IMM and regulate CI activity by assisting CI assembly and stabilizing CI structure. Based on this exhaustive structural information, we were able to analyze Q10 dynamics in the Q chamber, the active/deactive switching process, and the effect of NDUFA9 on CI activity.

Finally, we proposed a "Double Q model," where the Q10 molecule will not leave the Q chamber after being reduced but transfer the electrons to a second Q surrounding the entrance (Fig. 11.3d–f), which is a similar process as the electron transfer process between two Qs in Ndi1 (Feng et al. 2012), and the oxidation of Q at Site 2 could trigger the coupling of proton pumping. In addition, a recent docking and molecular simulation study also a Q-binding site on the surface of respiratory complex I (Djurabekova et al. 2022), which is consistent with our proposal. However, this model needs to be verified by further computational and biochemical approaches.

11.5 Cooperative Assembly of CI and SC

11.5.1 CI and SC Assembly Could Enhance Cellular Respiration

The composition of SC can be varied in a given species, and can even be more varied in different species (Braun 2020; Ukolova et al. 2020). Furthermore, as the level of energy requirement within the cell changes, the composition of the SC changes accordingly (Cuillerier et al. 2021; Signorile et al. 2022; Scrima et al. 2023). Though several mechanisms have been proposed to explain the assembly process (Fang et al. 2021), regulation pathways (Marx et al. 2024), and physiological properties of SC (Novack et al. 2020), as will be discussed below, we must admit most functional roles of SC are still enigmatic.

CI is the largest complex among the OXPHOS and is rate-limiting for cellular respiration. It has been reported by many groups that the majority of CI (around 80%) is found within SC (Schägger et al. 2004; Wittig et al. 2006), and SC is strictly required for the full assembly and stability of CI (Moreno-Lastres et al. 2012;

Tropeano et al. 2020). More than 18 chaperons have been recognized as assembly factors of CI (Signes and Fernandez-Vizarra 2018), and the assembly profile of CI was well-documented (Guerrero-Castillo et al. 2017). Although the dynamic complexome profiling approach has pictured the sequence of CI subcomplex clustering into holo-CI and suggests CI assembly is completed before SC formation starts (Guerrero-Castillo et al. 2017), many recent studies declared that a pre-CI around 830-kDa can interact with CIII and CIV to form a pre-respirasome and a cooperative assembly model of CI and SC was proposed (Fang et al. 2021; Moreno-Lastres et al. 2012; Alston et al. 2018; Protasoni et al. 2020). More accurately, the latest study proved that the P_D-a module alone can function as the scaffold to recruit CIII and CIV with the help of SCAF1, which can further recruit the P_P-b module, P_D-b module, P_P-a/Q module, and finally N module to form the complete SC, and it is the P_D module, rather than other CI modules, that is critical for the maintenance of CIII and CIV levels (Fang et al. 2021).

The biological significance of SC in promoting OXPHOS still waits to be verified; however, several advantages of forming SC are widely accepted (Novack et al. 2020): (1) assembly into SC can help to stabilize the structure of CI, CIII, and CIV, for numerous studies have demonstrated that the CIII and CIV association could favor the assembly and stability of holo-CI, and dysfunctional CI can in turn lead to lower activity of CIII and CIV. (2) Assembly into SC can reduce the rate of ROS generation. The matrix arm of CI and the Rieske domain of CIII is the major ROS sources in the cell, while their assembly into SCs can effectively isolate these ROS-producing sites from the outside environment. (3) Assembly into SC can prevent undesired aggregates of RCC subunits. The IMM is crowded with proteins; therefore, the correct assembly of SCs may be crucial in minimizing the possibility of nonfunctional aggregation or nucleation of their components. (4) Assembly into SC can favor ubiquinone and Cyt. c shuttling between RCCs by minimizing the distance between the coenzyme contacting sites. Though some enzyme kinetic studies introducing an alternative Q oxidase, AOX, into the OXPHOS membrane showed results against the compartmentation of Q pool, many other kinetic studies with different inhibitors and in vivo studies measuring the respiration parameters have provided evidence supporting that assembly into SC favors the OXPHOS process (Sabbir et al. 2023; Kobayashi et al. 2023; McKenzie et al. 2006).

11.5.2 The Regulation of SC Assembly

Several mechanisms to regulate the assembly of SC have been proposed, including the cAMP/PKA pathway (Signorile et al. 2022; Scrima et al. 2023; Parmar et al. 2024), CHRM1-β-arrestin-MAPK signaling cascade (Sabbir et al. 2023), spleen tyrosine kinase (SYK) pathway (Kobayashi et al. 2023), lipid composition (Leon et al. 2010), $\Delta\Psi m$ (Scrima et al. 2023), and CI or CIV dysfunction (Cuillerier et al.

2021; Marx et al. 2024). One group reported that the activation of the cAMP/PKA cascade resulted in an increase in SC formation associated with an enhanced capacity of electron flux and ATP production rate, which could probably be owing to the phosphorylation of CI subunit NDUFS4 (Signorile et al. 2022). On the contrary, another group following the metabolic flux theory reported that the activation of cAMP/PKA signaling could phosphorylate serine residues at the contact interfaces of the three complexes and hinder the assembly of SC. Also based on the metabolic flux theory, the authors suggest that elevated $\Delta\Psi m$ could lead to the separation of RCCs, and loss of $\Delta\Psi m$ could favor the assembly of SC (Scrima et al. 2023).

Mitochondria properties from cortical neurons were carefully analyzed in CHRM1 knockout (Chrm1−/−) and wild-type mice to identify mitochondrial abnormalities, and the authors claimed that loss of Chrm1 led to a significant reduction in cortical mitochondrial respiration (oxygen consumption) concomitantly associated with reduced oligomerization of CV and supramolecular assembly of SC (Sabbir et al. 2023). Using a FRET-based respirator assembly screen, one group screened over 1200 bioactive small molecules and identified SYK inhibitors as enhancers of MRC supercomplex formation, which was validated using BN-PAGE, and genetically confirmed with siRNA.

The IMM is highly enriched in phosphatidylethanolamine (PE) and cardiolipin (CL), the latter an evolutionarily conserved dimeric phospholipid that is precisely located in mitochondria. Studies on cells from patients with a hereditary disease known as Barth syndrome (BTHS), whose CL maturation is hindered, demonstrated that mitochondria from BTHS lymphoblasts show reduced abundance and instability of SCs which in turn results in a higher rate of CL degradation (McKenzie et al. 2006; Xu et al. 2016). Untargeted lipidomics of isolated mitochondria from neurons of McGill-R-Thy1-APP transgenic (Tg) rats observed a 60% decrement in mitochondrial CL and PE, and the authors speculate that at early stages of AD pathology, there could be a disassembly of the SCs of neuronal mitochondria due to alteration in CL and PE synthesis, which in turn may promote neuronal bioenergetics dysfunction (Leon et al. 2010).

Many studies have proved the activity and abundance of CI or CIV could influence SC formation. Inhibitors of CI were reported to be associated with increased ROS production, impaired calcium homeostasis of cells, reduced efficiency of ATP generation, and reduced levels of SC abundance (Marx et al. 2024). In hepatic Lrpprc knockout mice (Cuillerier et al. 2021), low levels of the mtRNA binding protein LRPPRC induce a global mitochondrial translation defect and a severe reduction (>80%) in the assembly and activity of CV, however, the animals show no signs of overt liver failure, and capacity of the ETC is preserved. Beyond stimulation of mitochondrial biogenesis, an enrichment of the residual CIV in supercomplexes (SCs) is observed, pointing to a role of deficient CIV in promoting SC formation.

11.6 Conformational Change in MCs

11.6.1 Verification of MC Existence

As discussed in the previous section, it is generally believed that RCC tends to assemble into various SCs, including $SCIII_2IV_1$, $SCIII_2IV_2$, $SCI_1III_2IV_1$, and so on. In 2017, we used Cryo-EM to identify an even larger complex containing two CIs, a CIII dimer, and two CIVs and defined it as $MCI_2III_2IV_2$ (Guo et al. 2017). Despite the tremendous efforts of recent studies mentioned above to show the regulating mechanism and physiological significance of SC, the picture of SC functions and properties is far from clear. As for MCs, the situation is even worse. In this section, we present our latest observation of MC structure and discuss why MC is structurally important in energy metabolism (Zhang et al. 2024).

After the initial discovery of the giant protein machine ($MCI_2III_2IV_2$), there was debate about its authenticity. In a previous study, cryo-ET images of bovine mitochondrial membranes failed to detect $MCI_2III_2IV_2$, probably because the authors destroyed mitochondria to obtain the final membrane fragments (Davies et al. 2018), which would certainly eliminate the transmembrane potential and could lead to mechanical disruption of the RCC assembly. However, more and more studies have shown that there are larger assemblies of RCCs than the classic SCI1III2IV1. In another cryo-ET study, the authors report the in situ structure of four kinds of larger assemblies of RCCs in mitochondria, which irrefutably verified our view that RCCs can assemble into $MCI_2III_2IV_2$ (Zheng et al. 2024). In addition, our latest research taking advantage of the powerful gSTED technology even detected intact $MCI_2III_2IV_2$ in live cells. gSTED images were obtained from rapidly fixed cells without disruption of cellular or mitochondrial structures to preserve the intact $MCI_2III_2IV_2$ giant complex (Zhang et al. 2024). We labeled the NDUFS1 subunit of CI and the Cox4I1 subunit of CIV, both located on the outer surface of $MCI_2III_2IV_2$, to visually screen for components containing both CI and CIV. These selected particles strongly suggest the presence of substantial amounts of $MCI_2III_2IV_2$ in live cells (Fig. 11.4).

Meanwhile, we reported the high-resolution cryo-EM structure of the porcine $MCI_2III_2IV_2$ in five different conformations (Zhang et al. 2024), including State 1, State 2, Mid 1, Mid 2, and Mid 3. More importantly, after careful analysis of our five structures, we concluded that the conformational changes of each complex in $MCI_2III_2IV_2$ are elegantly coordinated, so that $MCI_2III_2IV_2$ is an internally organized integral unit, rather than just a simple clustering of RCCs.

11.6.2 Conformational Change of CI in MC

$MCI_2III_2IV_2$ was classified into two conformations, expanded and constricted, based on the position of CIV. The two CIs found in constricted forms of $MCI_2III_2IV_2$ were all in the deactive state. By contrast, the two CIs in the expanded conformations of

Fig. 11.4 Super-resolution optical microscopy (gSTED) of fixed cell samples. (**a–d**) The immunofluorescence STED images of NDUFS1 (left, red) and COX4I1 (middle, green) in HPAEC, the merged image is shown on the right (merge). These two-fluorescence representing two subunits of megacomplex display similar distribution, which indicates the same mitochondria. Representative images of fluorescent immunostaining of MCI2III2IV2 with anti-NDUFS1 (red) and anti-COX4I1 (green), respectively. The four-point fluorescence pattern of MCI2III2IV2 was selected from the magnified merged image. Scale bar (**a**) 5 μm; (**b, c, d**) 50 nm

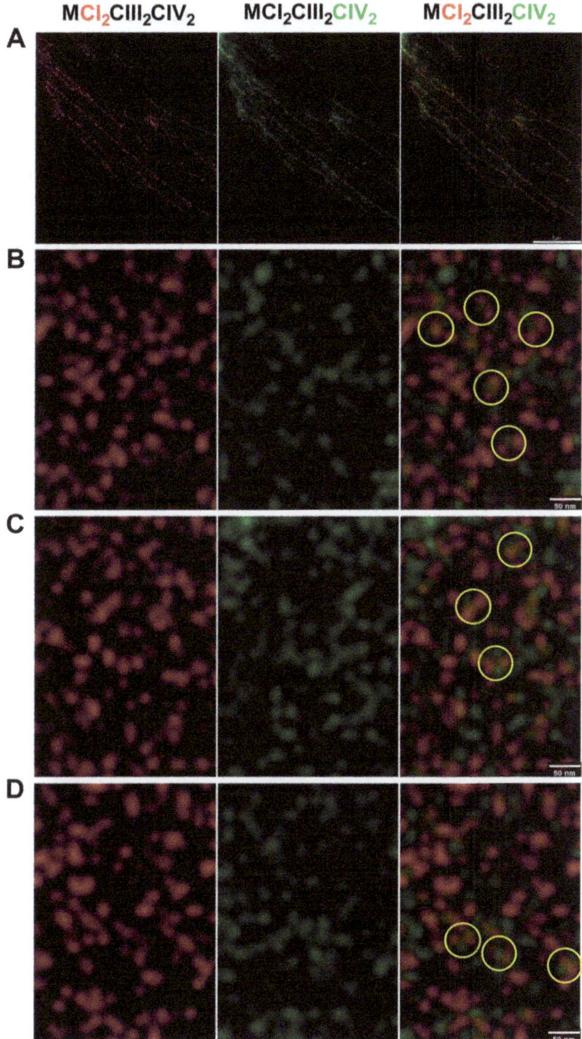

MCI2III2IV2 could be found in either active or deactive states. The expanded MCI$_2$III$_2$IV$_2$ is subdivided into three distinct states, including State 1, which contains two active CIs; Mid 1, containing 2 deactive CIs, and Mid 2, containing 1 active CI and 1 deactive CI. Both CIs in the constricted MCI$_2$III$_2$IV$_2$ were in the deactive state, but the position of the CIVs was found to shift between the two sites. Thus, the constricted MCI$_2$III$_2$IV$_2$ is subdivided into two states: State 2, where both CIVs are located at the same location, and Mid 3, the two CIVs are in different positions. Both CIs of State 1 are active and carry the UQ10 molecule bound to the Q cavity. In the structure of the State 2 MCI$_2$III$_2$IV$_2$, the two CIVs are in relative proximity to the CIII dimer. Both CIs in State 2 MCI$_2$III$_2$IV$_2$ are deactive. In both the

State 1 and State 2 structures, the CIII dimer exhibits a direct interaction with both CIs but forms a direct interaction with CIV only in State 2. The density of Cyt. c gradually breaks from State 1 to Mid 1 and then to State 2, indicating that Cyt. c binds more tightly to CIII in State 1 and more loosely to CIII in State 2.

From State 1 to State 2, the two interaction points between CI and CIII on the IMM side undergo significant conformational changes. (1) the long helix (C80-G126) of NDUFB7 extends from the distal end of the CI transmembrane arm to the Rieske domain that contacts CIII monomer A; (2) The C terminus of NDUFB10 protruded from the middle part of the CI transmembrane arm to contact the N terminus of the UQCRH subunit of CIII monomer B. The Rieske domain is essential for electron transfer within CIII, and the N-terminal region of UQCRH is adjacent to the Cyt. c binding site of CIII. Therefore, $NDUFB7_{A123}$ and $NDUFB10_{A176}$ of CI may act as two anchor sites to cause conformational changes in the CIII Rieske domain and Cyt. c binding site (Fig. 11.5).

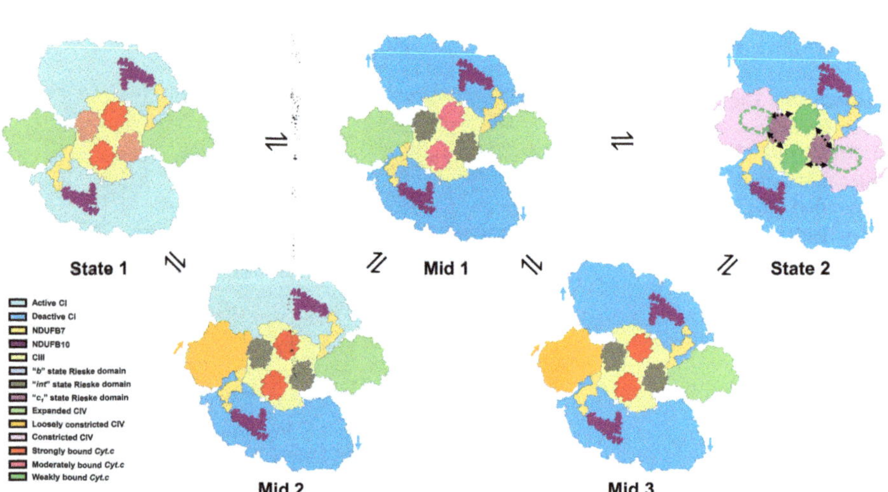

Fig. 11.5 Proposed activation/deactivation mechanism of MCI2III2IV2. We propose that State 1 represents the activated MCI2III2IV2, while State 2 represents the deactivated MCI2III2IV2. The transition between State 1 and State 2 via Mid 1 is the major path, while Mid 2 and Mid 3 are branches. CIs in State 1 are active, while CIs in Mid 1 and State 2 are inactive. There are two anchor points between CI and CIII, NDUFB7 and NDUFB10. The Rieske domain is the "b" state in State 1, the "int" state in Mid 1, and the "c1" state in State 2. The coordinated conformational changes of CI and CIII may be mediated by the two anchor points. CIV is away from CIII in State 1 and Mid 1 but tightly attached to CIII in State 2. Cyt.c is strongly bound to CIII in State 1, moderately bound in Mid 1, and weakly bound in State 2. The position of CIV and the stability of Cyt.c could be related to the availability of electrons. PDB accession codes: 8GOQ (State 1), 8GOV (Mid1), 8GOW (Mid 2), 8GOX (Mid 3), and 8GOR (State 2)

11.6.3 Conformational Change of CIII and CIV in MC

CIII also has three well-defined conformations, including the "b," "c1," and "int" states (Sect. 11.2.2). The dynamic range of the 2Fe-2S cluster of State 1 is the same as that of the "b" state. The dynamic range of Mid 1 coincides with that of the "int" state. The dynamic range in State 2 overlaps between the "int" and "c1" states. It is reasonable to speculate that the Rieske domain of CIII changes from the "b" to the "c1" state when $MCI_2III_2IV_2$ transitions from State 1 to State 2 and that the $NDUFB7_{A123}$ and $NDUFB10_{A176}$ localization sites can mediate the cooperative conformational change of CI and CIII. These findings suggest that the classification of CI and CIII based on the $MCI_2III_2IV_2$ state corresponds to the previously identified conformational changes of CI and CIII and that the conformational changes of CI and CIII are coordinated.

When $MCI_2III_2IV_2$ transitions from State 1 to State 2, the position of CIV changes significantly. In State 1, the CIV was located in the same position as previously reported porcine or sheep $SCI_2III_2IV_1$ structures. However, the location of the CIV in State 2 is different from that in any previously reported CIV-containing supercomplex. Given that both CIs in State 2 are deactivate, the unique position of CIV in State 2 suggests that this position may be involved in keeping CI in a deactivated state (Fig. 11.5).

11.7 How Can Mitochondrial Disorders Lead to Various Diseases?

Mitochondria play many key roles in the cell, the most fundamental of which are oxidative phosphorylation, metabolism of core carbohydrates, and biosynthesis of cell growth intermediates, which can largely determine cell function and fate (Akbari et al. 2019; Murphy and Hartley 2018). Thus, disruption of mitochondrial function caused by mutations in nuclear or mitochondrial DNA (mtDNA) can lead to devastating "primary" mitochondrial diseases, such as mitochondrial encephalomyopathies. Although the most classical function of mitochondria is to provide the energy necessary for cell survival through carbohydrate metabolism, mitochondria are far more than the "engine" of the cell. With the deepening of research, people's understanding of the biological functions of mitochondria has been greatly expanded, including lipid regulation, calcium homeostasis, thermogenesis, autophagy, apoptosis, oxidative stress, inflammation regulation, immune regulation, and so on (Jenkins et al. 2024; Liu and Birsoy 2023). It is now clear that mitochondria are involved in almost all aspects of cellular function, affecting processes not traditionally ascribed to organelles. Therefore, mitochondrial dysfunction can be a symptom and cause of many common diseases, including neuropathy, metabolic diseases, cardiovascular diseases, acute and chronic inflammation, tumors, and viral infections. In these "secondary" mitochondrial diseases, even if the direct cause is not mitochondrial abnormalities, the downstream processes of the disease also disrupt mitochondrial function, leading to further deterioration of the condition.

The assembly state and conformations of RCCs are closely related to cellular energy demand and many signaling pathways (Nesci et al. 2021). Generally speaking, the pathogenic factors (virus or bacterial infection, nuclear or mitochondrial genetic mutations, metabolic stress, oxidative damage, abnormal protein aggregates, etc.) could lead to SC disassembly, OXPHOS inhibition, ROS generation, cristae loss, and mitochondria scattering (Murphy and Hartley 2018; Xu et al. 2022). Furthermore, damaged mitochondria could release their contents (damage-associated molecular pattern, DAMP), which are danger signals for the cell and could lead to chronic inflammation (Al Amir Dache and Thierry 2023; Borcherding and Brestoff 2023). Taken together, the disrupted metabolic homeostasis, elevated ROS load, restrained energy supply, disordered cell survival signaling, and inevitable chronic inflammation could result in various diseases in almost every kind of organ of the human body. Now we will explain how the process of RCC structural damage leads to disease in detail to provide a theoretical basis for proper mitochondrial targeting therapies.

11.7.1 *Damaged RCCs Could Disrupt Membrane Potential, Release ROS, Lead to Abnormal Mitochondria Morphology, and Disturbed Quality Control*

Pathogenic factors could firstly cause damage to the RCC subunits, leading to the disassembly of RCCs and a loss of $\Delta\Psi m$ (or a gain of $\Delta\Psi m$ by stress, depending on different diseases) (Murphy and Hartley 2018; Nesci et al. 2021). The damaged RCCs could be a source of ROS, which could in turn cause more damage to RCC subunits, forming a forward feedback loop (Ulfig and Jakob 2024). Over 80% of ROS in the cell are generated from the RCCs, among which CI and CIII are the major sources (Tirichen et al. 2021). CI likely releases ROS in the matrix, while CIII releases ROS mostly in the IMS. The superoxide released at the IMS may be directly exported to the cytoplasm through an anion channel related to the Voltage-Dependent Anion Channel, VDAC (Han et al. 2003; Lustgarten et al. 2012). Mitochondrial ROS production increases when the membrane potential is high and the electron transfer rate decreases (Suski et al. 2018). ROS can act as second messengers by modulating the expression of several genes involved in signal transduction, influencing proliferation, autophagy, apoptosis, angiogenesis, and metastasis (Napolitano et al. 2021; Brillo et al. 2021). Many of the ROS-mediated cellular signals paradoxically protect the cell against oxidative stress, but above a given threshold ROS become harmful and induce oxidative stress. A slight increase in ROS could enhance mitophagy to remove impaired mitochondria and maintain homeostasis, but when over the threshold, ROS can damage the proteostasis network of the cell (Ulfig and Jakob 2024) (Fig. 11.6).

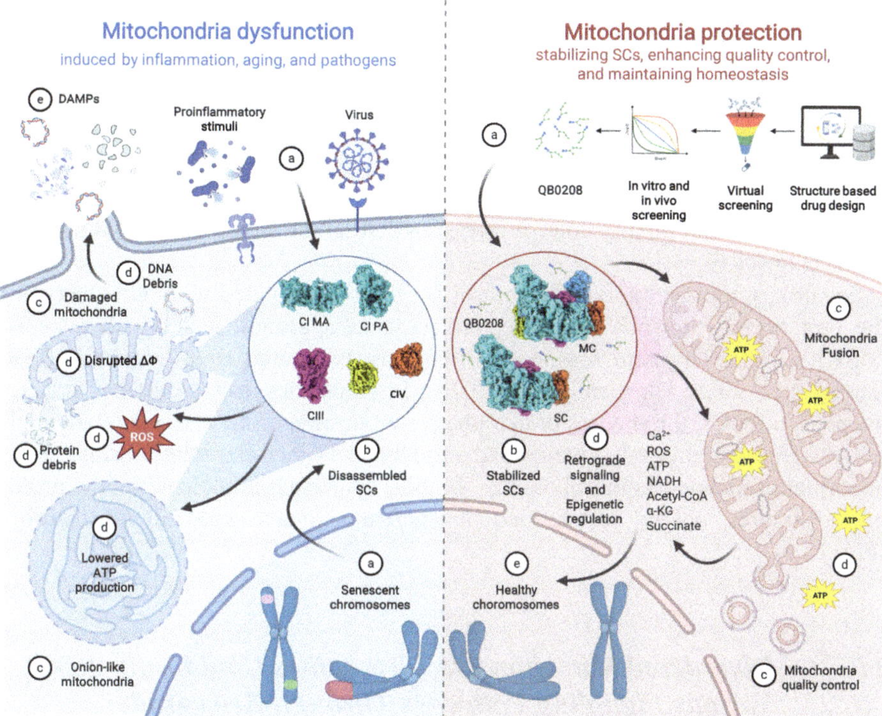

Fig. 11.6 The influence of mitochondria dysfunction and mechanism of mitochondria protection. The left panel shows the process of mitochondria dysfunction induced by inflammation, aging, and pathogens. The right panel shows the mechanism of mitochondria protection by novel mitochondria targeting drugs. In the left panel, (**a**) proinflammatory stimuli, viruses, and senescent chromosomes could cause the disassembly of SCs; (**b**) disassembled SCs could cause disrupted cristae formation, and lead to onion-like mitochondria and damaged mitochondria; (**c**) in onion-like mitochondria and damaged mitochondria, $\Delta\Psi$ disruption, ATP reduction, and ROS increase could happen; (**d**) damaged mitochondria could release protein and DNA debris into cytosol; (**e**) finally, these debris are even released into extracellular environments and recognized as DAMPs. In the right panel, (**a**) through the structure-based drug design, virtual screening, and in vitro and in vivo screening process, the lead compounds were selected; (**b**) these lead compounds could stabilize these supramolecular assemblies of SCs; (**c**) stabilized SCs could maintain the mitochondria in normal morphology, promote mitochondria fusion, and enhance mitochondria quality control; (**d**) healthy mitochondria could provide enough ATPs, and maintain the homeostasis of metabolites; (**e**) through retrograde signaling and epigenetic regulation, metabolite homeostasis could keep the chromosome in healthy state

The formation of SC and ATP synthase dimer or tetramer can influence the curvature of IMM, and the concentration of proteins is directly associated with the extent of membrane curvature, with protein crowding resulting in an accumulation of "steric pressure" that is alleviated by membrane bending, so the dissociation of SCs and ATP synthase polymers could lead to fewer cristae and even abnormal morphology of mitochondria (Johnson et al. 2024). As the progress in electron microscopy, many morphologies other than the canonical thread shape (kidney beans) were reported, including compact mitochondria, branching mitochondria, nanotunnel, mitochondrial donut, elongated mitochondria, large volume mitochondria, small volume mitochondria, and megamitochondria (Jenkins et al. 2024). These abnormal mitochondria morphologies, with different functional relevance, are generally considered to be related to diseases, like AD, PD, heart failure, and NASH, etc., and these morphologies are often caused by disturbed mitochondria fission, fusion, and quality control processes (Ulfig and Jakob 2024; Jenkins et al. 2024) (Fig. 11.6).

In turn, mitochondria quality control is closely connected with autophagy and the proteostasis network, including molecular chaperones, heat shock response (HSR), unfolded protein response (UPR), and integrated stress response (ISR) (Hong et al. 2024; Ulfig and Jakob 2024). In conditions where overload damage happens to RCCs, the process of autophagy and normal proteostasis network could fail to remove damaged proteins and mitochondria in time, which brings about abnormal mitochondria morphologies. In these abnormal mitochondria, the profile of metabolites, energy supply, and downstream signals could all be disturbed (Fig. 11.6).

11.7.2 Disordered Mitochondria Metabolites Could Influence Various Signaling Pathways Through Retrograde Signaling and Epigenome Modification

Mitochondria adopt feedback or feedforward strategies to suppress metabolic toxicity, achieve metabolic conservation, ensure stable levels of key metabolites, allow metabolic plasticity, and prevent futile cycles (Liu and Birsoy 2023). Typically, feedback mechanisms counteract the perturbation of the environment to achieve homeostasis. On the other hand, feedforward mechanisms aim not to attenuate perturbations but to rewire the metabolic network to adapt to such new perturbations. In some cases, both circuits can act simultaneously to enable downstream effectors to maintain homeostatic control and promote adaptive responses. When mitochondria morphology is damaged, the profile of metabolites (citrate, acetyl-CoA, α-ketoglutarate, glutamate, Ca^{2+}, ATP, NADH, etc.) is also disordered, which could lead to downstream disruption of retrograde signaling (Fig. 11.6).

Retrograde signaling initiates from the mitochondria and is proposed to be deployed in states of stress in which mitochondria require the nucleus to adjust gene expression to regain homeostasis (Butow and Avadhani 2004; Takeda and Yanagi 2019; Quirós et al. 2016). The canonical retrograde signals involve energetic or oxidative stress, calcium-dependent responses, and, more recently, protein toxicity. The IIS (insulin and insulin-like growth factor 1 (IGF-1) signaling), PI3K-Akt-mTOR pathway, and AMPK signaling pathway are the three major retrograde signaling pathways influenced by mitochondria (Fig. 11.6).

In addition, recent years have seen an explosion in the role of mitochondria metabolites to regulate cellular programs through the epigenome (Santos 2021). Many studies reported that changes in glucose or lipid metabolism impact post-translational modifications in transcription factors, kinases, and histones that in turn influence gene expression and cell differentiation programs. Levels of the metabolite acetyl-CoA or α-ketoglutarate (α-KG) were shown to regulate histone acetylation and DNA methylation, respectively, thereby impacting transcriptional outputs and cellular fate (Fig. 11.6).

11.7.3   Damaged Mitochondria Is a Strong Stimuli of Inflammation

The endosymbiotic origin of mitochondria makes components of this organelle (mtDNA, mitochondrial transcription factor A (TFAM), extracellular ATP, etc.), and the whole organelle itself (if outside of cells), danger signals that are particularly sensitive to immune pathways (Al Amir Dache and Thierry 2023; Borcherding and Brestoff 2023). Either secreted as intact-free, fragmented, or vesicle-encapsulated particles, these DAMP molecules could interact with the pattern recognition receptors (PRRs) and thus stimulate undesirable inflammatory signaling pathways. Besides, recent studies also report that whether intact or damaged, whole mitochondria discharged outside the cell show considerable pro- or anti-inflammatory effects in different conditions, highlighting the paradoxical interactions between these organelles and immune cells (Al Amir Dache and Thierry 2023; Ramirez-Barbieri et al. 2019). In general, damaged mitochondria could release their components into surrounding fluids and stimulate acute or chronic inflammation, which could be a critical factor in causing diseases (Wynn et al. 2013; Wang et al. 2022; Hotamisligil 2017) (Fig. 11.6).

11.8   Developing Drugs Targeting Mitochondria

The most successful mitochondria-targeting drug is metformin. Metformin is the first-line medication to treat type 2 diabetes mellitus (T2DM) in most guidelines and is used daily by >200 million patients (Foretz et al. 2023; Triggle et al. 2022). The classical mechanism of metformin is to be enriched in mitochondria due to a strong positive charge in the solution and inhibit CI reversibly (Bridges et al. 2014). Downregulated CI activity could result in higher NADH/NAD$^+$ levels and higher AMP/ATP levels, both of which can slow down gluconeogenesis in the liver (Foretz et al. 2023). Besides, higher AMP/ATP levels could activate AMPK signals, which can inhibit mTOR, STAT3, and NF-κB, to lower the production of inflammatory cytokines (IL6, TNF-α, IL1B, PRF1, CXCL1, etc.) (Foretz et al. 2023). In addition, reduced CI activity could also produce less ROS, which is also a known factor in generating protein/lipid damage and inflammation (Foretz et al. 2023; Bharath et al. 2020). Structures of CI with biguanide (IM1092, analog of metformin with stronger CI affinity) provided evidence for this theory to some extent (Bridges et al. 2023). However, more and more targets of metformin are proposed nowadays, such as CIV, glycerol-3-phosphate dehydrogenase, lysosome PEN2 2–ATP6AP1 axis, and gut microbiota, though the mechanisms are less clear (Foretz et al. 2023). Most recently, many groups worked cooperatively for over 40 months to evaluate the geroprotective effects of metformin on adult male cynomolgus monkeys (Yang et al. 2024). A comprehensive suite of physiological, imaging, histological, and molecular evaluations were applied, and pan-tissue transcriptomics, DNA methylomics, plasma proteomics, and metabolomics were leveraged to develop innovative monkey aging clocks. Their results suggested a significant slowing of aging indicators (including tumor necrosis factor-α (TNF-α), interleukin-1β (IL-1β), S100 calcium-binding protein A8 (S100A8), and matrix metalloproteinase 9 (MMP9)), and notably a roughly 6-year regression in brain aging.

The most densely tried direction of mitochondria targeting compounds is to kill tumors. Many mitochondrial pathways are altered in tumors, including OXPHOS, fatty acid oxidation, glutamine, and one-carbon metabolism, due to mutations in oncogenes, tumor suppressor genes, or metabolic enzymes (Sainero-Alcolado et al. 2022; Wu et al. 2023). These altered pathways lead to metabolic reprogramming that sustains the rapid cell proliferation and increases ROS, which can be used by cancer cells to maintain pro-tumorigenic signaling pathways while avoiding cellular death (Sainero-Alcolado et al. 2022; Wu et al. 2023). Consequently, many compounds were developed targeting these pathways to selectively kill cancer cells (Chung et al. 2021; Molina et al. 2018; Pujalte-Martin et al. 2024; Shi et al. 2019; Tsuji et al. 2020; Zou et al. 2023). Conditions could be much more complicated in clinical use, for cancer cells can reprogram their metabolism to counteract the blocking of one pathway by activation of another metabolic hub, and inhibition of metabolism may cause adverse effects in normal cells (Sainero-Alcolado et al. 2022). Nonetheless, the idea of targeting mitochondria to treat cancer is still very appealing because the combined use of several compounds as a cocktail could be promising, and many compounds are worth further developing, such as IACS-010759 and IACS-2858 inhibiting CI (Chung et al. 2021; Molina et al. 2018; Tsuji et al. 2020); Gboxin inhibiting CV (Shi et al. 2019; Zou et al. 2023); AGI-5198, AG-221, and AG-881 inhibiting mutated IDH (Rohle et al. 2013; Yen et al. 2017; Medeiros et al. 2017); CPI-613 inhibiting KGDHC and PDC (Pardee et al. 2014; Lycan et al. 2016); LY345899 targeting one-carbon metabolism (Ju et al. 2019); Etomoxir targeting fatty acid oxidation (Ratheiser et al. 1991). Besides, mitochondria-endoplasmic reticulum contacts (MERCs) are extracting growing attention recently, with their main roles in ion and lipid transport, ROS signaling, membrane dynamic changes, and cellular metabolism determined and are now considered potential targets for novel anticancer therapy strategies (Wu et al. 2023).

Except for cancer, metabolic syndromes, chronic inflammation, cardiovascular diseases, neurodegenerative diseases, and microorganism infection are all closely related to mitochondria (Murphy and Hartley 2018; Xu et al. 2022), and the logic of these therapies mostly focuses on the protection of mitochondria, rather than mitochondria inhibition as the case in cancer. Common mitochondria damage includes overload ROS production, lack of NAD^+, abnormal MPTP formation, and proteostasis network disruption (Murphy and Hartley 2018). S1QELs and S3QELs bind directly to the ROS-generating sites in CI and CIII (Brand et al. 2016; Orr et al. 2015), respectively, to inhibit ROS production. The depletion of NAD^+, which can lead to both bioenergetic defects and inappropriate protein acylation, can be counteracted by compounds such as nicotinamide (NAM), NAM riboside (NR), and NMN to replenish NAD^+ levels (Sorrentino et al. 2017; Gariani et al. 2016). CsA binds to the matrix protein cyclophilin D (CYD, also known as PPID) and thereby prevents cell death caused by the formation of the MPTP (Ottani et al. 2016; Cung et al. 2015). A disturbed proteostasis network, like a disrupted mtUPR response, could lead to abnormal mitochondria fission and mitophagy, a failure in mitochondria quality control, while Urolithin A could enhance mitophagy; clear damaged protein and organelles, to decrease inflammation; and restore muscle function (Ryu et al. 2016).

In recent years, based on our high-resolution structure of SC, our group also developed a novel mitochondria protective compound targeting complex I, QB0208, which is now in Phase II clinical trial. The result of the Phase I clinical trial suggested no toxicity in human bodies under the highest tested dose. Our pharmacological study suggests that QB0208 can help damaged cells and tissues return to normal basal homeostasis by reducing mitochondrial ROS, enhancing mitochondrial quality control, and increasing oxidative phosphorylation levels. At the same time, these processes of improving mitochondrial activity can also enhance the body's metabolic level, reduce the metabolic burden, and reduce the level of pro-inflammation factors. In addition, for immune cells, especially macrophages, the activation of the oxidative phosphorylation pathway can enhance the immune function, promote the M2 polarization of macrophages, enhance the ability of phagocytosis of foreign entities, and eliminate inflammation. Based on these processes, QB0208 could act as a neuroprotective, cardiovascular-protective, anti-inflammatory, and antiviral compound (Fig. 11.6).

11.9 Conclusion

In this chapter, we first discussed the hotly debated coupling mechanism of CI and explained our "Double Q model" in brief, which suggests Site 2, rather than the classic Site 1, could be the position that triggers the coupling of proton pumping, and a second UQ surrounding the entrance is needed to accept electrons from the UQ in the Q chamber, which is reminiscent of the electron transfer between two Qs in Ndi1 (Feng et al. 2012) and consistent with a recent molecular simulation study (Djurabekova et al. 2022). Then, we explained the mutual influence of CI and SC assembly process based on the latest reports and raised examples supporting that forming SC could be an adaptation to high energy demand and higher assembly of MC is a common situation in living mammalian cells.

Most importantly, we solved the high-resolution structures of porcine $MCI_2III_2IV_2$ in five conformations, State 1, State 2, Mid 1, Mid 2, and Mid 3, showing that the two CIs found in constricted forms of $MCI_2III_2IV_2$ were in deactive state, while the two CIs in the expanded forms of $MCI_2III_2IV_2$ were in either active or deactive state, suggesting that $MCI_2III_2IV_2$ shifts from the active state (State 1) to the deactive state (State 2), via the intermediates (Mid 1, Mid 2, and Mid 3). In addition, the dynamic range of the 2Fe-2S cluster in State 1 largely aligned with that reported in the "b" state; the dynamic range in Mid 1 largely coincided with that in the "int" state; and the dynamic range in State 2 overlapped between the "int" and "c1" states. So, when $MCI_2III_2IV_2$ shifts from State 1 to State 2, the Rieske domain of CIII changes from the "b" state to the "c1" state. Our results showed that the conformations of CI, CIII, and CIV in MC are coordinately changed so that their function could very likely be fine-regulated to make the whole MC work as an integral unit.

Finally, we discussed how mitochondria dysfunction can be translated into clinical syndromes. Mutation of or damage to RCCs could on one hand lead to overload ROS generation which could in turn cause more damage to RCCs and other cellular components in a forward feedback manner and, on the other hand, could result in assembly failure and dissociation of SCs and RCCs which could cause less cristae formation and abnormal mitochondria quality control. The damaged mitochondria could therefore release their contents and cause inflammation. Ultimately, decreased energy supply and disrupted signaling pathways could lead to various diseases almost in every part of the human body.

Extensive efforts have been made in developing drugs targeting mitochondria in the last decade, and several drugs are very successful, like metformin, Mito-Q, and NMN. Our group also developed a leading compound, QB0208, based on our high-resolution structure of SC to protect mitochondria morphology and reduce ROS production. Now, our molecule is in Phase II clinical trial, aiming to treat a series of diseases caused by mitochondria dysfunction.

Acknowledgments This wo r k w a s supported by funds for M.Y. from the National Key R&D Program of China (2022YFA1302701), the National Natural Science Foundation of China (32030056), and for R.G. from the National Natural Science Foundation of China (32100962).

Author Information The authors declare no competing financial interests.

References

Akbari M, Kirkwood TBL, Bohr VA. Mitochondria in the signaling pathways that control longevity and health span. Ageing Res Rev. 2019;54:100940. https://doi.org/10.1016/j.arr.2019.100940.

Al Amir Dache Z, Thierry AR. Mitochondria-derived cell-to-cell communication. Cell Rep. 2023;42:112728. https://doi.org/10.1016/j.celrep.2023.112728.

Alston CL, et al. Bi-allelic mutations in establish its role in early-onset isolated mitochondrial complex I deficiency. Am J Hum Genet. 2018;103:592–601. https://doi.org/10.1016/j.ajhg.2018.08.013.

Altmann R. Die Elementarorganismen und ihre Beziehungen zu den Zellen. Leipzig: Veit; 1890.

Bharath LP, et al. Metformin enhances autophagy and normalizes mitochondrial function to alleviate aging-associated inflammation. Cell Metab. 2020;32:44–55 e46. https://doi.org/10.1016/j.cmet.2020.04.015.

Blaza JN, Vinothkumar KR, Hirst J. Structure of the deactive state of mammalian respiratory complex I. Structure. 2018;26:312–319 e313. https://doi.org/10.1016/j.str.2017.12.014.

Borcherding N, Brestoff JR. The power and potential of mitochondria transfer. Nature. 2023;623:283–91. https://doi.org/10.1038/s41586-023-06537-z.

Brand MD, et al. Suppressors of superoxide-H2O2 production at site I of mitochondrial complex I protect against stem cell hyperplasia and ischemia-reperfusion injury. Cell Metab. 2016;24:582–92. https://doi.org/10.1016/j.cmet.2016.08.012.

Braun H-P. The oxidative phosphorylation system of the mitochondria in plants. Mitochondrion. 2020;53:66–75. https://doi.org/10.1016/j.mito.2020.04.007.

Bridges HR, Bill E, Hirst J. Mossbauer spectroscopy on respiratory complex I: the iron-sulfur cluster ensemble in the NADH-reduced enzyme is partially oxidized. Biochemistry-Us. 2012;51:149–58. https://doi.org/10.1021/bi201644x.

Bridges HR, Jones AJ, Pollak MN, Hirst J. Effects of metformin and other biguanides on oxidative phosphorylation in mitochondria. Biochem J. 2014;462:475–87. https://doi.org/10.1042/BJ20140620.

Bridges HR, et al. Structural basis of mammalian respiratory complex I inhibition by medicinal biguanides. Science. 2023;379:351–7. https://doi.org/10.1126/science.ade3332.

Brillo V, Chieregato L, Leanza L, Muccioli S, Costa R. Mitochondrial dynamics, ROS, and cell signaling: a blended overview. Life (Basel). 2021;11:332. https://doi.org/10.3390/life11040332.

Butow RA, Avadhani NG. Mitochondrial signaling: the retrograde response. Mol Cell. 2004;14:1–15. https://doi.org/10.1016/S1097-2765(04)00179-0.

Choi EH, Kim MH, Park SJ. Targeting mitochondrial dysfunction and reactive oxygen species for neurodegenerative disease treatment. Int J Mol Sci. 2024;25:7952. https://doi.org/10.3390/ijms25147952.

Chung I, et al. Cork-in-bottle mechanism of inhibitor binding to mammalian complex I. Sci Adv. 2021;7:eabg4000. https://doi.org/10.1126/sciadv.abg4000.

Cuillerier A, et al. Adaptive optimization of the OXPHOS assembly line partially compensates lrpprc-dependent mitochondrial translation defects in mice. Commun Biol. 2021;4:989. https://doi.org/10.1038/s42003-021-02492-5.

Cung TT, et al. Cyclosporine before PCI in patients with acute myocardial infarction. New Engl J Med. 2015;373:1021–31. https://doi.org/10.1056/NEJMoa1505489.

Davies KM, Blum TB, Kuhlbrandt W. Conserved in situ arrangement of complex I and III(2) in mitochondrial respiratory chain supercomplexes of mammals, yeast, and plants. Proc Natl Acad Sci USA. 2018;115:3024–9. https://doi.org/10.1073/pnas.1720702115.

Djurabekova A, et al. Docking and molecular simulations reveal a quinone-binding site on the surface of respiratory complex I. FEBS Lett. 2022;596:1133–46. https://doi.org/10.1002/1873-3468.14346.

Efremov RG, Sazanov LA. Structure of the membrane domain of respiratory complex I. Nature. 2011;476:414–U462. https://doi.org/10.1038/nature10330.

Ernster L, Schatz G. Mitochondria: a historical review. J Cell Biol. 1981;91(3 Pt 2):227s–55s.

Euro L, Bloch DA, Wikström M, Verkhovsky MI, Verkbovskaya M. Electrostatic interactions between FeS clusters in NADH:Ubiquinone oxidoreductase (complex I) from Escherichia coli. Biochemistry-Us. 2008;47:3185–93. https://doi.org/10.1021/bi702063t.

Fang H, et al. A membrane arm of mitochondrial complex I sufficient to promote respirasome formation. Cell Rep. 2021;35:108963. https://doi.org/10.1016/j.celrep.2021.108963.

Fedor JG, Jones AJY, Di Luca A, Kaila VRI, Hirst J. Correlating kinetic and structural data on ubiquinone binding and reduction by respiratory complex I. Proc Natl Acad Sci USA. 2017;114:12737–42. https://doi.org/10.1073/pnas.1714074114.

Feng Y, et al. Structural insight into the type-II mitochondrial NADH dehydrogenases. Nature. 2012;491:478–82. https://doi.org/10.1038/nature11541.

Foretz M, Guigas B, Viollet B. Metformin: update on mechanisms of action and repurposing potential. Nat Rev Endocrinol. 2023;19:460–76. https://doi.org/10.1038/s41574-023-00833-4.

Friedman J, Mourokh L, Vittadello M. Mechanism of proton pumping in complex I of the mitochondrial respiratory chain. Quantum Rep. 2021;3:425–34. https://doi.org/10.3390/quantum3030027.

Gandhi S, Sweeney HL, Hart CC, Han R, Perry CGR. Cardiomyopathy in Duchenne muscular dystrophy and the potential for mitochondrial therapeutics to improve treatment response. Cells. 2024;13:1168. https://doi.org/10.3390/cells13141168.

Gariani K, et al. Eliciting the mitochondrial unfolded protein response by nicotinamide adenine dinucleotide repletion reverses fatty liver disease in mice. Hepatology. 2016;63:1190–204. https://doi.org/10.1002/hep.28245.

Green DE, Hatefi Y. Mitochondrion and biochemical machines – mitochondria or their equivalents are principal energy transducers in all aerobic organisms. Science. 1961;133:13–9.

Gu J, et al. The architecture of the mammalian respirasome. Nature. 2016;537:639–43. https://doi.org/10.1038/nature19359.

Gu JK, et al. Cryo-EM structure of the mammalian ATP synthase tetramer bound with inhibitory protein IF1. Science. 2019;364:1068–+. https://doi.org/10.1126/science.aaw4852.

Gu J, Liu T, Guo R, Zhang L, Yang M. The coupling mechanism of mammalian mitochondrial complex I. Nat Struct Mol Biol. 2022;29:172–82. https://doi.org/10.1038/s41594-022-00722-w.

Guerrero-Castillo S, et al. The assembly pathway of mitochondrial respiratory chain complex I. Cell Metab. 2017;25:128–39. https://doi.org/10.1016/j.cmet.2016.09.002.

Guo R, Gu J, Wu M, Yang M. Amazing structure of respirasome: unveiling the secrets of cell respiration. Protein Cell. 2016;7:854–65. https://doi.org/10.1007/s13238-016-0329-7.

Guo R, Zong S, Wu M, Gu J, Yang M. Architecture of human mitochondrial respiratory Megacomplex I2III2IV2. Cell. 2017;170:1247–1257.e1212. https://doi.org/10.1016/j.cell.2017.07.050.

Guo R, Gu J, Zong S, Wu M, Yang M. Structure and mechanism of mitochondrial electron transport chain. Biom J. 2018;41:9–20. https://doi.org/10.1016/j.bj.2017.12.001.

Haapanen O, Djurabekova A, Sharma V. Role of second quinone binding site in proton pumping by respiratory complex I. Front Chem. 2019;7:221. https://doi.org/10.3389/fchem.2019.00221.

Han D, Antunes F, Canali R, Rettori D, Cadenas E. Voltage-dependent anion channels control the release of the superoxide anion from mitochondria to cytosol. J Biol Chem. 2003;278:5557–63. https://doi.org/10.1074/jbc.M210269200.

Hirst J. Mitochondrial complex I. Annu Rev Biochem. 2013;82:551–75. https://doi.org/10.1146/annurev-biochem-070511-103700.

Hoias Teixeira M, Menegon Arantes G. Balanced internal hydration discriminates substrate binding to respiratory complex I. Biochim Biophys Acta Bioenerg. 2019;1860:541–8. https://doi.org/10.1016/j.bbabio.2019.05.004.

Hong WL, Huang H, Zeng X, Duan CY. Targeting mitochondrial quality control: new therapeutic strategies for major diseases. Mil Med Res. 2024;11:59. https://doi.org/10.1186/s40779-024-00556-1.

Hotamisligil GS. Inflammation, metaflammation and immunometabolic disorders. Nature. 2017;542:177–85. https://doi.org/10.1038/nature21363.

Jenkins BC, et al. Mitochondria in disease: changes in shapes and dynamics. Trends Biochem Sci. 2024;49:346–60. https://doi.org/10.1016/j.tibs.2024.01.011.

Johnson DH, Kou OH, Bouzos N, Zeno WF. Protein-membrane interactions: sensing and generating curvature. Trends Biochem Sci. 2024;49:401–16. https://doi.org/10.1016/j.tibs.2024.02.005.

Ju HQ, et al. Modulation of redox homeostasis by inhibition of MTHFD2 in colorectal cancer: mechanisms and therapeutic implications. Jnci-J Natl Cancer I. 2019;111:584–96. https://doi.org/10.1093/jnci/djy160.

Kampjut D, Sazanov LA. The coupling mechanism of mammalian respiratory complex I. Science. 2020;370:eabc4209. https://doi.org/10.1126/science.abc4209.

Kampjut D, Sazanov LA. Structure of respiratory complex I – an emerging blueprint for the mechanism. Curr Opin Struct Biol. 2022;74:102350. https://doi.org/10.1016/j.sbi.2022.102350.

Kobayashi A, et al. A FRET-based respirasome assembly screen identifies spleen tyrosine kinase as a target to improve muscle mitochondrial respiration and exercise performance in mice. Nat Commun. 2023;14:290. https://doi.org/10.1038/s41467-023-35865-x.

Leon WC, et al. A novel transgenic rat model with a full Alzheimer's-like amyloid pathology displays pre-plaque intracellular amyloid-β-associated cognitive impairment. J Alzheimer's Dis. 2010;20:113–26. https://doi.org/10.3233/Jad-2010-1349.

Li B, et al. Structure of the Mrp complex reveals molecular mechanism of this giant bacterial sodium proton pump. P Natl Acad Sci USA. 2020;117:31166–76. https://doi.org/10.1073/pnas.2006276117.

Li M, et al. Supramolecular antagonists promote mitochondrial dysfunction. Nano Lett. 2021;21:5730–7. https://doi.org/10.1021/acs.nanolett.1c01469.

Liu Y, Birsoy K. Metabolic sensing and control in mitochondria. Mol Cell. 2023;83:877–89. https://doi.org/10.1016/j.molcel.2023.02.016.

Lustgarten MS, et al. Complex I generated, mitochondrial matrix-directed superoxide is released from the mitochondria through voltage dependent anion channels. Biochem Bioph Res Co. 2012;422:515–21. https://doi.org/10.1016/j.bbrc.2012.05.055.

Lycan TW, et al. A phase II clinical trial of CPI-613 in patients with relapsed or refractory small cell lung carcinoma. PLoS One. 2016;11:e0164244. https://doi.org/10.1371/journal.pone.0164244.

Margulis L. Origin of eukaryotic cells. Recherche. 1970;1(2):121–6.

Marx N, Ritter N, Disse P, Seebohm G, Busch KB. Detailed analysis of Mdivi-1 effects on complex I and respiratory supercomplex assembly. Sci Rep. 2024;14:19673. https://doi.org/10.1038/s41598-024-69748-y.

McKenzie M, Lazarou M, Thorburn DR, Ryan MT. Mitochondrial respiratory chain supercomplexes are destabilized in Barth syndrome patients. J Mol Biol. 2006;361:462–9. https://doi.org/10.1016/j.jmb.2006.06.057.

Medeiros BC, et al. Isocitrate dehydrogenase mutations in myeloid malignancies. Leukemia. 2017;31:272–81. https://doi.org/10.1038/leu.2016.275.

Mitchell P. Coupling of phosphorylation to electron and hydrogen transfer by a chemi-osmotic type of mechanism. Nature. 1961;191:144.

Mitchell P. Chemiosmotic coupling in oxidative and photosynthetic phosphorylation. Biol Rev. 1966;41:445–502.

Molina JR, et al. An inhibitor of oxidative phosphorylation exploits cancer vulnerability. Nat Med. 2018;24:1036–46. https://doi.org/10.1038/s41591-018-0052-4.

Moreno-Lastres D, et al. Mitochondrial complex I plays an essential role in human respirasome assembly. Cell Metab. 2012;15:324–35. https://doi.org/10.1016/j.cmet.2012.01.015.

Mühlbauer ME, et al. Water-gated proton transfer dynamics in respiratory complex I. J Am Chem Soc. 2020;142:13718–28. https://doi.org/10.1021/jacs.0c02789.

Murphy MP, Hartley RC. Mitochondria as a therapeutic target for common pathologies. Nat Rev Drug Discov. 2018;17:865–86. https://doi.org/10.1038/nrd.2018.174.

Napolitano G, Fasciolo G, Venditti P. Mitochondrial management of reactive oxygen species. Antioxidants (Basel). 2021;10:1824. https://doi.org/10.3390/antiox10111824.

Nesci S, et al. Molecular and supramolecular structure of the mitochondrial oxidative phosphorylation system: implications for pathology. Life (Basel). 2021;11(3):242. https://doi.org/10.3390/life11030242.

Nielsen SO, Lehninger AL. Phosphorylation coupled to the oxidation of Ferrocytochrome-C. J Biol Chem. 1955;215:555–70.

Novack GV, Galeano P, Castano EM, Morelli L. Mitochondrial supercomplexes: physiological organization and dysregulation in age-related neurodegenerative disorders. Front Endocrinol (Lausanne). 2020;11:600. https://doi.org/10.3389/fendo.2020.00600.

Orr DL, et al. Suppressors of superoxide production from mitochondrial complex III. Nat Chem Biol. 2015;11:834–U836. https://doi.org/10.1038/Nchembio.1910.

Ottani F, et al. Cyclosporine A in reperfused myocardial infarction the multi-center, controlled, open-label CYCLE trial. J Am Coll Cardiol. 2016;67:365–74. https://doi.org/10.1016/j.jacc.2015.10.081.

Padhy I, et al. Structure based exploration of mitochondrial alpha carbonic anhydrase inhibitors as potential leads for anti-obesity drug development. DARU J Pharm Sci. 2024;32(2):907–24. https://doi.org/10.1007/s40199-024-00535-w.

Pardee TS, et al. A phase I study of the first-in-class antimitochondrial metabolism agent, CPI-613, in patients with advanced hematologic malignancies. Clin Cancer Res. 2014;20:5255–64. https://doi.org/10.1158/1078-0432.Ccr-14-1019.

Parey K, et al. High-resolution structure and dynamics of mitochondrial complex I-insights into the proton pumping mechanism. Sci Adv. 2021;7:eabj3221. https://doi.org/10.1126/sciadv.abj3221.

Parmar G, et al. Accessory subunit NDUFB4 participates in mitochondrial complex I supercomplex formation. J Biol Chem. 2024;300:105626. https://doi.org/10.1016/j.jbc.2024.105626.

Protasoni M, et al. Respiratory supercomplexes act as a platform for complex III-mediated maturation of human mitochondrial complexes I and IV. EMBO J. 2020;39:e102817. https://doi.org/10.15252/embj.2019102817.

Pujalte-Martin M, et al. Targeting cancer and immune cell metabolism with the complex I inhibitors metformin and IACS-010759. Mol Oncol. 2024;18:1719–38. https://doi.org/10.1002/1878-0261.13583.

Qiang GF. Natural products targeting mitochondria: a promising strategy for metabolic syndrome. Chin J Nat Med. 2020;18:801–2. https://doi.org/10.1016/S1875-5364(20)60020-6.

Quirós PM, Mottis A, Auwerx J. Mitonuclear communication in homeostasis and stress. Nat Rev Mol Cell Biol. 2016;17:213–26. https://doi.org/10.1038/nrm.2016.23.

Ramirez-Barbieri G, et al. Alloreactivity and allorecognition of syngeneic and allogeneic mito-chondria. Mitochondrion. 2019;46:103–15. https://doi.org/10.1016/j.mito.2018.03.002.

Ratheiser K, et al. Inhibition by Etomoxir of carnitine Palmitoyltransferase-I reduces hepatic glucose-production and plasma-lipids in noninsulin-dependent diabetes-mellitus. Metabolism. 1991;40:1185–90. https://doi.org/10.1016/0026-0495(91)90214-H.

Roca FJ, Whitworth LJ, Prag HA, Murphy MP, Ramakrishnan L. Tumor necrosis factor induces pathogenic mitochondrial ROS in tuberculosis through reverse electron transport. Science. 2022;376:eabh2841. https://doi.org/10.1126/science.abh2841.

Rohle D, et al. An inhibitor of mutant IDH1 delays growth and promotes differentiation of glioma cells. Science. 2013;340:626–30. https://doi.org/10.1126/science.1236062.

Ryu D, et al. Urolithin A induces mitophagy and prolongs lifespan in and increases muscle function in rodents. Nat Med. 2016;22:879–88. https://doi.org/10.1038/nm.4132.

Sabbir MG, Swanson M, Speth RC, Albensi BC. Hippocampal versus cortical deletion of cholinergic receptor muscarinic 1 in mice differentially affects post-translational modifications and supramolecular assembly of respiratory chain-associated proteins, mitochondrial ultrastructure, and respiration: implications in Alzheimer's disease. Front Cell Dev Biol. 2023;11:1179252. https://doi.org/10.3389/fcell.2023.1179252.

Sainero-Alcolado L, Liano-Pons J, Ruiz-Perez MV, Arsenian-Henriksson M. Targeting mito-chondrial metabolism for precision medicine in cancer. Cell Death Differ. 2022;29:1304–17. https://doi.org/10.1038/s41418-022-01022-y.

Santos JH. Mitochondria signaling to the epigenome: a novel role for an old organelle. Free Radic Biol Med. 2021;170:59–69. https://doi.org/10.1016/j.freeradbiomed.2020.11.016.

Schägger H, Pfeiffer K. Supercomplexes in the respiratory chains of yeast and mammalian mito-chondria. EMBO J. 2000;19:1777–83. https://doi.org/10.1093/emboj/19.8.1777.

Schägger H, et al. Significance of respirasomes for the assembly/stability of human respiratory chain complex I. J Biol Chem. 2004;279:36349–53. https://doi.org/10.1074/jbc.M404033200.

Scrima R, et al. Mitochondrial sAC-cAMP-PKA Axis modulates the DeltaPsi(m)-dependent control coefficients of the respiratory chain complexes: evidence of respirasome plasticity. Int J Mol Sci. 2023;24:15144. https://doi.org/10.3390/ijms242015144.

Shi YF, et al. Gboxin is an oxidative phosphorylation inhibitor that targets glioblastoma. Nature. 2019;567:341–6. https://doi.org/10.1038/s41586-019-0993-x.

Signes A, Fernandez-Vizarra E. Assembly of mammalian oxidative phosphorylation complexes I-V and supercomplexes. Essays Biochem. 2018;62:255–70. https://doi.org/10.1042/Ebc20170098.

Signorile A, et al. cAMP/PKA signaling modulates mitochondrial supercomplex organization. Int J Mol Sci. 2022;23:9655. https://doi.org/10.3390/ijms23179655.

Sorrentino V, et al. Enhancing mitochondrial proteostasis reduces amyloid-β proteotoxicity. Nature. 2017;552:187–+. https://doi.org/10.1038/nature25143.

Steiner J, Sazanov L. Structure and mechanism of the Mrp complex, an ancient cation/proton antiporter. elife. 2020;9:e59407. https://doi.org/10.7554/eLife.59407.

Suski J, et al. Relation between mitochondrial membrane potential and ROS formation. Methods Mol Biol. 2018;1782:357–81. https://doi.org/10.1007/978-1-4939-7831-1_22.

Takeda K, Yanagi S. Mitochondrial retrograde signaling to the endoplasmic-reticulum regulates unfolded protein responses. Mol Cell Oncol. 2019;6:e1659078. https://doi.org/10.1080/2372 3556.2019.1659078.

Tirichen H, et al. Mitochondrial reactive oxygen species and their contribution in chronic kidney disease progression through oxidative stress. Front Physiol. 2021;12:627837. https://doi.org/10.3389/fphys.2021.627837.

Triggle CR, et al. Metformin: is it a drug for all reasons and diseases? Metabolism. 2022;133:155223. https://doi.org/10.1016/j.metabol.2022.155223.

Tropeano CV, et al. Fine-tuning of the respiratory complexes stability and supercomplexes assembly in cells defective of complex III. Biochim Biophys Acta Bioenerg. 2020;1861:148133. https://doi.org/10.1016/j.bbabio.2019.148133.

Tsuji A, Akao T, Masuya T, Murai M, Miyoshi H. IACS-010759, a potent inhibitor of glycolysis-deficient hypoxic tumor cells, inhibits mitochondrial respiratory complex I through a unique mechanism. J Biol Chem. 2020;295:7481–91. https://doi.org/10.1074/jbc.RA120.013366.

Ukolova IV, et al. New insights into the organisation of the oxidative phosphorylation system in the example of pea shoot mitochondria. Biochim Biophys Acta Bioenerg. 2020;1861:148264. https://doi.org/10.1016/j.bbabio.2020.148264.

Ulfig A, Jakob U. Cellular oxidants and the proteostasis network: balance between activation and destruction. Trends Biochem Sci. 2024;49:761–74. https://doi.org/10.1016/j.tibs.2024.07.001.

Vial G, Detaille D, Guigas B. Role of mitochondria in the mechanism(s) of action of metformin. Front Endocrinol. 2019;10:294. https://doi.org/10.3389/fendo.2019.00294.

Wang Q, et al. The role of microglia immunometabolism in neurodegeneration: focus on molecular determinants and metabolic intermediates of metabolic reprogramming. Biomed Pharmacother. 2022;153:113412. https://doi.org/10.1016/j.biopha.2022.113412.

Warnau J, et al. Redox-coupled quinone dynamics in the respiratory complex I. Proc Natl Acad Sci USA. 2018;115:E8413–20. https://doi.org/10.1073/pnas.1805468115.

Wittig I, Braun HP, Schägger H. Blue native PAGE. Nat Protoc. 2006;1:418–28. https://doi.org/10.1038/nprot.2006.62.

Wu M, Gu J, Guo R, Huang Y, Yang M. Structure of mammalian respiratory Supercomplex I(1) III(2)IV(1). Cell. 2016;167:1598–1609 e1510. https://doi.org/10.1016/j.cell.2016.11.012.

Wu H, Chen W, Chen Z, Li X, Wang M. Novel tumor therapy strategies targeting endoplasmic reticulum-mitochondria signal pathways. Ageing Res Rev. 2023;88:101951. https://doi.org/10.1016/j.arr.2023.101951.

Wynn TA, Chawla A, Pollard JW. Macrophage biology in development, homeostasis and disease. Nature. 2013;496:445–55. https://doi.org/10.1038/nature12034.

Xu Y, et al. Loss of protein association causes cardiolipin degradation in Barth syndrome. Nat Chem Biol. 2016;12:641–7. https://doi.org/10.1038/Nchembio.2113.

Xu J, et al. Mitochondria targeting drugs for neurodegenerative diseases—design, mechanism and application. Acta Pharm Sin B. 2022;12:2778–89. https://doi.org/10.1016/j.apsb.2022.03.001.

Yang Y, et al. Metformin decelerates aging clock in male monkeys. Cell. 2024;187(22):6358–6378.e29. https://doi.org/10.1016/j.cell.2024.08.021.

Yen K, et al. AG-221, a first-in-class therapy targeting acute myeloid leukemia harboring oncogenic mutations. Cancer Discov. 2017;7:478–93. https://doi.org/10.1158/2159-8290.Cd-16-1034.

Zhang L, et al. Structural basis for the regulatory mechanism of mammalian mitochondrial respiratory chain megacomplex-I2III2IV2. hLife. 2024;2:189–200. https://doi.org/10.1016/j.hlife.2024.03.003.

Zheng W, Chai P, Zhu J, Zhang K. High-resolution in situ structures of mammalian respiratory supercomplexes. Nature. 2024;631:232–9. https://doi.org/10.1038/s41586-024-07488-9.

Zou Y, et al. Cancer cell-mitochondria hybrid membrane coated Gboxin loaded nanomedicines for glioblastoma treatment. Nat Commun. 2023;14:4557. https://doi.org/10.1038/s41467-023-40280-3.

Index

A

Absent in melanoma 2 (AIM2), 43, 46, 48–50, 170, 178

Acetyl-CoA, 208, 210–213, 217, 230, 232, 236, 238–240, 320, 321

Adherens junction, 8–10, 13–19, 29

Allostery, 2, 58, 59, 61, 66, 68–69, 72, 74–76, 215, 228, 230, 262, 277, 278, 280–282, 287, 288, 306

Amide-hydrogen deuterium exchange mass spectrometry (HDXMS), 58, 59, 66–75

Anti-termination transcription complex, 190, 191

ATP-dependent protease Lon, 192, 193

B

Biomolecular condensates, 114, 133, 256, 263, 268

Bioorganic chemistry, 1

C

Cadherin, 10–12, 14–20, 22–24, 30

Cancer associated PCNA isoform (caPCNA), 167, 179–182, 184

Caspase, 42, 43, 164, 165

Cell adhesion, 10–13, 24, 29

Cell junction, 8–25

Citric acid cycle, 208, 239, 257

Clustered protocadherin, 11, 14, 24–30

Complex I, 300, 302, 304, 323

Condensates, 4, 114, 133–134, 139, 263, 264, 267–268, 277

Cryo-electron microscopy (cryo-EM), 14, 20, 21, 26, 30, 44–49, 76, 86, 89, 90, 93–98, 100–103, 117, 191, 209, 222, 224, 225, 279, 280, 282, 284, 287, 288, 308, 311, 314

Cryo-electron tomography (cryo-ET), 14, 20–23, 27, 28, 30, 31, 314

Cyclic adenosine monophosphate (cAMP), 55–76, 256, 312, 313

D

Damage-associated molecular patterns (DAMPs), 42, 45, 168, 318, 319, 321

Delta (δ) protocadherin, 25, 27, 28

Desmosome, 8–10, 13, 14, 19–23, 29, 30

DNA damage repair, 182

DNA replication, 114–119, 134–137, 139, 160–166, 169, 170, 178, 180–184, 262, 284

DNA replication and repair, 2, 158, 160, 161, 178, 180, 184

Droplet interface bilayer membranes, 99

E

Electron cryo-microscopy (cryo-EM), 86

Electron microscopy, 3, 21, 94, 193, 196, 200, 208, 278, 319–320

Electron transport chain, 257

Enzyme compartmentalization, 268, 269, 289

© The Editor(s) (if applicable) and The Author(s), under exclusive license to
Springer Nature Switzerland AG 2026
A. M. Pedley (ed.), *Supramolecular Protein Assemblies In Cells*, Advances in
Experimental Medicine and Biology 1514,
https://doi.org/10.1007/978-3-032-26629-3

Enzyme regulation, 256
Enzymology, 201, 208–241
Escherichia coli RNA polymerase (ERNAP), 190–192, 195–200

F
Filaments, 10, 45, 131, 219, 263

G
Gasdermin D (GSDMD), 42
Genome stability, 114, 131, 134, 136, 139, 162, 168–169
Glycolysis, 165, 167, 208, 210, 257, 264
G-quadruplex H-DNA, 114, 135

H
Histone acetylation, 208, 210, 236, 238, 321

I
Immune surveillance, 159, 168–169, 173, 176–179, 184
Inflammasome, 42–50, 170
Innate immunity, 50, 168, 171

K
Keto acid dehydrogenase complex family, 225
Krebs cycle/tricarboxylic acid (TCA) cycle, 208, 229, 233, 236, 237, 263, 264
    See also Citric acid cycle

L
Lambda N, 190–201

M
Mechanosensitivity, 98–102
Membrane assembly, 11, 18, 29
Metabolic diseases, 317
Metabolism, 3, 56, 114, 118, 119, 123, 124, 159, 183, 184, 192, 210, 211, 229, 232, 233, 238, 239, 256–289, 317, 321, 322
Metabolons, 2–4, 208, 209, 211, 212, 216, 218, 220–222, 224–228, 236, 240, 256, 263–275, 277, 285, 289
Mitochondria, 4, 86, 168, 192, 208, 257, 300
Mitochondrial disorders, 317–321

Mitochondrial pyruvate carrier (MPC), 229, 233–235
Multi-enzyme complexes, 1, 208, 228, 264

N
Native mass spectrometry, 3, 93–96
Neuronal apoptosis inhibitory protein (NAIP), 43, 46–47, 50
NLRC4, 43, 46–47, 50
NLRP1, 42–45, 50
NLRP3, 43, 45–46, 48, 50
NLRP6, 43, 46–48, 50
Non-canonical DNA structure, 114, 118, 134–139
Noncanonical inflammasome, 42
Nuclear function, 236–238, 317
Nucleotide, 42, 56, 117, 161, 190, 228, 256
Nucleotide-binding and oligomerization domain leucine-rich repeat containing (NLR) family, 44–46

P
Pathogen-associated molecular patterns (PAMP), 42, 45, 46, 168
Phosphodiesterase (PDE), 56–60, 62–65, 69–76, 126
Poly(ADP-ribose) polymerase 1 (PARP1), 114, 115, 124–126, 128–130, 132, 133, 137, 139
Post-translational modifications (PTMs), 4, 158, 160, 162, 165, 174, 181, 182, 185, 268
Proliferating cell nuclear antigen (PCNA), 114–117, 120–124, 131, 139, 158–185
Protein assembly, 3, 4, 91, 123, 263–285
Protein complexes, 42, 66, 87, 88, 100, 117, 173, 176, 177, 182, 199, 224, 264, 288, 300
Protein degradation, 193–196, 270
Protein dynamics, 66, 75
Protein import, 86–103
Protein interaction, 11, 167, 177, 179
Protein kinase A (PKA), 56–76, 312, 313
Protein-protein interactions, 10, 66, 114, 124, 158, 201, 210, 216, 219, 221, 289
Purine, 3, 256–260, 262–266, 268–273, 275, 276, 278, 280, 284–286, 288
Purinosome, 3, 264–273, 276, 277, 285–286, 289
Pyrimidine, 119, 127, 256–258, 260–264, 273–275, 284

Pyroptosis, 42
Pyruvate oxidation regulation, 208, 210, 211,
     229, 232, 233

R
Regulation, 4, 10, 43, 58, 101, 122, 159, 192,
     208, 262, 307
Replication stress, 120, 132–136, 139,
     169–170, 172–175, 177, 178, 180–183
Replisome, 3, 114–116, 139
Respiratory chain complexes (RCCs),
     234, 300–319

S
Secondary messenger signaling, 72
Signaling, 3, 8, 42, 57, 115, 159, 190, 232,
     256, 301

Single molecule fluorescence microscopy,
     100
Small molecule inhibitors, 180
Spatiotemporal regulation, 58, 75, 76, 237
Structural biology, 59–65, 101, 103, 216, 226,
     232, 240
Structure, 1, 8, 43, 57, 89, 114, 158, 190, 208,
     256, 300
Supercomplexes (SCs), 301, 308, 313

T
Translocase of the outer membrane (TOM)
     complex, 86–103

X
X-ray crystallography (XRC), 1, 2, 14, 17, 18,
     21, 22, 29, 30, 66, 76, 93

GPSR Compliance
The European Union's (EU) General Product Safety Regulation (GPSR) is a set
of rules that requires consumer products to be safe and our obligations to
ensure this.

If you have any concerns about our products, you can contact us on

ProductSafety@springernature.com

In case Publisher is established outside the EU, the EU authorized
representative is:

Springer Nature Customer Service Center GmbH
Europaplatz 3
69115 Heidelberg, Germany